H.A. Kramers

The Solvay Congress, 1927. From left to right, bottom row: E. Langmuir, M. Planck, Madame Curie, H. A. Lorentz, A. Einstein, P. Langevin, Ch. E. Guye, C. T. R. Wilson, O. W. Richardson. Second Row: P. Debye, M. Knudsen, W. L. Bragg, H. A. Kramers, P. A. M. Dirac, A. H. Compton, L. V. deBroglie, M. Born, N. Bohr. Back Row: A. Piccard, E. Henriot, P. Ehrenfest, Ed. Herzen, Th. DeDonder, E. Schrödinger, E. Verschaffelt, W. Pauli, W. Heisenberg, R. H. Fowler, L. Brillouin.

H.A. Kramers
Between Tradition and Revolution

With 16 Illustrations

Springer-Verlag
New York Berlin Heidelberg
London Paris Tokyo

M. Dresden
Institute for Theoretical Physics
State University of New York at Stony Brook
Stony Brook, NY 11794
U.S.A.

Library of Congress Cataloging-in-Publication Data
Dresden, Max
H. A. Kramers: between tradition and revolution.
Bibliography: p.
Includes index.
1. Kramers, Hendrik Anthony, 1894–1952. 2. Physics—History. 3. Physicists—Netherlands—Biography.
I. Title.
QC16.K69D74 1987 530'.092'4 [B] 87-4767

Softcover reprint of the hardcover 1st edition

Typeset by Asco Trade Typesetting Ltd., Hong Kong.

9 8 7 6 5 4 3 2 1
ISBN-13: 978-1-4612-9087-2 e-ISBN-13: 978-1-4612-4622-0
DOI: 10.1007/978-1-4612-4622-0

To come very near a true theory and to grasp its precise application are two very different things as the history of science teaches us. Everything of importance has been said before by somebody who did not discover it.

A. N. WHITEHEAD
The Organization of Thought,
Greenwood Press, Westport, CT, 1974, p. 127.

Acknowledgments

It is a pleasure to thank the many people who helped and supported me in the arduous task of writing this biography. It would have been impossible to even begin this biography without the approval and cooperation of Kramers' children. I had the pleasure of meeting all of them: three daughters—Suus (Perk), Agnete (Kuiper), Maartien—and one son—Jan. My very first preliminary interview was with Kramers' eldest daughter, Suus. She, in what turned out to be a pattern, invited me to her house. We had a rather long discussion and she fed me a lovely dinner. For some reason, almost all the interviews with the family were associated with gastronomic events and were, without exception, informative and pleasant. I feel privileged in getting to know the family so well. I deeply appreciate the confidence they placed in me by giving me access to Kramers' personal papers and private correspondence. Even though I doubt that they will learn a great deal new about Kramers from the scientific chapters, I hope that the personal and general chapters will give them a renewed admiration of the remarkable person who was their father.

Many people provided indispensable technical help, especially Dr. E. Rudinger of the Bohr Institute in Copenhagen, who gave me access to the fabulous files of that institute. Perhaps even more important were our private conversations, which were essential in clarifying the intricate personal relations between Kramers, Pauli, Heisenberg, and Bohr. I would like to thank Joan Warnow of the American Institute of Physics and the Bohr Library in New York for her hospitality and council about just how to collect the materials necessary for a systematic biography. There are many anonymous librarians all over the world who assisted and directed an inept scholar in locating necessary materials for a Kramers biography; they all have my warmest appreciation.

Three young historians of science were kind enough to send me their theses in advance of publication. By now all have made a name for themselves: Mara Beller (now in Jerusalem), Klaus Stolzenburg (now in Stuttgart), and Neil Wasserman (then at Harvard).

I want to thank two of my colleagues at Stony Brook (coincidentally, both have Danish wives) who were kind enough to translate some Danish materials into English: Dr. Jack Smith, for two important postwar letters from Kramers to Bohr; and Dr. A. Jackson, for an English translation of a sketch of a paper on the Bohr–Kramers–Slater theory that Kramers never published.

The beautiful and sensitive translations of Kramers' poems into English are due to Claire Nicholas White. Many aspects of this biography were shaped by an incisive study entitled *The Structure of the Biography*, by Professor S. Dresden of Leiden University, my brother. I thank him for introducing me to his penetrating analysis of biographical writing.

I obtained the photograph (p. xxiii) of Kramers in Princeton from two old friends from Evanston, Dr. and Mrs. Daniel Zelinsky. While spending a year (1947) in Princeton, they became close friends of Agnete Kramers and her husband, the mathematician Nico Kuiper. Kramers himself also spent a great deal of time in Princeton; consequently, the Zelinsky's had met Kramers quite often and the photograph was taken during one of those meetings. I want to thank Zelda Zelinsky for sending me that photograph.

I am grateful to my colleagues in the Institute for Theoretical Physics and the Department of Physics for their understanding and patience. With but very few exceptions, they accepted that I would spend a great deal of time on this biography. They also showed great forbearance in listening to innumerable stories about Kramers, Pauli, and Heisenberg.

Reading this book will show immediately how much it owes to the interviews. It is really impossible to thank all those interviewed, but I want to make an exception for Annie Romeyn–Verschoor. The wife of the noted historian Jan Romeyn, she was a major author, thinker, and historian in her own right. Jan Romeyn was certainly Kramers' oldest and best friend—a friendship that spanned well over 50 years. As a consequence, Romeyn knew and understood Kramers in a very special way, and his wife Annie, a most intelligent and perceptive woman, understood Kramers at least as well. It was a great privilege and a personal pleasure to have several lengthy, intense conversations with her. Her insight and knowledge added greatly to my understanding of Kramers. It was especially generous of her to give me free access to the Romeyn–Kramers correspondence. That most interesting, very personal correspondence covering some 40 years gives an unusual and illuminating insight into Kramers' character. Without access to this correspondence an important segment of Kramers' personal as well as scientific life would have been lost to posterity. I thanked Annie Romeyn profusely for her time, her many telephone conversations, her kindness. I am sorry that she will not be here to see this biography finished. I planned to present her with a copy. Perhaps the best way to thank her now is to call attention to her brilliant autobiography, *Omzien in Verwondering* (Looking Back in Amazement), which describes the Dutch political and academic ambience—and a little bit of Kramers' life—in sensitive detail.

A book must not only be written, it must also be typed, corrected, proofread, collated, and duplicated. In the initial stages of this work I had the help of an excellent secretary, Isabel Harrity. I want to thank her for all her efforts, which in fact went much beyond the routine activities. She corrected the English, complained about the length of some chapters, and reminded me of deadlines. Unfortunately, she had to leave for greener pastures well before the work was finished.

I want to express my appreciation to a very special Dutch physicist, H. B. G. Casimir. As soon as I told him about my plans to write a biography of Kramers, he was most encouraging and helpful. In numerous interviews and discussions he gave me a penetrating description of Kramers, his role in Dutch physics, and his personality. Since Casimir, like Kramers, is a physicist of world class, his comments were especially important. It was Casimir who, together with Bohr, summarized Kramers' scientific work at the memorial session for Kramers at the Dutch Physical Society Meeting in July 1952. To have his sympathetic support in this venture meant a great deal to me. If I had problems or got confused, I knew that I could count on his help.

For the first 20 years of my life, there was no person I knew better than Abraham Pais, now a distinguished physicist at Rockefeller University. Pais and I went to kindergarten, grade school, high school, and the first few years of college together. Then our paths diverged. Pais got to know Kramers extremely well during wartime, 1939–1945, while I was in the United States in Ann Arbor, Michigan. Kramers developed a very high regard for Pais' scientific ability. The first time I saw Pais again after my departure in 1939 was in Washington in the spring of 1947. It was a curious coincidence that Kramers, Pais, and I shared a room in the Willard Hotel (it no longer exists!) in Washington. I had reserved a room there for the Washington Meeting. I came from the far west at that time (Lawrence, Kansas). The only room available was a double room and I took it, even though I could not afford it (the University of Kansas did not reimburse its professors for anything at that time). Soon after I arrived in Washington I met Bram Pais, who asked if by chance there was some space in my room. Of course, I was overjoyed to have him share the room. A little later, Kramers met Pais and asked him whether he knew of any space in the hotel. So he joined us as well. That was the only time that the three of us spent any time together, but we had some long, animated, and lively discussions. Kramers and Pais knew and understood each other well; they had shared the war experiences and had evidently discussed quantum field theory at great length. I was a bit of an outsider, but I was intrigued and fascinated by their discussions. I often reflected on this chance meeting while writing this biography. In retrospect, it seems that it might well have influenced my decision to embark on this project. So it is not surprising that after I had decided to write the biography, I often went to Pais for information and suggestions. He was always helpful; he knew things about

Kramers and his physics that nobody else knew. His council and advice were always excellent, his suggestions always pertinent. It is an unusual pleasure to thank a friend from my earliest youth, a highly esteemed colleague, for helping me to complete a biography of a person he knew much better than I.

Although C. N. Yang, my friend and colleague for 20 years, did not contribute a great deal of specific information about Kramers, this is still the place to express my appreciation for our long and warm association. I often went to Yang to ask specific questions about physics, about the development and future of physics, or about personalities. I profited immensely from the ensuing talks; his trenchant insight, his immediate deep understanding, his almost frightening intellectual directness inevitably led to new insight and new understanding on my part. His off-hand, brilliant, and perceptive remarks are literally strewn all over this biography. It is not true that everthing that Yang says is necessarily true or correct, but it is true that he literally does not have a trivial bone in his body, so all his comments always led to further analysis and more profound understanding. I consider it an unusual stroke of luck that for the last 20 years I have had the privilege to have daily informal contact with one of the greatest theoretical physicists of the 20th century. For this humbling and exhilarating experience I am deeply grateful.

Finally, I dedicate this book to my wife, Bertha Cummins. She helped in thousands of ways, by accompanying me on innumerable interviews and by organizing and reading a good deal of the Kramers correspondence. Her reactions, as always of deceptive simplicity, revealed extraordinary insight and understanding, which helped me in dealing with this sensitive material. All this is true, yet it really misses the point of describing the importance of Bertha Cummins for this work. I am not sure that I know how to give an adequate description of her significance for this work or for my life. Research and scholarship, whether successful or not, require tranquility, serenity, and tacit understanding. These she has been able to give me in an unhappy, turbulent world, apparently without trying, without strain. It has always been a miracle to me how she managed to express extraordinary intelligence and unusual sensitivity in the most ordinary of matters. She created the conditions which made the completion of this work possible. I devotedly hope that the work is worthy of her.

Contents

Acknowledgments vii
Preface xv

Part 1 A Remarkable Person in a Very Special Epoch

Chapter 1
H. A. Kramers: Why a Biography? 3

Chapter 2
The Classical Synthesis: Lorentz 8

Chapter 3
The Unraveling of Classical Physics: Planck and Einstein 13

Chapter 4
Conflicts and the Basic Incompatibility 17

Chapter 5
Bohr: Success and Ambiguity in a Period of Transition 20

Chapter 6
Bohr, the Photon, and Kramers 28

Chapter 7
The Bohr Institute: Pauli and Heisenberg 34

Chapter 8
From Virtual Oscillators to Quantum Mechanics 41

Chapter 9
The Revolution in Progress 56

Notes for Part 1 79

Part 2 Living Through a Revolution

Chapter 10
A Hesitant Start with an Unusual Twist 87
I. Youth and General Background 87
II. The Leiden Years (1912–1916) 90
III. The Arnhem Interlude 94

Chapter 11
The Early Copenhagen Years: From Student to Apostle 97
I. Summary: Between Leiden and Copenhagen 97
II. The Initial Contacts Between Kramers and Bohr 100
III. Elaboration and Extensions of the Bohr Theory 102
IV. Kramers' Contribution to the Bohr Theory: The Thesis 107
V. The Changing Status of Kramers 110
VI. How Parsimonious the Emotions? 114

Chapter 12
From Apostle to Prophet: Hints of Trouble to Come 119
I. Helium: The First Failure 119
II. More Problems: What Is the Right Direction? 122
III. The "Virtual" Encounter of Kramers and Heisenberg 124
IV. The Gospel According to Kramers and Holst 132
V. The Great X-Ray Paper: The Last Success 134

Chapter 13
The Multifarious Consequences of a Desperation Revolution 139
I. The Struggle About the Photon 139
II. The Kramers Dispersion Theory: Not a Consequence but a Success 144
A. The Unreliability of Chronology 144
B. Dispersion Theory: Background and Basic Ideas 146
C. The Kramers Derivation 150
III. The Bohr–Kramers–Slater Theory 159
A. Background: Slater's Ideas 159
B. Bohr–Kramers–Slater: The Interplay of Personalities 163
C. The Bohr–Kramers–Slater Theory: Basic Ideas 171
D. The BKS Theory: Reflections and Second Thoughts 178
E. The Compton Effect in the BKS Theory 185
F. The Intellectual Framework of Kramers' Physics 191
G. Kramers' Personal Elaborations 195
H. The Reception of the Theory 199
I. The Crash 207
IV. The Road to Quantum Mechanics: The Changing of the Guard 215
A. The Kramers–Heisenberg Dispersion Theory Paper: Background and Chronology 215
B. General Approach Between Virtual Oscillators and Matrices 223
C. Technical Matters: The Old Method and New Puzzles 230
D. From Virtual Oscillators to Matrices: Heisenberg Goes It Alone 242
E. The Personalities of Pauli, Heisenberg, and Kramers: The Strange Path from Deference to Ridicule to Doubt 252

F. The Changing Patterns 276
G. Why Didn't Kramers Discover Quantum Mechanics? 285

Chapter 14
The Curious Copenhagen Interlude 289
I. The Events 289
II. Recollections and Some Documentation 291
III. Conclusions 295

Notes for Part 2 299

Part 3 Waiting for a Revolution That Did Not Happen

Chapter 15
The Search for Identity in Changing Times 311
I. The Move to Utrecht 311
II. The Change in Style and Mood 314
III. The Scope of Kramers' Activities 319
IV. Kramers' Contributions to Statistical Mechanics 321

Chapter 16
The Recurrent Theme: Electrons and Radiation 325
I. Kramers' Approach: The Legacy of Lorentz and Bohr 325
A. Lorentz's Influence 325
B. Kramers' Uneasiness with the Dirac Radiation Theory 332
C. The Chronology of an Obsession 338
II. The Struggle Toward a Consistent Theory 344
A. The Construction of the "First" Kramers' Hamiltonian 344
B. Further Elaborations of Kramers' Ideas 357
III. The Story of Two Conferences: Shelter Island and Solvay 372
A. The Pre-Shelter Island Status: The April 1944 Lecture 372
B. The Shelter Island Conference: Kramers' Presentation 379
C. The Solvay Report: Culmination and Termination of Kramers' Program 391
D. Culmination or Oblivion: Van Kampen's Thesis 402
IV. Retrospect: The Postponed Recognition 410
A. The Lamb Shift: How High the Stakes 410
B. The Convoluted Recognition 421
C. What Went Wrong? Can One Be Too Principled? 426

Notes for Part 3 433

Part 4 Kramers as a Person and a Scientist: Conflict or Harmony?

Chapter 17
Personality and Style 441
I. The Assessments: Public and Personal 441
II. The Conundrum of the Near Misses 446
A. The Role of Near Misses 446
B. The Litany of Near Misses 449

C. Interlude: Near Miss Number 9—The *S* Matrix Incident 453
D. Near Misses Concluded: Omissions and Style 458
III. Elements of Kramers' Scientific Style 461
A. The Role of Fear and Courage 461
B. Characteristics of Kramers' Scientific Style 464
C. The Role of Mathematics, Structures, Abstractions 466
IV. Personal Matters 470
A. Some Personal Traits 470
B. The Surroundings 474
C. The Interpersonal Relations 477
D. Fear and Decency 484

Chapter 18
Obligation and Duty 488
I. The Equivocal Role of Religion 488
II. Obligation and Duty 490
III. The Special Role of Teaching 491
IV. Duty: The War and Its Aftermath 498
A. The Wartime 498
B. Kramers' Work in the United Nations 501

Chapter 19
Kramers' Self-Image 506
I. Kramers and the Psychology of Physicists 506
II. Kramers' Lecture on Debye: What Is an Ideal Scientific Biography? 512
III. Fragments from Notebooks and Diaries 516
A. From the Notebooks: Science and Scientists 516
B. From the Diaries: The Person and His Demons 520
C. From the Diaries: Personal Entanglements 523
D. The Story of Camilla 528
E. The Poems 532

Chapter 20
Epilogue: Does One Know Better—Understand More? 536
I. The Objective Recognition: Prizes and Awards 536
II. Finale 539

Appendix: Poems 542

Notes for Part 4 546

Index 553

Preface

It is now a little more than 11 years since the idea of writing a personal and scientific biography of H. A. Kramers took hold of me. A few days earlier I had been lecturing, in a course on field theory, on the renormalization procedures of relativistic quantum field theory. Since the students had considerable trouble understanding the physical basis of the procedure, at the end of the lecture I explained that renormalization is not an exclusive quantum or relativistic procedure. A careful treatment of classical electron theory as started by Lorentz and developed in detail by Kramers also requires renormalization. The students appeared quite interested and I promised them that I would explain all this in more detail in the next lecture. I could have looked up this material in Kramers' book, but I remembered that Kramers had stressed this idea in a course I had attended in Leiden in 1938–1939. I did dig up some of these old notes and, although they were considerably less transparent than my recollection seemed to indicate, they reminded me forcefully of the thrilling days I had spent in Leiden with Kramers. Kramers' deep insight and originality were apparent even when distorted by my opaque notes. The students had never heard of these ideas of Kramers' and were totally unaware of his work in field theory. That his approach and understanding were no longer a common part of physics was deeply regrettable; that his heroic efforts were all but forgotten was intolerable. This was when I first realized that a biography was needed. But it was not at all clear just who should write it. Kramers had had many friends, students, and colleagues who had known him better and longer than I. Van Kampen, one of his last and most talented students, had written a brilliant thesis on Kramers' favorite topic: "the interaction of radiation and electrons." Wouthuysen was a student who Kramers had loved and respected; ter Haar was not only a devoted

Kramers receives the Lorentz medal, 1948.

student of Kramers but a close friend of the family; all of them were more likely authors of a Kramers biography than I.

There were many others equally or better suited to this task: Abraham Pais, the close friend of my youth, who had spent some of the most intense moments of his life with Kramers and who was thoroughly familiar with Kramers' work in quantum field theory; Jan Korringa, Kramers' last assistant; Fred Belinfante, another one of Kramers' faithful assistants who typed his great Volume II; and above all, H. B. G. Casimir, the great Dutch physicist, who knew Kramers well and was more nearly his scientific equal than anyone else. I, on the other hand, did not know Kramers very long and only in a rather superficial way.

I was a student in Amsterdam for about three and a half years. I had been fascinated from my early high school days by relativity and quantum mechanics. Bram Pais and I tried to learn these topics with limited success in high school; we did learn very soon that to get anywhere we needed calculus. I remember clearly my excitement when, in a swimming pool, Bram Pais told me that he had learned to take the derivative of a product. Another early but quite indelible memory was when Heisenberg received the Nobel prize for inventing matrix mechanics. Neither Bram Pais nor I had any idea what a matrix was, but we were almost mesmerized by these fields.

The classes in the University of Amsterdam were not inspiring and nobody ever mentioned either relativity or quantum theory. With perhaps a youthful mixture of arrogance and insight, I decided that no one in Amsterdam knew quantum theory. I had read a popular book on atomic structure written by

Kramers and Holst. It seemed clear that the only master of quantum theory in the Netherlands was Kramers in Leiden. Consequently, I wrote to Kramers asking him whether I could come to Leiden to continue and complete my studies, while living in Amsterdam. In those days, a commuting student was somewhat unusual. Kramers answered promptly. He was extremely kind and suggested the courses that I should attend in Leiden. He even invited me to participate in the "Wednesday evening Ehrenfest colloquium," something that I considered a great honor and an almost "holy" obligation. I never missed it.

Kramers felt that I should try to do some research and he gave me a very specific problem to try. I visited him regularly in his home in Oegstgeest to discuss my progress (which was painfully slow!).

During the period in which I attended Kramers' classes, he often had to go to Amsterdam for, I believe, committee meetings of the Dutch Royal Society. So very often Kramers and I walked (or bicycled) to the station after his classes to take the train back to Amsterdam. During these many walks and train trips, Kramers told me a great deal about physics—how he expected or hoped it would develop. He often talked about his years in Copenhagen and about his meetings with Bohr, Heisenberg, and Pauli. I was thrilled by these conversations—even though Kramers would often mention things I did not really understand—and the words stayed with me. He often asked, "How do you know that the mass in the Dirac equation is the same as the experimental mass?" (I only fully appreciated the profundity of this question after I completed this biography.)

I do not believe that I can adequately express the excitement I felt when I walked with Kramers in the Breestraat in Leiden and we passed a bookstore that had a copy of his great textbook on quantum mechanics in the window. To walk in Leiden with a person who not only knew quantum mechanics but could write a masterly book on it (so said the advertisement) was to me to be near heaven. Through Kramers I got my first glimpse of true science and learned first-hand about physicists and their special world.

The approaching war made serious long-range plans difficult; but through considerable luck and Kramers' very active help, I received a stipend to study in the United States with Fermi. Kramers urged me to seize this unusual opportunity. So, after about a year of reasonably close association with Kramers, I left. I saw Kramers a few times after World War II in Washington and in Princeton, but apart from an occasional letter we had no further contact.

It was on the basis of this very limited experience that I had to decide whether or not to write a biography of Kramers. To do it right was clearly a formidable task. I wrote to many physicists asking for help and suggestions. I was also interested in finding out whether anyone else was planning to write such a book. Although some people had thought about it in a general, rather vague way, nobody seemed anxious to undertake this enterprise. One person strongly advised me against doing this. He insisted, quite correctly, that a slipshod biography would not only be worthless but indeed an insult to Kramers. He further asserted that it would take 10 years to do justice to such a work. I would long remember his admonition. But the desire to try to write Kramers' biography became stronger and stronger, almost an obsession. And so, 11 years ago, I began what turned out to be an exciting adventure in a new field: This book is the result.

It may appear unfair, or unfortunate, but it nevertheless is true that the decisive advances in science can be traced back to the efforts of relatively few individuals. These outstanding scientists were, of course, influenced by their peers and by the times in which they lived. Indeed, the crucial developments can most often be associated with particular persons at particular times. Thus, a scientific biography should be viewed as a contribution to the history of science. It is important to stress that in science not all ideas—even if ingenious and original—are of equal importance, nor is it true that all practitioners—no matter how brilliant—are of equal significance. Within science there is a tortuous and indirect path to the present, and for better or worse, "success" is defined in terms of the contemporary criteria. The struggles, the mistakes, the false leads are rarely mentioned in the development of science. They obviously play an important role in the succession of scientific ideas; they play a crucial role in the scientific life of an individual. It is possible that by ignoring all mistakes, all false clues, all misdirections, an efficient, smooth presentation of a scientific field or the life of a scientist could be given. (Most textbooks follow exactly that method.) But it is not good history: It gives a false and misleading picture of the progress of science and it misrepresents the role of most scientists in the scientific evolution. A serious biography must address the background and motivation of an individual, his hopes, expectations, fears, ambitions, the relation of his science to his life, his relations to his friends, peers, and relatives. But a scientist neither lives nor works in a vacuum: The status of his science, the role of the science in the prevailing culture, the political and financial circumstances—all these will affect his science and his scientific productivity. It is only an analysis of that complex amalgam that can do justice to the richness, the variety, and the subtlety of a life devoted to science.

After reading a number of scientific biographies, and after reflecting on the general character of scientific biographies, I could not procrastinate any further: I had to start the actual preparations for the writing of the Kramers biography. I first read all of Kramers' papers (those printed in his *Collected Works* and a few more). I repeated most of the calculations, not because I expected that Kramers had made many mistakes (he actually made very few), but because actually performing the calculations gives an appreciation of Kramers' power and originality. My respect for Kramers' formal ability, always enormous, soared to new heights after merely imitating his calculations. It became quite clear while performing these computations that Kramers derived a great deal of pleasure from his formal power. I, in turn, was pleased when I began to recognize his special tricks and highly personal techniques.

Picture from a painting, late 1950.

Kramers at his best, 1935.

While reading his papers I also spent a considerable amount of time reading Kramers' public lectures, inaugural addresses, commemorative lectures, and obituaries. Kramers prepared these lectures with great care and delivered them candidly: In them he revealed more of his personal philosophy and personal expectations than in his purely scientific papers. These lectures and articles were very helpful in getting a large picture of Kramers' personal and scientific activities.

Kramers' life was totally dedicated to physics, and he belonged to that blessed set of individuals who made genuine contributions. No further justification for his biography is needed other than the recognition of the cultural role of physics so beautifully expressed by Oppenheimer: "I think that taken as a story of human achievement and human blindness, the discoveries in the science are among the great epics of all time and they should be available in our tradition ... It was a heroic time—it had its share of crooks and robbers ... the achievements in physics seem to me to stand with the high points of human knowledge—in the quality of insight—in the beauty of work."

About 6000–8000 of Kramers' letters are still available. I have read, I would think, about half of them, and I glanced at many more. The letters provide an invaluable source of quite specific information.

Kramers kept diaries on an irregular schedule; he also kept something like a daily appointment book, an extensive calendar in which he often jotted down "thoughts of the moment." The Kramers family was most generous in making these materials available to me.

Kramers also kept his lecture notes. Some of these are quite complete and polished, while others are very sketchy. But their study is especially illuminating in tracing the evolution of Kramers' thinking. Kramers' daughter, Maartien, kept most of these notes; it was most kind of her to allow me totally free access to them.

Perhaps the most important source of information about Kramers consisted of talks, conversations, formal interviews, and totally informal, unstructured discussions with his colleagues, students, and family. I conducted about 40 formal interviews and another 50 or 60 informal discussions. Although as far as I know no tapes of Kramers himself exist, he and his work are often mentioned in taped interviews with "quantum physicists"; so that the use of these tapes was especially important.

I was generally struck by the "soft" character of the information so obtained. When I replayed the tape of an interview with a well-known physicist, about one week after the interview had taken place, he found it hard to believe that he had indeed said what was on the tape. In fact, he promptly denied it. Of course, letters and diaries are hardly much more reliable; they strongly depend on the mood or the particular circumstance at the time of the writing, or on the impression the writer wishes to create at that moment. But even though each one of these separate sources, scientific papers, public lectures, diaries, and interviews is necessarily partial, incomplete, and replete with ambiguities and contradictions, it seemed that by putting them all together, I was able to arrive at a surprisingly complete, coherent image of a remarkably decent, brilliant, oversensitive, slightly melancholic, profoundly human scientist. Even admitting the doubtful nature of any and all detailed information available, this conclusion, I believe, stands unchallenged.

It is inevitable that an enterprise of this scope would have a profound influence on the author. In the past 11 years I have learned an enormous amount, and my understanding and appreciation of physics has changed substantially. I became aware of the strange, roundabout way physics advances, following misleading signs and misinterpreted experiments. I read more about physicists, more biographies, more history than I had in my whole life up until that time. I developed an extraordinary admiration and respect for historical scholarship. When I was first confronted by the Kramers file at the Bohr Institute in Copenhagen, with its many hundreds of letters, written in many languages in almost undecipherable handwriting, I was dumbfounded. Who could, would, read all this? This, to a historian of science, is a routine matter, a daily occurrence, but I was ready to give up the whole project. Fortunately, I recovered to such an extent that I eventually became quite familiar with Kramers' handwriting and Pauli's scribbles. I always knew that physics is difficult and subtle, that mathematics is sophisticated and deep, but I was not prepared for the forbidding difficulty and the erudition of the history of science.

This is a biography of a physicist written by a physicist. I frankly doubt that this study could meet all the exacting demands of rigorous historical scholarship. That is not exclusively because of my inability to meet these standards, but because I believe that the approaches and viewpoints of the physicist and the historian are characteristically and legitimately different. I speculate frankly and frequently about the remarkable number of "near misses" in Kramers' work. Similarly, I wonder why there were so many problems, so near his interest and well within his grasp, which he never touched. To a physicist these are interesting questions which can give unusual insight into the scientific style of a person. To a historian, these are probably undocumented speculations of doubtful value. I am also well aware that all the objective information I collected in the end is filtered through me. Ultimately I made the choice of what papers and what events to stress. Even though I believe that I could adduce excellent and compelling reasons for the choices made, I clearly was not an uninvolved bystander in this process. I worried a good deal about how I could be more objective, more even-handed. But I finally came to terms with recognizing that what is presented here is *my* view and *my* understanding of Kramers' life.

For many years I was so imbued by the intellectual power and beauty of theoretical physics that the realization that the creators of such beauty could be anything short of perfect human beings came as a jarring, unpleasant surprise. Now it is no longer surprising: It is of course obvious, even though it still seems incongruous to me.

A scientific biography, by its very nature, must deal with a number of distinct and, indeed, barely compatible matters.

(1) It certainly has to be concerned with the personal, everyday details of the scientist's life; for example, education and background need to be included. Even more important is a careful discussion of his personal hopes, dreams, expectations, and disappointments. The ebb and flow of his scientific aspirations form an integral part of any scientific biography. Such a discussion deals primarily with personal and social relations, rather than with mathematical or scientific ideas. As such, this part of a biography can be appreciated by a very general audience.

(2) Since science plays such a dominant role in all aspects—even the emotional aspect—of a scientist's life, it is only possible to appreciate this deep personal involvement in science, in terms of a perceptive appreciation of the status, the current intellectual ambience, of the science itself. Only then can the subtle and ever-changing interactions between the individual scientist and the field he so desperately tries to influence be fully understood. The role of research and the definition of success require a sensitive understanding of the relation between the scientist and his chosen field. The vacillating relations between the scientist and his peers, teachers, students, competitors, and critics must be examined with special care to obtain a balanced, coherent view of the scientist. This type of discussion

Kramers and Einstein at the Solvay Congress, 1927.

The last picture of Kramers.

requires some general understanding of the nature of the field but not too much specific technical knowledge. A very broad audience should have no trouble in appreciating the issues involved, although there might be occasional items whose relevance leaves something to be desired.

(3) Lastly, a scientific biography should consist of a thoughtful, illuminating analysis of the major technical contributions made by the subject of the biography. It is especially important to stress the many false starts, the futile pursuits of misleading clues, that inevitably precede or accompany great achievements. It is essential that enough technical details be supplied, so that the originality, power, and importance of the scientific contributions can clearly be discerned. An overall assessment of the total impact of the scientific work on the further development of science is, of course, indispensable. This type of discussion, in contrast to the earlier parts, demands a great deal of technical sophistication.

I tried to balance these conflicting demands as I wrote this biography. After a serious effort to be complete, clear, honest, and correct it came as a stunning shock that the manuscript produced ran between 2000 and 2200 pages! In print this would amount to about 1400 pages. Obviously, a book of that length would be too long to read and too expensive to produce. The alternative was

Kramers and his wife at his home, Oegstgeest, 1949.

to publish the work in two volumes, but this did not seem practical because it was difficult to find a natural division. Furthermore, one volume would likely concentrate on his pure science, while the other would cover his personal involvements. This was the dichotomy that I wished to avoid at all costs. Thus, it was decided to publish just one volume. This meant that large sections of Kramers' significant and beautiful scientific work could not be included. Needless to say, I made this decision with severe misgivings and an acute sense of disloyalty.

Perhaps some of the omitted material can be included in later collections, or in a series of individual essays, or papers in appropriate journals.

It is particularly galling that no serious technical discussions are included of Kramers' fundamental work in statistical mechanics; neither the Ising model papers nor the Kramers equation are discussed in any detail. The W.K.B. method, the Kramers–Kronig relations, are just mentioned. Kramers' great textbook, his "scientific confession of faith," is barely touched. His remarkable paper on the Kramers degeneracy, the important study on charge conjugations, his penetrating original "symbolic method," are all omitted. There are persuasive reasons to include them all. His long series of papers on magnetism is just mentioned in passing.

Chapters 1–9 set the scientific stage for Kramers' science. The remarkable four-cornered fluctuating personal relations between Bohr, Kramers, Pauli, and Heisenberg, which had such a potent, long-lasting effect on Kramers' scientific and personal life, are introduced in these chapters. These introductory chapters, and also Chapter 10, should be accessible to any reader. The early papers and his thesis, all written in Copenhagen, are treated in some detail in Chapters 11 and 12. Chapters 13, 14, and 16 are lengthy and technically demanding. These represent Kramers at his technical best. The topics selected for this detailed treatment are the dispersion relations, the Bohr–Kramers–Slater theory of 1924–1925, and Kramers' version of renormalization. As will be detailed, both were watershed events in Kramers' scientific life. Kramers certainly made many basic contributions to physics, but no individual contributions were more fundamental than these. Even though it is perhaps possible to argue, on purely scientific grounds, that this statement is not strictly true, there is no doubt that to Kramers these investigations were of overriding significance. He certainly considered them as the most important efforts of his career. Chapter 15 gives a partially technical analysis of Kramers' search for a scientific identity after his departure from Copenhagen. The last four chapters (17–20) all deal with Kramers as a human being and a scientist. Even in the technical chapters there are frequent sections easily understandable to a general reader. They are appropriately marked. All in all, 14 chapters should be generally understandable; for four other chapters (10–12, 15) some background in physics is surely helpful, while two chapters (13, 16) require a great deal of physics.

As structured, the book has considerable unity, coherence, and logic. It presents a clear, interesting picture of an extraordinary person and an unusual scientist, whose indispensable role is not now and probably never was fully realized. Even so, it is deeply sad and profoundly upsetting that so much of Kramers' work (and hence of his life) had to be omitted. It almost seems necessary to apologize to somebody for omitting so much of what was important and meaningful to Kramers. But perhaps the more focused presentation of Kramers' work and effort given here might still be preferable over a more complete, but necessarily more diffuse, meandering, and exhaustive treatment.

M. Dresden

PART 1

A REMARKABLE PERSON IN A VERY SPECIAL EPOCH

CHAPTER 1

H. A. Kramers: Why a Biography?

On April 25, 1952, almost all the major newspapers in The Netherlands carried news reports on the death of Professor Dr. H. A. Kramers, holder of the chair of theoretical physics at the University of Leiden. These obituaries recorded the major events in his life, including his birth, on December 17, 1894 in Rotterdam, and his education in Leiden. Kramers' contributions to science, which brought him national prominence and international eminence, were dutifully enumerated. Some of the highlights of his career were stressed: Kramers was the first scientific collaborator of Niels Bohr in the formulation and elaboration of the first modern theory of the atom. In 1946 Kramers was elected Chairman of the scientific and technological subcommittee of the United Nations Atomic Energy Commission. From 1946 to 1951 Kramers was president of the prestigious Union Internationale de Physique Pure et Applique. Clearly, Kramers was a major figure in physics and in the world of science.

But time so relentlessly erases the efforts and accomplishments of all but the greatest people that now—not quite 35 years later—his public service activities are effectively forgotten and even within the world of physics there are only a few results still associated with his name. This is more than a pity; it is a genuine loss. For not only was Kramers a sensitive and perceptive physicist, so that his reactions and impressions were of special interest, but he lived through one of the truly heroic periods in the evolution of science, and his struggles, conflicts, and doubts mirror those of physics in that painful transition from the classical period to the unfamiliar and uncomfortable quantum period. To Kramers, physics and science were integral elements of the culture; he had many genuine interests and areas of competence outside physics, but it was physics that formed the core of his being. From his student days and thoughout his distinguished career, Kramers was driven, not to say haunted, by a desire to make significant contributions to fundamental questions in physics. In many addresses, in his popular lectures, the beauty and fascination of theoretical physics and the thrill of discovery in spite of wrong directions and misleading clues are recurring themes. Kramers often expressed himself poetically on the significance he attributed to research: "To do original physics and to get some results is as if one is kissed by an angel." [1]

It is remarkable that although Kramers came tantalizingly close to epoch-making discoveries on numerous occasions, he never really succeeded in carrying the advances he had initiated to their ultimate, often revolutionary conclusions. But the search for new results was the perennial motivation for practically all his activities. To this search he devoted most of his energies and all his available time.

It is not surprising that Kramers was personally acquainted with almost all active physicists of that time. This was that unusual time in which Einstein and Bohr labored and worked, argued and struggled. It was the time in which scientists of an earlier generation such as Lorentz and Planck, with their mature and profound judgments, were baffled by the intractability of the problems; while the succeeding generation of Heisenberg, Dirac, and Schrödinger created new concepts that appeared strange and bizarre and were assimilated with dazzling speed by some and rejected as incomprehensible by others. It is doubtful that there ever was a period in the history of thought in which such a collection of original, brilliant minds all concentrated on one set of problems. Within an incredibly short period of 2 or 3 years, atomic physics was built into a new imposing scheme of extraordinary coherence and predictive power. Concepts that a few years before seemed outlandish and capricious became practically commonplace. "It was truly a heroic time—it had its share of crooks and robbers—but the achievements in physics seem to me to stand with the high points of human knowledge in the quality of insight and in the beauty of work." [2]

Kramers was literally immersed in all these dramatic events. He was an active participant, an acute observer, an incisive critic, and at times a troubled scientific conscience. The scientific and personal reactions of such a person to one of the major intellectual revolutions are worth recording and remembering.

When Kramers died, tributes appeared in many scientific journals. That their number was unusually large merely testifies to Kramers' standing in the international scientific community. It is more surprising that commemorations appeared in many places which had nothing to do with either science or universities. Perhaps the most perceptive, most extensive, and certainly the most informative eulogy was published in the 1951–1953 *Yearbook of the Dutch Literary Society*. It was an unusual honor for Kramers to be elected to that society and especially unusual to be commemorated in a yearbook—a distinction reserved for major literary figures. Where the discussions in physics journals understandably concentrate on Kramers' scientific accomplishments, these others show that Kramers was as unusual a human being as he was a remarkable physicist. This was not only because of his literary erudition, his unquestioned linguistic proficiency, his deep knowledge of theology, or his almost professional mastery of the cello and piano, but primarily because of his unique understanding of and concern for human beings.

He demonstrated that understanding in such apparently simple matters as making sure that his students, assistants, or visitors immediately received the financial support to which they were entitled; he perceived correctly that most of them would be too shy or embarrassed to ask a professor for money, even if they did not know from where their next meal was coming. His concern for people was expressed dramatically during the war years when he helped and supported students who had gone underground, or when he and another colleague distributed potatoes during the disastrous hunger winter of 1945. In less dramatic circumstances, the concern manifested itself in other ways; by helping students find positions, by answering innumerable requests from acquaintances and strangers, by patiently listening to grievances, by introducing humor into his day-to-day dealings with people. There was a strain of gentle irony that often masked the seriousness of the concern. Especially later in life a melancholy air was noticeable—neither morose nor self-pitying—but the melancholy of a man who has understood and accepted the human condition and the limitations of the human mind.

The activities of the mind were, to Kramers, the true essence of life. He was deeply interested in all facets of intellectual life and a good deal of his social life consisted of intellectual discussions, speculation, and arguments. All aspects of intellectual life were approached in a highly personal, original manner by continually asking unusual questions, by persistent probing, by changing his viewpoint in the process. It was often hard to decide whether Kramers considered the questions as serious or facetious, silly or profound, trivial or deep, or all of these. Most always an interpretation on all these levels was possible and very likely intended. Kramers loved to turn arguments around, often achieving surprising effects. "You theologists speak about the miracle when you no longer understand something, but I consider it a miracle when I sometimes understand a little of it."[3] This quotation is a typical example of the continual, almost playful, sense of wonder in Kramers' attitude, and of Kramers' penchant to speak in aphorisms. Perhaps because of this sense of wonder or mental flexibility, perhaps because of his incessant questioning, Kramers never developed an excessively strong set of convictions. In artistic or political matters he would, with equal facility, defend opposing views. He was intensely antidogmatic in all aspects of life. In physics, where the results of experiments and the demands of mathematics and logic were more stringent, Kramers' attitude was more codified, but even there never totally rigid. He certainly never had much sympathy for axiomatic schemes. In specific technical matters in science, Kramers, as everyone else, was bound by the rules of strict logic; his argumentation was precise and his use of mathematics was impeccable. But in broader matters, with issues involving philosophical attitudes or overall intellectual orientations, his extraordinary mental versatility allowed him the luxury of not rigidly adhering to a single philosophy or viewpoint.

It is remarkable that practically all of the elements so characteristic of Kramers' intellectual style—the probing, the sense of wonder, the gentle irony, the playfulness—were present in his personal life. More unusual is that he behaved much the same with all persons of all walks of life, be they students, professors, government officials, Salvation Army officers, artists, or athletic coaches. The same intellectual machinery, the same mental set, was always present. It was just adapted to the person at hand. Often his open-ended interest in people and their experiences, viewed from Kramers' intellectual perspective, led to unusual results, as when Kramers paid a visit to his friend the historian Jan Romeyn. Nobody was home except the maid, who was a member of the Salvation Army. Kramers became so interested in the Salvation Army, the function it fulfilled in the maid's life, that he stayed longer and longer and even had lunch. This was not the expected behavior of a professor in the Dutch culture at that time. At other times his continual probing, mixed with irony, was not appreciated and his intrinsic playfulness was misinterpreted as quixotic behavior.

Kramers was an especially private person. Even though he was willing to discuss anything, it often seemed that Kramers detached his personality from the subject and the discussion. It was mentioned in one obituary[4] that Kramers "loved his family, his science, his colleagues, his students." While this is all true, the quotation gives a flat, one-dimensional picture of a many-sided complex, person. As for most people, but for Kramers especially, life's wishes, dreams, doubts, fears, and expectations were deeply personal and were revealed only rarely if at all. His extraordinary mental agility enabled him to deflect any approach which he knew or sensed would come too near intruding on those domains he wished to keep private. This sense of privacy had also a counterpart in his scientific life, where his publications and public lectures possess all the impersonal objective features which physics demands. Only rarely did he give hints of his scientific hopes and aspirations.

The insistence on his own intellectual and personal privacy occasionally created the impression that Kramers was somewhat reserved and detached. Students often didn't exactly know what to make of Kramers. Especially in Leiden, where in 1934 Kramers succeeded Ehrenfest, who was deeply involved with his students, it seemed to many that Kramers was cold and disinterested. This was actually quite wrong, but it is easy to see how the combination of intense privacy, irony, and a brilliant, playful mind would present a confused picture to young students (and to some older professors as well). It took time to appreciate Kramers' profundity, wisdom, and sense of humor. It is, of course, to be expected that Kramers would produce quite different reactions in different people. He was, after all, a man of many talents and diverse interests who furthermore moved in a variety of scientific, social, artistic, and intellectual circles. Inevitably, such different reactions did occur, but it is surprising how unanimous the understanding and appreciation of Kramers' personality actually were. The Christian minister who gave the funeral oration described Kramers in terms strangely similar to those used by a Jewish refugee physicist (whom Kramers helped as a young man): "An old Jewish legend relates that God had compassion with the human race and took care that there would always be thirty-six just men on earth to make the human condition bearable. They were to be recognizable only by their deeds and themselves unaware of being one of the thirty-six. I think I have known one of them."[5]

Not many people obtain new knowledge or produce new insights into the understanding of Nature. There are not even too many who can appreciate, let alone enjoy, the progress achieved. Generally, such progress is slow and painstaking. Only rarely are there periods of rapid development, where each seminar lecture or each journal issue appears to announce new advances—so that the picture changes rapidly and understanding and insight increase with dramatic suddenness. Those active in such a period of turmoil must continually evaluate the significance of these claims; they may have to readjust their approach, and reassess their goals. This is a difficult and painful process. It was especially so for Kramers, who was more interested in fundamental understanding than immediate or cheap success. Furthermore, as a major physicist, Kramers was not merely interested in reacting to the new results, he wanted to influence that development himself: He expected to mold it and guide it along lines consistent with his views. The resulting struggles, disappointments, successes, heartbreaks, frequently missed opportunities, and rare moments of elation—all these are now hardly remembered. Yet it is only through a detailed understanding of these conflicts and struggles that a genuine appreciation of the significance of the advances can be obtained. The central role played by the personal relations between the scientists in this process is often overlooked. But their likes, dislikes, scientific prejudices, antagonisms, jealousies, fears, ambitions, and generosity often had a significant influence on the determination of the eventual physical theories—an influence that cannot be ignored.

It is the confluence of all these factors—a highly intelligent, highly intellectual person who matured at a time when a scientific revolution was in progress, who was sensitive to the personalities of physicists and nonphysicists alike—which makes Kramers such an especially interesting person. A description of his life is a description of humanity at its best and most human. Surprisingly—or perhaps not—all of us will see a partial personal reflection of our hopes, aspirations, and struggles in his life.

CHAPTER 2

The Classical Synthesis: Lorentz

About 1900—although this clearly did not happen in one day—radical and dramatic changes began to occur in the nature and character of physics. Although the profound character of the changes was not realized away, many new facts required a serious rethinking of the basic principles of physics without, however, arriving at anything resembling a systematic new set of concepts. When Kramers entered the field of atomic physics, as a beginning researcher in 1916, he entered a field in turmoil. There were basic disagreements about the goals of physics and there was considerable confusion about the appropriate methods of handling the existing problems.

At the end of the 19th century, the confidence of physicists, which was shared by the population at large, in their general approach was based to a considerable extent on the spectacular success of Newtonian mechanics. Based on beautiful measurements by Tycho Brahe and stimulated by the analysis of planetary orbits by Kepler, Newton inferred that there should exist an inverse square law of gravitational attraction between all material bodies everywhere in the universe. This extrapolation appears even now as surprisingly accurate. Combined with Newton's laws of motions, which predict the motion of a body once the forces acting on it are known, this provides a compact and powerful method to describe *all motions* in the universe. It is certainly impressive that the motions of the sun, the moon, the earth, and the moons of Jupiter or the motions of oceans, rockets, footballs, and cars are all described with great accuracy by the same laws expressed in the same types of equation. This surely creates confidence in the existence of universal laws. So overpowering was the influence of Newtonian mechanics that investigations in many different areas were patterned after mechanics. In fact, a common measure of success in a scientific area was the discovery of laws equal in universality and power to those of mechanics; another would be the *reduction* of the phenomena to mechanics.

During the 19th century, laws having the scope and generality of the laws of mechanics were discovered in two new important areas: thermodynamics and electrodynamics. In both fields there was a long history of incomplete experimental results organized under a number of separate rules which were then eventually integrated and summarized in a very few simple laws. These laws, like the Newtonian laws, could be expressed naturally and compactly as differential equations. Also as in mechanics, an enormous amount of empirical material could be understood and correlated on the basis of a small number of general laws.

There is another large class of phenomena, all having to do with light or optics, that appeared as separate and distinct from most other phenomena. Experiments on light, which were among the first precision measurements, demonstrated not only reflection and refraction, but also interference and diffraction. Precise empirical laws were found and the 19th century view was that light was a wave motion in the ether. That provided an excellent description of the phenomena but it still did not establish any connection between light and the rest of physics. It was therefore a genuine triumph when an incisive analysis by Maxwell (about 1870) of the equations of electrodynamics (the Maxwell equations) showed that light waves were electromagnetic waves; so light is indeed an electromagnetic phenomenon. Optical phenomena (refraction would be an example) could be derived from the Maxwell equations. Thus, a genuine unification was achieved. Once again a set of diverse and apparently unrelated observed regularities were seen to be consequences of a few basic universal laws.

With this impressive list of achievements, it is easy to understand why physicists at the end of the 19th century would have enormous and justified confidence in their general approach. Physics was a precise, well-defined, logical, and deductive scheme that seemed complete and understandable. It was indeed an intellectual structure of imposing classical beauty; it possessed coherence and an inner harmony. No one had any serious doubts about the general approach, the universal validity of the underlying laws, or the effectiveness of the mathematical descriptions. Optimism and confidence reigned supreme. "The more important fundamental laws and facts of the physical sciences have all been discovered and these are so firmly established that the possibility of their ever being supplanted in consequence of new discoveries is exceedingly remote .... Our future discoveries must be looked for in the sixth place of decimals."[6] It was recognized, as in this quote by Michelson, that there might be finer details, further elaborations, but the basic approach to physics was clear, its methodology was understood, and it was known to be correct and effective: The intellectual scheme was set. By the end of the 19th century, physics had come to a certain finished stage. There was confidence that, at least in principle, any phenomenon in physics should be capable of quantitative explanation in terms of known causal laws. This total structure is nowadays referred to as *classical* physics.

It seems that all the characteristics of physics at that time were personified in H. A. Lorentz (1853–1928), a professor of theoretical physics at the University of Leiden. There can be little doubt that Lorentz was one of the greatest Dutch physicists who ever lived. He certainly was one of *the* major physicists of his time; he made monumental contributions to all divisions of classical physics. In his researches he always showed a complete mastery of the subject matter, so much so that the results often made an effortless impression, as if they flowed naturally and inevitably from the basic laws. All his work possesses a basic harmony and unity characteristic of Lorentz's view of life and physics. Lorentz was revered in The Netherlands. He was lecturing at Leiden when Kramers was a student there, from 1912 to 1916. Kramers often commented that he was destined for physics because his initials H. A. were the same as those of Lorentz. On occasion there were slight hints that Kramers was somewhat bothered by the tendency to extend the comparison beyond the similarity of the initials. Lorentz's personality and his approach to and expectations of physics exerted a powerful influence on Kramers.

It is hard to realize that in a relatively short time most of the high hopes and optimistic expectations gave way to a sense of confusion and concern. In his opening address of the Solvay Congress in Brussels, October 30, 1911, Lorentz declared: "Black body radiation is a most mysterious phenomenon and a most difficult one to unravel." This tone is quite different from that of Michelson's statement (p. 9) made only 12 years earlier. Of course it is not true that this breakdown of classical physics occurred suddenly. There were a number of strong hints that the classical scheme had some troubles and weak points. Given the great successes of classical physics, it was only natural that these deficiencies would be considered as technical problems rather than indications of deeper difficulties.

The molecular description of matter gave an early indication, not so much that something was wrong, but that some physical objects had quite extraordinary properties. Today the idea of materials made up of molecules is commonplace and in no way paradoxical, but 100 years ago it was difficult to visualize and accept that matter consisted of an enormous number of extremely tiny molecules moving with typical speeds of 3600 miles per hour. Even after Boltzmann (1844–1900) and Maxwell (1831–1879), in the latter half of the 19th century, constructed a brilliant theoretical structure explaining many of the properties of matter on a molecular basis, there were many first-rate scientists who refused to take molecules seriously. This gave rise to intense discussions among Boltzmann, Planck (1858–1947), and Mach (1838–1916). The relations among these three protagonists are complicated and changed in time, with personal elements playing a significant role. But the existence and method of description of molecules were central to the discussion. In a lecture in Leiden in 1908, Planck took the molecular view very seriously with a corresponding anti-Mach position. Since Mach had an almost pathological aversion to molecules, it is not surprising that at times the discussion became quite acrimonious.

In spite of their unusual properties, molecules could be incorporated in the classical description. However, to do this a new element had to be introduced:

the notion of *probability*. (This was especially emphasized by Boltzmann.) Thermodynamic laws, as explained on this molecular basis by Boltzmann, emerged as describing the average behavior, or the most "probable" behavior, of a system. This was unpalatable to many physicists who believed and certainly wished that in the strict classical tradition laws should be precise and causal; they should deal with actual events and real phenomena and not with "averages" or "probabilities of occurrence." The further development showed that the Boltzmann view was the only tenable one, although several problems remained. The probability interpretation of thermodynamics was used with great success by Einstein in 1905. Thus, the classical scheme of physics could be enlarged to include the molecular description of matter by the explicit use of probability notions. This was an extension of the scheme which was perhaps unexpected and unwelcome to some, but it was not a breakdown of the system.

Another strain was put on the classical scheme by the discovery of the electron. Contrary to the molecular substructure, which for some time was restricted to theoretical discussions, the electron was a purely experimental discovery: It was unanticipated and unexpected. The discovery made by Thompson in 1887 was the existence of a particle much smaller than an atom which carried the smallest amount of electricity possible. If it was difficult to imagine a molecule moving at 3600 miles per hour, it is staggering to contemplate an electron in a typical experiment exhibiting a speed of 18,600 miles per second. Indeed, a number of physicists had their doubts and concerns about the reality of electrons. As late as 1920, Roentgen did not allow his coworkers to utter the word "electron" in his laboratory!

What is perhaps most remarkable about the electron is that the laws of electrodynamics were obtained and put in their final form in Maxwell's equations well before the electron was discovered. (Maxwell died 16 years before the discovery of the electron.) An obvious problem now arises: Can one relate and harmonize the existence of electrons with the already known laws of classical electricity and magnetism? It is indeed possible to construct a theory of electrons which not only leads to the electrodynamics of Maxwell but predicts a number of striking new effects as well. These major accomplishments were due to Lorentz. Once again the classical scheme was enlarged, this time by the introduction of the electron. It is true that classical physics did *not* predict the existence of the electron, nor its properties, but its existence did not appear to *contradict* classical physics in any way. The Lorentz theory of electrons stands as one of the impressive achievements of classical physics. It also stands as one of the "last" achievements of classical physics. Any efforts to enlarge it further or to analyze the nature of the electron itself or otherwise modify that theory inevitably led to new, no longer classical, domains. Kramers, who took classes in electrodynamics from Lorentz, was profoundly influenced by the Lorentz theory of electrons. Kramers' later approach to problems involving electrons shows unmistakably Lorentz's influence.

It was possible to extend the classical scheme of physics to accommodate the molecular theory of matter and the theory of electrons while maintaining the basic classical ideas of causality, universality, and continuity. In the succeeding period, it was no longer possible to maintain all three, which necessitated a far-reaching reexamination and scrutiny of many principles of physics. But it must be stressed that just as in the first instance, physical laws and concepts are justified by their effectiveness in organizing and ordering experiences and experiments; a breakdown in a law or limitation of a concept is justified only by repeated and irreconcilable disagreements with experimental results. Guidance for new laws or new concepts can only be initiated by new phenomena.

CHAPTER 3

The Unraveling of Classical Physics: Planck and Einstein

It is not a particularly unusual observation that hot bodies emit radiation. This can be thermal radiation, which can be measured and felt, or it can be visible light, which can be measured and seen. The type and amount of radiation emitted depend on the temperature and the kind of material. One can measure both the total amount of energy emitted and the amount of energy of a given frequency (color). It is more unusual that when matter in the form of hollow balls is heated and observed through a small hole, all materials exhibit the same thermal spectrum; this is *not* the case for solid spheres. Although perhaps unexpected, this particular behavior has been observed and can be understood on the basis of the laws of thermodynamics, a well respected division of classical physics. Early in 1900, the first really accurate measurements of that thermal spectrum became available. Surprisingly, there was no theoretical formula that agreed with these observations. There was a theoretical expression derived on the sound basis of classical physics, but it agreed with the data only for low frequencies. For higher frequencies the theory and experiment did not agree at all. On the basis of a rather different semi-theoretical picture, it was possible to obtain an expression for the high-frequency part of the spectrum, but then the low-frequency part did not fit at all. There was no formula of any kind that fit the complete experiment.

It is important to observe that a hot body emitting radiation is not a particularly unusual phenomenon. In terms of classical physics, it should be governed by the laws of heat (thermodynamics) combined with those of light (electrodynamics). As such, the thermal spectrum was calculated by Lorentz and by Planck. They used different methods but both obtained the same incorrect low-frequency answer. Not only was it wrong, it was nonsense, because if the *total* energy emitted at a given temperature was calculated on that basis it was found to be infinite, while experimentally it is a perfectly reasonable finite amount (proportional to T^4). How serious this result was taken can be inferred from the name it eventually received from Ehrenfest (1880–1933)—"the ultraviolet catastrophe."

To Planck, who had a deep abiding faith in the methods of classical physics, this situation presented a dilemma. He first obtained, literally by some inspired guesswork, a formula that interpolated between the high- and low-frequency parts of the thermal spectrum. It fit the data extremely well and Planck received congratulations (October 19, 1900) from the experimentalist (Rubens) who had carried out the delicate measurements. But now it became imperative to see how this empirical result could be obtained from the established laws of physics. After two months of work, Planck produced his formula. In so doing, it was necessary to introduce a new universal constant called h, which had the physical dimension of an action. The existence of this constant led to the result that the energy, in the form of a vibration of frequency ν, is an integer multiple of $h\nu$; that is, it is discrete, not continuous. In the course of time, Planck gave a number of derivations of his basic formulas based on different assumptions, using different methods. However, they all contained the result that the radiant energy is *not continuous* but instead is emitted in pieces or packets called quanta. The smallest amount of radiation energy of frequency ν which can, according to this assumption, be emitted is $h\nu$; h (Planck's constant) is the universal constant also called "the quantum of action." Thus, energy *cannot* be divided arbitrarily fine; it is *not* continuous. Planck introduced an essentially discrete element in physics. This was not all immediately clear, but after Einstein's analysis, the discrete energies emerged as the essential novelty. This was not only nonclassical, it explicitly contradicted classical physics, where energy and energy exchanges are known to be continuous. Since in classical electrodynamics the continuity of energy and of energy exchanges is often used and always implied, Planck's derivation of the experimentally observed thermal spectrum rested on two mutually incompatible ideas: the discrete character of the energy exchanges and classical electromagnetic wave theory, which presupposes continuity. There are hints (I owe this insight to Professor Jost) that Planck recognized the existence of contradiction. This is not an enviable position for a systematic physicist.

This fundamental conflict haunted Planck all his life. He was torn between his wish to have a quantitative explanation of blackbody radiation (which he had given) and the unconventional radical nature of the necessary assumptions. "I [Planck] considered the introduction of the quantization of energy an act of desperation—the phenomenon required a theoretical explanation, whatever the price."[7] The best assessment of the status of physics after Planck's work comes from Planck himself. "The floundering of all efforts to bridge the chasm [to obtain the thermal distribution] soon left little doubt .... Either the quantum of action was a fictional quantity, in which case the whole deduction of the radiation law was in the main illusory and represented nothing more than an empty non-significant play on formulas, or the derivation of the radiation law was based on a sound physical conception. In that case the quantum of action must play a fundamental role in physics and this was something entirely new, never before heard of, which required us to reverse all our physical thinking since the establishment of the infinitesimal calculus by Leibniz and Newton, upon the acceptance of the continuity of all causative connection."[8]

Planck realized perfectly well that he had taken an extraordinary step. He was a modest, completely honest person, but he knew what he had done. On December 14, 1900, when he presented his derivation to the German Physical Society, he is alleged to have mentioned to his son "today I have made a discovery as important as that of Newton."[9] Even so, Planck was deeply ambivalent about the quantum; he hoped all his life to be able to fit the quantum idea within the framework of classical physics. Since this was impossible, he naturally wanted to limit the significance of the quantum idea as much as possible. In fact, the role of the quantum varied in successive versions of his theories. Planck primarily invoked the quantum idea to regulate the way radiation and matter interact with each other. He was most cautious about any actual, independent existence of quanta.

Both Lorentz and Einstein were soon aware of the intrinsic *incompatibility* of the quantum idea and classical physics. But Einstein, who criticized the Planck derivation mercilessly (because of its contradictory assumptions), nevertheless took the quantum idea much more seriously than did Planck. On the basis of a most brilliant analysis of radiation, Einstein suggested that radiation behaved as if it were actually composed of independent, actually existing quanta. Radiation of frequency ν could be considered as a stream of quanta of energy $h\nu$. It was common to call these quanta "light quanta" to distinguish them from Planck's quanta. Later, as here, they were called "photons." Thus, Einstein conferred explicit existence and reality on quanta; they were not, as for Planck, entities that merely monitored the exchange of energy. On the basis of these ideas, Einstein predicted a simple formula for the photoelectric effect, whereby light shining on a piece of metal liberates electrons. With these properties, Einstein had come very close to ascribing a particle nature to light.

Nobody liked Einstein's photons. The suggestion of their existence caused an uproar in physics circles. The basic reason was that the interference properties of light were experimentally so well established and so beautifully explained by the electromagnetic wave theory of light that nobody wanted to contemplate the introduction of particlelike quanta of light. They would, with one blow, spoil both the experimental agreement and the fit with the Maxwell theory. Einstein's argument for the existence of photons or light quanta is a physical interpretation of some mathematical results. It is a beautiful argument but it was not considered compelling. Even Planck, who started with quanta, wanted no part of Einstein's photons. Planck was willing to tolerate quanta in the emission and absorption processes, which were poorly understood anyhow. He was *not* willing to entertain the existence of photons for a pure radiation field which was accurately described by the classical wave theory. In one of the most unusual letters of recommendation ever written, Planck wrote about Einstein: "That he [Einstein] may sometimes have missed the target in his speculations, as for example in his hypothesis about light quanta, cannot really be held against him for it is not possible to introduce fundamentally new ideas, even in the most exact sciences, without occasionally taking a risk." [9a]

The validity of Einstein's viewpoint about photons was tested in the best possible way by a careful, painstaking series of experiments carried out by Millikan (1868–1953), who had no sympathy for that viewpoint. Indeed, Millikan had a thorough dislike for Einstein's photons. "The suggestions of Einstein are bold, not to say reckless. They seem to violate everything we know about interference." [10] But after more than 10 years of the most meticulous experimentation, Millikan concluded in 1923: "The work resulted, *contrary to my own expectation*, in the first experimental proof of the Einstein relation and the first photoelectric determination of the Planck constant h." [11]

It is very important to recognize that in spite of this success of Einstein's view on quanta, the initial objections of Planck, Lorentz, and Millikan against the photon picture remained unanswered. The classical description of light in terms of electromagnetic waves, possessing the interference and diffraction properties, remains correct. But in addition to the experiments demonstrating this wave character, there are new experiments that are *incompatible* with the wave theory and require new features. The blackbody thermal spectrum requires quanta, while the photoelectric effect demands a photon description.

Most physicists were understandably reluctant to abandon the wave picture of light and for the first 20 years of the 20th century, very few did. Only Einstein took the photons seriously. But all physicists, one way or another, had to accommodate to a quantum notion of some kind: The Planck formulas agreed too well with the precision experiments to be ignored. Furthermore, especially through the work of Einstein, quantum ideas were instrumental in explaining serious deviations from the classically expected behavior in the specific heat of solids at low temperature. Again, quantum ideas produced a marked improvement in bringing theory and experiment together. Such improvements could not be obtained without the use of quantum ideas. There were certain phenomena for which quanta were essential, but the existence of quanta contradicts classical theory, and there are many phenomena that must be explained classically. The situation was confusing and not very satisfactory. A very typical assessment was given by Nernst (1864–1941) on January 26, 1911: "At present the quantum theory is essentially only a rule of calculation, of apparently very strange, one might say grotesque nature, but it has proven so fruitful by the work of Planck ... and Einstein that it is the duty of science to take it seriously and subject it to careful investigation." [12] This is indeed a far cry from the systematic understanding and well-defined starting points which classical physics had provided for the explanation of phenomena. And more conflicts were on the way!

CHAPTER 4

Conflicts and the Basic Incompatibility

The discussion in Chapter 3 shows that classical physics is incapable of providing an adequate description of blackbody radiation and the photoelectric effect. There is another set of experimental results that demonstrates the inadequacy of classical concepts even more dramatically. The results all concern atomic structure and the light (or radiation) emitted by atoms.

The analysis of light emitted by hot bodies had long been the subject of serious, mainly experimental and organizational, investigations. The experimental aspect dealt with the intensity of the light emitted, the ability of a body to absorb light, and especially the *spectrum* (the set of frequencies of colors) of the emitted light. After many years of experimentation, it was shown that *atoms* can, under appropriate circumstances, emit light. Their spectrum consists of a number of *lines*. The line spectrum is characteristic for the atom. If one wanted to phrase this result as a biblical injunction it would be: "By their spectral lines ye shall know the atom."

The organizational studies concentrated on relating the frequencies (or wavelengths) of the experimentally discovered lines. These line spectra could be very complicated, for example, the spectrum of iron in the visible region has easily 5000 lines. The simplest atom, hydrogen (H), has just four lines in the visible region, but other atoms have much more complicated spectra. Occasionally, lines occur in groups, such as doublets or triplets. The spectral pattern is affected by electric and magnetic fields, which usually increase the number of lines. To obtain any understanding from this mass of data it is necessary to have some order. This was accomplished bit by bit—by the recognition that the frequencies of a spectrum were not random numbers, but that within a spectrum there are definite patterns called "spectral series." A spectral series is a set of lines whose frequencies are related to each other. One of the earliest series was discovered by Balmer (1825–1898) for some of the lines in the hydrogen spectrum. The numerical relation he found is of the form

$$\nu = R\left(\frac{1}{4} - \frac{1}{n^2}\right)$$

where R is a constant. If n is given integer values, that is, 3, 4, 5, and so on, the corresponding ν's give the frequencies of lines. It is interesting that at the time Balmer obtained this formula (1885), he knew of only four lines in the hydrogen spectrum. He predicted, using his formula, the existence of a fifth line. Later, Balmer showed that his formula accounted for all 12 hydrogen lines known at that time.

The line spectra usually show a great deal of regularity, although this generally cannot be recognized as easily as in the Balmer case. These more complicated regularities can still be expressed as *numerical* relations between the frequencies.

Since the spectra produced by the atoms are characteristic of and determined by the atoms, it must be possible to understand these spectra on the basis of the atom's structure. According to classical physics, the structure is a geometric arrangement of positive charges and negative electrons. As such, it should be described by classical electrodynamics. The difficulty was that no one knew how to predict the observed line spectra of the atoms from that model. In fact, there were many indications that such a classical charge distribution could not produce the series of sharp spectral lines at all. If not an out and out conflict, there was at least a serious logical hiatus between the application of electrodynamics to the atom and the observed results.

A major insight came with the experimental results of Rutherford (1871–1937). In a celebrated paper (March 7, 1911), the result of 2 years experimentation, he announced that the positive charge in an atom is all concentrated in a very tiny region. This region, the nucleus, is about 10^{-12}–10^{-13} cm, a hundred thousand times smaller than the atom; it contains practically all the mass of the atom. This experiment, and many succeeding studies confirming it, settled the question of the geometric arrangement in an atom. This appears at first sight to provide a very simple picture, now known to everybody, of a neutral atom. It consists of a heavy positively charged nucleus with negative electrons moving around it at great speeds.

What laws of physics should describe an atom? To a classical physicist this is a silly question, because the laws are already known; they just need to be applied to this system. The motion of the electrons—the orbits and energies—should be described by classical mechanics, with the force between the electron and the nucleus being the familiar electrostatic force. Such an atom classically can emit light. It is a consequence of classical electrodynamics that *if* an electron is *made* to move in a circle by applying external forces, it emits electromagnetic radiation (light) of frequency $\nu = 1/T$ (where T is the period of the motion). The *classical* frequency emitted is directly related to the period of the *classical* motion.

In spite of these appearances there is a profound conflict. If an electron moves in a curved path around the nucleus, it is accelerated. This is a consequence of classical kinematics. According to classical electrodynamics an accelerated charge radiates energy away. Thus, the electron loses energy and will—by classical mechanics (the law of conservation of energy)—move closer to the nucleus, radiating more energy away and spiraling into the nucleus in a very short time. This whole process can be calculated in detail using the classical equations. The light emitted is not a series of sharp spectral lines of given frequencies but a continuous broad band of all frequencies. Thus, this mechanism cannot be the origin of the spectral series; that is one disagreement with experiment. Much more serious is the problem that, based on this picture, atoms cannot exist for more than 10^{-5} seconds, in flagrant conflict with experiment. This is one of the reasons that physicists in 1911 reacted rather quietly and with great caution to Rutherford's results. Nobody seriously doubted his experiments, but the resulting system with negative electrons moving around a positive charge could not be stable according to the accepted and well-understood laws of mechanics and electrodynamics. It is easy to understand that there would be great reticence to accept or even consider a picture in which the atoms, the basic constituents of matter, are fundamentally unstable. Unfortunately, the combination of experiments and classical theory led inevitably to that unacceptable conclusion.

By 1913 there was no *single* framework in physics which could accommodate the available information: the classical experiences, the experiments on atomic structure and electrons, the experiments on blackbody radiation and photons, and the information about line spectra. Not only was there no single framework, there were serious compatibility questions and a number of explicit conflicts.

This was the situation in physics when Niels Bohr (1885–1962) published his seminal papers in the English journal *Philosophical Magazine* in 1913. It took unusual genius and extraordinary courage to suggest a rethinking and reexamination of the phenomena along strange and unfamiliar lines. It took a special kind of intuition and insight to discern within the confusion the possibility for an effective and rational approach to atomic problems.

CHAPTER 5

Bohr: Success and Ambiguity in a Period of Transition

With the advent of Bohr's papers, a new period started in the development of atomic physics. Because of the important role of the quantum, the theoretical structure is often called quantum theory. Here the two terms are used as denoting more or less the same set of ideas. Bohr's ideas, general approach, and underlying philosophy dominated atomic physics to an unprecedented extent in the period from 1916 to 1925. This coincides exactly with the period Kramers spent in Copenhagen, so apart from the initial phase from 1913 to 1916, Kramers was directly involved in the formulation, elucidation, and application of quantum theory.

Bohr's approach to the conflicts and incompatibilities of atomic physics was radically different from that of Einstein and Planck. Although Planck was obviously aware of the problems which the quantum of action presented for classical physics, there are frequent indications that he would have been pleased with a partial reconciliation of classical and quantum notions, instead of an irreconcilable split. Einstein, by contrast, tried to adapt the foundation of theoretical physics to the new and contradictory facts. Bohr's approach, although his philosophical ideas were influenced by both Einstein and Planck, was operationally quite different. Bohr was totally convinced that the conflicts were genuine and irreconcilable. No minor modification of classical theory could produce the results of Einstein and Planck on blackbody radiation, nor could alterations of electrodynamics explain the line spectra. Furthermore, the quantum of action plays a central role in the description of atoms; it cannot be treated as a subsidiary concept. Thus, according to Bohr, there was no way in which the classical theory could describe atomic processes and therefore no point in even trying to explain the stability of atoms or the existence of the quantum of action on classical grounds.

The procedure advocated by Bohr is to start from experiments and observations—and perhaps some classical ideas which seem trustworthy—and then to construct theoretical concepts and rules and try to fit this all into a mathematical scheme. This is not a very precise prescription! To implement it, vague as it is, one has to be prepared to abandon classical laws and classical constructs although it is not clear just where to do this and what should replace them. The study of atoms takes us into a new domain with new laws and new concepts, but it is not known at this stage what concepts are appropriate and what the laws actually are.

Under these circumstances, Bohr suggests that a basic change in viewpoint is required. Instead of trying to reconcile quantum and classical approaches, the first task is to search for a scheme that can accommodate the basic facts. From these facts one can hopefully construct a scheme into which these facts fit. In the Bohr view, the concepts and the scheme that combines the necessary elements will provide a *language* for the handling of the new phenomena.

As a minimum, the scheme must allow for the existence of stable atoms, the existence of the quantum of action, and the line spectra. But sometimes classical theory is correct and that must fit into the scheme as well. The theoretical scheme sought starts from these general ideas. Also, to be useful, it should correlate data and predict new results. The ideas of Bohr are more in the nature of a program than a deductively derived theory. Experiments are at the heart of the program. Concepts must be introduced in such a way that they allow a convenient organization of the observations. Bohr's proposal is therefore to seek a minimum scheme that contains only the elements needed to describe experiments, even if, on a classical basis, these experiments are contradictory. The eventual underlying theory should be constructed almost by trial and error from that scheme. Thus, there has to be a continual exploration of theoretical alternatives to organize the experiments and to make the scheme more coherent. Theoretical physics in this period dealt with provisional constructs, with ambiguity, with incomplete theories, a striking change indeed from the formal deductive character of classical physics.

Although the ideas presented so far are vague, the implementation in Bohr's paper of April 1913 is extremely specific. It is most transparent to formulate Bohr's theory in terms of three separate postulates:

(1) An atom can exist in certain states only. In these states, called the stationary states, the atom does not radiate. In such a state an atom has a definite energy; the possible energies form a discrete set.

(2) If an atom exchanges energy with its environment, the atom changes from one stationary state to another. An atom can emit or absorb electromagnetic radiation only by performing a transition from one stationary state of energy E_1 to another state of energy E_2, so that

$$E_1 - E_1 = h\nu$$

where h is Planck's constant and ν is the frequency of the emitted or absorbed electromagnetic wave. This Bohr rule is a generalization of

Planck's idea that energy exchanges take place in quanta. It asserts that the *only* way an *atom* can exchange electromagnetic energy is in quanta $h\nu$.

(3) In a stationary state the electrons in an atom move according to the laws of classical mechanics and classical electrostatics; however, *no* radiation is emitted.

These Bohr postulates are well known, but their revolutionary character and great subtlety are not always sufficiently appreciated.

The concept of "stationary states" is genuinely new. It corresponds to a possible orbit of the system in classical physics. It is strange on two accounts: First, there is no radiation emitted in these states. That flatly contradicts classical electrodynamics, and Bohr effectively denies the validity of the Maxwell equations for electrons inside an atom, although the validity of Newtonian mechanics is maintained. Second, it is remarkable that certain orbits are singled out as stationary states. That means that the theory must contain a *selection* procedure which picks those states from all the possible classical states. Since the stationary states have definite energies, the selection mechanism picks out a discrete set of states having definite energies. This means that the energy of such a system can have certain discrete values only; that is, the energy is *quantized.* A stationary state is also called a quantum level or quantum state. The selection mechanism, which is usually called the quantization rule, can obviously not be classical; it is a totally new element in the scheme.

The method whereby a Bohr atom could emit or absorb energy is also totally nonclassical. In classical theory, the frequency radiated is determined by the properties of a *single* orbit. Specifically, the geometry of the orbit determines the frequency. In the Bohr atom, this frequency is determined instead by an energy difference between two orbits and has little to do with the period of an electron in a single orbit. This process of transition, called a quantum jump, is another novelty of the Bohr view. Bohr was explicitly clear, and mentioned numerous times, that he had not given and could not give a mechanism for a particular jump, nor was it sensible to ask when the electron jumped. He merely said that the emission of light by an atom could occur only if the atom made a quantum transition. The transition itself was not described any further. The transition notion itself is as *nonclassical* as it can be.

The selection mechanism, or the quantum rules, used by Bohr can be expressed in a variety of ways but most simply by the requirement that the angular momentum of an electron in a circular orbit can have only certain values allowed by the relation $mvr = nh/2\pi$, where m is the mass, v the velocity, r the radius of the circular orbit, $n = 1, 2, 3 \ldots$, and h is the Planck constant. This choice may seem very arbitrary or even capricious. This impression does little more than confirm that the quantization rules are outside the scheme of classical physics and must be supplied from quantum principles. The rule given is one way to accomplish this. One must, of course, demand that these rules *do* lead to results in agreement with experiment.

On the basis of these ideas, Bohr was able to account quantitatively for a surprising number of spectral observations. In fact, the successes were nothing short of spectacular. Bohr could explain the Balmer series of hydrogen; moreover, he obtained an explicit value for the Rydberg constant, which agreed within 6% with experiment. That is already something of a success, although one might argue that this success only came about because Bohr had picked the quantum condition very cleverly: So cleverly that with that choice Bohr had given a model that was based on the nuclear atom, gave spectral series, and related it to quantum ideas! No one had accomplished anything like that before.

Additional successes followed rapidly. There was a series of lines in hydrogen, the so-called Pickering–Fowler series, which appeared to disagree with the lines allowed by the Bohr theory. On the basis of his postulates, Bohr suggested that these lines should *not* be hydrogen lines, but instead should belong to ionized helium. There was understandable excitement when these lines were indeed found to be helium lines. The Bohr theory made a prediction (for the constant R) for that series which agrees well with the data, but it was not exact. There was enormous excitement when Bohr, by a slightly better treatment (taking the motion of the nucleus into account), obtained an agreement between calculation and experiment correct to 1 part in 100,000. As an additional consequence of this more precise calculation, Bohr predicted another series of helium lines, at that time unobserved. These lines should be slightly displaced from the Balmer lines. They were found a year later at the right places by Evans.

A direct verification of the energy level scheme in an atom and the frequency condition came from a series of experiments by Franck and Hertz. By carefully controlling the energy of particles colliding with the atoms, they showed that unless there was enough energy to cause the atom to make a transition, the particle and the atom could not exchange energy at all (all the collisions were elastic). It should be mentioned that although the Franck–Hertz experiment shows the quantum nature of energy exchange directly, the experiment was not designed for that purpose. In fact, Franck and Hertz were at first (for about 3 years!) unwilling to accept Bohr's explanation.

This is certainly an impressive set of successes. In spite of the curious mixture of classical and quantum ideas, and in spite of—or perhaps because of—the unusual ideas of stationary states and quantum jumps, it was evident that Bohr's ideas had something to do with the real world. They certainly could not be ignored because no other theory or scheme came anywhere near predicting and correlating as much spectral material as the Bohr theory.

The reactions of the world of physics were swift and intense. The first Bohr paper was published in the July 1913 issue of the *Philosophical Magazine*. One of the first reactions came from Sommerfeld (1868–1951), a leading German physicist in Munich in September 1913: "Although I am for the present rather sceptical about atom models in general, nevertheless the calculation of this constant [R] is indisputably a great achievement."[13] Bohr had been in close contact with Rutherford in England, who had been very helpful in getting Bohr's papers published. Rutherford clearly appreciated the incisive insights of Bohr, although he had some difficulties in grasping all the subtleties of the unconventional arguments. It was due to Rutherford's urging that Bohr was invited to give a lecture on the "theory of radiation" for the British Association of the Advancement of Science, which would be held in Birmingham, in September 1913. That was a considerable honor, but even so, Bohr was reluctant to accept. At the last minute, mainly because the great Lorentz would be present, Bohr decided to go. At that meeting, just a few months after Bohr's papers appeared, the work caused quite a stir. Jeans (1877–1946), one of the main speakers, referred to Bohr: "Dr. Bohr has arrived at a most ingenious and suggestive, and I think we must add, convincing explanation of the laws of spectral series .... The only justification put forward for Dr. Bohr's bold postulates is the weighty one of success."[14] In general there was, at least publicly, a very favorable acceptance of the new ideas; an article in *Nature* called Bohr's theory "convincing and brilliant."[15] Bohr's colleagues and friends (Oseen, a Swedish physicist, and Hevesy, a Hungarian) sent congratulatory letters.

Einstein took a great interest in Bohr's results. He immediately recognized their great importance, but he was also profoundly troubled. Einstein first heard of Bohr's paper at a physics colloquium in Zurich. At the end of the talk, V. Laue (1879–1960) a famous x-ray physicist, protested: "But this is nonsense! Maxwell's equations are always valid, an electron in a circular orbit must radiate!"[16] Einstein got up and said: "Very curious. There must be something to it. I don't believe that it is pure chance that the Rydberg constant can be expressed numerically, so accurately in terms of basic constants."[17] Einstein kept a close watch on the developments of the Bohr picture. When Einstein heard that the Pickering–Fowler lines were indeed helium lines, he was extremely astonished, and with that penetrating insight that was so characteristically Einstein's, he went right to the point: "Then the frequency of the light does not at all depend on the frequency of the electron [in its orbit] ... this is an enormous achievement."[18] Many years later (1949) Einstein reflected again on his initial reaction to the early days of quantum theory and the Bohr postulates. "All my attempts to adapt the theoretical foundation of physics to the new facts were completely unsuccessful. It was exactly as if the ground was slipping away from under our feet and we had no firm soil that we could build on. It always seemed a miracle to me that this vacillating and quite contradictory basis turned out sufficient to enable Bohr, a man of brilliant intuition and keen perceptibility, to find the predominant laws of spectral lines and the electron shells of atoms; including their significance for chemistry. This still seems a miracle to me. It is a manifestation of the highest form of musicality in the realm of thought."[19]

In spite of the immediate successes of Bohr, many physicists had serious reservations. There was the strange circumstance that while the quantization rules were nonclassical, once the stationary orbit was selected the behavior was again described classically. The quantization rules themselves seemed, and were, arbitrary. But the most intense criticisms and the most serious objections concerned the frequency condition and the quantum jumps from one level to another. There is something very strange about these quantum jumps. They are, of course, discontinuous events. Even Rutherford, who was generally sympathetic, wrote to Bohr about the quantum jumps: "I do not see how the electrons know where they are suppose to stop." But other people had much stronger feelings. For example, Schrödinger, in a letter to Lorentz, wrote: "The frequency condition in the Bohr theory seems to me, and has indeed seemed to me since 1914, to be something so *monstrous*, that I should like to characterize the excitation of light in this way as really almost inconceivable."[20]

Another person quite upset with Bohr's ideas was Ehrenfest, a professor at Leiden and the person with whom Kramers had the most direct contact as a student. In August 1913, Ehrenfest wrote: "Bohr's work in the quantum theory of the Balmer formula has driven me to despair. If this is the way to reach the goal I must give up doing physics."[21] Even 3 years later (spring 1916) after a number of impressive Bohr successes, Ehrenfest remained unconvinced, unimpressed, and unrepenting. In that year, Sommerfeld had given a more general and more persuasive version of the quantization conditions which defined their logical status in a much better way. This prompted an unusual reaction from Ehrenfest: "I am sending you congratulations .... Even though I consider it horrible that this success will help the preliminary, but still completely monstrous, Bohr model on to new triumphs."[22] It is important that the two major scientific influences in Kramers' early life, Ehrenfest and Lorentz, both took a dim view of Bohr's approach to physics. Ehrenfest disliked that approach, as is evident; Lorentz had doubts of such magnitude about the consistency and the logic of Bohr's argumentation that he never actively worked in atomic physics although he was thoroughly conversant with its status.

The reaction to Bohr's ideas split quite generally according to scientific generations. The up and coming generation saw much to admire in result, approach, and style. After all, Bohr did have a procedure and he obtained results nobody else had gotten before. The older generation was more restrained in its praise; more emphasis was placed on the problematic aspects and the logical inconsistencies. At the meeting in Birmingham, Lord Rayleigh, a venerable classical physicist who wrote the monumental *Theory of Sound*, reacted to Bohr's ideas by saying nothing at all. In private, he confided to Bohr that he, in his younger years, was convinced that a physicist over 60 should not participate in debates on new problems. Even though Lord Rayleigh added that he no longer held this view so firmly, he held it strongly enough not to take part in the discussions.

All these negative comments could not obscure the fact that Bohr's ideas were essential for an understanding of spectra. The ideas of the quantum level, quantum state, and quantum transitions provided the language and formal scheme for the organization and correlation of spectral properties. Equally important are the quantum conditions which eventually select the stationary states, and hence the spectrum. These concepts, and the accompanying language, provide the key for the description of atoms and spectra, although they surely could not be defined or justified classically; in Bohr's time they could not even be defined very precisely. But the picture presented by all these ideas had a coherence nevertheless, and the model was unmatched in its accurate description of spectra and atoms. (A more honest and accurate statement would be: To the extent that spectra could be understood at all, the Bohr theory provided the language and the concepts. The actual quantitative successes were pretty much restricted to H, He^+, Li^+.) So effective was this description that many physicists forgot its provisional character and considered the model as "the truth" or "the reality." Sommerfeld, who in 1913 "did not believe in 'atomic models'," took them very seriously in 1917. This was in part because Sommerfeld had succeeded in giving a much more significant version of the selection mechanism, the quantum rules, than Bohr. This enabled him to include some effects of relativity which led to the first quantitative understanding of the fine structure splitting of spectral lines. This was unquestionably a major success.

Bohr, however, emphasized time and time again the provisional character of his ideas. To Bohr the issue of overriding significance was always that the classical ideas could *not* be applied in atoms. For that reason, he continually stressed the *conflicts* between the classical theory and the needs of atomic theory. By focusing on this conflict, he hoped that some day a new theory encompassing both domains could be found, but the explicit recognition of the conflict was a necessary preliminary to that major step. "I hope I have expressed myself sufficiently clearly, that you appreciate the extent to which these considerations conflict with the admirably coherent group of conceptions ... the classical theory of electrodynamics. On the other hand, by emphasizing this conflict, I have tried to convey to you that it may be possible in the course of time to discover some coherence in the new ideas."[23]

The intellectual and, no doubt, emotional struggles to liberate one's thinking from classical precepts was a major theme throughout Bohr's life. A closely related struggle concerns the efforts to forge a new theoretical structure for the new circumstances that had the same coherence, predictive power, and beauty as classical physics. In Bohr's, and others, writings from 1913–1925, there are frequent references to a "future, more complete theory." At that time there was no such theory, but there were clearly hopes and expectations. The search for a theory that would contain classical physics as a special case and that would give *rational, systematic* explanations of the quantum conditions, stationary states, and quantum transitions became central in all theoretical efforts.

It is important to stress that the conflicts and the struggles left an indelible mark on Bohr and indeed on all those directly associated with the changes from classical to quantum physics. It is all well and good to emphasize conflicts, contradictions, and the need for changes, but it is extremely difficult and acutely painful to give up concepts that have provided an understanding of the world. It is particularly hard to abandon trusted ideas for the sake of rather indefinite, fleeting, and as yet, poorly defined speculations. The ensuing struggles, trials, conflicts have all the earmarks of an intellectual revolution. Some old ideas must be discarded, some kept, some modified; altogether, it was a fluid, ambiguous, but immensely exciting period with unusual promise for further development. This is certainly the way the future of physics was perceived by many young physicists.

Kramers arrived in Copenhagen in the fall of 1916 when the Bohr model had achieved considerable success but was far from universally accepted. Kramers' training in physics under the powerful teaching of Lorentz and the mesmerizing influence of Ehrenfest was primarily classical. As soon as Kramers started to work with Bohr he became immersed in Bohr's approach to quantum theory. This required a restructuring of Kramers' thinking, which he accomplished with incredible speed. He was personally involved in some of the further successes and some of the later failures. As for most theorists, the search for a coherent, organized, rational description of atoms, radiation, and quanta dominated Kramers' scientific life.

CHAPTER 6

Bohr, the Photon, and Kramers

It was inevitable that Kramers would be immediately affected by the subject matter of Bohr's physics. Of equal importance for Kramers' later development were the combined influences of Bohr's approach to physics, his style, and his personality. Kramers' scientific views were molded by his close contact with Bohr. Bohr's personality and his approach to physics were intertwined so closely that an analysis of Bohr's influence on Kramers' physics automatically must involve an examination of the interplay of the personalities.

One particularly important characteristic of Bohr was his single-minded pursuit of an issue. He had phenomenal powers of concentration; he was capable of, and insisted on, a complete and totally exhaustive analysis of any question. Partial or confused answers were anathema to Bohr. Many aspects of a question would be examined, then reexamined and various reformulations would be considered, tried, and rejected. The problems were considered from many viewpoints in many different contexts, so that in the end an extremely precise, thoroughly thought-out result emerged. This was a very slow and painful process. Bohr would demand clarity and precision with an almost fanatical fervor. He was well aware of the enormous intellectual burdens this imposed; every day he suffered the agonies of the struggle for perfection. One of Bohr's favorite poems by Schiller expresses his feelings regarding the struggle to arrive at the truth: *Nur die fülle führt zur klarheit und im abgrund wohnt die wahrheit* ("Only complete understanding can lead to clarity and the truth resides in an abyss").* Bohr had not a trivial bone in his body and he demanded from himself and those around him a total uncompromising intellectual honesty.

Bohr's method of work was also unusual. Even though he spent many hours

* This translation is rather free!

in solitary thought, his actual intellectual activity was most often carried out in the presence of others, by argument or by discussion. He needed the stimulus of other minds, or the type of response a Greek chorus would provide, to function most effectively. Only rarely did he write himself; usually he *dictated*—his Ph.D. thesis to his mother, his early papers to his wife. Later, Bohr's scientific collaborators, starting with Kramers, would write down notes while Bohr dictated and cogitated. There would be discussions and arguments; Bohr's method of thinking was one of continual verbalization. This demanded the utmost concentration on the part of Bohr, but also on the part of the collaborator. This intense intimate method of personal cooperation—debating, arguing about every word, groping for the correct ideas—also produced a peculiar emotional strain on all participants. When Bohr started a sentence he often did not know where it would lead or how to finish it. This prompted Dirac (1902) to say: "When I was a boy I was always taught never to start a sentence without knowing the end of it."[24]

The same process of thinking via articulation persisted in Bohr's public lectures. Although he prepared them with meticulous care, he was not a very successful public speaker. While speaking, Bohr was in deep and intense thought, and the rethinking and rearrangement of the arguments while giving a lecture did little to enhance the cogency of the presentation. When Bohr's brother, Harold Bohr, a world famous mathematician, was asked why he (Harold) was one of the greatest mathematical lecturers in the world while Niels was so poor as a public speaker, he explained: "This is because at each place in my lecture I speak only about those things I have explained before, but Niels usually talks about matters which he means to explain later."[25]

This somewhat unusual procedure, where the creative thought processes are stimulated by conversations (in a sense these discussions are the thought processes), emphasizes the significant role of the verbal interchanges and the language in which these discussions are carried out. This same insistence on the appropriate use of language characterized Bohr's writing style from his very earliest papers. To Bohr, the language, problems, concepts, and physics were so entangled that every word, innuendo, and equivocation had to be exactly right to express the physics correctly.

When Rutherford, in Manchester, received Bohr's first paper for publication, he thought that it was most significant, but he commented numerous times that it was too long and ought to be cut down. In Rutherford's letter to Bohr he added a rather off-hand postscript: "I suppose you have no objection to my using my judgment to cut out any matter I may consider unnecessary in your papers. Please reply."[26] Bohr did not just answer—he immediately set out from Copenhagen to Manchester to "fight it out." If it is recalled how complicated and expensive such a trip was at that time, the immediate decision to go to Manchester becomes especially remarkable. Although the details of the succeeding meeting are not known, the event made a profound impression on Rutherford. Even 20 years later, Rutherford recalled—in words that are perhaps the best and most compact characterization of Bohr's style: "I could see that he had weighed up every word in it, and it impressed me how determinedly he held on to every sentence, every expression, every quotation, everything had a definite reason and although I at first thought that many sentences could be omitted, it was clear when he explained to me how closely knit the whole was, that it was impossible to change anything."[27]

It is important to recognize the great contrast between the manner in which Bohr expressed himself in writing and in oral discussions in a small group. In the former case, the greatest care was exercised to obtain the exact meaning, but with all needed equivocations. He was careful not to offend opposing views nor to provoke confrontations. In the privacy of his office, however, his expression might contain drastic imagery and quite explicit and uncompromising criticism.

Mathematics played a minor role in Bohr's creative activities. He certainly was acquainted with contemporary mathematics, in part because of his closeness to his brother Harold. But to Bohr, mathematics was a tool to be used for calculations; it was not an intellectual instrument to be employed as a means to investigate a theory. Such matters as formal structure, mathematical beauty, logical simplicity, or mathematical elegance meant little to Bohr. "I cannot understand", he used to say, "what it means to call a theory beautiful if it is not true."[28]

Bohr had an incredible intuitive insight into the nature of phenomena and their interrelation. It seemed almost as if he were a personal friend of electrons and atoms and could sense their behavior without the use of formal mathematics. Instead, his verbalization of the logical connections between the phenomena and their description enabled him to arrive at surprising insights. In this process his insistence on a precise language caused him continually to scrutinize and refine the logical chains that had led to his conclusions. So much so, that Bohr, in practically every lecture, reviewed the considerations that had led to his decisive break with classical physics and eventually to the Balmer formula.

This particular combination of qualities—a superb physical intuition, an unmatched verbal facility to handle and manipulate concepts, the ability to pursue logical implications with single-minded fanaticism, and inexhaustible intellectual and physical strength—made it very difficult to maintain or originate dissenting views in Bohr's presence. It was not so much that Bohr was a priori opposed to new or different ideas, but rather that new ideas or contrary suggestions were subjected to the same intensive scrutiny as were Bohr's own views, and few ideas could withstand that penetrating analysis. It is not surprising that although Bohr was open-minded, it was difficult, if not impossible, for other scientists to generate opposing views very effectively or hold them for very long. The combination of all these qualities made Bohr's persuasive powers just too great.

The struggles and conflicts surrounding the photon notion afford an outstanding example of the tenacity and effectiveness with which Bohr defended his views and stuck to his ideas. These same struggles, and the investigations resulting from them, also had a particularly strong influence on Kramers' thinking and subsequent scientific development.

The views of Einstein and Planck concerning the quantums of radiation were quite different. Planck, it will be recalled, thought of *quanta* as controlling the interactions between matter and radiation but believed free fields to be described correctly and totally by classical electromagnetic waves. Einstein, on the other hand, attributed considerable reality to the quanta of even the free field. As time went on, Einstein became more and more convinced about the existence of quanta (photons) and their possession of *both* energy and momentum. Einstein wrote in 1917, after he finished an important paper on blackbody radiation: "... with that [the existence] of light quanta is practically certain."[29] Two years later he wrote: "I no longer doubt the *reality* of photons—although I still stand quite alone in this conviction."[30]

In this controversy Bohr unquestionably sided with Planck. Bohr was quite willing to attribute all kinds of nonclassical features to the interactions between matter and radiation; after all, he did introduce quantum jumps. But Bohr was deeply impressed by the precision of the classical electromagnetic theory of light, so he wanted to maintain the wave description at practically all costs. Bohr often stressed that the frequency ν (which defines the energy of a quantum) is itself derived from the wave theory. Bohr at one time even contemplated giving up the conservation of energy in order to keep the wave character of radiation. Even *after* the Compton effect (April 1923) demonstrated the existence of a photon momentum rather conclusively, Bohr remained unconvinced. Both before and after that time, Bohr insisted on the *exclusive* validity of wave concepts in the description of light. When many persons, including Einstein, criticized this single-minded insistence on the exclusive use of wave concepts by pointing to the particle (photon) properties of light, Bohr allegedly replied: "Even if Einstein had found an unassailable proof of the existence of photons and would want to inform me [Bohr] by telegram, this telegram could only reach me by radio because of the existence and reality of radio waves."[31] Even so, the experimental discovery of the Compton effect caused Bohr profound troubles, and he expressed these worries in a letter to Rutherford (January 27, 1924): "You can understand my concern—for a person for whom the wave theory is a creed."[32] Only a little later (February 7, 1924), Bohr expresses the same sentiments in a letter to Michelson: "I hope that it will be possible using Slater's idea to harmonize the essential reality of the quantum theory with the continuous character of the radiation theory."[33] The unusual ideas of Slater will be discussed later, but for the present, Bohr's insistence on the continuous nature and wave character of radiation must be especially noted. Bohr's objections to the photon concept were deep; he was unable, or unwilling, to consider the photon as a viable option. He admitted that Einstein's photoelectric equation (based on the photon idea), which was confirmed by Millikan's experiments, yielded the most accurate value of Planck's constant. He admitted further that this phenomenon was "entirely unexplainable in the classical theory." But admitting that these results were in favor of a photonlike explanation brought Bohr no closer to taking that explanation seriously. For example, he concluded his Nobel lecture by expressing grave reservations about the photon: "In spite of its heuristic value [Einstein's] hypothesis of light quanta, which is quite irreconcilable with the so-called interference phenomena, is not able to throw light on the nature of radiation."[34]

Kramers, at that time, was in daily contact with Bohr. From the description of Bohr's didactic methods of thinking and communicating, it is clear that Kramers must have been totally familiar with, and convinced of, Bohr's views on photons. In the first published comments on this issue, Kramers himself is in fact *plus royaliste que le roi* ("more of a royalist than the king"), and he rejects the photon notion explicitly and forcefully. (See Chapter 13, Section I.) In a later study (see Chapter13, Section III), Kramers and Bohr had high hopes for a new theory of radiation based on some original ideas of Slater. In Slater's suggestion, the photon played an important role, which of course neither Bohr nor Kramers liked, although they were intrigued by the other aspects of Slater's work. In a joint paper by Bohr, Kramers, and Slater, the photon idea was eliminated altogether. Under the persistent criticism of both Kramers and Bohr regarding the reality of photons, Slater gave up and agreed to eliminate the photons. There was a substantial price to be paid, however, and the resulting theory no longer possessed strict conservation of energy and momentum, the conservation laws being satisfied only in an average sense. Slater, in a letter to van der Waerden, recalled the mood at that time: "As you suspected, the idea of the statistical conservation of energy and momentum was put in the theory by Kramers and Bohr quite against my better judgment."[35] As if this was not radical enough, the theory also demanded that strict causality had to be abandoned in individual processes. This is not yet the place to assess the merits of the proposal, but it is clear that Bohr and, at that time, also Kramers would go to practically any length to avoid the Einstein photon. When the Bohr–Kramers–Slater theory was shown to conflict with experiment, Bohr of course recognized that the theory was no longer tenable but was still not convinced that photons actually existed. He concludes a letter to Geiger (one of the experimentalists involved) by expressing doubts about the photon: "Conclusions concerning an eventual corpuscular nature of radiation lack a satisfactory basis."[36]

These events played a central role in Kramers' scientific life. However, that was *not* the only reason to recall them at this juncture. They were mentioned here to illustrate the enormous strength and power of Bohr's convictions. These convictions could be shaken and modified only by exhaustive analysis and by the most compelling arguments. Forced by events Bohr eventually shifted his attitudes on photons—or more precisely, evolved a scheme in which there was room for photons. In fact, in one of these curious rearrangements which occur so often, Einstein and Bohr appear to have crossed their intellectual allegiances. But this happened much later (1926–1928). In this earlier period (1920–1924) Einstein believed in photons and Bohr did not, and as the quoted literature showed, Bohr was not easily dissuaded.

Another reason to record these events now is that even this small fragment of the tangled history shows not only the strength of Bohr's personal convictions but also his extraordinary persuasive powers. Bohr's persuasive powers were not derived from brilliant performances in public debates or lectures, but rather through painstaking, detailed analyses, in which no argument was left incomplete, no alternative was left unexplored, and no evidence was omitted. Through such exhaustive treatment, Bohr's ideas and scientific ideology permeated the minds of his colleagues and collaborators. Opposition was eventually ground down. This was quite a common occurrence. It was most pronounced, of course, for those who were in close contact with Bohr because they were exposed daily to Bohr's uncompromising searching analysis of all basic issues in physics.

Kramers met Bohr for the first time in September 1916. He came unannounced, as a young student of not quite 22 years old. He had studied classical physics in Leiden, The Netherlands, and wanted to continue his studies. Nothing else was very definite. Bohr at that time had just returned from Manchester to accept the position of Professor of Theoretical Physics in Copenhagen. Bohr was about 10 years Kramers' senior and he had already published his epoch-making papers on atomic structure; his fame was spreading rapidly. Bohr, with all his intellectual and personal power, introduced Kramers to the problems and promises of the quantum world. Kramers assimilated the new ideas with incredible speed. Although Kramers rapidly became an independent thinker and researcher—in part because of his superior mathematical skill—it was inevitable that his philosophical outlook and approach to fundamental physics would be shaped and formed by Bohr. From the very start there was a very special personal and scientific relation between Kramers and Bohr. As with all human relation, it changed in time, but the initial differences were never totally erased. Kramers' approach to physics, his judgment as to what was important and what was not, and his hopes for physics were in the beginning of his career dominated by the ideas of Bohr. This influence remained strong for a considerable length of time and only began to fade when Kramers moved into new, more formal aspects of physics. But throughout Kramers' scientific life, significant remnants of Bohr's influence were clearly discernible.

CHAPTER 7

The Bohr Institute: Pauli and Heisenberg

When Bohr returned to Copenhagen in 1916 to accept the first chair in theoretical physics in Denmark, he probably did not realize that he was on the threshold of establishing one of the most outstanding and most unusual institutions in the history of science. It had taken the Danish authorities a good two years to establish a professorship in the first place, and it was not clear that they would be willing to do anything more—this in spite of the meteoric rise of Bohr and the frequent hints that he might be lured away to positions in other countries. By any standards, the conditions in Copenhagen in 1916 were abysmal. Bohr had only one small room which he had to share with his assistant, and there was no laboratory for experimentation at all. Clearly, something had to be done to make scientific activity possible and to keep Bohr in Copenhagen. After considerable thought, Bohr proposed to the faculty that a new "Institute for Theoretical Physics" should be established. Bohr submitted a report to the faculty on April 18, 1917 in which he suggested that the institute would provide teaching and research facilities and, in addition, facilities for numerical computations and experimental investigations. It was further suggested that the residence of the director (the professor of physics) would be right in the institute because "the supervision of research stretches over all hours of the day and night." [37] After the usual ups and downs, strikes, delays, changes, and modifications, a modest three-story building was built. The institute was inaugurated on March 3, 1921.

Soon after Bohr arrived in Copenhagen, and well before any suggestions for an institute were made, he received a curious note from a totally unknown Dutch student:

> Copenhagen, August 25, 1916
> Prof. N. Bohr:
> To begin let me introduce myself, by telling that I am a Dutch student in physics and mathematics. I've studied for 4 years in Leiden and followed the lectures by

Prof. Kuenen and later of Prof. Lorentz and Prof. Ehrenfest in physics and of Prof. Kluyver in mathematics. I passed all examinations, so that I am now what we call in Holland "Doctorandus"; I want to get the doctors title by writing a dissertation. These four years I always hoped and wanted to go to a foreign university, in order to go on with my studies. As I didn't like to go to a country that is in war now, I decided to go to Copenhagen and hope to study now mathematical physics. I've not yet specialized myself much in that branch but only worked out something, together with Prof. Ehrenfest. Of course I should like very much to come in acquaintance with *you* in the first place and also your brother Harold. Therefore I would be very glad, if you would permit me to visit you one of these days. Perhaps you'll be so good to write me a card or telephone to my hotel *when* I may come and see you. My address is ...."[38]

The visit which Kramers requested in this letter did take place, and Kramers asked whether Bohr would take him on as a student and assistant. (That meant that Bohr would have to find some financial support.) Bohr was a bit nonplussed by it all, but his brother Harold suggested that if the young Dutchman was so keen on working with him, he might as well be given the chance. Bohr agreed to do just that. Kramers' small stipend came from funds made available to Bohr by the Carlsberg Foundation, which later would support the Institute for Theoretical Physics. Thus, apart from Bohr, Kramers became the first member of the Institute for Theoretical Physics—well before it existed, even before it was conceived.

Kramers' entry into Bohr's institute, and thereby into the inner circle of quantum physicists, is a remarkable story from beginning to end. It is not altogether usual for a practically penniless student to take a trip in the middle of a world war just to go study at a different university. Since Kramers had no way of knowing or guessing how Bohr would react to his wish to complete his studies there, he might well be on his way back before he learned a thing. Money and assistantships were not abundant enough to count on getting financial help in a foreign country where no one knew him. Equally unusual is that while Kramers, in his note, refers to Professor Lorentz and Professor Ehrenfest—both well-known physicists—he brings no letters of recommendation or letters of introduction from them to Bohr. (It is doubtful that he could have gotten such letters!)

Perhaps most surprising of all is how well it all worked out. Kramers' formal ability in classical mechanics was immediately useful in helping Bohr with his great study "on the Quantum theory of line spectra." This study was stimulated by some important papers of Sommerfeld on generalizations of the quantum rules. Kramers' control of the intricate mathematics was essential for the successful completion of these studies.

It is also amazing that a short seven months after Kramers came to Copenhagen, Bohr trusted his understanding and knowledge sufficiently to have Kramers visit the physics department in Stockholm to explain quantum ideas to a not very sympathetic audience. Kramers continued to make rapid progress, so that when Oskar Klein, the next (pre-institute) visitor, showed up in May 1918, Kramers initiated Klein in the ideas of quantum theory as well as in Bohr's unconventional style of working. By the time Kramers finished his

Ph.D. thesis (1919), which was a brilliant technical implementation of one of Bohr's general suggestions, he was recognized as one of the few experts in the Bohr version of quantum theory. After the Institute for Theoretical Physics was established, Kramers became its first assistant. Soon afterward, in 1924, he was appointed as a lecturer at the institute, which formalized the teaching obligations he had carried out for some time. He was a superb expositor and brilliant lecturer. His rapidly expanding reputation was recognized by his election, at an uncommonly young age, to the Danish Academy of Sciences and Letters in 1925.

The gradual expansion of the institute paralleled the increasingly significant role of Kramers in physics and in the institute. It was natural that visitors to the institute would spend a great deal of time with Kramers as well as with Bohr. It was known and accepted that Kramers was thoroughly conversant with Bohr's ideas—and he often could explain them better. However, the scientific orientation—the delineation of research directions—was throughout determined by Bohr.

Bohr's style of working also set the tone for research at the institute. Collaboration between people and groups, rather than working in isolation, was the accepted mode of research. There were innumerable dialogues, discussions, and arguments—often until deep into the night. Of special importance for the scientific interchanges were the many seminars, which were informal, with many interruptions, and without time limit. Heisenberg (1901–1977) wrote (much later) that "Science is rooted in conversations,"[39] which is an accurate description of the manner in which science was carried out at the Copenhagen Institute at that time. The fact that Bohr lived right in the institute added greatly to the ease of communication, the informality of the atmosphere, and the intensity of the debates.

Bohr had an uncanny talent for picking the most gifted, original, and perceptive young physicists. Since he controlled funds available for visitor's stipends, he could invite promising talent. For the record, his first four selections were Kramers, Klein, Pauli, and Heisenberg. It is hard to imagine how one could do much better. Usually, the physicists assembled around Bohr were young. They generally came from similar social backgrounds, and they themselves formed a rather close-knit social group. All were passionately interested in physics and felt that atomic physics was exceedingly important. An aura of continued intellectual excitement pervaded the atmosphere. They all (at least in the period from 1916 to 1925) acknowledged Bohr as the intellectual and scientific leader. In this period (1920–1925) Kramers often acted as Bohr's spokesman; he was referred to as "Bohr's lieutenant."[40] Pauli, with his sharp wit, described Kramers with the phrase, "Bohr is Allah and Kramers is his Prophet."[41] This tied Kramers and *his* physics directly to the status and success of Bohr and *his* physics. The Institute for Theoretical Physics at Copenhagen was not only a physical location for physicists to gather but a "school": Its physics had a well-defined style and character, primarily determined by Bohr. There were other schools also possessing equally definite, but of course distinct, character. Under Arnold Sommerfeld, a long and strong tradition in theoretical physics had developed in Munich. Sommerfeld's physics was quite different from Bohr's: He was perhaps unkindly (and probably incorrectly) referred to as a "mathematical mercenary."[40] He used his considerable mathematical ability in *any* theory that would yield quantitative predictions for realizable experiments. He worried less about the rationale, the logical structure of the theory, than he did about clear-cut procedures, specific numerical results, and comparisons with experiment. Sommerfeld wrote the following to Einstein (June 11, 1922): "I can only contribute to the technology of quantum theory, you have to create the philosophy of quantum theory."[42] As this quote shows, Sommerfeld was interested primarily in the mathematical elaboration of a theory and shied away from fundamental questions. He would almost go as far as to say that the calculation rules *are* the theory.

Sommerfeld's insistence on specific and concrete problems, combined with his mathematical mastery, exerted a major influence on atomic physics. His approach to physics was also ideally suited to introduce young students to physics. Consequently, it is not surprising that Sommerfeld's school produced a very large number of extraordinary students. Sommerfeld took a great interest in his students, even in the early phases of their careers. It was his custom to involve his students very early in some kind of research. The problems were often minor, consisting of numerical interpretation of data, but Sommerfeld created an atmosphere of excitement and involvement with fabulous results. In Munich, the Bohr-type models, especially as modified by Sommerfeld, were taken more or less literally as the description of atoms. Several of Sommerfeld's students gave detailed and numerical analyses of these models, and the results were often compared with experiments. The results were taken quite seriously.

Göttingen also had a "school" of theoretical physics. The University of Göttingen had a long tradition as *the* major mathematical center of Germany. In 1910, the great mathematician Hilbert (1862–1943) continued that tradition. Physics in Göttingen was good; especially because of the presence of Max Born (1882–1970), a first-rate, all around physicist. The style in Göttingen was considerably more formal and mathematically oriented than in Munich. The physicists in Göttingen were more aware of general mathematical results than were those in Munich or, surely, those in Copenhagen. The atmosphere in Göttingen tended to be stiff and formal. Pauli often referred to *Göttingen Gelehrsamkeit* ("Göttingen's [excessive] learnedness"), which definitely was *not* intended as a compliment. The initial reception of Bohr's ideas in Göttingen had been quite cool. As the theory began to show more and more successes, the appreciation increased, but the initial doubts lingered and were never totally overcome. Of course, it was recognized here, as everywhere else, that Bohr's viewpoint had shown itself so powerful that in spite of its problems it could not be ignored. It is most interesting that what might appear as a fortuitous meeting of these three schools led to a series of events which had a decisive influence on the direction the theory of quanta was to take.

The Academy of Sciences, Göttingen, obtained in 1908 a bequest from a mathematician, Paul Wolfskehl, to reward 100,000 marks for the first person to publish a proof of Fermat's last theorem. By a reasoning which is perhaps not totally clear, the Wolfskehl committee decided to use the interest of the bequest to invite prominent scientists to give lectures at Göttingen. Whatever the reason, the lectures were a smashing success. The selection committee picked a remarkable set of lecturers including Poincaré, Lorentz, and Sommerfeld. The last lectures in the series* were given by Bohr from June 12 to 22, 1922. It was at these lectures (sometimes called the "Bohr Festival") that persons from the Copenhagen, Göttingen, and Munich schools met face to face for the first time.

In the audience were two of Sommerfeld's most brilliant students, W. Pauli and W. Heisenberg. It is no exaggeration to say that Bohr's lectures caused a major reorientation of the scientific outlook of both. On July 25, 1921, Pauli obtained his Ph.D. degree with Sommerfeld in Munich. If anything, his thesis was something of an anticlimax, for Pauli burst forth on the scientific stage as a 19-year-old prodigy who understood more about relativity than any person, with the exception of Einstein. The book on relativity which Pauli wrote at the request of his professor (Sommerfeld) while still a student, still stands as one of the most thoughtful and incisive treatments of relativity. Pauli became interested in atomic physics, and his thesis dealt with the hydrogen molecule ion. He went to Göttingen as Born's assistant and was there when Bohr gave his lectures in June. Pauli recalls his impressions of Bohr: "A new phase of my scientific life began when I met Niels Bohr for the first time.... It made a strong impression on me that Bohr at that time and in later discussions was looking for a *general* explanation ... in contrast to Sommerfeld's approach."[43]

Not only was Pauli deeply impressed by Bohr, but Bohr was also impressed by Pauli. He immediately invited Pauli to spend some time at the institute in Copenhagen. Bohr informed Kramers by letter (July 15, 1922): "Today I just want to tell you, what I think will interest you, that Pauli is coming up here in October to stay for about half a year. He is an excellent man in all respects and will certainly be most helpful.... he and Born have solved the problem of the helium orbits and they have in all details verified all our results.... He would now like to try his hand on the helium atom and is, I believe, at the moment busy looking at it by a method different from yours. He emphasized strongly that he was doing it only for his pleasure and that he, of course, would not publish anything before you have published your results."[44]

Pauli did spend the fall of 1922 in Copenhagen and later became a regular visitor at the institute. He and Kramers were very congenial, talked a great deal, and became fast friends. Pauli's relation to physics was very much like that of a man to his mistress: He usually loved it, but on occasion wished he never heard of it. There was an enormous involvement and great passion, but also deep depression, rejection, and anger. His extraordinary creative gifts

* The inflation in Germany at that time terminated the lecture series.

were superseded only by his critical faculties, and he became the scientific conscience of a whole generation of quantum physicists. In his ruthless critical appraisals he spared no one, with the notable exceptions of his teacher Sommerfeld and (almost of course) Lorentz; but not Bohr, not Einstein, not Heisenberg, and certainly not Kramers.

Heisenberg, at the time of Bohr's lectures, was a student just in his second year. Sommerfeld knew how fascinated Heisenberg was with atomic theory and took him along from Munich to listen to Bohr in Göttingen. During these lectures Bohr talked about a number of topics: One of them was the quadratic Stark effect in hydrogen,* and he reported on a calculation made in Copenhagen by Kramers. This calculation, as all calculations at that time, was a mixture of classical ideas, quantum ideas, and inspired guess-work. Although the results were unusual, Bohr felt that one could have confidence in Kramers' results and he expected that they would be confirmed by experiment.

There is more than a touch of irony in the circumstance that Heisenberg publicly entered the world of physics by an incisive criticism of a calculation by Kramers. In the discussion after Bohr's lecture, Heisenberg explicitly disagreed with Bohr. This was most unusual! To say anything after a big lecture is not too common, for a second year student to ask a question is practically unheard of, and to argue with Bohr on even terms should be impossible, but this is precisely what Heisenberg did.

Bohr was obviously worried by Heisenberg's remark. He was equally deeply impressed by Heisenberg. He invited Heisenberg first for a walk, then for a visit to Copenhagen. Heisenberg recalled later how important this first meeting with Bohr had been for his further development."[45] It was a great surprise to Heisenberg that Bohr, in marked contrast to Sommerfeld, had not calculated anything. His insight and understanding came from an intense preoccupation with the actual phenomena rather than from calculations and mathematics. Furthermore, Bohr was much more skeptical about the validity of atomic models than many other physicists, including Sommerfeld. It was a revelation to Heisenberg that one could do physics in the manner of Bohr. After that meeting, the conceptual and philosophical aspects of theoretical physics began to play an increasingly important role in Heisenberg's thinking.

Heisenberg was an authentic genius. His rapidity of comprehension and assimilation were phenomenal, allowing him to do an enormous amount of work in practically no time at all. He was an indefatigable worker, an excellent sportsman, and an accomplished pianist, in addition to being very strong minded and very ambitious. His initial training with Sommerfeld was mathematical and pragmatic, and he developed a taste for the fundamental questions of physics with Bohr. The combination of those approaches led Heisenberg to some of the most original and daring innovations in physics.

It took until Easter of 1924 before Heisenberg's visit to Copenhagen could be arranged. Among other things, Heisenberg had to finish his Ph.D. thesis.

* This effect and Heisenberg's objection to the result will be discussed in Chapter 12.

By that time, Heisenberg—and also Pauli—had become increasingly critical of the Bohr theory. No doubt Heisenberg hoped to discuss these matters with Bohr. During the Easter vacation of 1924, Bohr took Heisenberg for a walking tour through Denmark where they discussed many things: history, philosophy, nationalism, and physics. As a consequence, Heisenberg was converted —if not to the details of Bohr's physics, certainly to the philosophical approach underlying Bohr's physics.

The first extended period Heisenberg spent in Copenhagen was from September 1924 until April 1925. During that visit, Heisenberg spent a great deal of time with Kramers, and they had lengthy discussions practically every day and every night. Not infrequently, Bohr would join them. That period was especially productive for Kramers; he originated and published some of his most significant ideas at that time. Those discussions were primarily concerned with extensions and generalizations of Kramers' ideas toward a more systematic theory of atoms and radiation.

By this time, the contradictions and internal conflicts of the Bohr theory had become so severe that it was generally agreed that altogether new and different ideas seemed to be required for a viable theory of atomic phenomena. It is probably fair to say that all persons actively working in atomic physics searched quietly, or perhaps frantically, for ideas, principles, or methods that would lead to the construction of a rational and consistent theory of atoms. The time for the formulation of "the future more complete theory," so often invoked in the discussions of the Bohr theory, seemed to be at hand—although most physicists believed this would be a prolonged and stepwise process.

After this period of intense cooperation with Kramers, Heisenberg left Copenhagen in April 1925 to return to Göttingen. By June 1925 he had, by a single stroke of genius, discovered the ideas, which within a few months produced such a complete theory.

CHAPTER 8

From Virtual Oscillators to Quantum Mechanics

The early years of the Institute for Theoretical Physics (1920–1925) coincided with very turbulent times in physics. Everyone was aware of the successes of the Bohr theory—publicly recognized by the award of the Nobel prize to Bohr in 1922. But the initial objections, such as the renunciation of classical ideas and the nonsystematic aspects of the Bohr theory, remained serious difficulties. In the period 1920–1925, a large number of contradictory and incomplete features appeared. For example, the Bohr theory gave a beautiful explanation of the hydrogen spectrum, but in spite of many efforts, it got nowhere with the helium spectrum. So the criticism that was muted by the successes of the theory became louder with its failures.

The manifest problems of the Bohr theory caused a curious split in expectations. Those physicists, pretty much the older generation, who never liked the Bohr theory in the first place (they described its practitioners as "quantum jumpers" and "atom mystics") expected that a theory along more classical and traditional lines would eventually emerge. They generally admitted that there was as yet no such "neoclassical" theory, but they hoped that the future would bring one. The physicists actively working in atomic physics (usually the younger generation) were quite aware of the contradictory and incomplete aspects of the Bohr theory, but they fully expected that a future theory of the same coherence and precision as classical physics could be found. It would probably entail new features and even further renunciation of the classical concepts per se. They also admitted that there was no such theory at the time. So both opponents and supporters looked toward the future, albeit with quite different anticipations.

Lorentz, who was probably as close to an impartial arbiter as any one, gave this assessment (December 10, 1923, Sorbonne, Paris): "All this [quantum theory] has great beauty and extreme importance but unfortunately we do not really understand it. We do not understand Planck's hypothesis concerning oscillators, nor the exclusion of non-stationary orbits, and we do not understand how, after all according to Bohr's theory, light is produced. There can be no doubt, a mechanics of quanta, a mechanics of discontinuities, has still to be made."[46] Apart from expressing the consensus, this statement by Lorentz also coins the name for the future theory—"mechanics of quanta." A little later Bohr used the term "quantum mechanics" in an important paper to denote this same set of ideas. Thus, there was a name before there was a theory.

Several physicists actively working in quantum theory were attempting to delineate the precise domain of validity of the theory. In 1921 Born wrote to Einstein: "Quanta are a hopeless mess, I am making a systematic search of places where the Bohr theory fails."[47] One of the curious observations especially stressed by Heisenberg was that even if the Bohr theory failed, it did not fail by much, and a relatively simple modification (which was ad hoc) would reestablish a surprising agreement with experiment. One of the first problems that Heisenberg tackled was treated that way. The problem was the splitting of spectral lines in a magnetic field. Heisenberg was asked by Sommerfeld to account for the data. Heisenberg managed to do this, using 1/2 integer quantum numbers—for no other reason than that it worked. Sommerfeld was impressed, Bohr appalled, and Pauli scoffed. Pauli told Heisenberg: "First you introduce 1/2 integers, then, 1/4, 1/8, ... etc. and quantum theory will crumble to dust in your capable hands."[48] Later, Heisenberg and Born showed that the usual Bohr quantization does *not* work for helium. However, if one uses the half-integer quantization it does work. Heisenberg pointed out that in many places where the Bohr theory uses integers, J, one can regain agreement with experiment merely by replacing J by $\sqrt{J(J+1)}$.

Bohr was extremely unhappy about the half-integer quantization and preferred to give up the validity of mechanics. The difference in viewpoints is illustrated in a letter from Heisenberg to Pauli (January 15, 1923): "Bohr does not believe in half integer quantization. He cares more for general theoretical principles than agreement with experiment."[49] This letter was written before Heisenberg visited Copenhagen; the conflicting approaches are very evident.

Shortly thereafter, Heisenberg again wrote to Pauli. There was general agreement that the "atomic theory" was not very good and needed major revision: "We both agree that fundamentally all the models are wrong."[50] There was even agreement about what this revision should achieve: "It must be taken as absolutely certain that both the electromagnetic theory and the quantum theory are valid in their respective fields, and equally certain, that the two descriptions are incompatible. We can only conclude that they are parts of an overriding system, which would give rise to *mathematical formulae identical* with those of the present theories."[51] There was no agreement, however, about the specific nature of the "overriding scheme," nor was there agreement about the path to be followed to arrive at the overriding scheme or a new more complete theory.

In such circumstances, it is to be expected that the personal predilections of the scientists would play an especially important role in determining the research directions. Heisenberg and Pauli investigated the organization and classification of spectral lines. Their theoretical efforts stayed pretty near the experimental reality while they attempted to infer general principles from their analysis.

Kramers, together with Holst, wrote a popular exposition of the Bohr theory, which indicated that they had confidence in the general approach—if not in all the details—of the Bohr theory. Born tried to settle once and for all where the Bohr theory broke down and where it gave correct results. Such continual efforts without a well-defined direction and without clear-cut successes led inevitably to disappointments and frustration. These are clearly noticeable in the many letters being exchanged. For example, Bohr writes to Sommerfeld: "In the later years I myself have often felt very lonely as a scientist because I had the impression that my endeavors to develop the principles of quantum theory systematically to the best of my ability were received with little comprehension. For me it is not a matter of petty didactic details, but a serious attempt, to reach such an inner coherence that there could be hope of obtaining a more secure foundation for further constructive work .... I know how unclear matters still are and how clumsy I am, in expressing my thoughts in easily understandable form .... During a stage in science where everything is still in ferment, it cannot be expected that everybody has the same views about everything."[52] The personal reactions to these frustrations also varied; for example, Pauli wrote: "Physics has run into a blind alley again. In any case it is too difficult for me and I'd prefer to be a comedian or a movie actor—and never hear about physics again."[53]

For Kramers, who was so very close to Bohr and so personally involved with the formalism of the Bohr theory, these times of vacillation and changing fortunes of the Bohr theory, were very difficult, but also very important. In this period (1920–1925) Kramers achieved some of his greatest successes and suffered some of his deepest disappointments. He was as aware as any person of the shortcomings of the theory and, like others, was preoccupied with applications, improvements, and possible modifications. He was less willing than others (especially Heisenberg and Pauli) to look for radical departures from the Bohr theory, but he was anxious to amend, alter, and clarify it, always, however, leaving the basic structure intact.

The failures and frustrations of physics produced in Kramers, as in most physicists, large swings of mood. In his correspondence with other physicists, these moods were not so explicit, and Kramers appeared reasonably hopeful about the success of a modified Bohr theory. If he expressed more negative feelings, these were generally tempered by irony. However, in letters to his lifelong friend Romeyn, Kramers is much less guarded, and the depressions and the occasional euphoria are much more evident: "I work on so many things at the same time, mainly together with Bohr, that for the time being there is no harvest .... I imagine that there is now starting a third period in my life, devoted to work, with a Capital W."[54] Romeyn answers—quite a bit later—advising Kramers to return to Holland: "Come to Bergen, where you can rest and reflect on the new phase in your life, for you cannot remain an appendage to Bohr forever."[55] This letter was answered by Kramers: "I am extremely busy—a little tired—I have no time and opportunity for desires and dreams .... I will for the time being remain in Copenhagen, there was a possibility of a position in the U.S.A. ... but that was not a permanent position."[56] These letters are typical: There is a great deal of work, many obligations, and frequent allusions to being tired.

The following year (1924–1925) was one of the most important in Kramers' scientific life. It had been known for a long time that the dispersion of light could be described exceedingly well by classical considerations going back to the work of Lorentz. The underlying physical picture was classical: An atom consists of electrons bound to a center and moving harmonically, obeying classical laws and radiating accordingly. This classical picture gives results that agree with experiments. But this picture cannot be maintained in the Bohr theory. So the question arises: How can one reconcile spectra (described by the Bohr theory) with dispersion (described classically)? This question was first asked by Lorentz when he heard Bohr's talk on quantum states in 1913. It was first treated by Ladenburg in 1921 and then reconsidered by Kramers in a short but brilliant paper in 1924. By using a mixture of classical and quantum considerations, Kramers managed to "translate" the classical expression for the dipole moment into a quantum expression that described dispersion in purely quantum terms. The expression did contain frequencies of the atomic transitions, as well as the probabilities of the transitions, but there were no references to the *orbits* or to quantum rules: All classical features had disappeared from the formulas. Thus, Ladenburg, and especially Kramers, had demonstrated that as far as its response to radiation is concerned, an atom is equivalent to, or may be replaced by, a set of harmonic oscillators whose frequencies are equal to the absorption frequencies ν_i of the atom. To reobtain the classical results in the limit of large quantum numbers, however, it is necessary to introduce *negative* absorption as well as the usual absorption. This was one of Kramers' key observations. Kramers' derivation made use of an important technical trick: the replacement of a differential quotient in classical physics by a difference quotient in the quantum translation. Kramers evidently had these results on January 17, 1924, for he showed them to Slater in Copenhagen at that time.[57]

It should be stressed that Kramers did not say that an atom *is* a set of oscillators, but rather that, in certain respects, it behaves as *if* it were. It is therefore understandable that a theoretical idea due to Slater, also involving oscillators, was of special interest to Kramers and also to Bohr. Slater suggested, actually as a means to reconcile the photon and wave characters of light, that while an atom is in a stationary state it is emitting electromagnetic waves of all frequencies corresponding to transitions to lower states. This field would not be an ordinary field and would not be directly observable. The corresponding waves would not carry energy but the field intensity is presumably related to the probability of finding photons. This field can be called a virtual field, and the oscillators producing it are virtual oscillators. Bohr and Kramers both liked the idea of a virtual field and virtual oscillators and in fact invented the names. These ideas fit naturally with those of Ladenburg and Kramers on the dispersion relation. But Bohr and Kramers did not like the photon idea (as was mentioned earlier), and they eliminated any reference to the photon in the paper which was published.

In this theory, called the Bohr–Kramers–Slater theory, the idea of a virtual set of oscillators—or a virtual radiation field—was explicitly introduced. As a result of the elimination of the photon notion, energy and momentum are conserved statistically, and *not* in each individual process.* The easiest way to understand the necessity of this radical step is to recognize that in the Bohr–Kramers–Slater theory the energies of atoms change discontinuously, as the Bohr theory demands, while the energy of the field is changing continuously. In that case there can be no conservation for the *individual* events.

Thus, Kramers was actively involved in the derivation of the dispersion relations and the elaboration of the Bohr–Kramers–Slater theory. It is important to note that there is, in principle, no direct logical relation between these two results. The dispersion relations suggest a *formal* virtual oscillator picture, but this picture by itself does not require the absence of photons and hence the statistical nature of the conservation of energy. The absence of photons is a separate, unrelated assumption, and the Kramers dispersion formula would be valid in the same form even if photons were assumed to exist. The importance of both contributions was recognized very rapidly.

Kramers himself stressed the most interesting and important feature of the dispersion formulas: their exclusive dependence on quantities that allow a direct physical interpretation on the basis of the fundamental laws of quantum theory and the structure of atoms. All references to the classical mathematical theory of orbits (the theory of multiple periodic orbits) have disappeared.

Several authors (e.g., van Vleck, Born) tried to apply the Kramers procedure to obtain corresponding quantum transcriptions of other systems. Born's specific idea was to treat the interactions between *mechanical* systems by the same methods by which Kramers had treated the interactions between a radiation field and an atom. When Bohr later commented on Kramers' activities in Copenhagen, he referred to the derivation of the dispersion relations as Kramers' greatest achievement.[58] Pauli, not given to exaggerated appreciation, wrote the following to Kronig (October 9, 1925): "Many regards to you and also to Kramers, whom I basically like a lot, especially when I think about

* A more detailed, more technical description of the Bohr–Kramers–Slater theory is contained in Chapter 13, Section III.

his beautiful dispersion formula."[59] Heisenberg was also very positively impressed by the Kramers dispersion relations. In fact, while in Copenhagen (until April 1925), Heisenberg and Kramers extended the original Kramers dispersion formulas to include the Raman effect. It was Heisenberg's plan to use the Kramers procedure, or a variation thereof, to obtain the intensities of the hydrogen lines, but this proved to be too difficult.[60] These comments clearly show the great significance attributed to the Kramers dispersion relations by the world of physics.

It is not easy to disentangle the various strands of Kramers' work at this period. Although he no doubt was aware of the logical independence of the description in terms of virtual oscillators and the statistical character of the conservation laws, his published work indicates that he considered the Bohr–Kramers–Slater theory as one unified structure. Furthermore, he appeared pleased with the results already obtained and the future prospects. He writes to Jan Romeyn on January 8, 1924, just a week before he tells Slater about the technique used in deriving the dispersion formulas: "I am lately having a great deal of success with my work .... I have had the good luck to be doing some new things."[61] All during 1924 Kramers worked seriously, and perhaps even feverishly, on the Bohr–Kramers–Slater theory. This in spite of, or perhaps because of, the gathering objections to the theory. In a typewritten manuscript written late in 1924,[62] Kramers reacts to an observation made by Einstein that in the Bohr–Kramers–Slater theory the fluctuations in total energy increase proportional to the time. Instead of taking this as an objection (it was so intended by Einstein), Kramers proposed to take this seriously and speculated that this energy loss might be observable for large stars. There are other manuscripts attempting various extensions, all indicating that Kramers took the Bohr–Kramers–Slater theory, in all its aspects, including the statistical conservation of energy and the *intrinsic probability* character of the transitions induced by the virtual field, extremely seriously. He might even have hoped that the "theory of the future" would contain significant elements of the Bohr–Kramers–Slater theory. On October 28, 1924 he wrote the following to Romeyn: "I fight in my field for the broadest possible viewpoints. Shall I ever have the good fortune to climb a new mountain? Maybe, just maybe."[63] For a person who expresses himself in an habitually understated manner, this is surprisingly explicit.

If Kramers' papers on the dispersion relations created a significant stir among responsible scientists, the Bohr–Kramers–Slater paper created an uproar. The reactions were immediate and came from all quarters. Schrödinger, who firmly believed in classical continuous pictures, was particularly pleased that theory allowed the serious discussion of conservation of energy as a statistical law. This was something he had been trying to do for some time. Born wrote to Bohr: "I would like to tell you how much pleasure I derived from the direction you have given this problem.... Here we have had similar thoughts.... I am quite convinced that your theory is fundamentally correct and in a certain sense it is the final word."[64] But while some people were ecstatic, others were doubtful, and still others had severe misgivings. Einstein objected vigorously and Pauli was very critical of the whole approach. While the main features of the Bohr–Kramers–Slater theory stemmed from the fact that Bohr and Kramers would not accept Einstein's photon, Einstein was even more unhappy with the alternative that the Bohr–Kramers–Slater theory provided. Einstein wrote to Hedwig Born: "The idea that an electron ejected by a light ray can choose of its *own free will* the moment and direction in which it will fly off is intolerable to me. If it comes to that I would rather be a shoemaker, or even an employee in a gambling casino, than a physicist."[65] Thus, there was a fundamental difference between Einstein and Bohr which even made its way to the newspaper. The Danish and the German newspapers tried to provoke Bohr and Einstein, respectively, into a public debate, but neither agreed to any public statement. The world of physics was, for a short time, divided into two camps, since the two architects of modern science, Einstein and Bohr, fundamentally disagreed.

Luckily, it was possible to test the validity of the statistical character of the Bohr–Kramers–Slater theory by direct appeal to experiments. Two experimental groups—Bothe and Geiger in Germany and Compton and Simon in the United States—carried out different experiments using different techniques. The results were unambiguously in conflict with the predictions of the Bohr–Kramers–Slater theory. The experiments were in harmony with the predictions of a theory based on Einstein's photon. On April 17, 1925, Geiger wrote to Bohr: "Our results are not in accord with Bohr's interpretation of the Compton effect ... it is recommended therefore to retain until further notice the picture of Compton and Debye.... One must therefore probably assume that the light quantum concept possesses a high degree of validity as assumed in that theory."[66] In a letter written to Fowler on April 21, Bohr adds a postscript saying that he just heard from Geiger that the experiments exclude the Bohr–Kramers–Slater theory. He ends his postscript with the following: "It seems therefore that there is nothing else to do than to give our revolutionary efforts as honorable a funeral as possible."[67]

The three originators of the Bohr–Kramers–Slater theory reacted characteristically differently to the definitive experimental disproof of their theory. Slater, who felt that he was coerced into a position of which he was not convinced, was not surprised. In his later writings, Slater claims that his initial opposition was in fact vindicated by the experimental findings. He did not appear unhappy with the results and it affected his subsequent approach to physics only to the extent that he remained antagonistic to Bohr.

Bohr took it all in good grace. Because of the unrelenting critique by Pauli and Einstein of the Bohr–Kramers–Slater theory, he was prepared for the (to him) negative outcome of the experiments. With the intellectual resilience and toughness so characteristic of Bohr, he started to think right away about the next step to be taken. From the Geiger and Compton experiments it followed that for the description of radiation the (Einstein) photon is every bit as necessary as the classical wave, and Bohr tried to deal with that reality. On

April 21, 1925, four days after Bohr received the experimental results, he wrote to Franck: "We will try in as painless a manner as possible to forget our attempt at revolution. Our goals we cannot forget quite that easily. In the last few days [4 days] I have suffered through all kinds of wild speculations to find an adequate basis for radiation theory. I have discussed this intensively with Pauli—who is visiting, he was for a long time unsympathetic toward our Copenhagen 'Pocket Revolution' [Putsch]."[68] It is interesting and characteristic that within the first 4 days after receiving the news, Bohr was already exploring new directions with Pauli. Bohr summarizes his reactions in a note (July 1925): "One must be prepared for the fact that the required generalization of classical electrodynamic theory demands a profound revolution in the concepts on which the description of nature has until now been founded."[69]

Kramers had much more difficulty accommodating to the experimental results than either Slater or Bohr. He thought that some of his most original and successful scientific efforts were an integral part of that theory. It is not surprising that he had taken the theory seriously and literally, and to have that structure destroyed by incontrovertible experimental evidence must have been a major disappointment. Interestingly enough, Kramers, in a letter to Born, still expresses some reservations about the interpretation of the experiments (something Bohr never did): "I can unfortunately not survey how convincing the experiments of Bothe and Geiger actually are for the case of the Compton effect."[70] At that time, a few weeks after the "news" had arrived in Copenhagen, Bohr was already looking for all kinds of new directions, while Kramers still appeared to hope for an interpretation of the experiments consonant with the Bohr–Kramers–Slater theory. This hope evidently disappeared during the summer. All during the spring and summer of 1925, Kramers was deeply depressed, in part because of the failure of the Bohr–Kramers–Slater paper. Oskar Klein, who was a close friend of Kramers, tried to console and encourage Kramers numerous times during that period.[71] Klein tried to cheer him up by referring to the pleasure children get out of their games, even if ultimately they are not successful,[72] but nothing was especially effective. Even as much as a year later, Kramers was still upset and a little defensive about the Bohr–Kramers–Slater theory. On February 26, 1926, Kramers wrote a letter to R. de L. Kronig,[73] in which the matter came up again. Kronig had written to Kramers, criticizing certain features of the Bohr–Kramers–Slater theory, in a rather unpleasant and aggressive manner. Although by this time Kramers no longer believed in it, he still felt compelled to explain the rationale of the Bohr–Kramers–Slater theory.

Kramers' disappointment in the demise of the Bohr–Kramers–Slater theory influenced him profoundly, and it may well have made him hesitant to pursue speculative ventures later. From that point on, his research deals primarily with specific, very difficult computational questions. He seems to shy away from very general considerations. If generality is to be achieved, it must come from very specific examples. However, in his popular lectures, Kramers remains a good deal more speculative and adventurous than in his technical papers.

There was a second, perhaps even more important, reason for the depressed mood of Kramers in the spring of 1925. This does not deal with something Kramers did which failed, but rather with something he *did not* do which succeeded.

To appreciate this properly, it must be recalled that the search for a systematic organized theory of atomic phenomena was the main preoccupation of all physicists. The precise form and character of such a theory was, of course, not known, but it was possible to list the desirable characteristics of such theories. They should preferably *not* contain classical quantities which describe the interior of atoms; they should not contain arbitrary quantization rules; and they should use mechanics and electrodynamics only in those domains where experiments assured their validity. Kramers, and later Kramers and Heisenberg, obtained *results* of *just* this character for the dispersion of light. (In fact, Kramers considered these relations as a support for the Bohr–Kramers–Slater theory.) The manner in which Kramers and Heisenberg obtained their results, however, was not systematic, but rather involved guesswork, interpolations, and considerable artistry. This was not unusual at that time. In much of quantum theory the starting point was classical with quantization conditions being applied and the classical solutions recast in quantum terms. Just what was "the correct translation" was not clear and became a matter of intuition, often based on a familiarity with the phenomena. Quantum theory was not an autonomous scheme. Quantum and classical ideas were mixed. It was not a well-defined scheme either; the rules were neither explicit nor a priori, with success depending greatly on the skill of the practitioner.

The *result*, the Kramers disperson formula, was much more impressive than the method used to obtain it. Kramers' formula depended only on the transition frequencies and their probabilities. Both could directly be related to observable quantities.

Upon his return from Copenhagen to Göttingen in April 1925, Heisenberg concentrated on this point: A very sensible and reasonable result was obtained by very unreasonable and doubtful methods. For example, the result did not depend on the orbits, but the derivation did; the result did not contain the quantum conditions, but they were used in the transcription process. The result also appeared to describe the atom as a set of oscillators, but the starting point did not. It now makes immediate sense that a proper procedure should not contain elements, for example, orbits, that are of no importance in the final physical description. As a start, Heisenberg looked for a scheme in which the *orbit* of an electron in an atom does not occur. This orbit is unobservable; therefore, the theory should not contain it.

This point had been discussed previously. Pauli wrote to Bohr: "I do not know to what extent one can talk about specific orbits of electrons in the stationary states. Kramers *never* admitted this doubt as sensible."[74] On December 12, 1924, Pauli repeated that comment: "I am sure of the necessity of the modification of the classical orbit notion, in spite of our good friend

Kramers' [colored] pictures*.... I believe that energy and momenta of stationary states are a good idea more real than orbits."[75] Whereas Pauli considered possible modifications of the orbit notion, Heisenberg proposed to eliminate it altogether. What is measurable, after all, are the transitions between states and their probabilities. Single states roughly correspond to single classical orbits; thus, transitions between states need information from two orbits. It is understandable that Kramers, an expert in classical mechanics, would not be anxious to give up the orbit notion.

But Heisenberg, looking at the Kramers dispersion formula, observed that only quantities referring to *transitions* occurred, hence calling for a description in terms of *two* orbits. Succinctly stated, classically one needs *all* the information about a single orbit; in an atom one needs *some* information about *pairs* of orbits. Specifically, one needs the frequencies and the transition probabilities between orbits. Heisenberg set himself the task of describing the behavior of atoms in terms of frequencies ν_{ij} (between states) and probabilities for such transitions, P_{ij}. In the Kramers dispersion formula, these probabilities are always the square of the classical amplitudes, so actually the description sought is in terms of the frequencies ν_{ij} and these amplitudes a_{ij}. Heisenberg knew beforehand that, at least for dispersion phenomena, such a description must be possible because the Kramers dispersion relations can be expressed in terms of just these quantities. Heisenberg then claimed that the *complete* behavior of all atoms must be so described. There just remained the problem of finding what laws these quantities satisfy. Stated differently, one looks for the physical laws—in the first instance the laws of mechanics—in terms of these quantities. These laws must also include the known information on atoms. In a letter to Pauli dated June 24, 1925, Heisenberg writes: "I would like to know the precise meaning of the equations of motion, considered as a relation between amplitudes."[76]

In the month of June, Heisenberg completed the first phase of his program. He had constructed an autonomous theory of a simple model which simulated the behavior of atoms. Quantum conditions disappeared as did orbits. The quantities occurring in the theory always refer to pairs of orbits and were unfamiliar mathematical objects to Heisenberg. (They are actually matrices, rather well-known objects, but Heisenberg didn't know that.) In this construction he leaned heavily on the Kramers–Heisenberg dispersion relations and also used the Kramers trick of replacing differential quotients by difference quotients. On June 17, 1925, Heisenberg returned from Helgoland, where he had gone to recover from an attack of hayfever. His quantum mechanics paper was now more or less complete. After stopping by to see Pauli in Hamburg, Heisenberg returned to Göttingen. He wrote to Pauli on June 21 to thank him. In a letter to Pauli dated June 24, 1925, he explained his scheme of matrix mechanics further.

* A reference to the popular book by Kramers and Holst in which the orbits of electrons are represented in beautiful pictures.

Coincidentally, Kramers had visited Göttingen at Born's invitation from June 8 to June 22. It would be expected that Kramers and Heisenberg would talk in great detail about Heisenberg's matrix mechanics, since that is about all that Heisenberg would talk about at the time. That such a discussion indeed took place can be inferred from a letter Heisenberg sent to Pauli: "I am convinced in my heart that this quantum mechanics is actually correct—for this reason Kramers accuses me of optimism."[77] On July 9 1925, Heisenberg sent his completed paper to Pauli and on July 11 showed it to Born in Göttingen.

With this paper by Heisenberg, the Bohr period of quantum theory came to an end. The formalism set up by Heisenberg—significantly extended and eventually completed by Heisenberg, Born, and Jordan in Göttingen—formed the basis for the "future more complete theory" of the Bohr period, the theory for which everyone was looking. It was true then, and is still true now, that any systematic treatment of quantum theory, whether applied to atomic, molecular, or nuclear problems, *can* be phrased in terms of the language and concepts of Heisenberg's original matrix mechanics.

Born was immediately impressed by Heisenberg's paper. On July 15, 1925 he wrote to Einstein about Heisenberg's new paper, "which might appear mystical ... but it is very deep and certainly correct."[78] Pauli, in numerous letters, was very encouraging, and at times even enthusiastic about Heisenberg's approach. He wrote to Kronig on October 9, 1925: "Heisenberg's quantum mechanics has once again given me joy in my life."[79] Bohr was also enthusiastic. In the first published reference to Heisenberg, he states: "... a step of extraordinary scope ... through which the difficulties associated with mechanical models can hopefully be avoided."[80] Later, on October 14, 1925, Bohr wrote to Ehrenfest: "I am extremely enthusiastic about this development and the prospects for the future."[81] On January 27, 1926, Bohr wrote to Rutherford: "In fact due to the last work of Heisenberg, prospects have been realized with one stroke, which although only vaguely grasped, have for a long time been the center of our wishes."[82] Perhaps the best assessment of Heisenberg was given by another genius, Dirac (1902): "Heisenberg's contribution was so great that it allowed second rate people to do first rate work."[83]

A very rapid development followed Heisenberg's paper in which many of the ambiguities of Bohr's quantum theory were removed and advances followed each other with dizzying speed—"quantum mechanics fever" seemed to spread through physics. Einstein, however, had very mixed feelings. He wrote Ehrenfest, on a postcard: "Heisenberg has laid a large quantum egg. In Göttingen they believe it. (I don't.)."[84] Later, Einstein wrote to his close friend, Michele Besso: "The most interesting theory is the Heisenberg–Born–Jordan theory of quantum states. A veritable witches brew where infinite matrices take the place of Cartesian coordinates."[85] This was the first instance of Einstein's discomfort with the further development of quantum theory. Einstein's insistence on the photon picture of light as a coequal of the wave picture had precipitated a profound conflict. When resolutions to that conflict began to appear, as with Heisenberg's quantum mechanics, Einstein was hesitant to accept this as the ultimate solution.

Heisenberg achieved these monumental results after he left Copenhagen in April 1925. His paper was finished about July 1 and he saw Kramers about June 21. It would clearly be of great importance to know Kramers' reaction to this unquestioned breakthrough. It might be anticipated that, because of his formidable formal talents, Kramers would play a dominant role in the necessary mathematical elaboration of Heisenberg's scheme. It is remarkable and surprising that Kramers did not appear to react at all. There are many letters by Born, Bohr, Einstein, and Pauli commenting on Heisenberg's achievement. Kramers' first recorded reaction is several months later.

It is also curious that Heisenberg, who had just spent a semester in Copenhagen in intense cooperation with Kramers, did not keep Kramers informed about his progress in constructing the quantum mechanical scheme, especially since the programmatic idea was an extension of the Kramers–Heisenberg cooperation. This must be contrasted with the blow-by-blow description that Pauli received; even Kronig, not a physicist in the inner circle, got rather detailed information. Equally strange is that Bohr was not kept up-to-date either, especially because at that time Heisenberg still possessed a hefty dose of hero worship for Bohr. For example, Heisenberg wrote to Bohr in connection with some other matters: "Today I have a terribly bad conscience and I don't know whether you won't be angry with me. Yesterday without waiting for a message from you, I have submitted my paper."[86] This letter—written almost in the style of an errant son to a father—still makes no mention of Heisenberg's major results in quantum mechanics. Evidently, the source of information available to the Copenhagen circle about quantum mechanics would have been the conversation between Kramers and Heisenberg, on June 21 in Göttingen. But strangest of all, it appears that Kramers did not tell *anyone* anything about that conversation. In the prevailing atmosphere of the Institute for Theoretical Physics that would almost appear to be impossible, but it is certain that as late as August 31, 1925 (long after Kramers had returned to Copenhagen—and a month and a half after Born wrote to Einstein), Bohr did not know about Heisenberg's work. Heisenberg wrote to Bohr on August 31 and asked Bohr to get a visa: As far as physics is concerned I haven't thought about it for a month and I don't know whether I still understand it at all. Earlier, as Kramers has probably told you, I produced a paper on quantum mechanics. I would very much like to hear your opinion."[87] Bohr answered this letter on September 4: He got the visa for Heisenberg and was happy Heisenberg was coming, but there was not a word on Heisenberg's quantum mechanics paper. Furthermore, Bohr had to give a lecture on mechanics and atomic theory on September 2 and 3 at the mathematical congress in Copenhagen. He described the situation as "highly unsatisfactory and provisional with little future." From what is known about Bohr's rapturous reaction to Heisenberg's work, it is clear that when Bohr gave the lecture he did not know about quantum mechanics. Clearly, Kramers did *not* tell him about Heisenberg's results.

On July 16, 1925, Kramers wrote a letter to Harold Urey, an American physicist, who had visited Copenhagen and who wanted to know from Kramers "what was new." In that letter, five typewritten pages, Kramers mentions all kinds of things, including his visit to Göttingen: "About ten days ago my wife and children moved to a small villa which we built ourselves. I myself was in Holland that day, with my father who had his Doctor's degree for 40 years. Before that time I stayed some days in Göttingen where I had never been before and where I had a good time with Franck, Born and the other physicists there. Now I am quite happy in our new home."[88] In this long letter, specifically intended to bring Urey up-to-date on physics, there is not a word or even a hint, of quantum mechanics and Heisenberg.

It is sometimes argued that this indicates that Kramers at that time did not fully appreciate the significance of Heisenberg's contribution* and hence neglected to tell Bohr and Urey about it. It would appear more reasonable that to the contrary, Kramers, as soon as he saw Heisenberg's results, understood their full impact immediately. It must be remembered that a good share of the formalism was taken over from the Kramers–Heisenberg paper, so Kramers knew it inside out. The manipulations with Fourier series were also very familiar to Kramers, since he had used similar procedures for many years. The systematic character of Heisenberg's scheme must have been evident to Kramers, who had a quick appreciation of the structure of a theory.

Although Kramers no doubt understood all this right away—and no doubt recognized its far-reaching significance—it was probably clear to him, or in any case he sensed, that he could not have taken this major step by himself. To eliminate the orbit notion would have been foreign to Kramers, steeped as he was in the theory of Bohr orbits, and he would have resisted it. To make a jump from a physical model into a mathematical abstract scheme, something natural for Heisenberg, would have been difficult and artificial for Kramers. Already in the earlier discussions in Copenhagen, Heisenberg was more willing than Kramers to *free* the theory (and himself) from physical models and take a step into mathematical abstraction. Heisenberg had sensed early that the Kramers–Heisenberg relations were the first in a set of relations which together would provide a formalism that would comprise the whole of quantum theory. It therefore seems overwhelmingly likely that Kramers recognized the epoch-making significance of Heisenberg's work, an organized and systematic theory, and a goal Kramers also pursued.

It must be remembered that Kramers was already depressed because of the failure of the Bohr–Kramers–Slater theory. It is understandable that when he was confronted with a theoretical structure for which he had assembled all the pieces, but which he failed to put together, that his depression would deepen to such an extent that he was unable to talk about Heisenberg's work. All of this was certainly not helped by an incisive and cruel letter from Pauli which coupled the failure of the Bohr–Kramers–Slater theory with the success of Heisenberg's new approach: "In general I think it was a magnificent stroke

* K. Stolzenberg, Ph.D. thesis, Stuttgart, 1977.

of luck that the theory of Bohr–Kramers–Slater was so rapidly refuted by the beautiful experiments of Geiger and Bothe, as well as the recently published ones by Compton. It is certainly true that Bohr himself, even if these experiments had not been carried out, would not have maintained his theory. But many excellent physicists (e.g. Ladenburg, Mie, Born) would have maintained this theory and this ill-fated work of Bohr–Kramers–Slater would perhaps for a long time have become an obstacle to progress in theoretical physics.... I have especially been impressed by Heisenberg's bold new ventures (you have certainly heard about them in Göttingen)... Heisenberg's methodology pleases me a great deal.... I wish him great success in all his efforts."[89] This letter ends with an admonition to Kramers' wife—that whenever Kramers judges those people who are not true believers in the Copenhagen (Bohr) style of physics, she must sing him a song about spring, beauty, and love. It is of course impossible to know how Kramers reacted, but it is obvious that the failure of the Bohr–Kramers–Slater theory and the redirection of physics must have left their mark. Kramers' actual state of mind during that summer is probably accurately described in a letter written to Romeyn, perhaps the only person in whose presence Kramers felt that he did not need to, and probably could not, pretend: "That I didn't answer earlier can be blamed on this miserable mood of tiredness and overwork, from which I have suffered the last three months. I lacked the strength to overcome it.... About July 1st I was for 10 days in Göttingen, I saw Dirk Struik and his wife.... I myself have a busy fall coming up, *even* if *I would forget* about my scientific aspirations.... In the past few months, I have often had the desire to be back in Holland.... I would think seriously about returning to Holland if I could get a good position there. ... My friends here say that I suffered from 'Melancholia Theoretica', the last few months. There is some truth in that, no theoretician is never entirely free from that disease."[90]

In view of all this, it is not so surprising that for one summer Kramers appeared to withdraw from the center stage of quantum theory. It is a sign of remarkable intellectual resilience that in September he was back at work. For the Golden Jubilee Celebration of Lorentz, December 11, 1925, Kramers contributed an interesting paper entitled "Some Remarks About the Quantum Mechanics of Heisenberg." It was finished in Copenhagen in November 1925, so by that time Kramers was back at the frontier of physics. The remarks contained in this paper relate Heisenberg's ideas to classical considerations and they demonstrate Kramers' control of the formal apparatus of quantum theory. The paper was written in Dutch, however, and it was not widely read. Furthermore, Kramers just made the remarks and did not develop the ideas further. Almost simultaneously, and certainly independently, Dirac made the same remarks, but he made them the starting point for a highly original, transparent new version of quantum mechanics.

In the midst of all this turmoil, another problem having both personal and scientific ramifications preoccupied Kramers: Just how long should he stay in Copenhagen and, if he left, where should he go? On April 11, 1925, two weeks before the disproof of the Bohr–Kramers–Slater theory, Kramers heard that there was a possibility of a professorship in Holland. He commented on that in the letter to Romeyn and also mentioned it in the letter to Urey. After some indecision, Kramers decided to apply for the position in Utrecht. The four persons writing letters of recommendation had shaped and formed the physics of the 20th century more than any others: Planck, Lorentz, Einstein, and Bohr. Understandably, Kramers had misgivings about leaving, and he wrote to Fowler: "The idea of leaving Copenhagen and Bohr after almost ten years is not very agreeable, but on the other hand I feel it is a wholesome change to obtain an independent post."[91] On February 25, 1926, Kramers gave his inaugural address in Utrecht. He actually stayed in Copenhagen until May 1926 and then moved to Utrecht.

Heisenberg was Kramers' replacement in Copenhagen.

CHAPTER 9

The Revolution in Progress

The time from September 1925 until about March 1927 was one of the most exciting, productive, and ususual periods in the history of physics. Although not all the significant developments took place there, Copenhagen was very much the nerve center. It was there that new results were continually analyzed and scrutinized, and it was there, strongly influenced by Bohr, that new results were eventually synthesized into a coherent, complete scheme. Thus, Kramers spent his last year in Copenhagen in the midst of this exuberant scientific environment.

The events during this hectic period had a profound effect on all participants, including Kramers, and the developments in that period had a decisive influence on physics and eventually all fields of science. In physics, quantum theory separated those who understood and appreciated it, such as Heisenberg, Pauli, Bohr, Born, and Kramers, from those who did not understand it too well and appreciated it even less. But all physicists were profoundly affected by this avalanche of results. It is no exaggeration to describe the rapidly changing outlook as a revolution in science. One way or another—in teaching, research, applications, or the selection of problems—by action or by inaction—all physicists had to take a position vis-à-vis that revolution. Kramers was present in Copenhagen until May 1926 and directly experienced the changes and surprises. Even after he had moved to Utrecht in the summer of 1926, he maintained close individual contact with many of the major figures, and so continued to have first-hand information of all that transpired. To appreciate Kramers' later role and subsequent activities in physics, it is essential to realize just how profoundly this revolution affected physics. For Kramers, as for all physicists of his generation, the events of these few years required a major intellectual reorientation. It required nothing less than a complete reassessment of the goals and methods of theoretical physics. It required the introduction of new concepts, the extensive use of new and largely unfamiliar (to physicists) mathematics. Most difficult, it required the relinquishing of old, trusted, and familiar ideas—such as the notion of an orbit of an electron in an atom. The new concepts were more abstract, more formal, and less tangible than those of classical physics. Succeeding generations tend to accept the quantum ideas as a matter of course without realizing how difficult and painful the renunciation of the old concepts had been. Nor do they fully recognize the profundity of the changes in the conceptual basis of physics occasioned by quantum ideas. An examination of the concerns, conflicts, and contradictions during this revolutionary period is an excellent way to appreciate the struggles each serious scientist had to endure in order to understand and to contribute to the further development of physics. (For Kramers, the successive stages of these struggles are accurately reflected in the several versions of the courses he taught in Utrecht starting in 1926.) There actually were a number of unrelated factors all contributing to the intense intellectual ferment accompanying this revolution.

One important element was the rapid and successful development of Heisenberg's matrix mechanics. Although to some people, such as Einstein and Fermi, its extreme formal mathematical nature was objectionable, others rapidly recognized the great promise and power of the scheme, especially after Born and Jordan demonstrated, only a short time after Heisenberg's article, that Heisenberg's mysterious noncommunicating quantities were nothing but ordinary matrices (then known to mathematicians, and now known to all physicists, chemists, engineers, and economists). Many problems and questions which could not be treated previously now became amenable to discussion. The excitement and surprise are clearly noticeable in the reaction of E. Wigner who had just obtained his Ph.D. degree in Berlin: "One day a fantastic article appeared in the Zeitschrift für Physik, claiming that the purpose of microscopic physics must be solely the determination of energy levels and the transition probabilities between them.... The Born–Jordan article was a great surprise and it electrified me deeply. I felt it represented, coming from the blue sky, a sudden profound recognition ... that it gave the prescription for the quantisation of an arbitrary system was what impressed me most deeply."[92] Klein wrote: "I remember my own astonishment when Kramers in the autumn of 1925 sent me a postcard saying that there now was a new formulation of quantum theory without explicit quantum conditions, but in their place, a mysterious algebraic relating with noncommuting quantities."[93] Beside its status as an autonomous system, and not a strange mixture of classical physics and quantum rules, the abstract and formal scheme of Heisenberg soon began to show its superiority over the Bohr theory. Perhaps the most auspicious example was a calculation made by Pauli. During September and October of 1925, Pauli, who had liked Heisenberg's first paper, had become more and more upset by what appeared to him as excessive mathematical elaboration of Heisenberg's ideas. He continually lambasted *Göttinger Gelehrsamkeit* ("Göttingen's learnedness"). Heisenberg, in a strange outburst of irritation, wrote to Pauli: "Your eternal ridiculing of Copenhagen and Göttingen is a flagrant scandal. You will have to admit that we are not seeking to ruin physics by malicious intent. If we are donkeys having never produced anything new, you are an equally big jackass, for you haven't done anything either."[94] It is not known whether this angered or encouraged Pauli, but it is true that in an incredibly short time he derived, using matrix mechanics, not only the energy-level scheme of hydrogen, which was already known from the Bohr theory, but also the behavior of a hydrogen atom in crossed electric and magnetic fields. That problem, if handled by the Bohr theory, or any of its variants, gave rise to insurmountable difficulties and in fact was a famous unsolved problem in the context of the Bohr theory. Pauli's treatment was a technical tour de force, showing unequivocally that the Heisenberg theory was superior to the Bohr theory—a very good reason for the excitement that the Heisenberg scheme engendered. Evidently, Heisenberg's anger was totally mollified by Pauli's brilliant physics. He wrote after receiving Pauli's calculations: "I need not tell you how delighted I am about the new theory of hydrogen and how pleasantly surprised I am about the speed with which you produced this theory."[95] Pauli informed Kramers in Copenhagen, who in turn informed Bohr. Both were deeply impressed by this feat of Pauli (Heisenberg had tried to calculate the hydrogen atom according to matrix mechanics but gave up after a short while) and by the evident power of matrix mechanics.

In September–October of 1925 it was beginning to become clear that with the discovery of Heisenberg's matrix mechanics a quite extraordinary new development in physics had started. Whether this had anything to do with the entry of an unusual number of brilliant young people into this field is not so clear, but it is certain that a surprising number of original and powerful intellects were drawn to quantum theory. In turn, they contributed to its rapid growth. Many of them came to Copenhagen for long or short periods, and all of them had connections with somebody in Copenhagen so that everyone there was in continual contact with ongoing developments. This influx of brilliance, most of it directed toward the solution of related problems, is another feature that gives this period such a special character. It can be legitimately argued that at no other time in physics (and presumably in all of intellectual history) has as great an array of talent been focused on a single set of problems. Outstanding, even among this unusual group of scientists, was the young British physicist, Dirac. There is, in writing about this period, something of a risk to declare reasonably intelligent persons, who happen to live and work in unusual and fertile times, as brilliant, or to refer to merely brilliant persons as geniuses. There is, however, no danger of exaggeration in referring to Dirac as a genuine genius. Dirac's contributions to the systematic theory of quanta, from 1925 to 1928, starting as they did with a paper based on Heisenberg's ideas submitted on November 7, 1925 but fiercely independent from then on, are hardly equaled and never surpassed in the history of quantum physics.

Whereas the early workers in quantum theory, including Heisenberg himself, were not totally clear why Heisenberg's scheme worked, or what its basic structure was, Dirac recognized this right away. He inferred from an incisive reformulation of Heisenberg's paper that the crucial point of matrix mechanics was that the quantities describing the dynamics do not commute as they do in classical mechanics. It was something Heisenberg in fact knew, but which rather bothered him.[96] Kramers, in a short note written in Dutch, recognized that appropriate classical quantities also exhibit a similar noncommutativity, but it remained for Dirac to discover that the noncommutativity was the key to the whole scheme. Even though Dirac and Kramers made similar observations, they used them in characteristically different ways—Kramers to establish the connection with the former classical theory, Dirac to construct an autonomous new quantum theory. Dirac's work in this period is nothing short of astounding. Apart from major contributions to the formalism of matrix mechanics, he was the first to formulate a rational quantum theory of the emission and absorption of radiation. Recall that in the Bohr theory the *mechanism* of the transition from one level to another was a complete mystery. Dirac, by a brilliant application of the new formalism to atoms as well as to electromagnetic fields, derived many known but previously disconnected and unrelated results in a completely systematic and unified way. In so doing, he resolved the contradictions and paradoxes that had plagued radiation theory since Planck first postulated the quantum of action. This was achieved using the formalism of quantum mechanics* and was obviously a tremendous success of that formalism. It did no harm to Dirac's reputation either. In late 1927–early 1928, Dirac published the first version of a quantum theory that also satisfied the principle of relativity (special relativity) thereby initiating one of the most fruitful developments of modern physics. (Among other things, the notion of anti matter was an outgrowth of the Dirac equation.)

Dirac was not only a genius, he was unusual and exceptional in every way. There can be little doubt that Pauli was a genius: He had great power, profound understanding, and extraordinary sense for the deductive structure of theories. There is no doubt that Heisenberg was a genius: His speed of comprehension was phenomenal, he was totally unafraid of mathematics, and he had a talent for abstracting physical pictures into mathematical schemes which he then handled with consummate skill. Dirac was very different. Even though very few people could accomplish what either Pauli or Heisenberg could, once it was achieved, many people could understand what was done and how they went about doing it. This was not the case with Dirac. He had a particular personal way of looking at a theoretical structure. His papers—although written with great lucidity and totally without circumlocution—are full of unexpected viewpoints and unusual exploitations of the logical possibi-

* The formalism actually used is mainly Dirac's version of quantum mechanics, although it contains some ideas of Schrödinger's method as well.

lities. Ehrenfest used to refer to Dirac's work as *unmenschlich* ("inhuman").[97] Ehrenfest did not really mean inhuman; more probably he meant "superhuman": No ordinary mortal could produce those arguments.

Another important difference between Dirac and the other physicists was the way in which he approached and used mathematics. Dirac was certainly a physicist and not a mathematician, but mathematical beauty played a major role in his approach to physics. Dirac, perhaps more than any other physicist, had an unusual talent for sensing what kind of mathematical structure was necessary and appropriate for the formulation of physical laws. Dirac's papers, almost without exception, have that typical stamp of genius which is manifested by possessing the apparently contradictory properties of being both inevitable and unexpected.

During that period in theoretical physics, perhaps best characterized as the "heroic" period, there were many other outstanding figures. Certainly Fermi, the young Italian physicist, made many epoch-making contributions. His style was quite different from Pauli's, Heisenberg's, or Dirac's. The physical phenomena spoke directly to Fermi; he is reputed to have said[98] that he could calculate the numerical value of any physical effect, accurate to 20%, within an hour. If an accuracy of 10% was needed it would take him a little longer. This is genius of quite a different kind from Dirac's or Heisenberg's, but equally impressive and, for the development of physics, equally important.

These were only some of the people contributing to the scientific atmosphere. Their work dominated the field and their style defined the scientific ambience. Naturally, other investigations would automatically be judged by those standards. For example, Slater of the Bohr–Kramers–Slater theory returned to M.I.T. in late 1925. There he listened to lectures by Born on the new matrix mechanics of Born, Jordan, and Heisenberg and tried to understand, clarify, and expand this work as Dirac had. He had made interesting progress and decided that he was ready to write up his results for publication. However, he recalls somewhat plaintively: "A paper by a hitherto unknown genius appeared, P. A. M. Dirac's first paper on quantum mechanics. It included not only the small points I had worked out, but much, much more besides. But the interlude wasn't a total loss, at least I got to know Born well."[99] A few years later Kramers would have a very similar experience.

An altogether different development (initially independent of Heisenberg's matrix mechanics) started when Ehrenfest wrote to Lorentz[100] to make an appointment and discuss what Ehrenfest called a "very cute" idea of two of his young students, Goudsmit and Uhlenbeck. The idea in question is that electrons possess an intrinsic spin, that is, they rotate around their axis to produce an angular momentum. In addition, electrons possess a magnetic moment (they act like miniature magnets). Because of this new spin magnetic moment, electrons are capable of new magnetic interactions. This has the observable consequence that many spectral lines, which are single lines according to the Bohr theory, will in fact be doublets (two closely spaced distinct lines). Such pairs (or multiplets) of lines had experimentally been known for a long time and their existence had remained yet another irritating riddle within the Bohr theory. The specific Goudsmit–Uhlenbeck assumptions, although easy to phrase, are rather unusual. They first assumed that the electron rotates—a somewhat doubtful assumption for a point particle. They further assumed that this intrinsic angular momentum would be quantized (i.e., sensible within the context of the Bohr theory), but instead of a component of the angular momentum having an integer value as in the Bohr theory, (integer) $\times$ $(h/2\pi)$, they assumed as the only possible values $\pm 1/2 \times h/2\pi$. The miniature magnet could be aligned only parallel or anti-parallel to a field. This clearly does not follow from the Bohr theory. The third Goudsmit–Uhlenbeck assumption is a particular numerical relation between the angular momentum and the magnetic moment. Classical physics gives an unambiguous number for that ratio, but Goudsmit and Uhlenbeck assumed that for their spinning electron the value was twice as large. They were led to their picture, and the subsequent assumptions, by a study of a paper by Pauli in which Pauli associated *four* quantum numbers with the state of the electron in an atom. Since three of these numbers had a fairly direct physical significance—such as energy and angular momentum—their suggestion amounted to providing a physical interpretation of Pauli's quantum numbers. The other assumptions enabled Goudsmit and Uhlenbeck to obtain the correct splitting of a spectral line in a magnetic field. This was the idea Ehrenfest wanted to discuss with Lorentz. Lorentz was friendly but skeptical. The conclusion was that Lorentz "would think about it." At a later meeting, Lorentz expressed the view that in classical physics the spin idea made little sense. But even before that meeting Ehrenfest encouraged Goudsmit and Uhlenbeck to write a short note about their suggestion and send it to the journal *Naturwissenschaften.* It was published in the middle of November 1925, and the paper was in print before Lorentz's criticism was known.

It is fair to say that it caused a mild sensation. Many people were interested. It should be emphasized that this spin idea, although explaining many features of spectra, had at this point nothing to do with the new matrix mechanics of Heisenberg, the photon, or radiation theory; it was an additional new attribute of electrons which evidently was needed to account for spectra. The language used to describe it was classical with a little quantum admixture of the Bohr type. Spin was another ingredient exerting a powerful influence on physics and on the Copenhagen school during the last half year Kramers was there.

Heisenberg, who it must be remembered entered physics with a proposal to use half-integer quantization, was immediately interested in the spin but very skeptical. He was skeptical in part because the spin proposal was unusual and unconventional, but mainly because he believed that his matrix mechanics was (at least in principle) a *complete* theory of atomic phenomena and no new additional physical assumption should be needed. All this did not keep Heisenberg from calculating the doublet splitting using the Goudsmit–Uhlenbeck ideas. He found the effect but it came out wrong by a factor 2.

Pauli believed nothing of the spin: He did not trust the picture of an electron as a rotating sphere; he knew, or in any case suspected, something about the factor of 2; and he felt the necessity to introduce a fourth quantum number was an essential quantum feature, perhaps hidden somewhere in the matrix mechanics formalism—but it certainly could not be described classically. Pauli's objections were particularly important, for as early as January 8, 1925, Kronig, a young Dutch physicist, had made substantially the same suggestions as Goudsmit and Uhlenbeck. He had also made the calculation Heisenberg made later: He also found the factor of 2. However, Pauli, who admitted that it was an amusing idea, also considered it nonsense, having nothing to do with reality. Kronig was obviously disappointed; he later discussed his idea with Heisenberg (on the ferry going to Copenhagen), who was indifferent. In Copenhagen, Kronig tried to discuss the spin with Bohr but Bohr seemed to have little time and less interest. In the few brief discussions Kronig had with Bohr, Bohr made it quite clear that he considered that idea total nonsense. Kramers had somewhat longer discussions with Kronig but he was not very impressed either. He did not spend enough time with Kronig to look at any of the details. (This was at exactly the same time that matrix mechanics was developed so everyone had plenty to think about without considering new, wild ideas.) It would appear that both Bohr and Kramers were very unclear about the coupling between the electron spin and the remainder of the atom (the spin–orbit coupling). This is needed to calculate the doublet splitting. In the face of strong opposition by Bohr and Pauli and indifference by Kramers and Heisenberg, Kronig published nothing and returned to the United States.[101] Because of the strong encouragement of Ehrenfest, Goudsmit and Uhlenbeck published their work.

At the time Goudsmit received Heisenberg's letter with the calculated doublet splitting and the correction factor of 2, Goudsmit and Uhlenbeck had not calculated this doublet splitting—in fact they did not know how to do it. They, like Bohr and Kramers, did not know the coupling of the spin to the rest of the atom. As luck would have it, this was just at the time of the Lorentz Jubilee, November 19, 1925, and Einstein was visiting Leiden, He explained to Goudsmit and Uhlenbeck that this coupling is a direct consequence of the relativity theory. This enabled Goudsmit and Uhlenbeck to make the calculation. In agreement with Heisenberg, they too found that the result was a factor of 2 larger than the observed value.

Perhaps more important than this calculation was the fact that Bohr was also at the Lorentz Jubilee. Bohr, who had dismissed Kronig's spin idea because he did not see how the spin would be coupled to the orbit, was completely convinced (and surprised) by Einstein's argument. Bohr then made a complete about-face and became extremely enthusiastic about the spin. In a letter to Kronig, Bohr writes: "Einstein's remark that the coupling was an immediate consequence of relativity came as a complete revelation to me and I have since never faltered."[102] That there were classical difficulties (as pointed out by Lorentz) did not bother Bohr at all: The factor of 2 bothered him a little, but he felt that the physics of the spin was basically correct and the factor of 2 would take care of itself, perhaps by a correct quantum mechanical calculation. Bohr in fact became an evangelist for the spin idea. He wrote to Ehrenfest: "On my travel [from Leiden] I felt completely as a prophet of the electron-magnet gospel and I believe I have succeeded to convince Heisenberg and Pauli that their present objections are not decisive and that it is very probably that a quantum mechanical calculation will give all details correctly."[103]

Heisenberg was grudgingly convinced, while Pauli was not: The factor of 2 remained a puzzle. Bohr's intuition was correct. On February 20, 1926, L. H. Thomas showed that the factor of 2 was actually the result of a subtle mistake in the usual relativistic calculation. Everyone was surprised, including Einstein.[104] Thomas' note was very difficult to read so it took a considerable time before it was understood. Heisenberg, who had corresponded with Bohr about the electron spin, accepted the conclusion, although he writes a postcard to Pauli saying that he is "too dumb" to understand Thomas' argument, Pauli still did not believe it; he hints that the factor of 2 is related to the structure of the electron. On several occasions Pauli expresses strong reservations about Thomas' arguments. He writes to Bohr on February 26, asking Bohr either to delay the publication of Thomas' note or to alter the text substantially. On March 5, in a letter to Bohr, Pauli states that he would be interested in knowing what Kramers thinks about his objections to Thomas' paper. He refers to Kramers as a "learned relativist and responsible scientist." As such, Kramers' reaction is certainly worth knowing. On other occasions Pauli refers to the spin idea as "the false doctrine." On March 8, Pauli writes to Kramers to settle the spin question and Thomas' derivation: "I appeal to you as a significant scientist ... to destroy this false doctrine.... This after all has nothing to do with quantum mechanics or politics or religion, it is just classical physics."[105] Kramers and Pauli evidently had discussed Thomas' paper numerous times without coming to any very definite conclusions. After Pauli received a letter from Bohr, which also contained a number of explanatory remarks of Thomas, he wrote[106] on March 12 to Bohr that he now believed the Thomas factor. On the same day he wrote to Kramers asking him to think about some models which would allow a pedagogical explanation of the factor of 2. With Pauli's acquiescence, the electron spin became an established part of physics.

It soon became clear that all objects—protons, neutrons, nuclei—possess a spin. It should be stressed that as far as atoms, electrons, and their quantum (nonrelativistic) description are concerned, the spin and the spin interactions are new additional features. They are not consequences of quantum rules, the nonexistence of orbits, or dispersion relations. The spin and spin properties are just extra, new properties. Although Kramers did not invent the spin and was, as all others, rather skeptical, the properties of the mathematical quantities describing the spin later became one of Kramers' serious interests.

There is an interesting postscript to the spin story which had a profound effect on Kramers. With the new spin properties, several questions came up. The most obvious one is whether the spin properties can be combined with matrix mechanics, that is, whether it is possible to construct a formalism that in some coherent way describes the usual properties together with the spin properties. It is illuminating that on February 5, 1927, Heisenberg wrote a letter to Pauli in which he mentions that he made a *bet* with Dirac that it would take 3 years to understand the spin fully. Dirac claimed this would be clear in 3 months. On May 3, 1927, Pauli submits a paper in which he succeeds in *combining* quantum mechanics and the spin. Of course he does not explain the spin, he just incorporates its description in quantum terms. This was an important advance, albeit a partial one. Heisenberg mentions the bet with Dirac in a letter to Jordan (May 1927); he feels that the insufficiencies in Pauli's paper will guarantee that he will win his bet. Pauli and others were dissatisfied with Pauli's paper because it did not meet the requirements of relativity. In fact, Pauli claimed that a consistent relativistic spin theory was in principle impossible. Kramers did not believe this; so Kramers and Pauli made a bet that one could *not* construct a relativistic spin theory. Pauli issued a direct challenge to this effect to Kramers.[107] Both Dirac and Kramers lost their bets—but Dirac was certainly the moral victor and Kramers demonstrated eventually, even to Pauli's satisfaction, that he had effectively met Pauli's challenge.

On January 2, 1928, about 6 months after the expiration of his bet with Heisenberg, Dirac submitted one of the most significant papers in physics in which he obtained a formalism satisfying both the requirements of relativity and quantum theory. That is what Dirac set out to do. He obtained as a gigantic bonus that his equations predicted the spin, with all its couplings, including the Thomas factor of 2, which had caused so much confusion and difficulty when first introduced and computed. At about that time (January 1928), Kramers had obtained, by a very intricate, complicated, and somewhat contrived method, a set of equations that had exactly the same mathematical and physical content as Dirac's equation. (It actually takes some work to demonstrate that precise equivalence.) Kramers was very disappointed that even though he (Kramers) had achieved his goal, Dirac was ahead of him by just a few months. Dirac's derivation was penetratingly original and very simple but Kramers never quite overcame his initial doubts, not to say suspicions, about Dirac's procedure. Perhaps because his derivation was somewhat cumbersome and Dirac's derivation was so elegant, he did not publish any of this until about 10 years later. All during his life, Kramers was bothered by the fact that he missed the relativistic spin description by only a few months;[107] and while there is no recorded evidence in which Kramers expressed his disappointment about missing matrix mechanics, he frequently referred to his dismay at missing the spin.[108] When Kramers eventually published his results, Pauli was again very critical, but after a long correspondence Pauli agreed that Kramers' procedure was legitimate "but not elegant." Thus, Kramers joined that long list of eminent physicists beaten out for a major discovery by Dirac. There is no evidence that anyone ever paid off any of the bets.

It might appear that with the successful start of a consistent quantum theory

(Heisenberg's matrix mechanics), with the discovery of a new particle attribute (the spin, the first in hundreds of years), and with the entrance of such great talent into physics—all crammed within a period of less than a year—nothing else could have happened during that time period. As it was, these developments produced enormous excitement in physics, so that new discoveries followed each other in rapid succession. It would seem that neither the field nor its practitioners could adapt to still more surprises, more new ideas, more controversy. This period, however, is so unusual that within the same time span (January 1926–July 1926) yet another absolutely spectacular development took place.

The Austrian physicist Schrödinger—who in contrast to Heisenberg, Pauli, and Dirac was not in his early twenties (he was at the venerable age of 39)—published a series of monumental papers that provided another systematic description of atoms, including the quantization of energy and the transition probabilities between energy levels. In short, all the problems that Planck, Bohr, Heisenberg, Pauli, and Dirac had been struggling with for so long were approached from a new, totally different viewpoint. Strangely enough, the philosophical and conceptual motivations for that viewpoint were radically different, even opposed to that of Bohr and Heisenberg; but the numerical results that this new approach yielded were identical to those of the Heisenberg–Born–Dirac theory.

The investigations of Schrödinger were the culmination of studies initiated by Louis de Broglie in France and by Einstein. It is interesting and possibly significant that *none* of the originators of this theory, usually called wave mechanics, participated in the Bohr type of analysis of spectra. It was Einstein who, by introducing the photon idea in 1905 and later more explicitly in 1917, had to confront the dual set of properties, which had to be ascribed to electromagnetic radiation. They were the wave properties, beautifully verified by classical interference and diffraction experiments, and the particle properties, as suggested by the photoelectric effect and forcefully demonstrated by the Compton effect. There is good evidence that this necessary assignment of dual—indeed contradictory—properties was continually on Einstein's mind.*

It was the unusual insight of de Broglie that a similar assignment of dual properties might be a general characteristic and not restricted to light. Thus, in the preface of his thesis, de Broglie writes: "After long reflection in solitude and meditation I suddenly had the idea, during the year 1923, that the discovery made by Einstein in 1905 should be generalized in extending it to all material particles and notably to electrons."[109] Stated differently, de Broglie asserted that just as light waves can, under appropriate circumstances, exhibit particle characteristics, so can particles under appropriate circumstances exhibit wave properties. Thus, de Broglie associated with each electron

* For a recent, incisive analysis of this concern, see the beautiful book by A. Pais, *Subtle is the Lord*, Oxford University Press, New York, 1982.

a wave, which he gave the unfortunate name "fictitious wave." Since the existence of these waves is to have *observable* consequences, there is nothing fictitious about them. Later, de Broglie made specific suggestions about methods whereby these waves might be detected.

Einstein, in his profound analysis of quantum theory, thought along very similar lines. As early as 1917, Einstein introduced the idea of the "guiding field" (*Gespenster Feld* or "ghost field"). Stated briefly, Einstein's idea was that light consisted of photons, but that the motion of these photons was guided by the electromagnetic field. Only slightly later did Einstein extend the same ideas to matter, so that the motion of particles is guided or determined by a matter field.[110] It is evident that these ideas have a great similarity to those of the Bohr–Kramers–Slater theory. Einstein never published these ideas, most likely because he knew that this picture is not compatible with the conservation laws of energy and momenta. Bohr, Kramers, and Slater were willing to give up these laws, something Einstein was never willing to do. That Einstein thought along these lines may further be inferred from a letter Einstein wrote to Ehrenfest in connection with the Bohr–Kramers–Slater theory: "Recently I gave a colloquium discussion of the paper of Bohr, Kramers, Slater. The idea is an old acquaintance of mine—which however I don't consider as a real possibility. Main reasons—conservation laws seem to hold rigorously. Why should actions at a distance be an exception."[111] This shows that Einstein seriously thought about wave fields associated with particles even though in this instance he did not publish it. Later, however, Einstein did publish a significant paper on the material quantum gas.[112] In this paper, formulas occur (for the fluctuations) that have exactly the same structure as those Einstein had obtained for light 20 years earlier. In both formulas there are two terms. In the case of light, the origin of these terms is easily identified: One term refers to the wave attributes and the other to particle attributes. In the case of the material gas, the identical terms appear once again: One term refers to the particles of the material quantum gas and the other by an obvious analogy should refer to the matter waves. Einstein clearly took this seriously. In the paper he mentions that "I pursue this interpretation further since I believe that here we have to do with more than a mere analogy."[112]

It was presumably this remark that stimulated Schrödinger to study and extend what he called "the de Broglie–Einstein wave theory." In a paper that precedes the actual formulations of wave mechanics, Schrödinger describes the ideas of de Broglie and Einstein in terms very reminiscent of those of the guiding field: "In the de Broglie–Einstein wave theory ... a moving corpuscle is nothing but foam on a wave radiation in the basic substratum of the universe."[113] This conveys a picture alright, but its precise physical meaning is certainly not clear. However, one knows from classical physics that if there is a wave, there is a wave equation, and Schrödinger set out to find what equation would describe these de Broglie–Einstein matter waves—whatever they might be. Actually, this equation was known for a free particle, but Schrödinger, by a brilliant argument, found the correct equation for a particle that experiences a force. Very soon he also found an equation for a system containing many interacting particles. "In this article I should like to show—first for the simplest case of the hydrogen atom that the usual rule for quantization can be replaced by another requirement, in which there is no longer any mention of 'integers'. The integral property follows rather in the same *natural* way that, say the number of nodes of a vibrating string must be an integer. The new interpretation can be generalized and, I believe strikes very deeply into the true nature of the quantization rules."[114] This was the opening phrase of the epoch-making series of articles by Schrödinger. In this series of articles he obtained a description of atoms in terms of wave equations (differential equations), the same mathematical machinery describing most of classical physics, elasticity, electromagnetism, and fluids. Most physicists were familiar and comfortable with this form of mathematics, in contrast to the abstract, unfamiliar mathematical character of matrix mechanics. In principle, by using the Schrödinger equations all problems in atomic physics could be reduced to more or less conventional problems in mathematical analysis. In practice, not all problems could be solved, but certainly many more than before. Consequently, the application of Schrödinger's equation did not require a great deal of new mathematics. Compared with the Heisenberg matrix method, this was an enormous mathematical simplification. For example, the hydrogen atom, which needed Pauli's tour de force to yield to matrix mechanics, was solved using routine mathematics in Schrödinger's very first paper. Both theories gave the identical, correct answer for the hydrogen atoms and indeed for all systems studied. The Schrödinger method was generally simpler because it was a more familiar method.

However, at least as important as the use of familiar mathematics in the Schrödinger scheme was the absence of arbitrary quantization conditions (as in the Bohr theory). Schrödinger felt uncomfortable with discrete elements such as quanta; he had disliked the quantum conditions from the outset and detested the notion of quantum jumps. Schrödinger's approach to physics was a classical approach where continuity played a fundamental role. If discrete features were to emerge at all, as, for example, in classical wave theory, then this should be a consequence of the particular constraints imposed on the system rather than a property of the basic equations. For many physicists the hope that a neoclassical theory that did away with quantum rules and quantum jumps would eventually be found was evidently never totally suppressed.

It is understandable that Schrödinger's papers caused a sensation. They were brilliantly written, they used mathematics that was understandable, they appeared to eliminate quantum rules, they seemed to herald a return to classical concepts, and they obtained with relatively simple means a number of stunning results: Clearly, one could not reasonably ask for much more (or indeed that much). In the euphoria of these first papers, it appeared that there was only one tiny remaining question. What is the physical meaning of the waves? In classical theory, equations referring to classical quantities either have a direct physical significance or can be related to quantities that have such a significance, such as an electromagnetic field strength, the tension in a string, or the density in a fluid. However, merely to call the waves de Broglie–Einstein waves identifies the subject of the discussion but does not provide enough information to extract the complete physical content from the formalism. For the calculation of the energy spectrum, such an interpretation is not so necessary. To Schrödinger and to many others, the accomplishments of wave mechanics were so evident that there was general confidence that the remaining questions would soon be cleared up. Schrödinger writes in his first paper: "It is hardly necessary to point out how much more gratifying it would be to conceive a quantum transition as an energy change from one vibrational mode to another than to regard it as a jumping of electrons. The variation of vibrational modes may be treated as a process *continuous* in space and time and enduring as long as the mission process persists." It was *this* particular feature which made the Schrödinger theory so attractive to Einstein. On April 16, 1926, Einstein makes a few comments in a letter to Schrödinger but in the margin he writes: "The idea of your article shows real genius."[115] No higher praise exists. On April 23, Schrödinger replies to Einstein: "The whole thing would certainly not have originated yet and perhaps never (at least by me) if the importance of de Broglie's ideas had not been brought home to me by your second paper."[116] Planck writes on April 2, 1926, to Schrödinger: "I read your article the way an inquisitive child listens in suspense to the solution of a puzzle that he has been bothered about for a long time."[117] And so the chorus of praise continued; Sommerfeld was overjoyed, Fermi immediately took to Schrödinger's approach, and a general air of optimistic expectation pervaded physics. Here was a theory that one could understand and use.

The Copenhagen and Göttingen reactions (Bohr, Kramers, Heisenberg) to Schrödinger's work were considerably more subdued. Actually, the first reaction came from Pauli who wrote to Bohr on February 9, 1926, only a few weeks after Schrödinger's first paper was submitted: "Schrödinger's paper has interested me a great deal, perhaps his method isn't all that crazy."[118]

The two descriptions were not only different, but appeared contradictory. The wave mechanical theory stressed continuity, a classical description in terms of waves. The Heisenberg scheme insisted on discontinuities (quantum jumps), the impossibility of a complete classical description, with the particle notion and quantum states as basic concepts. These were not just vaguely felt biases—these ideas were explicitly articulated and used.

Very soon afterward, Schrödinger demonstrated that the two theories, which started from such different physical assumptions and which had such different goals, nevertheless are formally identical. Thus, as far as the mathematical treatment was concerned, either Heisenberg's or Schrödinger's theory could be used. Very early and undoubtedly independent of Schrödinger, Pauli also proved the equivalence of the two theories. Actually, Pauli never published this result: He communicated it orally to several of his collaborators.[119] He also wrote it in a letter he sent to Jordan, and he took (for Pauli) the unusual step to type the letter and keep a copy.[120]

Schrödinger never made a secret of his intentions to substitute simple classical pictures for the strange concepts of quantum theory. He had a deep aversion to the abstract character of quantum theory. He writes to Planck: "I have the deepest hope that it now will be possible to construct a harmonious quantum theory free from all problems and arbitrariness and not in the sense that everything becomes more discontinuous with quantum conditions imposed but rather in the opposite direction. The *beautiful classical* methods themselves yield all the necessary quantum numbers. There is nothing mystical about these quantum numbers. We are well familiar with them in connection with Hermite functions, Laguerre polynomials. Sommerfeld's comparison with the number mysticism of the Pythagoreans is exactly right. The integers in the atom have about the same origin as the overtones of a vibrating string."[121]* This approach is in marked contrast to that expressed by Born and Jordan: "The new mechanics appears as a pure theory of discontinuities. The physical phenomena do not in any way order the quantum states. The quantum numbers are nothing but indices which characterize the quantum states, which can be ordered and normalized according to any viewpoint."[122]*

It is hard to imagine more contrasting viewpoints, and it is certainly a minor miracle that all calculations based on such contradictory viewpoints and different formalisms nevertheless give the same results. Even though there could not be any argument about the numerical results, there was a major conflict about the physical interpretation of the two formalisms and the degree and the necessity of physical visualization of the two schemes. To Schrödinger it was clear that his wave field had to be a totally classical field which could exchange energy and momentum; but as in a genuine classical theory, there were no discontinuous changes and no quantum jumps. In the Heisenberg theory, discontinuous transitions between states played an essential role. Of course, neither theory at that time possessed a complete consistent physical interpretation; this became the central area of discussion and controversy.

Since the mathematical results of the two theories were identical and the physical interpretations of both were incomplete in the spring of 1926, it could be anticipated that individual reactions to the two theories would strongly depend on the personal expectations for theoretical physics. Schrödinger, in discussing Heisenberg's matrix mechanics, writes: "I was of course familiar with his work. ... I was discouraged if not repelled, by what appeared to me a rather difficult method of transcendental algebra defying any visualization."[123] However, according to Heisenberg, the classical visualization of the wave field as suggested by Schrödinger was neither consistent with known physical quantum laws (the photoelectric effect, the Planck law) nor all that intuitive. Actually, Schrödinger himself recognized the difficulty of providing an acceptable classical interpretation of the wave field. After some indecisiveness, he eventually interpreted the square of the wave function ($\psi^*\psi$) as the actual electrical charge density which then, like the waves, is spread over all

* This is a rather free translation by M. Dresden.

space. That interpretation more or less works if one deals with just *one* particle (in that case the ψ waves exist in a three-dimensional space); however, it becomes more tenuous for many-particle systems or particles with spin, for then the charge is distributed in a many-dimensional space. So Heisenberg wrote to Pauli: "The more I think about the physical part of the Schrödinger theory, the more disgusting it appears to me. Imagine the spinning electron whose charge (according to Schrödinger) is spread out over all space with its axis in a fourth or fifth dimension. What Schrödinger writes about 'visualizable' ... I find a deep fog.... The only great contribution of the Schrödinger theory is the calculations of the matrix elements."[124] This criticism is not unrelated to Heisenberg's often expressed view that matrix mechanics was already a complete theory and that its physical interpretation would follow inevitably from an examination of its mathematical structure without introducing any new elements. Whatever the motivation of Heisenberg's objection to the Schrödinger theory might be—he pointed out immediately and forcefully that with the elimination of quantum states, quantum jumps, and photons, the clock would be turned back and all the conflicts and problems that led Planck and Einstein to the introduction of quanta in the first place would now be unresolved and would require new explanations.

After listening to Schrödinger give a seminar in Munich in the summer of 1926, Heisenberg wrote to Pauli; "When listening to Schrödinger one feels 26 years younger ... he throws out all quantum effects, photoelectric effect, Franck Hertz experiments ... then it isn't hard to make a theory."[125] This seminar was evidently the first time Schrödinger and Heisenberg met face to face. Schrödinger discussed his beautiful and elegant derivation of the hydrogen spectrum. Heisenberg had no objection to Schrödinger's mathematics; in fact, he used it with his customary brilliance to obtain new and interesting results for the helium atom. But he had profound objections to the physical interpretation where the physical charge density is given by the square of the wave field. Heisenberg observed in particular that if this charge density is now used as a classical source of radiation (as Schrödinger suggests), one does *not* and *cannot* get the Planck radiation formula. This was known to many people and certainly to Planck, who more than anyone else was aware that one had to introduce a discontinuity, a quantum, to obtain his empirical formulas. Even so, Planck was very hopeful that Schrödinger might, on the basis of his neoclassical pictures, eventually produce a consistent theory. When Heisenberg raised these objections to the Schrödinger interpretation at the seminar, he was taken to task by W. Wien, who told Heisenberg rather sharply that he understood Heisenberg's disappointment that quantum mechanics with its nonsensical quantum jumps was now finished, but undoubtedly Schrödinger would solve the problems signaled by Heisenberg very soon. Schrödinger recognized Heisenberg's objections as significant; he could not answer them at that time, but he too was convinced that their resolution was a matter of time.[125,126] Even Sommerfeld, who was well aware of Heisenberg's brilliance and who certainly recognized the cogency of Heisenberg's objections, nevertheless felt that Schrödinger's formalism was so powerful and elegant that the objections would sooner or later be overcome or perhaps just disappear. Whatever was wrong with the interpretation, the Schrödinger formalism was eminently usable and as such directly in Sommerfeld's line.

The experience at the seminar in Munich was rather unsettling to Heisenberg. His objections had not been answered, but at least in Munich no one seemed to care very much as long as there was a calculational scheme. However, Heisenberg had analyzed the necessity of discontinuous elements in the description of atoms so thoroughly with Kramers, Bohr, and Pauli that he was convinced that the physical interpretation of the Schrödinger theory could not be correct. He wrote that same night to Bohr, and a little later to Pauli, about the discussions.[125] As a consequence, Schrödinger was invited by Bohr to Copenhagen for further discussions. Pauli, who was enthusiastic about the formal opportunities provided by Schrödinger's theory, had no use at all for the Schrödinger interpretation or the attempt to return to a classical wave description. Pauli and Schrödinger had some correspondence on the interpretation question, but Schrödinger did not know how to deal with Pauli's ferocious criticism, and Pauli never took Schrödinger's interpretation seriously. In a letter to Heisenberg, Pauli refers to Schrödinger's interpretation as *Züricher lokal aberglauben* ("The provincial superstitions of Zürich").[127]

Dirac heard about the Schrödinger theory in a curious manner. In the spring of 1926, Dirac was engaged in applying his version of Heisenberg's matrix mechanics to the hydrogen atom. On April 9, 1926, Heisenberg wrote to Dirac and asked specifically: "Have you considered how far Schrödinger's treatment of the hydrogen atom is connected with quantum mechanics? These mathematical questions interest me especially because I believe that one can gain a great deal for the physical significance of the Theory."[128] Dirac describes his reactions with disarming frankness: "Well my anwer was that I had not considered Schrödinger's theory. I felt at first a bit hostile towards it. The reason was that I felt we have already a perfectly good quantum mechanics, which I believed would be developed for handling all the problems of atomic theory. Why should one go back to the pre-Heisenberg stage when we did not have a quantum mechanics and try to build it up anew. I rather resented this idea of having to go back and perhaps give up all the progress that had been made recently on the basis of the new mechanics and start afresh. I definitely had a hostility to Schrödinger's ideas."[129] Dirac's assessment of the physical significance of Schrödinger's theory was much the same as that of Heisenberg. Dirac argued that just as a pure wave theory of light (without discontinuous elements such as photons) cannot explain all experiments, a pure wave theory of matter (without particles) cannot explain all observations either. After Heisenberg explained the formal structure of wave mechanics to Dirac, and after Dirac found out that there was nothing to "unlearn" but that on the contrary Schrödinger's mathematical scheme provided an additional technique for the solution of problems, Dirac became very enthusiastic about the Schrödinger formalism and he used it with great success. After a while, Dirac and Schrödinger became good friends. Dirac and Schrödinger both were always concerned with the mathematical beauty of a theory, but they never reached an agreement on the interpretation of quantum theory.

Perhaps the most remarkable reaction to Schrödinger's work was Lorentz's. It was a custom of long standing that any new development eventually would be subjected to the incisive scrutiny of Lorentz. The world of physics always was anxious "to hear what Lorentz would say about it." Lorentz understood the coherence and unity of physics as few others did; he had an extraordinary facility to recognize essential physics even in entirely new situations. At age 72 (almost 73) he wrote two long letters to Schrödinger, containing a detailed and perceptive analysis of that work. He recognizes its great significance—but also its problems; he is unwilling to allow arbitrariness and confusion just to obtain a neoclassical description. He separates what is demonstrated from what is conjectured; he distinguishes what is accomplished from what is merely a hope. Quite different from Wien, Sommerfeld, or even Planck! In his letter, Lorentz comments on various aspects of the Schrödinger theory: "Reading these [papers] has been a real pleasure to me. Of course the time for a final judgment has not yet come and there are still many difficulties ... one would venture the hope that your efforts will contribute in a fundamental way to penetrating these mysterious matters."[130] These opening phrases of Lorentz's letter have something reasonable; they immediately seem to guarantee that Schrödinger's contribution will be judged fairly and put in its proper perspective. "If I had to choose now between your wave mechanics and the matrix mechanics, I would give the preference to the former because of its greater intuitive clarity so long as one only has to deal with the three coordinates *x*, *y*, *z* (i.e., one particle). If however there are more degrees of freedom then I cannot interpret the waves and vibrations physically and I must therefore decide in favor of matrix mechanics." Lorentz, with unerring instinct, points to one of the weakest points in the Schrödinger interpretation. Heisenberg had earlier made the same observation: "If I have understood you correctly then a 'particle' an electron for example, would be comparable to a wave packet which moves with the group velocity. But a wave packet can never stay together and remain confined to a small volume in the long run.... The slightest dispersion in the medium will pull it apart ... because of this unavoidable blurring a wave packet doesn't seem to me to be very suitable to represent things to which we want to ascribe a rather permanent individual existence." This comment of Lorentz goes to the heart of the Schrödinger interpretation. If, as Schrödinger suggests, the ultimate constituents of matter, for example, electrons, are to be described by a classical wave field and therefore are in fact waves, from where do the objects called particles and having particle properties come? Schrödinger suggested that packets of waves would simulate particles, but Lorentz argues that such packets would diffuse too rapidly to represent particles adequately. "If we decide to dissolve the electron completely so to speak, and to replace it by a system of waves this is both an advantage and a disadvantage. The disadvantage and it is a serious one is this: Whatever we assume about the electron in the hydrogen atom we must also assume for all electrons in all atoms: we must replace them by a system of waves. But then how am I to understand the phenomenon of photoelectricity and the emission of electrons for heated metals? The particles appear here quite clearly and without alterations, once dissolved how can they condense again? I do not mean to say that there cannot be many metamorphoses in the interior of atoms. If one wants to imagine that electrons are not always little planets that circle around the nucleus and *if one can accomplish* something by such an idea, then I have nothing against it. But if we take a wave packet as a model of the electron, then by doing so we block the way to restoring matters. Because it is indeed a lot to ask to require that a wave packet should condense itself again once it has lost its shape." With these simple words and direct images Lorentz effectively shows that the known experimental properties of particles are not compatible with the naive direct identification of particles with wave packets. These may be related, but the connection must be more subtle. For example, in connection with the serious objection of Lorentz that wave packets spread in time, Schrödinger showed that for a harmonic oscillator the packet does stay together. But Heisenberg showed later that this is true just for oscillators, so that generally Lorentz's objection is valid. The comments by Lorentz convey the mastery that Lorentz possessed over the field of physics at the age of 72. There is something intensely rational about his approach; his control of the subject is evident in the simplicity of his arguments. Schrödinger was deeply impressed by Lorentz's comments; he answered some, but other points remained unsettled. Perhaps most important in Lorentz's reaction was its balance. Clearly, this was a most significant development; it opened up novel ways of viewing and treating the problems in atomic physics, but many questions and puzzles remained. The Schrödinger theory was neither a return to a purely classical description nor a finished quantum theory: Much profound thought and a great deal of further work was needed to assess its precise significance.

Interestingly, Lorentz's conclusions were very similar to those of Bohr and Pauli who arrived at it "from the other side." Although Bohr was very much aware of the massive direct empirical support for energy levels and quantum transitions which Schrödinger wanted to eliminate via his wave theory, Bohr nevertheless felt that the wave formalism and the wave notion that this formalism implied had an important role to play in the physical interpretation of the quantum theory. It is interesting that Bohr, who had been a lifelong supporter of the wave theory of light and who accepted the photon character of light only reluctantly and grudgingly, accepted the wave character of matter rather easily. It is more accurate to say that with the discovery of the wave character of matter, it, like light, must be assigned both wave and particle characteristics. Bohr recognized that the central problem had become the physical description of objects possessing two distinct, perhaps contradictory, categories of properties. Bohr hoped to obtain the proper interpretation by a meticulous scrutiny of experiments. Heisenberg, on the other hand, believing as he did that the matrix scheme of quantum mechanics was already complete, felt that a deeper understanding of the formal abstract structure would provide him with the unique correct physical interpretation. As in the discussion of the dispersion relations, Heisenberg was willing and tempted to forego physical models of atomic phenomena and to be guided by abstract mathematical schemes, while Bohr argued that the description and interpretation of physical experiments were needed to arrive at any intelligible interpretation. Thus, Bohr felt that the wave features inherent in the Schrödinger formalism had to play an essential physical role. Heisenberg had little use for the waves and at first did not want to ascribe any physical meaning to the wave field. The only role of the wave field, according to Heisenberg, was to provide an effective way to calculate the elements in Heisenberg's matrices. The precise meaning—if indeed any—of the Schrödinger and de Broglie–Einstein waves was very unclear at that time.

A decisive advance was made by Born[131] and simultaneously and independently by Dirac.[132] Born studied a scattering problem. He was strongly influenced by the experiments of Franck on electron scattering. In the experimental situation one deals with *particles* that are scattered by a foil. Born treated this problem using the formalism of Schrödinger, which deals with waves. Thus, Born had to confront the problem already mentioned by Lorentz: How are these *particles* in the experimental situations to be related to the *waves* in the theoretical treatment? Through an incisive analysis, Born came to the startling conclusion that the Schrödinger formalism *does not* and *cannot* answer the question: "What is the state of the particle after collision?" Instead, it can only answer the question: What is the *probability* of a particular state after the collision? Thus, wave mechanics, in almost shocking contrast to classical mechanics, does not allow the prediction of future events with absolute certainty; it merely allows the prediction of probabilities. In his famous paper, Born summarized the interpretation: "The motion of the particle conforms to the laws of probability, but the probability itself is propagated in accordance with the laws of causality."[131]

The origin of this revolutionary interpretation can be traced to the (unpublished) ideas of Einstein on the "guiding field," the Bohr–Kramers–Slater theory, and the de Broglie–Einstein waves.[133] Einstein's ideas were developed for the electromagnetic field. To incorporate the photon in that scheme, Einstein conjectured that the local electromagnetic field would determine the *path* of the photon considered as a particle. The number of photons (particles) would be determined by the *intensity* of the field. Nothing more was said about the "reality" of this guiding field or ghost field as Einstein occasionally called it. Inasmuch as photons could be absorbed or emitted by atoms, the *number* of photons had to be understood as a statistical average. In the Bohr–Kramers–Slater theory, very similar ideas occurred: The field was produced by virtual oscillators which had no direct physical significance, but the field could produce real physical effects; that is, it could stimulate emission in distant atoms.

Born applied the Einstein ideas directly to the Schrödinger wave field. The intensity of the field ($\psi^*\psi$) at a point determines the probable number of particles at that point in exactly the same way as the intensity of the guiding field determined the number of photons. But Born emphasized particularly that the de Broglie–Einstein–Schrödinger wave field determines the number of particles in a statistical sense. Consequently, the field determines probabilities and not certainties. The matter field in the statistical interpretation of quantum theory is therefore not a real physical field like an acoustic field or an electromagnetic field, but it does numerically determine the probabilities of real physical events. This is reminiscent of, but distinct from, the Bohr–Kramers–Slater field.

Heisenberg in a letter to Born complains that "You have gone over to the other [Schrödinger's] side."[134] However, Heisenberg soon recognized the significance of Born's work. His only remaining criticism was a methodological one; from Born's analysis it might appear that there was a freedom of interpretation of the formalism, while Heisenberg throughout insisted that the interpretation was a necessary consequence of the mathematical structure. It was the reinterpretation of the Einstein and Bohr–Kramers–Slater ideas via the formalism of matrix mechanics which led to the probability interpretation of quantum mechanics. Heisenberg commented numerous times on the close interrelationship between the Bohr–Kramers–Slater ideas and the eventual interpretation of quantum theory.[135] In the summer of 1926, Born developed his theory of collisions and interpreted the wave in the many-dimensional space as a probability wave: "The *natural* continuation of an earlier idea of Bohr–Kramers–Slater." Later, in an interview (1963), Heisenberg was even more explicit about the crucial role the Bohr–Kramers–Slater attempt had played in arriving at the correct interpretation of the wave field: "Still I felt that the central step was made by Bohr–Kramers–Slater in the earlier paper—in that sense that the waves got that strange kind of reality. The Bohr–Kramers–Slater waves were something which was just in the middle between an actual reality and something which was only mathematics. And this intermediate position of waves, which were a physical reality in the sense that they *produced probabilities* for decay or emission and at the same time were not completely real like the electromagnetic waves, that I found so extremely interesting and attractive."[136] Thus, the Born interpretation the wave field achieved "an intermediate position" (in Heisenberg's terms) as determining the probabilities of events and configurations. With this interpretation, Born introduced probability as a *genuine* and *essential* element in the physical description, which further elaborations and refinements cannot remove. It is interesting and more than a little ironic that where Born believed that he had carried the Einstein notion of the guiding field to its natural conclusion, Einstein himself never accepted the probability interpretation of quantum theory as more than a temporary expedient.

Heisenberg, Pauli, and Bohr were impressed by the probability interpretation of quantum theory, even if it meant the renunciation of strict causality.

The notion that not all events in physics are causally describable had been "in the air" ever since Bohr introduced the quantum jumps in the description of atoms. It was one of the successes of the Born interpretation that he could *calculate* the probability of such transition but that the theory as a matter of principle *could not give* a description of what transpired during a quantum jump. Dirac independently arrived at the statistical interpretation. The probability interpretations scored a number of major successes, the calculation of the processes occurring during the passage of slow and fast particles through matter is but one example.

However, Schrödinger himself was deeply disturbed by the turn of events. His wave field was a classical field—which might have an *unknown*, but knowable, classical interpretation much like the classical electron fields. Furthermore, to Schrödinger, as to Einstein, physics had to be causal; probabilities should *not* play an intrinsic role. Thus, the Born and Dirac interpretation of his own wave field was anathema to Schrödinger. His concerns and worries are expressed in a letter to Planck: "I should like so much to know how the quantum situation is judged in Berlin and especially by yourself. Is it true what the matrix physicists [Born] and the q-number physicists [Dirac] are saying, that the wave equation describes only the behavior of a statistical ensemble?... I believe I am right that you yourself wrestled with the first and most basic assumption of discontinuity.... I believe that one is obliged to take up this struggle anew with the same seriousness among today's newly emerged points of view. I do not have the feeling that this is happening on the part of those who today already announce categorically: the discontinuity of energy *must* be adhered to.... What seems most questionable to me in Born's probability interpretation ... the most remarkable things come forth naturally ... the probability of events that a naive interpretation would consider to be independent do not simply multiply when combined, but instead 'probability amplitudes' interfere in a completely mysterious way.... Well as God wills I keep quiet. That is if one really *must*, I too will become accustomed to such things."[137]

Actually, Schrödinger never became accustomed to the probability interpretation, and he did not keep quiet either. Instead, he challenged the probability interpretation from its inception by pointing out logical shortcomings, by general philosophical objections, and by brilliantly conceived thought experiments. Schrödinger was engaged in a running battle with the exponents of the probability interpretation which lasted all his life and which started with Schrödinger's visit to Copenhagen in October 1926.

Bohr, in part at the instigation of Heisenberg, invited Schrödinger for a visit to Copenhagen. The official part of the program consisted of a lecture given by Schrödinger on October 4, 1926 on "The foundations of wave mechanics" for the Physics Club. After Schrödinger's usual brilliant performance, there was an official banquet and no serious discussion took place. The discussion started in earnest the next day when Schrödinger gave a seminar at the Institute for Theoretical Physics. Schrödinger stayed in Bohr's house at the institute so that nothing would interrupt the conversations. Bohr, Schrödinger, and Heisenberg were the main participants, but Klein was present most of the time. The discussions were long, at times heated, and of an extraordinary intensity.* All the participants, especially Bohr, Heisenberg, and Schrödinger, believed passionately in the essential correctness of their viewpoints. Even though everyone agreed that no scheme had been worked out in complete detail and that much work remained to be done, the opposing parties clung tenaciously to their respective approaches. So the debate depended as much on strength of personality, the conviction of ultimate correctness, intellectual toughness, and dogmatic adherence to a viewpoint, as it did on the actual issues. Bohr, who was usually a kind and considerate person in all his dealings with people, appeared in these discussions as a "remorseless fanatic, one who was not prepared to make the least concession or grant that he could be mistaken."[138] Heisenberg and Bohr, both superb logicians, relentlessly probed the inevitable consequences of Schrödinger's interpretation. Heisenberg had shown earlier that the Planck radiation formula could not hold. During these discussions it was demonstrated further that, based directly on Schrödinger's own ideas, the probability of spontaneous emission would be proportional to the product of the number of atoms in the upper and lower states. This is quite contrary to experiment and contrary to all known theories.[139] This is one sample of the problem Schrödinger faced; Heisenberg with his extraordinary speed could deduce inevitable consequences from Schrödinger's approach which then had to be acknowledged.

Schrödinger had an intense dislike of quantum jumps; his concepts were classical. He had produced a remarkable number of quantum results but, of course, not all. He held to the view that it was reasonable *to hope* that a further elaboration of these same ideas would yield the remaining results. This was very much the central point of the argument. Bohr and Heisenberg argued persuasively that there was *no hope at all* to eliminate the discontinuous elements in physics everywhere. One could, as Schrödinger had done, eliminate the discontinuities at some places, but then they would show up some place else. The already existing mathematical formalism and the empirical data together provided a structure which, although incomplete, already showed that somewhere in the interpretation discontinuities and quantum jumps had to occur. Schrödinger was (at least temporarily) convinced, for he said in a fit of exasperation: "If all this damned quantum jumping were here to stay, I would be more than sorry that I ever got involved in this quantum theory."[140] Bohr answered: "But the rest of us are extremely grateful that you did, your wave mechanics has contributed so much to mathematical clarity and simplicity that it presents a gigantic advance."[140] And so the argument raged day and night without letup. Perhaps as a consequence of this intense debate, perhaps as a defense, Schrödinger became ill. He stayed in bed for a few days

* There is no written record of these discussions, just letters written later by the participants. There are some written recollections of Klein and Heisenberg.

in Bohr's quarters in the institute. While Mrs. Bohr brought him tea and cakes, Bohr would sit on his bed and continue the argument: "But Schrödinger you must at least admit that...." After a few days Schrödinger left. Bohr and Heisenberg, each in his own way, were now preoccupied with the interpretation questions of quantum theory. Bohr summarized the discussion with Schrödinger in several letters to Fowler,[141] to Kramers,[142] and to Darwin.[143] These letters are all similar in content: "Klein has shown how it is possible on the basis of [Schrödinger's] wave mechanics to build a complete theory ... leaning on the quantum postulate. In this way wave mechanics may indeed by considered as a self consistent theory just as Heisenberg's matrix mechanics and it is very interesting to see how the notion of a corpuscle or of a wave presents itself as the more convenient one quite according to the place in the construction where the feature of discontinuity involved in the postulates is explicitly introduced."

Kramers had left Copenhagen a few months before the Schrödinger–Bohr–Heisenberg debates, but he received a blow-by-blow report of the final outcome. Like all other physicists of his generation, Kramers had to come to terms with the radically altered nature and novel viewpoints of physics. This is never an easy task: For one as scientifically and emotionally involved as Kramers, it was especially hard. In Utrecht, at that very time, Kramers was teaching what was very likely the first course in quantum mechanics in the world; this made thoughtful consideration an absolute necessity.

The scientific events which Kramers witnessed exerted a deep, continuing influence on him. The scientific and personal interactions were forever part of his personality. It would be totally impossible to have any appreciation of Kramers as a scientist and a human being if he were not viewed against the background of this changing period in physics, culminating in the quantum revolution.

Notes for Part 1

[1] Interview with J. Korringa by M. Dresden, July 6, 1976.

[2] J. R. Oppenheimer, "Reflections on the Resonances of Physics History," lecture given at the dedication of the Niels Bohr Library. American Institute of Physics, New York, September 1962.

[3] As quoted at the funeral oration on April 24, 1952. See also J. A. Wheeler, *Yearbook of the American Philosophical Society*, pp. 355–360 (1953).

[4] As quoted at the funeral oration, April 24, 1952.

[5] Letter from A. J. F. Siegert to M. Dresden, dated December 26, 1976.

[6] A. A. Michelson, Lecture 1899. See also Taylor, *Physics, the Pioneer Science*, Houghton Mifflin, Boston, 1941.

[7] Letter from Planck to R. W. Wood, dated October 7, 1931. Archive for the History of Quantum Physics.

[8] Planck's Nobel Lecture, Stockholm, June 2, 1920.

[9] Heisenberg, in *Physics and Philosophy*, Allen and Unwin, London, 1959, p. 35.

[9a] Letter signed by Planck, Warburg, Nernst, and Rubens to ensure the election of Einstein to the Prussian Academy of Science, June 12, 1913. See C. Seelig, *Albert Einstein*, Staples Press Ltd., London, 1956, p. 45.

[10] R. A. Millikan, *Rev. Mod. Phys.* **21**, 343 (1949).

[11] R. A. Millikan, Nobel Prize Address, Stockholm, 1923.

[12] W. Nernst, in a paper presented to the Berlin Academy of Sciences, January 26, 1911.

[13] Postcard from Sommerfeld to Niels Bohr, dated September 4, 1913, in Bohr Institute Files. Reprinted in *Niels Bohr*, S. Rozental (ed.), Wiley, New York, 1967 (English translation).

[14] Jeans' comparison of L. Rosenfeld and E. Rüdinger, in *Niels Bohr*, S. Rozental (ed.), Wiley, New York, 1967, p. 59.

[15] *Nature*, **92**, 304 (1913). See also *Niels Bohr*, S. Rozental (ed.), Wiley, New York, 1967, p. 60.

[16] Letter from Professor F. Tank to M. Jammer, dated May 11, 1964, as quoted in *M. Jammer: The Conceptual Development of Quantum Mechanics*, McGraw-Hill, New York, 1966.

[17] Letter from Professor F. Tank to M. Jammer, dated May 11, 1964, as quoted in

M. Jammer: The Conceptual Development of Quantum Mechanics, McGraw-Hill, New York, 1966.

18 Letter by G. de Hevesy to E. Rutherford, dated October 14, 1913, quoted in A. S. Eve, *Rutherford*, Cambridge University Press, Cambridge, 1939, p. 226.

19 A. Einstein, Autobiographical Notes, in *Albert Einstein: Philosopher Scientist*, P. A. Schilpp (ed.), Library of Living Philosophers, Evanston, Il, 1949, p. 3.

20 Letter from Schrödinger to Lorentz, dated June 6, 1926, in *Letters on Wave Mechanics*, Philosophical Library, New York, 1967.

21 Letter from Paul Ehrenfest to H. A. Lorentz, dated August 25, 1913, quoted in M. J. Klein, *Ehrenfest*, North-Holland, Amsterdam, 1970.

22 Letter from Paul Ehrenfest to A. Sommerfeld, dated May 30, 1916, quoted in M. J. Klein, *Ehrenfest*, North-Holland Amsterdam, 1970.

23 N. Bohr, Lecture to the Danish Physical Society, December 1913; also in *Fys. Tidssk.* **12**, 97–114 (1914).

24 Dirac, as quoted by J. Mehra, "The Golden Age of Theoretical Physics," in *Aspects of Quantum Theory*, Salam and Wigner, Cambridge University Press, 1972, pp. 17–52.

25 Harold Bohr, as quoted by Richard Courant in *Niels Bohr*, S. Rozental (ed.), Wiley, New York, 1967, p. 303.

26 Rutherford, as quoted in *Niels Bohr*, S. Rozental (ed.), Wiley, New York, 1967, p. 54.

27 Rutherford, as quoted in *Niels Bohr*, S. Rozental (ed.), Wiley, New York, 1964, p. 54.

28 L. Rosenfeld, *Essay Dedicated to Niels Bohr*, October 7, 1945, North-Holland, Amsterdam, 1961.

29 Letter from Einstein to M. Besso, dated September 6, 1916, as quoted from the Einstein–Besso correspondence, Pierre Speziali (ed.), Hermann, Paris, 1972.

30 Letter from Einstein to M. Besso, dated July 29, 1918, as quoted from the Einstein–Besso correspondence, Pierre Speziali (ed.), Hermann, Paris, 1972.

31 Interview with W. Heisenberg as recorded in *Archive for the History of Quantum Physics* (*AHQP*, Feb. 15, 1963). Interview with W. Heisenberg by M. Dresden, August 19, 1975.

32 Letter from Bohr to Rutherford, dated January 9, 1924, in *Archive for the History of Quantum Physics.*

33 Letter from Bohr to Michelson, dated Febrary 7, 1924, in *Archive for the History of Quantum Physics.* Compare the thesis of Klaus Stolzenberg, "The development of Bohr's thoughts on complementarity 1924–1929," Stuttgart, 1977.

34 N. Bohr, Nobel Lecture 1922, published in *Nobel Lectures II*, Elsevier, Amsterdam, 1967.

35 Letter from Slater to van der Waerden, dated November 4, 1964, quoted in van der Waerden, *Sources of Quantum Mechanics*, Dover New York, 1967.

36 Letter from Bohr to Geiger, dated April 21, 1925 (in the Bohr Institute files). See also thesis by Klaus Stolzenberg, "The development of Bohr's thoughts on complementarity 1924–1929," Stuttgart, 1977, p. 48.

37 Peter Robertson, in *The Early Years: Akademisk Forlag*, Universitetsforlaget in København, 1979, pp. 20–22.

38 *Niels Bohr Collected Works*, Vol. 3, p. 652, North-Holland, Amsterdam, 1976.

39 Heisenberg, *Physics and Beyond*, preface, Harper & Row, New York, 1971.

40 J. L. Heilbron, "Lectures on the History of Atomic Physics," p. 79 in *Proceedings of the International School of Physics, "Enrico Fermi,"* Vol. 57, Academic Press, New York, 1977.

41 Pauli in the Heisenberg–Pauli correspondence; J. Kramers in an interview with M. Dresden, August 14, 1975.

42 Letter from Sommerfeld to Einstein, dated June 11, 1922; *Correspondence of Einstein and Sommerfeld*, Hermann, Basel, 1968, p. 97.

43 W. Pauli, Nobel Prize Lecture, December 13, 1946, Stockholm.

44 Letter from Bohr to Kramers, dated July 15, 1922 (Bohr–Kramers correspondence

in the Bohr Library). Also in *Niels Bohr Collected Works*, Vol. 3, North-Holland, Amsterdam, 1976.

45 W. Heisenberg, in *From a Life of Physics*, evening lectures at the International Centre for Theoretical Physics, Trieste, IAEA Bulletin, 1968.

46 H. A. Lorentz, Lecture, Sorbonne, December 10, 1923, "The Old and the New Mechanics." Reprinted in *Lorentz*: *Collected Papers*, Vol. 7, pp. 82–302, Nyhoff, The Netherlands, 1934.

47 Letter from Born to Einstein, dated October 21, 1921, in the Born–Einstein correspondence.

48 In W. Heisenberg, *Physics and Beyond*, Harper & Row, New York, 1971, p. 35.

49 Letter from Heisenberg to Pauli, dated January 15, 1923 (in the Pauli correspondence).

50 Letter from Heisenberg to Pauli, dated March 26, 1923 (in the Pauli correspondence).

51 C. G. Darwin, *Nature* **III**, 771 (1923).

52 Letter from Bohr to Sommerfeld, dated April 30, 1922 (Bohr Institute files). Also quoted in *Niels Bohr Collected Works*, Vol. 3, North-Holland, Amsterdam, 1976.

53 Letter from Pauli to R. de L. Kronig, dated May 21, 1925 (in the Pauli correspondence).

54 Letter from Kramers to J. Romeyn, dated September 22, 1922. Original kept in The Institute for Social History at the University of Amsterdam. This correspondence will be referred to as R. All translations are by M. Dresden.

55 Letter from Romeyn to Kramers, dated February 20, 1923 (R).

56 Letter from Kramers to Romeyn, dated April 29, 1923 (R).

57 J. C. Slater, *Solid State and Molecular Theory, A Scientific Biography*, Wiley-Interscience, New York, 1975, p. 15.

58 N. Bohr, *Ned. Tijdschr. Naturrkd.* **XVIII**, No. 7 (July 1952).

59 Letter from Pauli to Kronig, dated October 9, 1925 (in the Pauli correspondence).

60 W. Heisenberg, in *From a Life of Physics*, evening lectures at the International Centre for Theoretical Physics, Trieste, IAEA Bulletin, 1968.

61 Letter from Kramers to Jan Romeyn, dated January 8, 1924 (R).

62 Manuscript in the Kramers' file at the Bohr Institute, Copenhagen.

63 Letter from Kramers to Romeyn, dated October 28, 1924 (R).

64 Letter from Born to Bohr, dated April 16, 1924 (Bohr Institute files, AHQP); see also Kramers files, Copenhagen, p. 19.

65 Einstein letter to Hedwig Born, dated April 29, 1924 (Born–Einstein correspondence).

66 Geiger letter to Bohr, dated April 17, 1925 (Bohr Institute files, AHQP). See also thesis by Klaus Stolzenberg, "The development of Bohr's thoughts on complementarity 1924–1929," Stuttgart, 1977, p. 47. See also Appendix, N. Bohr, *Z. Phys.* **34**, 142 (1925).

67 Letter from Bohr to Fowler, dated April 21, 1925 (AHQP).

68 Letter from Bohr to Franck, dated April 21, 1925 (AHQP). See also thesis by Klaus Stolzenberg, "The development of Bohr's thoughts on complementarity 1924–1929," Stuttgart, 1977, p. 51.

69 N. Bohr, *Z. Phys.* **34**, Appendix (July 1925).

70 Letter from Kramers to Born, dated May 13, 1925, SHQP microfilm.

71 Interview with O. Klein, October 29, 1975, by M. Dresden.

72 O. Klein, in *From a Life of Physics*, evening lectures at the International Centre for Theoretical Physics, Trieste, IAEA Bulletin, 1968.

73 Letter from Kramers to R. de L. Kronig, dated February 26, 1926 (in Kramers correspondence).

74 Letter from Pauli to Bohr, dated February 21, 1924 (Bohr Institute files). See also the Pauli correspondence.

75 Letter from Pauli to Bohr, dated December 12, 1924 (Bohr Institute files). See also the Pauli correspondence.

[76] Letter from Heisenberg to Pauli, dated June 24, 1925 (in the Pauli correspondence).
[77] Letter from Heisenberg to Pauli, dated June 29, 1925 (in the Pauli correspondence). See also thesis by Klaus Stolzenberg, "The development of Bohr's thoughts on complementarity 1924–1929," Stuttgart, 1977.
[78] Letter from Born to Einstein, dated July 15, 1925 (Born–Einstein correspondence).
[79] Letter from Pauli to Kronig, dated October 9, 1925 (in the Pauli correspondence).
[80] N. Bohr, in "Atomic Energy and Mechanics," Supplement to *Nature* **110**, 845 (Sept. 1925).
[81] Letter from Bohr to Ehrenfest, dated October 14, 1925 (Bohr Institute files). See also thesis by Klaus Stolzenberg, "The development of Bohr's thoughts on complementarity 1924–1929," Stuttgart, 1977.
[82] Letter from to Rutherford, dated January 27, 1926 (Bohr Institute files). See also thesis by Klaus Stolzenberg, "The development of Bohr's thoughts on complementarity 1924–1929," Stuttgart, 1977.
[83] Dirac, in *From a Life of Physics*, evening lectures at the International Centre for Theoretical Physics, Trieste, IAEA Bulletin, 1968.
[84] Postcard from Einstein to Ehrenfest, dated September 20, 1925, A. Einstein Archives, Princeton, New Jersey.
[85] Letter from Einstein to Besso, dated December 25, 1925 (in Einstein–Besso correspondence, pp. 215–216; Pierre Speziali (ed.), Hermann, Paris, 1972).
[86] Letter from Heisenberg to Bohr, dated June 8, 1925 (Bohr Institute files). See also thesis by Klaus Stolzenberg, "The development of Bohr's thoughts on complementarity 1924–1929," Stuttgart, 1977.
[87] Letter from Heisenberg to Bohr, dated August 31, 1925 (Bohr Institute files). See also thesis by Klaus Stolzenberg, "The development of Bohr's thoughts on complementarity 1924–1929," Stuttgart, 1977.
[88] Letter from Kramers to Urey, dated July 16, 1925 (Kramers' file); also SHQP microfilm.
[89] Letter from Pauli to Kramers, dated July 27, 1925 (in the Pauli correspondence); also in SHQP microfilm.
[90] Letter from Kramers to Romeyn, dated August 15, 1925 (R).
[91] Letter from Kramers to Fowler, dated December 9, 1925 (Kramer's file at the Bohr Institute).
[92] E. Wigner, in "Recollections and Expectations" lectures, October 13, 1975, inaugural address, Kramers' Chair.
[93] O. Klein, in *Niels Bohr*, S. Rozental (ed.), Wiley, New York, 1967, p. 87.
[94] Letter from Heisenberg to Pauli, dated October 12, 1925 (in the Pauli correspondence).
[95] Letter from Heisenberg to Pauli, dated November 3, 1925 (in the Pauli correspondence). See also *Theoretical Physics in the 20th Century—A Memorial to W. Pauli*, M. Fierz and V. F. Weisskopf (eds.), Interscience, New York, 1960, p. 43.
[96] Heisenberg in interview, February 15, 1963. Recorded in *Archive for the History of Quantum Physics*.
[97] As reported to M. Dresden by Professor G. E. Uhlenbeck.
[98] Private communication to M. Dresden.
[99] J. C. Slater, *Solid State and Molecular Theory, A Scientific Biography*, Wiley-Interscience, New York, 1975, p. 21.
[100] Letter from Ehrenfest to Lorentz, dated October 16, 1925, from the Kramers files.
[101] R. de L. Kronig in *Theoretical Physics in the 20th Century*, M. Fierz and V. F. Weisskopf (eds.), Interscience, New York, 1960, p. 26.
[102] Letter from Bohr to Kronig, dated March 26, 1926, quoted in B. L. van der Waerden, "The exclusion principle and spin," p. 215, in *Theoretical Physics in the 20th Century*, M. Fierz and V. F. Weisskopf (eds.), Interscience, New York, 1960.

[103] Letter from Bohr to Ehrenfest, dated December 22, 1925 (Bohr Institute files).
[104] Private communication from Professor G. E. Uhlenbeck. The widely held belief was that relativistic effects were always of the order v^2/c^2, so it appeared that a factor of 2 would be out of the question.
[105] Letter from Pauli to Kramers, dated March 8, 1926 (in the Pauli correspondence).
[106] Letter from Pauli to Bohr, dated March 12, 1928 (in the Pauli correspondence).
[107] Personal communication from Professor G. E. Uhlenbeck.
[108] M. Dresden interviews with N. H. Hugenholz on May 9, 1977, with J. J. Korringa on July 6, 1976, and with F. J. Belinfante on March 25, 1976.
[109] L. de Broglie, preface to Ph.D. thesis (reedited) *Recherches sur la théorie der Quanta*, Masson & Cie, Paris, 1963, p. 4.
[110] As reported by E. Wigner "Recollections and Expectations" lectures, October 13, 1975, inaugural address, Kramers' chair. See also Max Born, *My Life*, Charles Scribner & Sons, New York, 1970, p. 232.
[111] Letter from Einstein to Ehrenfest, dated May 31, 1924, Einstein Archives, Princeton, New Jersey.
[112] *A. Einstein*, S. B. Preuss, AK, Berlin, 1925, p. 3.
[113] E. Schrödinger, *Phys. Z.* **27**, 95 (1926).
[114] E. Schrödinger, "Quantisierung als Eigenwert problem", *Ann. Phys.* **79**, 527 (1926).
[115] Letter from Einstein to Schrödinger, dated April 16, 1925; reproduced in *Letters on Wave Mechanics*, Philosophical Library, New York, 1967, p. 24.
[116] Letter from Schrödinger to Einstein, dated April 23, 1926; reproduced in *Letters on Wave Mechanics*, Philosophical Library, New York, 1967, p. 26.
[117] Letter from Planck to Schrödinger, dated April 2, 1926; reproduced in *Letters on Wave Mechanics*, Philosophical Library, New York, 1967, p. 3.
[118] Letter from Pauli to Bohr, dated February 9, 1926 (in the Pauli correspondence).
[119] K. Lanczos, *Z. Phys.* **35**, 812 (1926); G. Wentzel, *Z. Phys.* **38**, 578 (1926), footnote 2.
[120] Letter from Pauli to Jordan, dated April 12, 1926. Available in the Pauli Archive, Zürich. Compare also van der Waerden in *The Physicist's Conception of Nature*, Jagdish Mehra (ed.), Reidel Publishing, Dordrecht, Holland, 1973.
[121] Letter from Schrödinger to Planck, dated February 26, 1926. Compare thesis by Klaus Stolzenberg, "The development of Bohr's thoughts on complementarity 1924–1929," Stuttgart, 1977, p. 103.
[122] M. Born and P. Jordan, *Z. Phys.* **34**, 879 (1925).
[123] Schrödinger, *Ann. Phys.* **79**, 734 (1926).
[124] Letter from Heisenberg to Pauli, dated June 19, 1926 (in the Pauli correspondence).
[125] Letter from Heisenberg to Pauli, dated July 28, 1926 (in the Pauli correspondence).
[126] W. Heisenberg, in *Niels Bohr*, S. Rozental (ed.), Wiley, New York, 1967, p. 103.
[127] Letter from Pauli to Heisenberg, dated July 11, 1926 (in the Pauli correspondence).
[128] Letter from Heisenberg to Dirac, dated May 26, 1926. See also P. A. M. Dirac, "Reflections of an Exciting Era," in *Proceedings of the International School of Physics, "Enrico Fermi"*, Vol. 7, Academic Press, New York, 1977.
[129] P. A. M. Dirac, "Reflections of an Exciting Era," in *Proceedings of the International School of Physics, "Enrico Fermi,"* Vol. 7, p. 131, Academic Press, New York, 1977.
[130] Letter from Lorentz to Schrödinger, dated May 27, 1926; reprinted in *Letters on Wave Mechanics*, Philosophical Library, New York, 1967, p. 43.
[131] Born, *Z. Phys.* **38**, 803 (1926), submitted July 21, 1926.
[132] P. A. M. Dirac, *Proc. R. Soc. London* **112**, 661 (1926).
[133] Max Born, *My Life*, Charles Scribner & Sons, New York, 1970, p. 232.
[134] Letter from Heisenberg to Born reprinted from *Max Born, My Life*, Charles Scribner and Sons, New York, 1970, p. 233.
[135] Heisenberg, "A Memoriam to W. Pauli," in *Theoretical Physics in the 20th Century*, M. Fierz and V. F. Weisskopf (eds.), Interscience, New York, 1960.
[136] Interview with Heisenberg on February 19, 1963, SHQP microfilm.

[137] Letter from Schrödinger to Planck, dated July 4, 1927; reprinted in *Letters on Wave Mechanics*, Philosophical Library, New York, 1967.
[138] W. Heisenberg, in *Physics and Beyond*, Harper & Row, New York, 1971, p. 73.
[139] O. Klein, interviews on February 20, 1963 and July 16, 1973 in SHQP.
[140] Heisenberg, interview on November 30, 1962; Heisenberg, in *Physics and Beyond*, Harper & Row, New York, 1971, pp. 75, 76.
[141] Letter from Bohr to Fowler, dated October 26, 1926 (Bohr Institute files). See also thesis by Klaus Stolzenberg, "The development of Bohr's thoughts on complementarity 1924–1929," Stuttgart, 1977.
[142] Letter from Bohr to Kramers, dated November 11, 1926 (Bohr Institute files). See also thesis by Klaus Stolzenberg, "The development of Bohr's thoughts on complementarity 1924–1929," Stuttgart, 1977.
[143] Letter from Bohr to Darwin, dated November 24, 1926 (Bohr Institute files). See also thesis by Klaus Stolzenberg, "The development of Bohr's thoughts on complementarity 1924–1929," Stuttgart, 1977.

PART 2
LIVING THROUGH A REVOLUTION

CHAPTER 10

A Hesitant Start with an Unusual Twist

I. Youth and General Background

Hendrik Antonie Kramers was born on December 17, 1894 in the city of Rotterdam, the largest commercial center and the second largest city in The Netherlands. He was the third son of what eventually became a family of five children, all sons. His father was a medical doctor who, in the time honored Dutch Calvinist tradition, was an imposing authoritarian figure. His mother, Suzanna Breukelman, by contrast was very easy-going, unusually kind, and gentle. She reputedly had a special talent for ameliorating conflicts. The combination of a strong, pragmatic father and a sensitive, kind mother produced a warm household, leading to a close-knit family life.

By any reasonable standards, the Kramers' sons were successful, at least two of them achieved unquestioned eminence. Two became physicians, another a chemical engineer. Jan Kramers, the oldest son and to whom Hendrik Kramers was probably closest, became a most outstanding scholar of Arabic. Jan Kramers the linguist and Hendrik Kramers the physicist were professors at the University of Leiden at the same time, their acknowledged international status providing added luster to a venerable institution.

The Kramers family was not especially wealthy, but they were comfortable and certainly had no serious financial troubles. This can, among other things, be inferred from the fact that all five sons received a university education, as their father had before them. Attendance at the university at that time was far from automatic, even for intelligent persons. It was a privilege restricted to a small affluent fraction of the population. The social milieu in which Hendrik Kramers grew up was that of upper middle class professional people. It had a rather strict Calvinist background, with a strong propensity for intellectual pursuits.

Hendrik Antonie early got the nickname Hans and the name stuck. It was by that name that he wished to be known by his friends and family. Later, when Kramers' interest in physics had become definite, he often referred to his initials H. A. as predisposing him to a career in physics. Holland's greatest physicist, Lorentz, had the identical initials—and this knowledge was never terribly far from Hans' thinking.

Already as a young child Hans had to wear glasses; this kept him from participating in the more active sports and rowdier games. There is no indication that this bothered him a great deal although he often got into fights. He evidently enjoyed a certain amount of privacy and liked to go for solitary walks. He showed a very early interest in reading; perhaps more remarkable, he started to write stories and poems at a very early age.[1] These stories show a surprising feeling for form and structure. Thus, the first indications of Kramers' special talents were revealed in his understanding and control of language, not in mathematics or physics. Some of these stories show a child's lively imagination, such as fantasies about gremlins and fairies. In the story "How Thor Got His Hammer Back," written when Hans Kramers was 10 years old, he showed a surprising knowledge and understanding of Norse mythology. The story "The Waiting Room of Professor Hans" is particularly original. In this story many persons are gathered in the waiting room of an enigmatic brilliant Professor Hans to ask for his counsel and advice. Actually, the professor rarely shows up and gives no advice. It is interesting as an indication of how the 12 year old hoped that he eventually might be seen by others. Probably most unusual is a story describing a Dutch–Belgian war in 1929. (Remember the story was written in 1906.) In this fantasy, apart from describing war-related activities, Hans also introduced a peace movement with "Mrs. Kramers of Rotterdam as the leader of the peace movement."

In addition to writing poems and stories, Hans Kramers was also an avid letter writer. It is a little hard to estimate, but a conservative guess would be that Kramers wrote at least 6000 letters. The personal letters, as well as the scientific letters, are all of great significance. In these letters Kramers expresses hopes, fears, and expectations with an abandon and freedom from constraint which sheds an interesting light on his personal thoughts and his scientific and intellectual development. Probably the earliest letter written by Hans Kramers is in the form of a poem, "On an Infected Toe."[2] It was written to Jan Romeyn, the son of a very close friend of Kramers' mother. Jan Romeyn and Hans Kramers had a close friendship which began when they were barely 5 years old. It lasted throughout their early school years (although they went to different schools), their joint student years in Leiden, and their different professional careers of historian and physicist, until Kramers' death. The correspondence between Romeyn and Kramers reflects the development of two intelligent persons from adolescent preintellectuals to mature scholars. The inevitable ebb and flow of personal relations recorded in the correspondence gives an unusual insight into the persons they were and the periods through which they lived.

As all children in Holland, Hans Kramers attended the primary school for 6 years. He then went on to a high school with an especially rigorous program in physics and mathematics for 5 years. During this period his unusual talents began to manifest themselves. Both in school and out of school his interest began to broaden. He retained and deepened his interest in everything connected with writing and literature, but his concern expanded from literature to literary criticism, to philosophy, to cultural history. Associated with his interest in philosophy and ethics was a deep preoccupation with theology (perhaps a remnant or recasting of his Calvinist upbringing). At the same time, natural phenomena and some rather fantastic engineering projects began to intrigue him. In high school, Kramers handled mathematics and physics with playful ease, and during the last few years in high school he began to develop an increasing interest in physics and chemistry. But he approached scientific questions from the logical, philosophical, or perhaps theological side. His interest in phenomena per se did not develop until much later, and it never really became the center of his interest in physics.

By the time that Kramers took his final examination from high school (he was then 17 years old), he had done a prodigious amount of reading and had been exposed to a vast amount of material. Kramers developed an early interest in music, both as a listener and a participant, and was a good pianist and an excellent cellist. Music played an exceptionally important role in his life. It provided an emotional outlet in a life that was otherwise controlled and rational. He was well on his way toward becoming a broadly cultured, intellectually oriented person. Although by this time his interest in science dominated, he did not wish to lose or neglect his other interests. The shift to science was gradual and not easy; it caused Kramers much soul searching. A letter to Romeyn in 1911, written when Kramers was visiting a relative in London, expresses his interest and preoccupation with literary scholarship, especially Byron, Shelley, and Goethe.[3] But about half a year later, he refers somewhat sadly to the demands science makes (or will make): "A man of science must sacrifice his individuality for his field."[4]

Kramers had by this time made a preliminary decision to enter the university and to study physics. However, it was not possible in Holland at that time to enter a university having graduated from the kind of high school Kramers had attended. Either one had to graduate from a gymnasium, which took 6 years and where the curriculum included a hefty dose of Latin and Greek, or one had to pass an Examination of State, which included a considerable amount of Latin and Greek (but other topics as well). It was evidently felt that only those people knowledgeable in Latin and Greek were suitable prospects for a university education. (This was the law of the land; it was changed many years later by an act of Parliament.)

It was rather common for students who had finished high school and who wished to go to the university to take private tutoring in Latin and Greek for a couple of years so they could pass the examination. Often several students would study together for that examination; so Hans Kramers, together with some friends, spent the next year (1911–1912) learning Latin and Greek. His friend, Jan Romeyn, was also a member of that group. (Actually, Romeyn had been attending the gymnasium and was taking Latin and Greek, but he had gotten into a fight with the principal, so he left the gymnasium and joined Hans Kramers and his colleagues to prepare privately for the examination.) It is impressive enough that it took Kramers but 1 year to prepare for a rather stiff exam in two new languages, but it is even more impressive that he learned these languages so well that from that time on he read them for his private pleasure. He especially enjoyed Cicero, Horace, and Homer and he read them all through his life.

II. The Leiden Years (1912–1916)

In September 1912, Hans Kramers enrolled at the University of Leiden. At that time in the university system, there was no well-defined time schedule which the students were compelled to follow. Presumably, they were to take classes which led to a degree, but the rate at which they took them was pretty much left to the students. They could take classes in the subjects of their choice and attend classes as they saw fit; the whole structure was exceedingly loose. It was required to take examinations in a number of subjects (for which the classes presumably provided the training), but the exams could be taken at any time. The examinations were all oral and consisted of a private discussion between student and professor. After a number of these exams (called tentamina), there was a somewhat more formal exam, again oral, but now with several professors present. Passing this exam made the student officially a "candidate." The whole process was then repeated, but now the final formal exam was called the doctoral examination; it corresponds rather closely to a Ph.D. preliminary examination in the United States. It would typically take 3 or 4 years to pass the examination for a candidate and another 2 or 3 years for the doctoral examination. There were, of course, wide variations. Very little or no research was done before the doctoral examination. There was (and is) a specific degree associated with that examination—a Drs. degree. This is an accepted useful degree; a fair number of students do not go beyond this degree. Many high school teachers have a Drs. degree, and for some strange reason, a number of politicians also hold that degree. To obtain a Ph.D. degree, a considerable amount of research has to be done; the writing of a thesis can take many years. Since there were no fixed schedules, a first indication of the interest and quality of a student was the speed with which he or she obtained the degrees. Kramers passed his doctoral examination in a little less than 4 years, which is extremely fast.

Upon entering the university, Kramers also joined the Student Corps, an exclusive social fraternity devoted to drinking, partying, and fooling around —all those activities which, at least in operattas, make student life so memorable and worthwhile. Kramers participated in some of these activities; he literally suffered through the initiation, but he soon became disgusted and quit. He was acutely uncomfortable in that environment.[5–7] He was a very social person but he intensely disliked the coerced, contrived conviviality of the traditional student social clubs and preferred discussions, talks, arguments, and debates.

Kramers engaged in many extracurricular activities: He was an editor of *Minerva*, a student literary magazine, and one of the organizers of an international association for students' interchanges. He evidently enjoyed these activities and the social contacts they brought. A not inconsiderable amount of time went into these efforts, and he performed the associated tasks with ease and skill.

Kramers was a serious student. Since there were no particular limitations on what classes to take or when to take them, Kramers very rapidly completed most of the available courses. These included the traditional classical fields of science: mechanics, electrodynamics, wave theory, statistical mechanics, and van der Waals theory (thermodynamics). Kramers also attended courses in mathematics. In addition to the customary classes in analysis and differential equations, he took mathematics of surfaces (really an introduction to two-dimensional differential geometry), a course in modular functions. Kramers later became an expert in complex variable theory and elliptic functions, his interest in these areas being first stimulated by the courses he took in Leiden.

Evidently, the classes were not too demanding; not only did Kramers master the material in short order, but it appears from his lecture notes[8] that his mind often wandered. Occasionally, in the middle of the lecture notes, there are stories, poems, and illustrations having no obvious connection with the lecture material. This did not seem to detract from Kramers' understanding or performance. It became evident quite early that Kramers was a superb analyst. He enjoyed calculations and manipulations. He had a special talent for inventing formal tricks. If something was computable at all, Kramers could invent an approximate or exact method to do it.

Unquestionably, the most important experiences during Kramers' Leiden years were his contacts with Ehrenfest and Lorentz. Kramers saw (but did not meet) both of them for the first time at the inaugural address by Ehrenfest on October 17, 1912. Kramers recognized Lorentz from the description his father had given him. (Kramers' father had taken physics from Lorentz many years earlier.) Kramers later recalls that he gazed on Lorentz "like a little, little boy who stares at a real Queen for the first time."[9] Later, Kramers took the famous "Monday morning" classes from Lorentz. This was a special topics course with the material changing from year to year. Lorentz gave beautifully organized, masterly descriptions (frequently his own) of many aspects of contemporary physics. So polished were the presentations that often students could not reproduce much after the classes were over. So effortless was the treatment that few listeners appreciated the struggles necessary to arrive at the results.

The routine (or so it was called) teaching and the immediate responsibility for students (certainly for students wishing to get a Ph.D.) was Ehrenfest's. Ehrenfest was a most unusual person and physicist.[10] His interests, both in physics and in the transmission of physics to young students, were of unparalleled intensity—amounting to a genuine passion. The people in Holland, especially the students, did not know exactly what to make of it. Kramers later tried to recall his initial reactions: "Such an intelligent enthusiasm was a strange phenomenon in these low lands, with its phlegmatic inhabitants. To evaluate and experience what humanity and the world had to offer with such a degree of intensity was bewildering to us (the students). Those who saw and heard Ehrenfest could not escape the feeling that they were subject to a whirlwind which would bring promise and novelty to all the corners of their soul."[11]

To say that Kramers was deeply impressed by Ehrenfest would be an understatement. The incisive way in which Ehrenfest could summarize the major points of an already existing theory, exposing its structure, was a revelation to Kramers. Kramers was continually amazed by the way Ehrenfest could focus on essentials, leaving out everything else and thereby exhibiting its physical features with utmost clarity. But he was literally mesmerized by the way Ehrenfest could guide his students unerringly to the burning questions at the frontier of physics. Ehrenfest's vision of research in physics as a relentless battle against "the great unknown" shaped and formed Kramers' attitude toward the role of research in physics. Ehrenfest, in turn, recognized the great promise, the already considerable formal power of Kramers. When Ehrenfest had to be absent from lectures, he asked Kramers, sometimes together with another of his most talented students, J. M. Burgers, to give some lectures to his class. Considering the enormous significance Ehrenfest attached to each and every presentation, this was an unmistakable sign of confidence. Of course, the two students had to give a trial lecture a few nights before to Ehrenfest to see that they did everything correctly and in the proper order!

It was very much in Ehrenfest's style to push his capable students as fast as possible toward independent, original work. He did this with both Kramers and Burgers; Kramers was about 2 years Burgers' senior. As is to be expected, this required a considerable degree of concentration on the part of the students. Kramers was quite willing and happy to make these efforts (so was Burgers), but Ehrenfest really demanded more: He insisted on a total commitment to physics in what amounted to a tripartite treaty between the individual, Ehrenfest, and physics. For Kramers, the increasingly explicit demands for exclusive concentration on physics with the accompanying renunciation of all other interests created a great deal of trouble. He was definitely interested in and fascinated by physics and by Ehrenfest, but he was apprehensive and uncertain about giving up everything else. There seemed to be too many unexplored possibilities. This created a great restlessness in Kramers which manifested itself in many ways. Some were quite silly, such as continually changing the furniture in his rooms or in changing rooms altogether, sudden suggestions for trips of all kinds, and capricious proposals to eat in out-of-the-way places. More seriously, he often did not give his full attention to physics; he frequently skipped Lorentz's class, which was very early in the morning. This was both unforgivable and incomprehensible to Ehrenfest. To Ehrenfest, nothing in the world could be an excuse to miss Lorentz's lectures. (Lorentz himself also took a somewhat dim view of Kramers' absences.) But most objectionable to Ehrenfest was that Kramers occasionally missed the seminars. The seminars were, to Ehrenfest, the purified essence of physics; it was there that "the battle against the great unknown" was in progress. Missing a seminar could only be a lack of interest—or worse, a shirking of a responsibility conferred by intellectual capability to participate in this struggle. Ehrenfest was quite serious about his total devotion to physics and to his students; but so intense were his relations that Ehrenfest had to be involved in everything, no matter how private or personal. The case of Burgers provides a telling example. Ehrenfest took Burgers under his wing and because of this and Burgers' own talents, Burgers developed rapidly. Burgers idolized Ehrenfest. While Burgers was working on his thesis he got engaged. He was amazed that Ehrenfest did not approve of either the girl or the engagement. He was even more amazed that Ehrenfest would deem it his prerogative to make such judgments at all.[12] Since Burgers married his fiancée, this caused considerable tension between Burgers and Ehrenfest and eventually led to an estrangement. Whatever his intentions, Ehrenfest was too oppressive, too stifling in his personal relations. He himself knew that. He writes to Burgers: "I am losing all contact with younger people and growing old.... I am reproached for hurting people by interference."[13]

Kramers already, or perhaps especially, at that early age insisted on his privacy. He had plenty of personal problems and his emotional life had all the intense ups and downs of a sensitive person of 22. But he kept these matters to himself. If he discussed them at all, he did so with persons such as Romeyn, of his own generation. To Kramers, such personal matters were separate and distinct from scientific matters, and his relation with Ehrenfest was limited to science.

In a fake newspaper produced on the occasion of his older brother Jan Kramers' marriage in 1915, Hans Kramers is depicted as very studious, reading Ehrenfest all the time, giving many excellent talks, traveling extensively in an airplane, a detached person with no wife or family and whose only possession is a cat. This clearly indicates that to his peers Kramers appeared totally serious about physics and devoted to Ehrenfest's physics. However, he was unable, or unwilling, to make a decisive break with all his other interests. To Ehrenfest, who could only conceive of an all-consuming passion for physics, this was a fatal flaw in Kramers. In addition, the beginnings of a different approach to physics between Kramers and Ehrenfest began to be noticeable. Of course, Kramers' style was by no means formed, but indications were already there. Ehrenfest was exclusively interested in very sharply formulated problems that dealt with physical principles of great import. Kramers was interested in these questions, but, in addition, he enjoyed solving problems and making very specific calculations using physics to answer questions. He did this all during his life, irrespective of the basic significance of the questions themselves. He evidently was thrilled by the power that mathematics and physics conferred on him in elucidating very special questions, even if they were unimportant. To Ehrenfest, this must have appeared as squandering formal power on insignificant physics.

A final element in the estrangement of Kramers and Ehrenfest was Kramers' early interest in the Gibbs formulation of statistical mechanics. In the Gibbs abstract formulation, the notion of the thermodynamic equilibrium state is central. This came very near being heresy to Ehrenfest, to whom irreversibility and the approach to equilibrium as formulated by Boltzmann were the key problems. Kramers, of course, did not know much about this at that time, but even his verbal interest in the Gibbs approach would not have endeared him to Ehrenfest. It is noteworthy that even before Kramers finished his thesis in Copenhagen, he did some work with Oskar Klein in Stockholm using Gibbs' ensemble method to study the behavior of strong electrolytes.[14] This indicates that his interest in the Gibbs formalism was more than childish curiosity. It is further interesting that Kramers' preference of the more abstract Gibbs approach over the more concrete Boltzmann approach was the *only time* in Kramers' professional life that he preferred a higher level of abstraction for the description of physical systems. Kramers maintained and deepened his interest in the Gibbs formulation of statistical mechanics throughout his life.

All these differences and disenchantments became explicit when Kramers took his doctors (Drs.) examination with Ehrenfest. It must be understood that taking this examination was not a success—it was a failure. No one, that is, no good student, ever took a Drs. examination with Ehrenfest.[15] Ehrenfest usually just worked with his students until they had carried out enough research to get a doctor of philosophy (Ph.D.) degree. Students had to arrive at a level of intellectual competence to become independent scientists; they then would get their Ph.D. So usually Ehrenfest dispensed with a Drs. examination; because of the generally loose administrative structure and the powerful role of a professor, this could easily be arranged.

It is certain that Kramers did take a Drs. examination. Although he formally passed it, he did not do particularly well.[21] Ehrenfest made it very clear that he did not think Kramers had a future as a physicist. According to Ehrenfest, Kramers lacked the motivation, the intense desire, and the concentration to become a physicist working at the frontier of physics. Ehrenfest suggested, or certainly intimated, that Kramers should take the Drs. degree, become a high school teacher, and give up any further attempt at becoming an active physicist.[16,17]

III. The Arnhem Interlude

Shortly after the traumatic Drs. examination, in the early spring of 1916, Kramers became a high school teacher of mathematics and physics at the Gymnasium in Arnhem. This was a particularly difficult period in his life. He was understandably upset and unhappy about the past, dissatisfied with the present, and confused and uncertain about the future. Exacerbating everything was a very unsettled and unsettling personal life. It is not surprising that in this period of tension and turmoil, Kramers' restlessness and idiosyncracies were quite conspicuous. In spite of his worries, he still projected a picture of a self-confident—not to say arrogant—young man. When the principal of the gymnasium rebuked him for coming late to class with considerable regularity, Kramers answered, totally seriously, that the principal should realize how much more he taught the students in half an hour than most teachers did in an hour. His tendency to make sudden capricious trips was especially pronounced at that time. One time he inveigled Burgers' brother, a prospective physics student, to bicycle all the way from Nymegen to Apeldoorn, a distance of some 50 miles, just to visit an aunt who allegedly made excellent fruit candy (which could in fact be bought in every neighborhood store). This was a typical example of what was called "a Kramers joke." All these signs of restlessness hid deep concerns and a considerable depression.[18]

As time passed, Kramers slowly and hesitantly considered the possibility of returning to physics in spite of Ehrenfest's dire pronouncements. He began, ever so vaguely, to see the sense—but also the limitations—of Ehrenfest's criticisms. Kramers recognized that he had difficulties in making the restrictions in his intellectual life necessary to become a productive scientist. He did not like to make decisions of any kind, and decisions reducing his possible options were almost impossible for him to make; this often caused him to appear as vacillating and undirected. But even admitting his problems with finding a single focus for his life did not justify Ehrenfest's categorical dismissal of Kramers' future as a physicist. Ehrenfest's extreme reaction was more a reflection of his own attitudes toward physics than a realistic assessment of Kramers' eventual role as a scientist.† These not very well articulated and rambling considerations[18] slowly led Kramers to explore ways to return to the world of research. He was strongly encouraged by his father who, as a pragmatic Dutchman, took a dim view of Ehrenfest's advice and suggestions.[19] Furthermore, Kramers had confidence in his unusual formal powers and this encouraged him to continue. At first, Kramers thought of going to Born, but to go to Germany in the middle of World War I was clearly not feasible. The choice, pretty much, had to be a country that was not at war and this suggested Denmark. Kramers by that time had certainly heard about Bohr and his unusual and brilliant ideas about atomic structure. It is hard to guess whether Ehrenfest's explicit antagonism to the Bohr theory at that time was an inducement for Kramers to go to Copenhagen. As it happened, the international student organization planned a conference in Copenhagen in the summer of 1916 and, as an officer, it was possible for Kramers to attend

† It took Ehrenfest a surprisingly long time to come to terms with Kramers as a physicist. He recognized the spectacular work Kramers did with Bohr in Copenhagen but even then had some misgivings. Only after Kramers became the acknowledged master of quantum theory, and especially after quantum mechanics, did Ehrenfest lose all qualms about Kramers' abilities, although he never completely got over his doubts of Kramers' motivation.

that conference. In addition, the principal of the gymnasium in Arnhem, who felt that Kramers—in spite of his idiosyncracies—was particularly able, encouraged Kramers to go to Copenhagen. The principal knew some people in Copenhagen and gave Kramers some letters of introduction.

So by the time the summer vacation came, Kramers had decided to go to Copenhagen. He prepared by trying to learn Danish in 2 weeks at a Dutch summer resort. (He succeeded surprisingly well!) It did not seem to worry Kramers particularly that he was going unannounced to a place where no one knew him; his real concerns were still about himself. He frequently thought and worried about the events in Leiden: "Have I been too selfish, am I really a physicist?"[20] There are a number of entries in his diary[20] in which he tries to forget an examination. He often chides himself for not working hard enough; but what he needs for the future is a "deepening of thought and effort—with that will come increasing self-confidence."[21]

Without having any idea of what the future would bring and with a curious mixture of confidence and doubt, Kramers left Holland to go to Copenhagen—and to Bohr.

CHAPTER 11

The Early Copenhagen Years: From Student to Apostle

I. Summary: Between Leiden and Copenhagen

It is surprising that a venture so haphazardly conceived as Kramers' trip to Bohr led to such unquestioned successes. The motivation for Kramers' departure from Holland was general unhappiness with Leiden and Ehrenfest, and severe doubts and concerns about his ability to do genuine physics. He came to Bohr without introduction, help, or recommendation from those who made Leiden renowned. He came at a time when his personal problems were severe and he was in the midst of great uncertainty and confusion about his scientific goals. Yet on this most tenuous basis Kramers proceeded to accomplish a prodigious amount of significant work in an incredibly short period of time.

A scant 8 months after Kramers' arrival in Copenhagen, Bohr had enough confidence in Kramers to send him to Stockholm to present his ideas on atomic physics. If it is recalled how jealous Bohr was of the precise and meticulous formulation of his ideas, this is an unusual sign of confidence in a young student. A little more than a year (October 3, 1917)[22] after Kramers' arrival, Bohr and Kramers achieved a major success by providing for the first time a qualitative explanation of the *intensities* and polarization of the Stark effect of hydrogen. On May 1, 1919, not quite 3 years after Kramers had taken his first tentative trip to Copenhagen, Kramers defended his thesis in Leiden. By that time the thesis was written, printed, and published. Its technical virtuosity and profound analysis of the physical ideas established Kramers at once as one of the masters of the quantum theory. This surely is an impressive productivity in the span of 3 years. But these signal achievements still do not exhaust Kramers' activities during these years. Kramers worked actively on Einstein's general theory of relativity. His dissertation contains an important proposition dealing with general relativity,† which later (1921) led to several papers in this area. Early in 1918, Kramers worked with Klein on the theory of electrolytes, an altogether different part of physics.

This veritable burst of scientific ideas is even more remarkable if it is realized that it was carried out in circumstances which were far from ideal. The physical facilities in Copenhagen were minimal. There were only occasional secretarial services. For a number of years (1916–1920), Bohr and Kramers shared one room about 15 square meters in size.[23] Efforts to obtain new facilities, especially earmarked for theoretical physics, took a great deal of time and energy and were hampered by frustrations and delays. When the new building was nearing completion, Bohr and Kramers had to move the books and papers themselves (January 1921).

But apart from these external demands, further augmented by a variety of teaching and educational obligations, Kramers' personal life was in a state of continual turmoil. It must be stressed again that Kramers left Holland not only because of confusion about his scientific abilities, but also because he was in the throes of an adolescent, indecisive love affair. The sudden transposition from Leiden, where he had many friends and many acquaintances, to Copenhagen, where he knew no one, inevitably caused feelings of loneliness, isolation, and even abandonment. These confused and conflicting emotions are quite explicit in Kramers' diary and personal letters. Thus, Kramers writes in his diary on August 25, 1916: "If you really want things to go differently, remember that you should *do* something, likely working can bring you toward personal deepening, consequently to peace and confidence."[24] These problems and concerns did not let up very rapidly, and he writes on September 25, 1916: "Now little Hans so hard you haven't worked and you have not struggled with deep questions at all, you have behaved like a coward.‡ This is discouraging if you know it yourself."[25] Even at that stage Kramers was occasionally wondering about somewhat fanciful questions in physics such as "What would gravity be like in spaces of higher dimensions?" and "What can be said about the temperature distribution of a universe filled with iron?"[26] But most of his diary entries and especially the letters to Romeyn express depression, concern, and confusion. "It cannot be of interest to any one, who I speak to and just what I do ... I occasionally ask myself what I am doing in Denmark.... I could have stayed in Holland that would have been just as cheap ... I could have studied there as well, but Ehrenfest did not find it all that wise ... and now I must confess, it is not just the thrill of travel, the wish

† In the Dutch tradition, each dissertation has to contain, in addition to the major research project, a number of original assertions (thesis), statements, or claims about a number of topics. These assertions are printed separately, and the public oral examination is actually devoted to a defense of these propositions. Usually, these propositions are related to the main topic, although not all have to be. In Kramers' thesis, one of these propositions dealt with general relativity.

‡ Although written in Dutch, Kramers uses the German word *Feigling*, which roughly translates to "coward." When applied to oneself it is a person who shies away from conflicts and confrontations.

to leave Leiden and all that was unpleasant there which drove me, but an escape from Waldie whom I love so much.... The fact is this. I don't hear a thing from Waldie and that makes me as unhappy as the time in Arnhem and I entered my room every morning with anticipation to see whether or not there was a letter from her."[27]

Thus, the trip to Copenhagen and to Bohr appears not just as a thoroughly cogitated scientific choice, but as a means to escape and avoid conflicts and confrontations. The Copenhagen trip is a direct continuation of Kramers' earlier quixotic behavior which was so prevalent in the difficult period in Arnhem. Kramers wrote his postcard, introducing himself to Bohr on August 25. Bohr must have answered rather promptly, for on September 13, 1916, Kramers sends a postcard to Romeyn, in which he mentions among other things that he had coffee with Bohr.[28] Bohr was somewhat nonplussed by Kramers' postcard but decided, especially at the urging of his brother Harold, that anyone so anxious to work with him should at least have a chance to show what he could do. Bohr evidently suggested to Kramers that he investigate the helium spectrum on the basis of Bohr's theory. This had been a famous unsolved problem in the context of Bohr's ideas. Bohr and Kramers must have plunged almost immediately into this study. In the Bohr Institute files, there are at least 200 handwritten pages on this problem. Also, Bohr in a letter to Sommerfeld[28] refers to results on the helium atom which were obtained in the fall of 1916.

This investigation, the collaboration with Bohr, and the association with colleagues at the University did a great deal to ameliorate Kramers' worries and confusion. His letters and diary entries, although retaining a touch of melancholy and loneliness, are more positive and less contorted. However, in late 1916 and early 1917, there is an abrupt cessation of all diary entries and practically all letters. This coincides in time with a shift in research topic from the study of helium to the investigation of the intensities and polarization of spectral lines. The helium study with which Kramers started his career did not then, and in fact *never* did, lead to significant results. However, the collaboration with Bohr on intensities and polarization yielded spectacular results in a very short period of time. Kramers worked extremely hard with total concentration and commitment in the spring of 1917. During that period he literally "caught fire," and he achieved, or came near to achieving, what all during his life remained a cherished and often elusive goal: a truly original and significant contribution to physics.

To put this accomplishment in its proper perspective, it must be recognized that even though Kramers' scientific life was acquiring direction and purpose, his personal life was still complicated and confused. The always tenuous relation with Waldie diminished in intensity, but he exchanged the attendant difficulties for even more involved problems with his future wife Storm. (Her actual name was Anna Petersen.) Kramers, during the early period in Copenhagen, had some moments of peace and tranquility, but most of the time his personal life was in a turmoil—fluctuating from exuberance to depression.

That under such circumstances Kramers was capable of the quantity and quality of work he did is nothing short of amazing. It is a true testimonial to his innate talents and intellectual power.

One of the contributing factors which enabled Kramers to accomplish so much in such a short time is the thorough, if somewhat unconventional, training he obtained in Leiden. In the true Dutch tradition, the students would take rough notes in class, which then should be rewritten neatly and in proper grammar in a second notebook, with important formulas underlined. Kramers never reached that neat notebook stage (many of his original notebooks are still available);[29] he usually appended brief remarks and comments on the material in the rough notes. Kramers' courses were in part traditional: differential equations, Maxwell theory, and Fourier series. In part they dealt with special topics such as dispersion theory and crystal optics. But what makes Kramers' background atypical is that in addition to the regular courses he selected certain particular topics for personal study. There appears to be no obvious reason for the topics picked; one notebook is devoted to many-dimensional waves, while several notebooks deal with the extinction theorem. These notes give a very clear picture of the manner in which Kramers approached a new topic. It could be said that he met the subject head on. He does not try to avoid difficulties, but rather seeks them out and attempts (and usually succeeds) to overcome them by detailed calculations. Most important for Kramers' later development and the collaboration with Bohr are his extensive and detailed studies of classical Hamiltonian mechanics. It is possible that this material originated from a course of lectures (although there is no record to that effect), but Kramers' notes go far beyond that. Specifically, he took the 1904 (first) edition of Whittaker's celebrated book, *Classical Mechanics*, and literally "slugged" his way through it. From the notes it appears that he was learning the topic: He not only works out most of the problems in detail, but he supplies details of derivations. There are, of course, mistakes, eventually corrected, a frequent retracing of steps, and redirections. The enormous amount of material in these three notebooks on classical mechanics shows that Kramers was familiar with a surprising number of subtle points in Hamiltonian mechanics. He acquired this detailed knowledge by his own independent efforts. Kramers' understanding and control of mechanics was a significant factor in his later ability to implement Bohr's ideas in atomic physics.

II. The Initial Contacts Between Kramers and Bohr

When Kramers' postcard arrived, Bohr had been back in Copenhagen for just about a month. Bohr returned to Copenhagen from Manchester in August to take up his duties as Professor of Theoretical Physics. He had applied for that (not yet existing) post more than 2 years earlier, and he officially received the appointment in April 1916. In his application for that position, Bohr had the strong support of many Danish and foreign physicists. This, as much as anything, showed that varied and tentative as the reaction of the physics community toward Bohr's ideas might be, there was universal recognition that with Bohr's studies a new era in physics had begun. The concepts of quantum level and quantum jump, perhaps uncomfortable and only partially understood, were of monumental importance and were in physics to stay. The curious combination of quantum ideas with classical notions provided a potent mix which, in the hands of Bohr, allowed the correlation of hitherto unrelated facts and phenomena.

In spite of all these sucessess and the accompanying recognition, Bohr himself was fully aware of the incomplete and provisional character of his theory. He worked incessantly at the precise delineation of his assumptions, at proper formulation of his theory, and at further applications. In January 1916, while still in Manchester, Bohr had finished a basic paper, "On the Application of Quantum Theory to Periodic Systems," which was submitted to *Philosophical Magazine*. It was scheduled to appear in April 1916,[30] and the page proofs arrived in early March. A few days later, Bohr received copies of two new papers written by Arnold Sommerfeld.[31] These papers contained a most significant extension of the scope of quantum theory. Bohr was deeply impressed by the work of Sommerfeld, so much so that he withdrew his own paper to *Philosophical Magazine* in order to revise and reconsider his own work in the light of Sommerfeld's new results. The need for this revision did not come at a particularly good time for Bohr. He was in a rather unsettled state, having to arrange a move from Manchester to Copenhagen, and he had to deal with the inevitable disruptions such moves cause. As a new professor in a new institute, there were many administrative demands and he had the full responsibility for the educational program of the new institute. Perhaps most important, Sommerfeld's new treatment of quantum theory, which necessitated the rethinking of Bohr's ideas, depended heavily on the Hamiltonian formulation of classical mechanics.

Bohr's revision started slowly because of lack of time and the need for a painstaking classical analysis of atomic mechanics. Kramers' sudden unexpected appearance in Copenhagen promised to alleviate two of Bohr's most pressing problems. Kramers was steeped and experienced in classical mechanics, and his expertise in this area could be used directly by Bohr to incorporate Sommerfeld's ideas in a comprehensive scheme of quantum theory. Kramers very rapidly showed himself to be a superior expositor. (He fully met the high standards Ehrenfest had set for all aspects of communication.) Kramers soon became a great help to Bohr in meeting his educational obligations, and he frequently substituted for Bohr in presenting his latest thoughts in seminars, conferences, and lectures. Very soon after Kramers' arrival, Bohr, with Kramers' help, embarked on a serious revision of Bohr's withdrawn paper. This revision was so thorough and so complete, that the original single paper expanded into four separate papers. It took about a year and a half to publish parts I and II[32] (they appeared in 1918) under the title "On the Quantum Theory of Line Spectra." Part III was not published until 1922 and part IV was never published at all.

The scope and style of these papers show that they were the outgrowth of the most careful scrutiny of the physical and formal basis of quantum theory at that time. The papers primarily contain Bohr's ideas, but as can be inferred from Bohr's method of working, these ideas were formalized and articulated in lengthy discussions with Kramers. It is in these discussions that Kramers was exposed to Bohr's ideas and approach to physics, and they in turn became part of his own scientific methodology and scientific expectations. There can be no doubt that in this development Bohr was the master and the guide, with Kramers the bright and eager pupil. But Kramers very rapidly made significant contributions, especially in the mathematical elaborations of the physical ideas. In Bohr's paper I of the series, there are many developments specifically attributed to Kramers. In turn, in Kramers' thesis, the style, approach, and type of argumentation are similar to those in Bohr's paper. Kramers' important role in the completion of Bohr's revision (especially of part I) was generally known and acknowledged. Oskar Klein writes in this connection: "Without the devoted and active help of Kramers this [rapid completion] would hardly have been possible."[33] In addition, Rosenfeld comments: "Bohr made full use of the powerful methods of Hamiltonian systems.... It was a fortunate circumstance that Kramers, a master in these questions formed in Ehrenfest's school, was on hand to help him."[34]

Thus, very soon after Kramers' arrival in Copenhagen, Kramers and Bohr established an effective working relation. The approach to quantum theory was all Bohr's. Kramers learned and applied these ideas almost as soon as he assimilated them; his formal powers allowed an almost immediate check on the implementability of Bohr's ideas. But at the same time, Kramers absorbed Bohr's methods, approach, and expectations and made them his own. As time went on, a very strong personal relation evolved between the two men. Even though they eventually became very close, a certain imbalance always remained. Kramers at 22 was a young, inexperienced, and unknown student, while Bohr at 31 already had a world reputation (and a permanent professorship). Bohr had very strong views and overpowering convictions, and Kramers enthusiastically adopted most of Bohr's views, but he was often doubtful and skeptical about his own. As time went on, there were slight shifts, but Kramers throughout his life retained an effusive admiration for Bohr and his physics.

III. Elaboration and Extensions of the Bohr Theory†

To appreciate fully the contributions Kramers made in the early years in Copenhagen, it is necessary to present his accomplishments in terms of the atomic physics of that time. The basic theoretical problem was always to

† This section is somewhat technical, as is Section IV. The more narrative description resumes in Section V.

calculate *all* (or as many as possible) features of the spectra of the various elements. Since Bohr's entry into that field, a fairly well-defined pattern for such calculations had been established.

(1) The starting point was always to view the atom as a classical system of positive and negative charges. To most physicists, this was not merely a picture but the *reality*. For purely calculational purposes, this pictorialization does not really matter.

(2) The next step beyond this visualization was an analysis of the classical motions of that particular system. Clearly, this is a purely classical mechanical problem, but for complicated atoms, especially when influenced by external fields, a complete control of all the classical motions is neither simple nor straightforward.

(3) The next step introduces the quantum ideas. Of all the classical motions, the Bohr theory selects certain *quantized orbits* as the only ones actually realized in nature. The selection principle is one of the key new elements in Bohr's theory. It represents a profound alteration of the classical ideas. Needless to say, the selection principle—sometimes expressed in terms of a set of rules (the quantum rules)—cannot be *derived* from classical physics.

Planck and Bohr obtained these quantum rules for very simple systems (the harmonic oscillator and the hydrogen atom) by remarkably intuitive arguments, by inspired guesses, and by (indirect) appeal to experiments. But no procedure was known which would extend these quantum rules to more complicated systems. It was not a question of deriving or justifying the quantum rules; it was not even known what the relevant rules were. Sommerfeld's studies solved this problem by providing a brilliant and beautiful generalization of the quantum rules of Bohr and Planck. Sommerfeld's conditions apply to any periodic system of s degrees of freedom. Such a system is described by s quantum numbers: n_i $(i = 1, \ldots, s)$. The orbits are selected by the conditions

$$\int p_k \, dq_k = n_k h, \qquad k = 1, \ldots, s$$

The integrals are over the periods of the coordinates q_k, and the p_k are the usual conjugate momenta. In principle, the calculation of the energies of stationary states of atoms is now reduced to an exercise† in classical mechanics (with the Sommerfeld conditions adjoined). Sommerfeld considered his quantum conditions as the ultimate foundations of quantum theory and "unproved and perhaps incapable of being proved." [37] These were the Sommerfeld results that inspired Bohr's thorough revision of his 1916 paper. The Sommerfeld quantum rules were incorporated and subjected to a profound analysis in Bohr's paper I. Even though these rules were not proved or derived, Bohr's analysis did much to remove the apparent arbitrary character from the quantum rules. In this analysis, Kramers was a constant help.

† Although an exercise, it is not a particularly simple exercise. Two other physicists, Wilson[35] and Ishiwara,[36] suggested the same quantum rules as Sommerfeld; however, they did not succeed in obtaining new results for more complex systems.

With the Sommerfeld quantum rules—however justified—the discussion of the energies of atoms was systematized in a formal scheme, now called "the old quantum theory," and Kramers soon became one of the masters of that theory.

(4) To obtain the actually observed spectrum, one key ingredient was still lacking. The Bohr theory postulated that in transitions between quantum levels n' and n'', light would be emitted of frequency $\nu_{n'n''}$ given by

$$h\nu_{n'n''} = E_{n'} - E_{n''}$$

It might appear that with the calculation of the energies E_n (now possible via Sommerfeld's scheme) the determination of the frequencies is merely a trivial subtraction. The frequencies, however, are only a small part of what is experimentally observed. Not all spectral lines are equally intense, some frequencies allowed by the Bohr theory do not occur at all, and the presence of external fields alters the frequencies. To understand this type of detail, it is necessary to supply information that allows the theoretical calculation of the *intensity* of spectral lines. This is the necessary additional ingredient, and the original Bohr theory did not contain that information. It is precisely in this area that Kramers made his fundamental contribution.

It is important to recall that the need for this additional information arose because Bohr's quantum theory had *separated* the emission process (and the emission frequency) from the mechanical properties of the classical orbit. Classically, the properties of the orbit determine *all* the properties of the emitted radiation, both the frequency (or frequency distribution) and the intensity. However, by insisting that the emission is due to a *transition* between two stationary states—which as a discontinuous event is *not* classically describable—Bohr lost all information about the classical intensity. To progress further required that, at some stage, methods had to be developed which would give information about the intensities. It could be anticipated that since the emission process intrinsically involved *two* stationary states, the intensities, as indeed the frequencies, would depend on the characteristics of two levels. However, just how the intensity was to be related to the properties of the two orbits was far from obvious. This basic question of the Bohr theory was brilliantly resolved by the collaboration of Bohr and Kramers. Bohr suggested the general physical ideas and the basic approach, while Kramers implemented the ideas, carried out the involved calculations, and performed the comparison with experiment. This successful comparison of course provided the ultimate justification of Bohr's physical ideas. The suggested method used all elements of the old quantum theory and introduced quantities that later would become fundamental in Heisenberg's formulation of quantum mechanics.

It is almost ironic that the fundamental clue leading to the calculation of the intensities in the Bohr theory was provided by Einstein. In a paper, which in spite of its brevity is as remarkable as any he ever wrote, Einstein provided the first systematic derivation of the Planck formula.[38] This derivation was based on the Bohr formula $h\nu_{n'n''} = E_{n'} - E_{n''}$ and on a novel idea—the existence of a numerical *probability* for the transition from a state $\{n'\}$ to a state $\{n''\}$. This process, a transition that takes place without external influences causing it, was called a "spontaneous emission" by Bohr (Einstein did not use that term). The probability for such a transition is denoted by $A_{n'n''}$ (the Einstein coefficient for spontaneous emission). The intensity of a line produced in a transition $n' \to n''$ is related directly to that transition probability. To connect this idea with the Bohr picture, $A_{n'n''}$ must be related to the *orbits* described by the quantum number $\{n'\}$ and $\{n''\}$. These orbits in turn are obtained from the Sommerfeld conditions

$$\int p_i\,dq_i = n_i'h \quad (\equiv I_i')$$

$$\int p_i\,dq_i = n_i''h \quad (\equiv I_i'')$$

Bohr, by a very ingenious argument, suggested a relation between calculable properties of these orbits and the Einstein A coefficients. His argument was an almost magical mixture of classical and quantum ideas. Classically, an orbit is determined by the constants I_i (which, of course, in quantum theory assume the special values n_ih). The classical motions occurring in atomic systems are multiple periodic, and they can be characterized by a number (usually s) of periods, $T_1, \ldots, T_2, \ldots, T_s$, or what amounts to the same thing by a set of mechanical frequencies $\omega_1, \omega_2, \ldots, \omega_s$. The explicit dependence of the frequencies of the orbits on I_i can be seen from the relations $\omega_i = \partial E/\partial I_i$. Consequently, every component of the displacement of a particle in an atom can be developed in a Fourier series, with Fourier coefficients C, which will depend on the orbit through the set of I_i values. For a classical orbit (with given I values), the intensity of the radiation of a particular harmonic component (a particular frequency) is proportional to the square of the Fourier coefficient corresponding to that frequency. Now apply these classical considerations to the particular orbit which corresponds to a stationary state with quantum numbers $\{n_i\}$ (thus $I_i = n_ih$ for this case). Then the Fourier development is given by

$$x(t) = \sum_{\{n\}} C_{\{n\}} \exp[i(n_1\omega_1 + \cdots + n_s\omega_s)t + i\delta_{n_1,\ldots,n_s}]$$

The intensity of the radiation of frequency $n_1\omega_1 + \cdots + n_s\omega_s$ which this orbit emits on classical grounds is proportional to $(C_{n_1,\ldots,n_s})^2$. However, the frequencies ν_i emitted by an atom according to the Bohr theory do not depend on a single orbit, but on two orbits. Two distinct orbits have different Fourier coefficients C' and C'' and two sets of frequencies; the observed frequencies ν_i depend on the energy differences and have a priori no simple relation to the mechanical frequencies ω. To use the Bohr frequency condition to obtain ν, it is necessary to relate the energies of neighboring motions. This can still be done within the classical context; it is only necessary to study two orbits—one described by a set of values I_i' and the other by a set I_i''. If I_i' and I_i'' are near each other, the differences δI_i are small numbers. The energy differences between two such orbits is then given by†

$$\delta E = \omega_1 \delta I_1 + \cdots + \omega_s \delta I_s$$

This result is still classical. In quantum situations the different I values would be given by $n_i'h$ and $n_i''h$, respectively. Bohr next considers two states where n_i' and n_i'' are large numbers. (This means that the difference $n_i' - n_i''$ is small compared to either n_i' or n_i''.) Then in the classical expression for δI_i, the difference between I values for neighboring motion is transcribed to $(n_i' - n_i'')h$. The corresponding energy difference is then

$$\sum_i (n_i' - n_i'')h\omega_i$$

This leads to the Bohr frequency for the transition of frequency:

$$\nu = \omega_1(n_1' - n_1'') + \cdots + \omega_s(n_s' - n_s'')$$

This is an instance of the correspondence principle which relates the Bohr frequency ν to the mechanical frequency ω, in the case of sufficiently large quantum numbers. This observation is now combined with the previous remark that classically the intensity of a frequency $(n_1\omega_1 + \cdots + n_s\omega_s)$ is given by $(C_{n_1,\ldots,n_s})^2$, where $C_{n_1,\ldots,n_s}$ is the Fourier coefficient in the Fourier expansion of the displacement. This combination leads to the following results:

(1) For large n' and n'', the frequencies emitted by an atom are given by $\nu = \omega_1(n_1' - n_1'') + \cdots + \omega_s(n_s' - n_s'')$, where $\omega_1, \ldots, \omega_s$ are the mechanical frequencies.
(2) The intensity of that particular frequency is proportional to $[C_{(n_1'-n_1''),(n_2'-n_2''),\ldots,(n_s'-n_s'')}]^2$, where C is the Fourier coefficient in the expansion of one of the components of the displacement.

Bohr's suggestion to use classical ideas in the quantum theory was based on the knowledge that in certain domains, such as low frequencies, the classical radiation theory worked well (this is even true for the Planck formula). However, Bohr boldly suggested that the *formal* connections uncovered by his arguments would be generally valid. For example, it should be possible in general to obtain the intensity of spectral lines by calculating quantities such as $C_{n'-n''}$. It should be stressed that these quantities are completely determined by the classical motion together with the quantum conditions. The connection between the intensities and the Fourier coefficients was first formulated by Bohr in his great paper I.[32] This connection was unusually fertile and yielded many surprising insights. In paper I, Bohr gives a very thorough, qualitative discussion of many examples of this connection. For example, if for certain states and frequencies C is zero, the corresponding line will not occur in the spectrum. Very interesting is the observation that the Fourier coefficients C may be zero (for certain systems and frequencies), in the expansion of the x

† This is one of the results in Bohr's paper I,[32] specifically attributed to Kramers.

and y components, but not equal to zero in the expansion of the z component of the displacement. This signals that the radiation is polarized. Thus, an examination of the C coefficients also gives detailed information about the state of polarization of the emitted radiation. The quantitative significance of these suggestive examples of the power of Bohr's ideas, however, could only be assessed through Kramers' detailed calculations.

IV. Kramers' Contribution to the Bohr Theory: The Thesis

It is striking and typical that although Bohr was intensely preoccupied with the relation between the Fourier coefficients of the motions and the intensity of spectral lines, he never explicitly wrote down† the precise formula that connects these quantities. This connection shows up for the first time in Kramers' thesis.[39] It is based on a simple argument. For a single harmonic component $x = C \cos \omega t$, the radiation energy emitted per second is proportional to $(\ddot{x})^2$ and thus to $C^2\omega^4$. Therefore, the number of quanta of frequency ω emitted per second is proportional to

$$\frac{C^2\omega^4}{h\omega} = \frac{C^2\omega^3}{h}$$

This quar tity must be proportional to the Einstein A coefficient for that harmonic co nponent. The general connection is given by

$$A_{n'n''} = \omega^3 (C_{n'-n''})^2$$

(some factors, e.g., 2π, h, and c, are omitted in this formula). It is typical that Bohr analyzed the implication of this formula without ever writing it down, while Kramers in his thesis did write it explicitly and made frequent use of it.

That Kramers used this formula so often is actually not surprising. Bohr suggested to Kramers that he use the relation proposed by Bohr to investigate the *intensities* of the fine structure component of the hydrogen lines and intensities of the Stark splitting of hydrogen and helium.[39] Once calculated, it would be possible to compare these results with experiment. For such a comparison the precise numerical relation is essential; a verbal connection is not enough. It must be emphasized that such intensity calculations had never been made before, so Kramers' investigations were ventures in a new, uncharted area. These studies were carried out in great detail in Kramers' thesis.

The systems actually considered by Kramers were all *one-body* systems: the hydrogen atom treated by relativistic mechanics (to obtain the fine structure splitting), the hydrogen atom in a strong electric field treated nonrelativis-

† At least this formula does not appear in the writings of Bohr published before Kramers' thesis was written.

tically (for the Stark effect), and the hydrogen atom in a weak electric field treated relativistically (to obtain the Stark effect of the fine structure lines). It might appear at first glance that the proposed study was a relatively routine exercise. After all, the intensities by Bohr's ideas and Kramers' formula are related directly to the Fourier coefficients of the classical motions, so all that needs to be done is to calculate these coefficients for the classical one-body problems considered. Although logically correct, the actual situation is not nearly that straightforward. It must be recognized that the whole quantization scheme using the Sommerfeld conditions works only for *separable* systems. These are, roughly speaking, systems for which the Hamiltonian is additive (in some coordinate system) in the s degrees of freedom.

The systems Kramers dealt with are separable only in rather complicated coordinate systems (general elliptic coordinates), so it was necessary to employ such coordinates to apply the method at all. The necessity of using elliptical coordinates complicates the formalism. It is possible, however, and this was carried out in the first chapter of his thesis, to reduce the needed coefficients C to integrals over quantities dependent on the mechanical motion. The main difficulty in the analysis is that, except for very simple systems, this cannot be done exactly; "appropriate" approximations have to be devised. Furthermore, what is appropriate depends on the system considered. The approximations are different for an atom in a strong electric field from those for a relativistic atom. The resulting calculations are lengthy and at times appear as a nightmare in special function theory.

The calculations were carried out by Kramers but evidently Bohr was in close touch with this work. Bohr must have been sympathetic to a geometrical interpretation, for in a letter to Bohr, dated January 1, 1918, Kramers writes: "I have seen that you were right when you said that one can calculate the Fourier coefficients for a motion which is a perturbed Kepler ellipse, provided one knows accurately the slow changes in the orbital elements; in the case of the Stark effect the whole thing reduces again to the determination of some simple integrals—this is probably also the case for the Stark effect with relativity; only the integrals become rather complicated."[40]

With the calculation of the C coefficients for the systems under consideration, half of Kramers' problem was solved (these calculations took 40 pages in the thesis). There remained the calculation of the A coefficients, or the intensities, using the Bohr connection. In his thesis,[41] Kramers gives a more detailed derivation of the energy relation between neighboring stationary states. With Bohr, he considers two stationary states, one initial state described by n_i' ($I_i' = n_i'h$) and a final state n_i'' ($I_i'' = n_i''h$), but he considers in addition a sequence of intermediate states given by $I_i(\lambda) = [n_i' + \lambda(n_i'' - n_i')]h$, where λ varies between 0 and 1, so that $I_i(0) = n_i'h$, $I_i(1) = n_i''h$. In these intermediate states the frequencies are $\omega_i(\lambda)$ and the Fourier coefficients are C_λ, also dependent on λ. Using the relation $\delta E = \omega_1 \delta I_1 + \cdots + \omega_s \delta I_s$, the Bohr frequency is then the average over all these intermediate states:

$$\nu = \frac{1}{h}\int_0^1 \delta E\, d\lambda = \int_0^1 d\lambda[(n_1' - n_1'')\omega_1 + \cdots + (n_s' - n_s'')\omega_s]$$

If the mechanical frequencies are independent of λ, the original relation used by Bohr is recovered. In more general situations, the average of the mechanical frequencies occurs in the emitted frequency. This more general relation (with the frequencies dependent on λ) is especially pertinent if external fields are considered because in that case, the mechanical frequencies $\omega_i(0) = \omega_i'$ of the initial motion are different from $\omega_i(1) = \omega_i''$, the mechanical frequencies of the final motion. It is even possible that the initial and final motions are not separable in the same coordinate system. This requires a rather careful treatment of the approximations. It is true that generally, at least for weak fields, $\omega_i' - \omega_i''$ is equal to the order of the field strength. Physically, this gives rise to a displacement of the spectral lines and to the appearance of new lines.

It is also reasonable to conjecture that the intensities or A coefficients are some average of C_λ over λ. However, this conjecture is difficult to check experimentally since the nature of the average to be performed over λ is not known. There is, however, one case in which one can expect to obtain reliable information about the intensities based on these considerations. There are many cases in which a single spectral line, emitted in the transition between levels, splits into a number of components under the influence of external fields or other perturbations. It is reasonable that in computing the *relative* intensities (the ratios of intensities), the method of averaging over λ to obtain the original intensity plays a minor role. Thus, Kramers suggests that the *ratios* of the intensities of these components should be given by the ratios of the squares of the C coefficients corresponding to the respective classical motions. With this basic assumption and Kramers' earlier calculations of the C coefficients, it is now possible to carry out an explicit numerical calculation of the theoretical predictions for these intensity ratios. The needed calculations and analysis make up the second half of Kramers' thesis. The agreement between theory and experiment is surprisingly good. Kramers describes this impressive agreement in his customary guarded manner: "On the whole it will be seen, that it is possible in Bohr's theory to account in a convincing way for the intensity of the Stark components."[42] "It will be shown that it is possible to account in a suggestive way for the observations especially for helium."[43] (The observations referred to are on the intensities of the fine structure lines of ionized helium.) It should be added that this agreement not only supports the basic ideas of Bohr connecting the Fourier components with the Einstein A coefficients, but it also demonstrates Kramers' skill in organizing the necessary approximation scheme, so that the essential physical aspects are maintained at each stage of the approximations. It is interesting, and again completely in character, that in spite of the unquestioned success of the thesis, Kramers would yet "emphasize the incomplete and preliminary character of the underlying considerations."[44] The world of physics, by contrast, considered these results as yet another example of the almost magical power of Bohr's mode of reasoning and acknowledged Kramers as an accomplished collaborator. It is curious that as time went on, Kramers' appreciation of his thesis effort seemed to diminish more than is usual. He later often expressed unhappiness and even embarrassment about his results.[45] In this instance, the judgment of the physics community was by far the better of the two.

V. The Changing Status of Kramers

Bohr very rapidly recognized Kramers' talents. As early as February 28, 1917, just about 6 months after Kramers' arrival, Bohr writes to Oseen in Stockholm, introducing Kramers: "I am sending you the greetings with Dr. Kramers, a young Dutch mathematician, who has worked with me here in Copenhagen.... I have been very pleased with my collaboration with Dr. Kramers, who I think is extremely able and about whom I have the greatest expectation.... We have worked together on the helium spectrum and have done a rather big job on the purely mechanical side of the problem."[46] Bohr expresses himself in a similar vein in many later letters to Rutherford[47] and to Richardson.[48] It is hardly surprising that with such support by Bohr, Kramers' reputation grew rapidly. The publication of Bohr's great paper[32] in late 1917, in which Bohr acknowledges Kramers' help and attributes specific derivations to Kramers, did a great deal to make Kramers' name widely known in the physics community. It is amusing that Epstein, in a comment to Bohr on paper I, refers to a "cute" proof Kramers supplied for one of Bohr's results. "About the footnote on page 29, I have several things to say. The proof of Kramers is very 'cute' and simple. It presupposes however knowledge of a theorem of Jacobi—which I did not possess two years ago."[49] This remark is interesting because it indicates once again the knowledge and control Kramers *possessed*, at age 24, of what surely was considered advanced and sophisticated mechanics, as exemplified by Jacobi's theorem.

Kramers also became known through his many lectures. These lectures were almost always on Bohr's ideas about atomic physics. Since Kramers was a brilliant expositor who possessed an excellent sense of humor, his lectures were instrumental in disseminating Bohr's ideas to broad audiences. On his first trip to Sweden (March 1917), Kramers lectured on the quantum theory and was received with rather considerable skepticism. Reporting to Bohr on that talk, Kramers writes: "There were many who criticized the frequency condition $h\nu = E' - E''$, the electron would need an information bureau to be able to calculate the frequency to be emitted."[50] However, the reception on his second trip to Sweden in December 1918 was a good deal better. Not that everybody understood everything, or even liked it, but Kramers could report to Bohr what Oseen felt: "It is all mysterious, but one cannot deny that all the reasoning is sound and it is at any rate a fruitful mysticism."[51] During these visits to Sweden, Kramers met Oskar Klein who was intrigued by Bohr's approach to atomic physics. In May 1918, Klein came to Copenhagen (he stayed until 1922) to study with Bohr and Kramers. Klein was one of the earliest "converts" to atomic physics in Sweden and did much to make Bohr's work known and understood in that country. However, Klein's entrance into the "Bohr World" was made possible through acquaintanceship with Kramers. Kramers and Klein remained close friends: They worked together in 1918, and Kramers played an official role at Klein's thesis defense. (When Klein met Kramers for the first time, Kramers' Danish was so fluent, that Klein thought for many months that Kramers was a native Dane.[52])

Through Bohr's influence, personal contacts, and especially his papers and lectures, Kramers began to acquire a reputation in the world of physics; he began to be known in Scandianvia, Germany (Sommerfeld), and England (through Bohr's contact there). However, during these years of increasing scientific stature in Copenhagen, Kramers had no contact with Leiden—neither with Ehrenfest nor with Lorentz. He did have an active correspondence with his friend Romeyn and some correspondence with others. So it was *possible* to communicate with scientific circles in Holland but Kramers did not choose to do it. Even when he had achieved his first great success in explaining the intensity and polarization rules, he asks his friend Romeyn to tell another friend, Coster, about it, but he neither asks him to inform Ehrenfest, nor does he do so himself. "Tell Dirk, as something new in physics, that we [Bohr and I] can explain ... the polarization and intensities of the Stark effect components."[53]

Not only did Kramers not contact Ehrenfest, but in his own way Ehrenfest also avoided any possible contact with Kramers. It was an established custom in The Netherlands that The Teyler Foundation in Haarlem would sponsor prize essays on assigned topics in physics. The Teyler Foundation was independent of the University but nevertheless had a definite impact on scientific life in The Netherlands—in part because the curator of that foundation was usually an eminent scientist (Lorentz held that position for many years) and in part because of the interest generated by the prize competitions organized by the foundation. In 1917 the topic for the competition chosen by the Teyler Foundation was "The Model of the Atom According to Rutherford and Bohr." In principle, any interested person could compete, provided of course they were aware of the competition and the topic. It is understandable and reasonable that Ehrenfest, who of course knew about the competition, encouraged his brilliant student J. M. Burgers to write an essay for that competition. It is indicative of the estrangement between Kramers and Ehrenfest that Ehrenfest not only did not inform Kramers of the competition (the topic would have been a natural one for Kramers) but specifically instructed others not to tell Kramers anything about the competition.[54] Burgers won the competition, and the prize essay was later expanded into his thesis. It was an important and beautiful piece of work. There is no hint that Kramers even knew about the prize essay competition. Ehrenfest deliberately kept that information from him.

Bohr was an essential intermediary in reestablishing the contact between Kramers and Ehrenfest. Actually, Bohr and Ehrenfest had not met nor had any contact before May 1918. As late as 1916, Ehrenfest still had misgivings about Bohr's physics. But in early May 1918, Bohr mailed a few copies of his paper I to chosen physicists, and Ehrenfest was one of them. Ehrenfest was picked by Bohr because Bohr's analysis was closely related to previous investigations by Ehrenfest. There can be little doubt that Ehrenfest must have been overjoyed. His incisive studies on the "adiabatic invariance" and its role in quantum theory had been largely misunderstood or ignored. (For an extensive analysis see M. Klein's beautiful biography of Ehrenfest.[55]) To see an influential figure like Bohr elevate his results from obscurity to fundamental principle must have been deeply satisfying to Ehrenfest. At about the same time that Bohr sent his paper to Ehrenfest, he drafted a letter to Ehrenfest. In the version of May 18, 1918, Bohr writes: "It has been my intention to write to you for a very long time and especially to tell about Kramers, whose stay here has been a great pleasure to me in these sad times. I consider him a very promising young physicist and he has been a great help for me in my work. We have worked a good deal together on the mechanical aspect of the problem of the helium spectrum and he is at present on his own account preparing for publication two papers on the intensity of the components of the fine structure and the Stark line of the hydrogen lines and on the simultaneous effect on these lines of the relativity modifications and of the external electric field."[56] This is a lengthy and explicit recognition of Kramers' ability and promise. Bohr ends this letter with a wish: "I hope very much to meet you sometime when the war is over. Kramers has told me very often about the interesting scientific life in Leiden and I should like so much to come to Holland when it is possible to travel again."[56] This letter marks the start of a close friendship between Ehrenfest and Bohr. On January 13, 1919 (2 months after the armistice), Ehrenfest invited Bohr to come to Leiden to deliver a lecture on April 25, 1919. Bohr would also be present at Kramers' thesis defense scheduled for early May. Ehrenfest's appreciation of the Bohr approach to physics also changed substantially during that period. As little as a year later, Ehrenfest was completely convinced of the fundamental significance of Bohr's investigations† (as can be inferred from the extensive Ehrenfest–Bohr correspondence concerning Bohr's participation in the Solvay Conference[56]).

These events had a direct influence on the tenuous relation between Ehrenfest and Kramers. Ehrenfest, in answering Bohr, wrote that he was overjoyed that Kramers was doing so well for he had always been aware of his talents but at the same time he expressed serious reservations about Kramers' commitment to research. Ehrenfest stressed that it would be necessary for Bohr to keep on pushing Kramers once he had started to work on a problem. At the same time, Ehrenfest could not help but be impressed by the importance

† It is clearly impossible to decide the causal order of the developing warm friendship between Bohr and Ehrenfest, and Ehrenfest's increasing admiration for Bohr's physics.

and sophistication of Kramers' work. Ehrenfest was much too perceptive a physicist to deny or denigrate the importance of Kramers' contributions. The technical virtuosity of the accomplishments was evident. Also, Kramers was the physicist who knew most about Bohr's ideas. His daily association with Bohr gave him immediate access to these ideas, sometimes before they were completely formed. The increasing recognition of Bohr's work automatically led to a greater appreciation of Kramers and his work. Under the influence of these factors, a gradual scientific role reversal began to take shape. Kramers from the hesitant student became the better, more up-to-date, and more skilled scientist; Ehrenfest from an unerring guide became more of an inquiring critic. This role reversal had become total when 10 years later (1926) Kramers returned to Holland. In spite of all this, Ehrenfest never quite got over his misgivings of Kramers' motivation, and several years after these events (1923), Ehrenfest still writes to Bohr that he thinks Kramers is insufficiently motivated and not committed like Pauli. Even as late as 1929, when Ehrenfest was completely aware of Kramers' mastery of quantum mechanics, he writes in a letter of recommendation for Casimir: "Casimir seems better than Kramers as a student. I think that he will be the true successor to Lorentz."[57] Although such comments are not unusual and these were certainly not intended for Kramers, they appear to echo, perhaps dimly, Ehrenfest's early disenchantment with Kramers.

Although this role reversal was a gradual process, it undoubtedly started with Kramers' triumphant return to Leiden for his thesis defense in May 1919. The thesis defense was (and is) an official public ceremony lasting exactly one hour. Since part of that hour is taken up by official and stylized rituals, it is not possible to go deeply into the scientific merits of the thesis. Furthermore, by tradition, the questions must deal with the specific propositions that accompany the thesis. More important than the specific questions are the persons who ask them. In Kramers' case, his "promoter" (thesis advisor) Ehrenfest asked questions (this is required), but Bohr and Lorentz did also. By all accounts Kramers handled himself superbly. Given his detailed knowledge of the subjects and his great verbal facility, this is not surprising. He had 13 propositions: Seven dealt with technical details of the quantum theory, two propositions were mathematical, one contained a generalization of a theorem of Poincaré, and another was a geometrical comment. The other two propositions dealt with the theory of strong electrolytes, referring to calculations that Kramers had carried out with Klein. Particularly noteworthy was the observation that the explanation of magnetic properties of materials would require a radical change of the usual laws of electrodynamics and mechanics. One proposition contained a result in general relativity which showed that, at that early time (1918) about 3–4 years after Einstein's publication, Kramers knew the theory in detail. He had mastered and could use the technical apparatus for detailed calculations. Kramers must have done this all by himself, since no one else knew the theory, and this was no mean feat.

After the successful thesis defense, there was a party in the room of Jan

Romeyn. Not only was he Kramers' best friend, he also had the largest room. Several students (old friends) came but, as was reported with great pride,[58] so did three Nobel prize winners—Lorentz, Kamerlingh Onnes, and Bohr. No mention is made of Ehrenfest and it is likely that he was not there. The presence of some fraternity types (members of the student "Corps") alone would keep Ehrenfest away. In the evening there was the traditional dinner at the best restaurant in Leiden, "The Gilded Turk" (it no longer exists). Kramers' fiancée sang Schubert songs. Both Bohr's visit and Kramers' promotion were a huge success.

Soon after the promotion, Kramers became ill and remained in the hospital in Rotterdam for a rather long time. After his recovery and the burst of scientific activity of the previous year, he resumed his various investigations. At the same time, the formal organization of the Bohr Institute progressed, not rapidly but steadily. In September 1919, Sommerfeld visited Copenhagen, supporting, most magnanimously, the creation of an Institute for Theoretical Physics. At that time, the first staff positions at the Institute were authorized. Kramers was made Scientific Assistant to Bohr. It is certain that Bohr never thought about another candidate for that post, and it is completely clear that Kramers never contemplated turning it down. The cooperation between Bohr and Kramers was so harmonious that it was natural that it would continue—if only resources were available. With the establishment of the Institute for Theoretical Physics, *some* financial support was guaranteed; this not only kept Kramers in Copenhagen but it also defined the scientific direction his career was to take. As an official and conscientious member of the institute staff, Kramers rapidly assumed a most varied set of administrative responsibilities, writing letters, ordering books and materials, and attending meetings. The official opening date of the institute was postponed again and again, due to strikes and shortages of materials. Bohr and Kramers actually moved to the new building in January 1921, and the date for the official opening was set for March 3, 1921. The day before the opening Kramers gave a tour of the new facility for a number of reporters. On the opening day the official inauguration took place, and the main address was given by Bohr. In this address, he publicly and explicitly thanked Kramers for his research and his teaching contributions. It was one of the few occasions where the teaching functions of the institute and Kramers' role in that endeavor were singled out.

In the not quite 2 years which had elapsed since Kramers returned from his thesis defense, he was an integral part of Bohr's "quantum" community.

VI. How Parsimonious the Emotions?

Kramers had made a number of Danish acquaintances who were not physicists. Through these contacts, especially through a businessman, Kjelgaard (who was actually a friend of the director of Kramers' school in Arnhem), he eventually met Anna Petersen,[59] a young woman studying voice in Copenhagen. She was an outgoing, energetic, exuberant person. Born in the small provincial city of Helsingör, her formal education consisted of little more than grade school. She did receive some further education from the nuns in a nearby mission who taught her French and in the process converted her to Catholicism. She had a beautiful strong voice, and while working and singing in a resort hotel on the Danish coast she was noticed by an elderly gentleman who was so impressed by her personality and voice that he offered to pay for singing lessons. This is what brought her to Copenhagen. Anna Petersen was warm, emotional, and full of life. She used to run up and down stairs with such vigor and energy that her friends referred to her as continually "storming" up and down. This is what got her the nickname Storm.[59] She had little formal training (something that bothered her all her life), but she was infinitely curious. Her intuition about people was excellent; although spontaneous and generous she also possessed a healthy dose of common sense.

It was through music that Kramers and Anna Petersen met and got to know each other well. Kramers was not only an unusually gifted cellist, he was also an excellent pianist. There were many occasions (some informal, some semi official) where Anna Petersen gave song recitals, accompanied by Kramers. It is hardly surprising that based on this joint devotion to music a close and intense personal relation developed very quickly. The relation evolved rapidly, but not smoothly. Hans was intellectual, sophisticated, and erudite; Anna was unsophisticated, simple, and without a great deal of formal knowledge. Hans was guarded, detached, private, and controlled in his emotional life, whereas Anna was warm, spontaneous, open, and unrestrained. They fell in and out of love innumerable times, sometimes breaking off their official engagement and promising never to see each other again, only to break that promise time and again. Storm very soon developed a fierce loyalty to Hans and even a loyalty to physics. Hans managed, or tried, to separate his attachments to physics and to Storm. It is clear from a letter Kramers wrote to Romeyn on November 25, 1917 that he was engaged to Storm at that time. Still the letter describes a deep disenchantment with his fiancée: "She is so different from me."[60] Kramers does not know exactly what to make of the "engaged state." He is still quite aware of—and interested in—other girls; he would like help and advice from Romeyn: "When I am with Storm, I miss you my old friend."[60] Hans' engagement to Storm created a bit of an uproar in Holland. In the summer of 1918, Kramers' father remarried, a former teacher in a boarding school. She entered the family when the sons were grown up; evidently, she felt that she was the guardian of the social level of the Kramers' family. To have a son (a stepson) engaged to a lower-class girl whose father was a carpenter, who had merely a grade school education, and who, to top it off, was a singer was clearly intolerable. It was alright to have Storm as a "music girl friend"[†][61] but no one of that class could be seriously considered

† The term "music girl friend" in Dutch, used in this context, is best if clumsily translated as "music concubine."

as a future wife of a Kramers. So Hans' stepmother tried valiantly to keep Storm at a safe distance, while actively encouraging contacts with the right class of girls.

On February 4, 1919, Kramers writes a postcard to Romeyn in which he reports that he has broken his engagement with Storm. They had a scene: "She cried, I had an intense discussion with her."[62] Although no doubt emotionally wearing, operationally the breaking of the engagement had little effect, for a few months later, in May 1919, Storm accompanies Kramers to Leiden for the occasion of his promotion. By that time the engagement was evidently on again. She was considered to be Hans' fiancée by most people but not by his stepmother. She did her best to make Storm uncomfortable, for example, by speaking French at dinners, which she believed would cause Storm embarrassment.[63] When Kramers became sick and was in the hospital in Rotterdam for some length of time, his stepmother encouraged visits from other girls, but kept Storm from visiting him. This did not have too much effect, because upon his return to Copenhagen in the fall their relation seemed to have stabilized somewhat. However, some of Kramers' initial doubts about his work and his future recurred at that time. That is perhaps not surprising. After a period of intense scientific activity, it is quite common to have a letdown. In Kramers' case this post-Ph.D. depression was compounded by the confusion and unsettled state of his personal life. A letter to Romeyn, written in this period, describes his mood quite accurately. He starts out by commenting on his love for Storm, and then he writes: "If I can get to work regularly I may yet end up a decent person."[64] He ends with this remarkable passage hinting at the changing relations and the turmoil this causes: "I have a feeling that I get continuously more parsimonious with my emotions. Almost all people I meet, I feel are more worth-while than I am—although this cannot be deduced from the words I say. I feel less in the mood to write you all these things."[64]

The period of vacillation and uncertainty continued for quite some time. Eventually, about January 1920, a decided break occurred. In a letter to Kramers, Romeyn inquires about many things. He specifically asks: "How are you without Storm?"[65] Storm had gone to Holland at that time and was considering a visit to Paris for possible further study and a trip to Indonesia to give recitals, but everything was tentative, depending on the relation with Hans. Although there was yet another break in that relation, it was not terminated by any means. Kramers had a terrible time making up his mind. Later that spring (1920), Romeyn writes to Kramers: "What I wanted to say in this letter, tell Storm *Yes* or *No*, but *not both*. She puts on a good front, she wants to make a tour to Indonesia.... I repeat take a decision ... say something ... so that both of you know what you have to do."[66] In spite of this strong letter from his best friend, Kramers still hesitated. There is a striking similarity between Kramers' unwillingness or inability to commit himself to physics in his student years, thereby giving up all other intellectual pursuits, and his indecision in his relation to Storm—which would similarly involve a commitment, with a corresponding renunciation of other options. As if to emphasize the problematic aspects of personal involvements for scientists, another Dutch friend, Dirk Coster, a physicist who was married in the winter of 1918, wrote Kramers a note[67] of which the main thrust was that marriage was difficult for scientists.

All these on-again, off-again encounters were bound to have an effect on Storm. Although she was deeply, totally in love with Hans, she had worries and misgivings of her own. These centered on several matters: She was quite aware of the differences in social class and religion; both bothered her to some extent. Most important, she was genuinely interested in a singing career. She saw quite clearly that if she married Kramers she would have to give up—or severely limit—that career. In late June of 1920, she definitely decided to go to Paris for additional voice training. She did go to Paris and it appeared that a decision had finally been made—whereby Storm and Hans each would go their own way. The decision to separate was unquestionably Storm's.[68] However, unexpectedly to Storm (and perhaps to Kramers), Kramers followed her to Paris where they had a very intense emotional encounter. As a consequence of the Paris interlude, it became necessary to decide on specific actions. Decisions could no longer be postponed or avoided. The ultimate decision was specifically left to Kramers;[68] Storm wanted no part of it. The necessity to make a decision weighed heavily on Kramers. He wrote to Romeyn on July 24, 1920: "The last few days I am in a deep depression." It was in this state of mind, pulled in many directions by scientific ambitions, kindness, love, and a Calvinist background, that Kramers reached the decision to marry Storm. The marriage took place on Monday, October 25, 1920, with Klein and Bohr as witnesses. The ceremony took place in a Catholic church: This was a matter of importance to Storm, but a matter of indifference to Kramers. Kramers was amused that the nuns in the church prayed for him. Although getting married in a Catholic service was of no concern to Kramers, he knew perfectly well that his parents, Calvinists to the core, would object strenuously. So he did not inform his parents of either the marriage or the character of the ceremony. Instead, he asked his friend Romeyn to do that. Romeyn demurred, but on November 8 he reluctantly informed Kramers' parents. They were far from overjoyed, but faced with a *fait accompli*, they made the best of it. It is interesting that most of their anger and criticism was directed toward Bohr. Bohr was blamed by Kramers' stepmother for making him work too hard; this is also the first instance where someone expressed the feeling that Bohr was exploiting Kramers. This theme was to recur: Strangely enough, it would be taken up successively by Romeyn and later by Storm.

It would of course be naive to believe that all the difficulties, the differences that created the problems, and the ups and downs in the relation between Storm and Hans Kramers would be removed by a mere ceremony. However, the marriage had a calming effect, serving to define the mutual obligations, and they established a somewhat Bohemian lifestyle. In the pattern adopted, physics, Bohr, and the institute played an important role. In fact, the life of the young couple revolved around the institute. On December 12, 1920, about a month after the wedding, Kramers and Storm visited Lund, where Kramers gave some lectures on the Bohr theory.[69] Marriage settled some but not all of the problems for Hans; he writes a postcard[70] on January 13, 1921 to Jan Romeyn in which he complains about being "tired and listless." Considering the emotional upheavals of the previous months and the continuing demands, this is hardly surprising. All this occurred at the time of the actual move to the new institute facilities. Understandably, this took a great deal of energy and no doubt contributed to the unsettled atmosphere. Event followed event in this stage of Kramers' life. The institute was opened March 3, 1921 and about a month and a half later, Suus, a daughter was born.

Kramers now had a wife, a daughter, a thesis, a position, and a growing reputation in physics. The direction of his life's work was fixed: He would be a theoretical physicist with broad interests. He certainly had great hopes and strong desires to make deep original contributions, but he was unsure about his ability to do so. He had demonstrated, and he had justifiable confidence in, his computational powers, but he both wished for and shied away from conceptual breaks.

After the intense emotional experiences preceding his marriage, he turned more and more inward; as he said himself, "he became parsimonious with his emotions." There was a gradual shift from personal to scientific, his friendship with Romeyn became more a matter of the past than a factor in the present. The center of his life was physics, even the inevitable emotional aspects of life were transferred and transfigured to a physical context.

CHAPTER 12

From Apostle to Prophet: Hints of Trouble to Come

I. Helium: The First Failure

It is quite natural that after the successful treatment of the hydrogen spectrum via the Bohr theory, attention would be directed toward the next simplest element, helium. Experimentally, the helium spectrum was reasonably well investigated, so its analysis and eventual understanding could be hoped or anticipated to be yet another brilliant confirmation of Bohr's approach to atomic physics. Indeed, with the selection of the stationary states according to the Bohr–Sommerfeld quantization rules and the determination of the frequencies by Bohr's frequency conditions, it would appear that the discussion of the helium spectrum would be little more than a complicated, perhaps tedious, exercise in classical mechanics, which ought to give results in close agreement with experiment. Whether so phrased or not, Bohr must have had similar expectations when he proposed the helium problem to Kramers as soon as he had arrived in Copenhagen; otherwise it is unlikely that he would have proposed this problem to an inexperienced student.

This is not at all the way things worked out: In fact, the helium problem was never really solved within the "old quantum theory." This is not for lack of trying, or lack of originality; rather, it turned out that the Bohr approach was just not powerful enough. There were two basic difficulties. Helium with a nucleus and *two* electrons is a more complicated system, making the mechanical problems more difficult to handle and the quantum conditions harder to implement. But more important than this technical problem was the physical fact that the spectrum of helium was found to consist of two totally separate spectra. An element like sodium (in general an alkali metal) has a recognizable series of lines in the visible spectrum; helium exhibits a very similar series *but* an *additional* series as well, quite similar to the original one.

It appeared from the spectrum as if helium consisted of a mixture of two types of helium, each with its own spectrum; they were usually designated as *ortho* helium and *para* helium. The spectral lines of *para* helium were all triplets (they split into three lines in a magnetic field). These two systems were separate; that is, no transitions occurred between the energy levels of these two systems (so-called noncombining terms systems). The lowest state is a state of *para* helium with an ionization energy of 24.5 eV.

The question now was how to explain these features on the basis of the Bohr theory. Since the only function of the quantum conditions is to select the correct stationary states, which in turn are particular classical orbits, the only way two types of system can arise is from two distinct classes of orbits. It was not a priori clear how these orbits were to be picked, but there was no doubt in anyone's mind that the two types of helium corresponded to two different sets of orbits. This merely reinforces the idea that different quantum states must be identified or related to different classical orbits. It was generally believed that in *ortho* helium the two electrons moved in coplanar orbits, while in *para* helium the planes of motion were either perpendicular or made some angle with each other. Since there were no principles to decide which geometrical arrangement was correct, many calculations had to be performed to see which one gave the best answers. None of the calculations carried out over many years by Bohr and Kramers, and later by Born and Pauli and Born and Heisenberg, gave particularly impressive answers.

In addition to these physical questions, there were nontrivial technical problems, giving rise to many different approximations and treatments. The net result of the Bohr–Kramers efforts for many years is contained in a paper published by Kramers in 1922.[71] This paper also could not reproduce the experimental answer. It is a careful and thoughtful study, and the negative result convinced pretty much everyone that the standard Bohr theory failed for helium. (There was no unanimity about the origin of that failure.) The method Kramers used was typical for the unorthodox ingenious manner in which he habitually treated mechanical problems. In the picture adopted, the two electrons in the helium atom move in separate planes, making an angle of 60° with each other. Each electron is assumed to move in an S orbit; the phase difference in the orbits is assumed to be 180°. There was a long history of semiempirical arguments for this picture. Using this picture, believed to be reasonable, Kramers set out to calculate the binding energy of the atom. Of equal importance was the stability calculation of that system. From the *assumed* geometry and symmetry, it follows that the system can be described in terms of the motion of a single electron which moves under the influence of a potential:

$$V = -\frac{Ze^2}{R} + \frac{e^2}{r}$$

where R is the distance of the nucleus to the electron, r is the distance from the electron to a fixed straight line through the nucleus, and Z is the nuclear charge. The corresponding mechanical problem is still rather nasty, so that approximations must be made to obtain numerical results. One might be tempted to consider the e^2/r term as small (it is proportional to the electron–electron interaction) and carry out a perturbation approximation on that basis. However, especially for helium, e^2/r is not small compared to the first term. Now Kramers' trick consists of a splitting of V into two terms:

$$V = \left(-\frac{\alpha e^2}{R} + \frac{c}{r^2}\right) + \left(\frac{e^2}{r} - \frac{c}{r^2} - \frac{(Z-\alpha)e^2}{R}\right) \equiv V_1 + V_2$$

α and c are coefficients that can be picked, for example, to minimize the second terms. This is a curious nonobvious splitting. It is motivated by its success. It is rather easy to show that a system described by $V_1 = -\alpha e^2/R + c/r^2$ can directly be solved; furthermore, the added term V_2 is indeed very small. The resulting perturbation procedure yields a very rapidly converging algorithm. This is a typical Kramers' trick; it is perfectly adapted to this problem and gives excellent results, but it does not work anyplace else. It depends crucially on the peculiar and unexpectedly simple properties of the motion described by the potential $V_1 = -\alpha e^2/R + c/r^2$. It is very hard to see where Kramers got this insight;[†] there is surely no general theorem or underlying structure suggesting this splitting. It almost appears as a formal coincidence that V_1, which is physically a contrived potential (the corresponding force is a $1/r^3$ dipole force), is exactly and easily soluble. That V_2 in the range of interest is small is an unexpected bonus. But contrived or not the method is exceedingly efficient. Because the classical system can now be controlled, the quantizations, via the Bohr–Sommerfeld rules, are straightforward.

In this problem, as in his thesis, it was Kramers' mastery of classical mechanics that enabled him to solve the problems. Kramers' treatment of the helium problem was his typical virtuoso performance, but the results were disappointing. The excitation energy calculated is 3.9 eV too low (out of 24.5), not an acceptable result. Perhaps worse, the mechanical picture on which the whole derivation is based yields a configuration that is classically unstable, thereby casting serious doubts on the reliability of the starting point. Kramers concluded that classical mechanics lost its validity rather than that the model was wrong or that the quantum rules were inapplicable. "One is not allowed to conclude that this negative result of our calculations must be considered as an objection against the correctness of our model—as is known—there are strong general reasons for its validity—so we must draw the conclusion that already in this simple case—mechanics is no longer valid."[72] A number of independent calculations all confirmed the failure of the Bohr picture when applied to helium. In a letter,[73] Born reports to Bohr about a calculation he and Heisenberg have made. Their results are poor even if one of the electrons is far removed from the other. Their conclusion is that something is wrong with the quantum ideas. This same theme recurs time and again in the cor-

[†] It is not difficult, once the potential V_1 is given, to verify its properties, but it is not the most obvious thing in the world to guess. This almost whimsical use of a formalism was quite characteristic of Kramers.

respondence between Bohr and Born in 1923. On April 9, 1923, Bohr writes to Born: "We meet deep seated difficulties every time we go to systems of more than one degree of freedom."[74] Born answers: "I wonder why mechanics seems to fail as in the case of many electrons."[75] It was unclear at this time whether the striking agreement Bohr had obtained for hydrogen was fortuitous, so that more complicated systems would require an altogether different type of treatment, or whether a minor modification or perhaps a change in viewpoint for helium would yield comparable success.

II. More Problems: What Is the Right Direction?

If the helium problem would have been the only difficulty of the Bohr theory, it would have been serious enough. Unfortunately, there were many others as well. One of the first serious discrepancies between theory and experiment was the hydrogen molecule ion. The quantum theory for this highly nontrivial system was carried out in Pauli's thesis. It was a remarkable piece of work, but as in Kramers' calculation of helium, the ionization potential computed did not agree with the experimental value. That was of course a disappointment, but more than that, the further discussion of these results also revealed the rather incomplete character of the calculations performed. In the Bohr theory, the only orbits to which a basic significance is attributed are the stationary states. However, in the classical description there are a host of other orbits. In making an approximate calculation as specifically outlined in Kramers' computation, it is necessary to average a quantity over *all* orbits. But now, should one take only the stationary states or classically stable orbits, or all classical orbits? The quantum rules as formulated at that time gave no indication what procedure if any was to be preferred. This uncertainty, together with the incomplete and incorrect results, caused deep concern about the Bohr theory.

Another difficulty was the inability of the Bohr picture of the atom to produce a quantitative description of the anomalous Zeeman effect. A quantitative discussion of this effect did exist; it was based on an ad hoc coupled oscillator picture, yielding all the square root expressions that such a theory typically possesses. Sommerfeld asked Heisenberg as his very first problem to investigate whether the Bohr theory could obtain similar results. Surprisingly, Heisenberg very rapidly produced a scheme that yielded a remarkable numerical agreement with experiment. However, to achieve this, Heisenberg had to allow half-integer values in the Bohr–Sommerfeld quantization conditions. This was a most radical step, even at a time when cherished principles were trampled on, and results of long standing were considered unreliable: It was bold to tamper with the single most trusted tenet of quantum theory. Heisenberg was well aware of what he had done. He wrote to Pauli; "the end justifies the means."[76] Understandably, this created an uproar: Kramers and Bohr would give up anything else before they would be willing to contemplate an alteration in the Bohr–Sommerfeld rules. The half-integer quantization became quite a "cause celèbre" and gave rise to a number of acrimonious discussions. The somewhat arbitrary use of half-integers was further stimulated (encouraged) by the observation that when the Bohr theory gave an incorrect result, this could often be patched up by replacing an integer quantum number J by $\sqrt{J(J+1)}$, where J could assume integer as well as half-integer values. This was all done in a purely ad hoc and capricious manner, causing physicists such as Bohr, who were searching for general principles and overall viewpoints, no end of grief. Heisenberg's pragmatic approach and his initial success in obtaining an empirically correct formula made him skeptical about Bohr's approach. He writes to Sommerfeld: "Bohr cares more for general theoretical principles than agreement with experiment."[77] In another letter to Sommerfeld,[78] Heisenberg is even more explicit: "What I find most terrible is that Bohr considers all that is wrong to be right and conversely."[78] However, further studies showed that the half-integer quantization by itself could no more explain all the spectral information than could the original Bohr–Sommerfeld rules. It became clear that, especially for the anomalous Zeeman effect, new physical ideas were required. The situation was succinctly, if not kindly, summarized by Pauli[79] in a letter to Bohr: "There exist now two types* of physicists in Germany: those who start with half integer quantum numbers, then if it doesn't work they go to integers and those who do the reverse—(*Heisenberg is excluded he is smarter)."[79] By the end of 1924, the anomalous Zeeman effect was yet another phenomenon not described by the Bohr theory.

The tenuous state of atomic physics at that time was illustrated most forcefully by the loss of the predictive power of the theory. When Stern and Gerlach prepared to set up their famous experiment to demonstrate space quantization, there was no unanimity among very fine accomplished quantum physicists about the outcome. In that experiment, a well-collimated beam of silver atoms is passed through an inhomogeneous magnetic field, and the question was simply: What happens to the beam after it emerges? Debye and Stern thought that there would be no effect at all or perhaps a slight shift; but nothing of interest would occur. Sommerfeld could not really make up his mind: He alternately predicted a broadening of the beam or a splitting into *three* separate beams. Bohr predicted (correctly) a splitting into *two* separate beams; but his reasons for this particular prediction are not recorded. This incident shows that the theory at that time was a very uncertain guide for the prediction of new phenomena. There was even substantial uncertainty about the formal—deductive—consequences of the theoretical scheme. A case in point was the effect of *crossed* electric and magnetic fields on the hydrogen levels. That was not an easy case to calculate; almost everyone, including Sommerfeld and Bohr, believed that the levels would broaden. A calculation made by Epstein convinced Pauli that the levels instead would remain sharp; subsequently, Pauli convinced the world.

The general status was therefore one where promise and doubts were combined in one confusing scheme. If Bohr or the Bohr ideas had nothing to contribute, the theory would have been discarded rapidly. But that was not the case. The theory had great merits. Often, even when it was wrong, it was qualitatively right. Although its logical basis was doubtful, judicious use of contradictory elements could lead to significant results, as it has in Kramers' work. The general belief was that the too literal and too detailed use of the classical orbit notion was pushing the classical theory too far. It seemed more fruitful to concentrate on the general regularities in spectra in the hope that the empirical information would point the way toward a more appropriate model or mechanics of the atom. This was the path followed by Heisenberg and Pauli; Kramers, by contrast, concentrated on those areas in which the Bohr approach in its original form still gave new and significant results. It was not until several years later that these two distinct approaches were recombined.

III. The "Virtual" Encounter of Kramers and Heisenberg

It would be a slight exaggeration to say that Kramers was not present at the first meeting between Kramers and Heisenberg. The meeting in question took place between Heisenberg and Bohr,[80] but Kramers' work was the immediate cause of their meeting. The reaction to—and assessment of—his results was the subject of their intensive, soul-searching discussions. Thus, although not physically there, Kramers' ideas were so much a part of the Bohr–Heisenberg discussions that it would not be quite true to say that Kramers was absent either. Perhaps to call this a "virtual" presence is as good a description as any.

In 1922, Bohr was asked to give the Wolfskehl lectures in Göttingen. That was a great honor, and it also gave Bohr a chance to rework and refine his ideas on atomic physics. Altogether there were seven lectures delivered in a large lecture hall; they started on June 12, 1923. The general topic was the theory of atomic structure. These lectures give an excellent summary of Bohr's thinking at that time. They started with a discussion of the Balmer formula and ended with Bohr's "building up principle" of the periodic table. The prospect of Bohr's lectures caused a good deal of excitement in Germany. Sommerfeld, who was always interested in advancing the careers of his students, asked Heisenberg whether he (and also Pauli) would like to go to Göttingen and meet Niels Bohr. So concerned was Sommerfeld about the welfare of his students that when he surmised that Heisenberg did not have the money to go to Göttingen he immediately added, as further clarification, that he would take care of the expenses involved. This is how it happened that Pauli and Heisenberg were present at the Wolfskehl lectures.

On the third lecture, Bohr turned from the general principles of quantum theory to applications.† It is of course primarily through applications to

† This part of the section is moderately technical, but the second half of the section should be easily understandable. (It is also important for the later developments.)

specific physical systems that the theoretical ideas become believable. Bohr mentions the Stark effect of hydrogen as a success. In the process, he gives a simple, elegant derivation of the energies of an atom in an electric field. But Bohr reserves his most enthusiastic comments for Kramers' successful calculation of the intensities of the lines in the Stark effect. The lecture contains a brief and somewhat stylized version of the method Kramers used in his thesis. The procedure was outlined in Chapter 11. The first crucial step is the calculation of the stationary state energies, using the Bohr–Sommerfeld quantum conditions. The calculation gives those energies, in terms of the quantum numbers of the states, and the mechanical frequencies. Next, in the limiting case of large quantum numbers, the emitted Bohr frequencies are approximately given by Kramers' formula:

$$\nu = \frac{1}{h}\int_0^1 \delta E\, d\lambda = \int d\lambda[(n_1' - n_1'')\omega_1 + \cdots + (n_s' - n_s'')\omega_s]$$

In this formula, $\omega_1, \ldots, \omega_s$ are the frequencies associated with the classical motions; the quantum states are determined by the quantum numbers n_i which fix the actions $I_i = n_i h$. In the model used, the transition frequency is pictured as due to successive transitions from a state where $I_i = I_i' = n_i' h$ to a state where $I_i = I_i'' = n_i'' h$, via the intermediary values $I_i(\lambda) = [n_i' + \lambda(n_i'' - n_i')]h$. If the mechanical orbits do not change much as λ varies, the dependence of the frequencies ω and λ can be neglected and one obtains for the emitted radiation the formula used by Bohr and Kramers for the Stark effect: $n \approx (n_1' - n_1'')\omega_1 + (n_2' - n_2'')\omega_c$. In this case, the motion is determined by *two* quantum numbers and two frequencies associated with the classical motion (e.g., ω_c is the classical frequency of the center of charge of the Kepler ellipse).

The last step is the calculation of the Fourier C coefficients of the classical orbit. It is then made plausible that the intensities of the spectral lines are proportional to C^2. Here, $C_{n'-n''}$ is the harmonic component corresponding to the values $n_1' - n_1'', n_2' - n_2'', \ldots, n_s' - n_s''$ in the Fourier decomposition. As discussed earlier, the calculation of C yields both the intensities and the polarization rules. The first step, the calculation of the energies, has exactly the same validity as the Bohr–Sommerfeld quantum rules. The next step, the calculation of the emitted frequency and specifically its relation to the mechanical frequencies, is more approximate. It is justified intuitively by the (presumed) geometrical similarity of neighboring orbits. Since for complicated motions, neighboring values of I_i need not be geometrically very similar, this assumption is not all that secure. Also, Bohr often applied the resulting relation, $\nu = (n_1' - n_2'')\omega_1 + \cdots$, to situations where $n_1' - n_1''$ was not all that small, which was really a condition for its validity.

The third step, the calculation of intensities via C, is the most tenuous of all. Using this procedure in the most straightforward manner, Kramers calculated the intensity ratios with inordinate success, thereby providing tangible support for this somewhat contrived procedure. In a remarkable passage, Bohr does not hide his pleasure and satisfaction at the outcome of Kramers' calculations and their confirmation by experiment. He states in his lecture:

"Kramers has calculated the coefficients C in the expansions and has also found that the calculated intensities agree with experience. The Stark effect offers an enormous amount of experimental material for which the quantum theory can account in all details. It would be difficult to mention a field in which the classical theory has achieved something similar."[81] There is no record indicating that any special attention was given to these comments, nor was there a great deal of discussion later, perhaps because these results were several years old and were generally known.

At the end of the lecture, Bohr referred to some newer results of Kramers.[82] The paper containing these results was published in late 1920; later in 1921, Heisenberg had to give a presentation of this very paper in Sommerfeld's regular seminar. Consequently, Heisenberg was quite familiar with the contents of Kramers' new paper.

Kramers' paper dealt with a rather subtle refinement of the Stark effect; the system was again a hydrogen atom in an electric field ε, but it was to be treated according to the laws of relativistic mechanics. Several special cases had been treated before. Sommerfeld, in a celebrated paper, had shown that a relativistic treatment yields a splitting of the hydrogen lines—the fine structure of the spectral lines. Epstein, Bohr, and Kramers had all calculated the effect of an electric field in an atom using nonrelativistic mechanics. But Kramers wanted to discuss the more general case where both relativity effects and an external electric field were taken into account. If a measure of the relativistic correction energy is called E_R and the interaction energy with the field is called E_ε, then the case of interest is that where E_ε is smaller than E_R. Physically, this is the Stark effect of the fine structure lines. Qualitatively, in this regime the Stark splitting caused by the electric field is smaller than the fine structure splitting caused by the relativistic effect. Kramers set out to calculate that Stark splitting and the intensity ratios of the split fine structure lines. The case where E_ε is larger than E_R, which is the usual experimental situation, had been calculated before; it was this calculation that was so successfully compared with experiment. But the case where E_ε is smaller than E_R had been neither calculated nor measured.

The procedure employed by Kramers was a technically demanding extension of the methods he had used in his thesis. First, the classical mechanical problem had to be investigated, then the Bohr–Sommerfeld quantization condition had to be applied. The mechanical problem is much more difficult than problems previously studied because, strictly speaking, the system does *not* separate even in parabolic coordinates. This separability had been of major importance in previous studies.† Actually, Kramers, following a method proposed by Bohr, carried out an approximate separation in parabolic coordinates; he then found an approximate Hamiltonian formulation for the

† It is interesting that Kramers mentions in passing that there does not exist *any* coordinate system in which this problem is exactly separable. The proof of this statement was promised in a future joint paper with O. Klein. This paper never appeared.

deviation from the Kepler motion. The Bohr–Sommerfeld conditions were applied to that new Hamiltonian system, leading to a description of the total energy in terms of *three* quantum numbers. The calculations were of formidable difficulty, requiring many advanced analytical techniques, tricky contour integrals, and elliptic integrals of the third kind. None of this fazed Kramers. He obtained an explicit expression for the energy levels of this system, enabling him to obtain detailed information about the Stark effect of the fine structure lines. Kramers demonstrated that each fine structure line is split in one or more sharp lines; they are shifted from the original line by an amount proportional to ε^2. For this reason, the effect is sometimes called the quadratic Stark effect. New lines appear as well. As ε increases so that E_ε becomes more comparable and eventually larger than E_R, several lines coalesce again, yielding the nonrelativistic Stark effect, whose lines are displaced by an amount proportional to ε.† All the results mentioned so far depend just on the Bohr–Sommerfeld quantization; as usual, it is possible to relate the quantum numbers to the rather involved geometry of the orbits.

It was clearly Kramers' intention to write a second paper in which at least some of the extensive calculations would be exhibited more explicitly. This paper (which was never written or at least never published) would contain details of the intensity calculations and the calculation of the Fourier coefficients. In the published version, however, Kramers merely makes some qualitative remarks about the polarization of the Stark lines, the number of lines to be expected as ε varies. These results were summarized by Bohr in his lecture; he reformulates them so that the use of the correspondence principle becomes quite explicit. Specifically, he writes for the transition frequency from a state n', l', m' to a state n'', l'', m'':

$$\nu \approx (n' - n'')\omega_1 + (l' - l'')\omega_2 + (m' - m'')\omega_3$$

where ω_1, ω_2, and ω_3 are frequencies associated with the classical motion. In this formulation, the Bohr version of the correspondence principle is evident, but this formula and the frequencies ω_1, ω_2, and ω_3 never appear in Kramers' paper. Bohr ends his lecture on Kramers' results on an almost exuberant note, expressing great confidence in the quantum description: "No experiments have yet been performed on the transition of the fine structure to the usual Stark effect by the gradual increase of an electric field. The quantum theory yields very many details of the phenomenon to be expected. Even if we really should not be unprepared to find that the quantum theory is false, it would surprise us very much if such a detailed picture obtained from the quantum should not be valid; for our belief in the formal reality of the quantum theory is so strong that we should wonder very much if experiments were to give a different answer than what is demanded by the theory."[81]

† If ε increases still further, there will be other displacements, some again proportional to ε^2. This is what in the contemporary nomenclature is called the *quadratic* Stark effect. It is better to call what Kramers discovered the relativistic Stark effect.

It was right after this optimistic assessment that Heisenberg dissented, vigorously and publicly. Heisenberg† raised an extremely interesting point; that is, the weak field Stark effect (this is really a much better designation than quadratic Stark effect or relativistic Stark effect) is intimately related to the phenomenon of dispersion. In dispersion phenomena, an atom is irradiated by electromagnetic radiation of a controlled kind (it could be monochromatic with a given intensity) and the scattered radiation is studied. Such experiments provide information about the intensities of atoms with weak electromagnetic fields. In the original experiment by Stark, the electric field used was about 50,000 V/cm. (By contrast, the average electric field of a 100 W bulb, a distance of about 1 m away is 1 V/cm, a great deal smaller.) Thus, by shining light of defined intensity on materials, the effect of weak fields can be investigated. Of course, from studies about the dispersion of light, one obtains the effect of fields varying in time. Taking the limit as the frequency goes to zero (or the wavelength goes to infinity) reduces the varying field to a static field. Taking the limit of a dispersion calculation for zero frequency should give information about the effect of the interaction of an atom with a static field, and that is exactly the weak field Stark effect. This observation was the starting point of Heisenberg's objection. He remarked further that a *classical* dispersion calculation would necessarily yield the orbital frequencies in the scattered radiation. That is because a classical atom can only emit frequencies determined by the classical motions. But this cannot be a correct quantum result, for the scattered frequencies in quantum theory are determined by the Bohr frequency condition, which depends on the energy difference between quantum states. This same situation ought to persist in the limit of zero frequency; consequently, Heisenberg reasoned that a classically computed or classically based Stark effect calculation must contain the classical orbital frequencies. It seemed to Heisenberg that Kramers' result did *not* contain these classical frequencies, so he considered them as wrong. He writes: "I did not believe Kramers' results were perfectly correct.... The quadratic Stark effect could be thought of as a limiting case of the scattering of light with very large wavelength ... one knew in advance that a calculation of scattering in a hydrogen atom by the methods of classical physics must lead to a wrong result—the characteristic resonance effect would occur with the electron's orbital frequency—Kramers' calculation could hardly be expected to give a correct result." [87]

Bohr was enormously impressed—by the question itself and by the cogency of the argument. He was also obviously worried by the objection. It appears that neither Bohr nor Kramers had thought about the connection between the dispersion calculation and the Stark effect. Perhaps most striking (certainly to his contemporaries)[86] was that in the ensuing discussion Heisenberg did not retract anything at all; in fact, he kept on challenging Bohr. Bohr first

† This is a paraphrase of what Heisenberg actually wrote. (In an interview,[83] Heisenberg confirmed that this was the kind of thing he had in mind. Other discussions[84, 85] are extremely sketchy about these points.)

tried to argue that in calculating the dispersion one should take into account the action of the radiation on the atom, but it was not exactly clear how this was to be done. Bohr also felt that the great success of the explanation of the linear Stark effect should give confidence in a similar explanation of the quadratic (weak field) Stark effect. But Heisenberg would have none of it. He pointed out that Kramers' results did *not* appear to contain the classical orbital frequencies; thus, either the correspondence principle was wrong or Kramers' calculation was incorrect. Bohr was not anxious to contemplate either alternative. After the discussion, Bohr invited Heisenberg to go for a walk so that they might discuss the whole topic more leisurely and more completely. It is noteworthy that the details of Kramers' work were not discussed in the lengthy conversation between Heisenberg and Bohr.[83] Instead, their interest was centered on the general setting of atomic physics, on the underlying philosophy. Only occasionally and obliquely did the discussion return to the status of Kramers' result. Bohr was most careful and tentative about what to make of Kramers' calculations: "A theory cannot explain anything in the usual strict scientific sense of the word. All it can hope to do is to reveal connections and for the rest leave us grope as best as we can. That is precisely what Kramers' calculations were intended to do."[87]

In retrospect, it appears that incisive as Heisenberg's line of argument was, it was still somewhat misdirected. Heisenberg's objections were directed against the *general* procedure rather than against Kramers' calculation in particular. Heisenberg, quite correctly, called attention to the vague and often ambiguous manner in which the Bohr theory related the Bohr frequencies ν, for the transitions between states, and the mechanical frequencies ω. The Bohr frequencies are the quantities eventually measured in the spectrum; the mechanical frequencies are computed from the model, so the connection between them is crucial for the theory. For large quantum numbers, the relation is approximately $\nu \approx (n' - n'')\omega_1 + (l' - l'')\omega_2 + (m' - m'')\omega_3$, where $n', l', m', n'', l'', m''$ characterize the quantum states. This relation is derived or suggested by the Kramers connection:

$$\nu \approx \int d\lambda[(n_1' - n_1'')\omega_1(\lambda) + \cdots]$$

Here the $\omega(\lambda)$ are the mechanical frequencies of the intermediate motions. These connections are all very tenuous and it is hard to make them precise. It is probably best to interpret them as guides from which the nature of the correct connection can be inferred. It is especially worrisome to use these formulas when the quantum numbers are not large (although this is frequently done). It takes additional confidence in the whole approach to transfer these connections to the Fourier components for the eventual intensity calculation. All these objections of Heisenberg are valid and pertinent. But once the Bohr procedure, with all its admitted problems and uncertainties, is considered a legitimate means of obtaining physical results, there is *no* difference from the *method* used for the linear Stark effect, which was so severely criticized. Furthermore, in neither case can the classical frequencies, $\omega_1, \omega_2, \ldots$, occurring in the formulas be interpreted as orbital frequencies in a straightforward manner. For these more complicated motions, the situation is more involved. The frequencies might refer to the oscillation frequency along a particular axis of the center of charge of a rotating Kepler ellipse, so that no simple connection remains between the mechanical frequencies occurring in the formulas and the geometrical picture of a particle rotating in a simple orbit. In Kramers' detailed calculations, a number of approximations were made, which makes the identification of the classical frequencies quite indirect. In fact, Kramers just derived the expressions for the energy; he never gave explicit expressions for the classical frequencies. But it must be emphasized that Kramers followed the standard procedure, quantizing the system via the Bohr–Sommerfeld rules and analyzing the results. His calculation was not rigorous (it was a systematic approximation procedure), but it was certainly not classical, as Heisenberg seems to imply. Kramers merely applied the usual Bohr procedure to the relativistic Stark effect. In so doing, he discovered a number of qualitatively new phenomena. It is interesting that for quite some time the qualitative predictions of Kramers were the only guide toward the interpretation of the relativistic Stark effect. As late as 1950, Kramers' results were routinely quoted[88] as providing a satisfactory description of the experimental data. Better experiments and a more definitive quantum mechanical treatment, including spin and the Lamb shift, have long since superseded Kramers' original work, but Kramers' specific results, in spite of Heisenberg's criticism (and Bohr's implied acquiescence), were valid consequences of the quantum theory of that time.

It took almost 2 years from the "virtual" meeting in the summer of 1922, until Easter in 1924, for Kramers and Heisenberg to meet face to face. They of course had known about each other before this first, actual meeting, which although congenial and pleasant was not especially remarkable. Later, their scientific careers would become closely intertwined; at this early stage their contacts and interactions took place with Bohr as the intermediary. For that reason alone, it is instructive to compare the initial reactions of Kramers to Bohr with those of Heisenberg to Bohr. It will be seen that even at this early stage there were characteristic and ultimately important differences.

There are certain superficial similarities. Kramers and Heisenberg were about the same age when they met Bohr for the first time. Both came from major centers of physics. Both almost immediately started to do research in atomic physics. Of course, Heisenberg came highly touted (by Sommerfeld) while Kramers came by himself. The important difference was that Kramers not only had the greatest respect for Bohr's physics (Heisenberg had that too), but he was deeply deferential toward Bohr's approach. His efforts were always to try to understand Bohr and, if absolutely necessary, to reinterpret him; but fundamentally he wanted to accommodate his thinking to Bohr's. Heisenberg fully recognized Bohr's importance, but from the very start he was more skeptical and much less accommodating. To write as Heisenberg did

—"Bohr cares more about theoretical principles than agreement with experiment"—certainly indicates independence and a critical attitude. In an earlier letter, Heisenberg in commenting on an idea of Bohr wrote to Pauli: "I am not particularly happy [with that idea] but one can of course not doubt that Bohr is correct."[84] This again shows the mixture of respect and independence which characterized Heisenberg's attitude. This was quite different from Kramers who always felt that it was not only a privilege to work so closely with Bohr, but that it was an equal privilege to pursue and develop physics according to Bohr's tenets. That of course did not mean that Kramers would automatically accept all of Bohr's suggestions, but his first reaction would always be that he tried to interpret Bohr's ideas into his thinking. This occasionally led to differences of opinion. Kramers was never very comfortable with conflicts and controversies; he tended to avoid them, often by incorporating Bohr's ideas or by making Bohr's approach his own. Heisenberg, although not especially pugnacious, did not shy away from confrontations, as his early exchange with Bohr illustrates so clearly. Heisenberg was self-confident and independent at an incredibly early age; he wished to construct atomic physics according to his own views (unformed and partial as they were at that time), unencumbered by Bohr's biases.

Apart from the differences mentioned, Kramers and Heisenberg also tended to tackle problems in quite different ways; it is perhaps better to say they perceived problems in different ways. The relativistic Stark problem was seen by Kramers as a computational question. If one calculated with skill and care and used the agreed upon interpretation, results would be forthcoming. Heisenberg, on the other hand, tried to visualize the nature of the result well before he had gone through the detailed analysis; he tried to infer the type of answer expected from general considerations. Heisenberg's approach tended to be more global, with more emphasis on universal physical principles, while Kramers' approach was more detailed, specific, and computational. Heisenberg (and in this he was strongly influenced by Bohr) was somewhat inclined to investigate *structures*; Kramers never really developed a taste for such broad inquiries and felt more at home with well-formulated, specific problems.

Kramers' intuition in quantum problems was based to a considerable extent on his experience with the geometrical description of the Bohr orbits. Heisenberg thought in terms of physical models, and later in terms of the formal mathematical character of the theory. At no time did Heisenberg engage in the elaborate geometrical analysis of the Bohr orbits as Kramers did in many of his papers. These differences in style and approach developed gradually. The early careers of both Heisenberg and Kramers were strongly influenced by Bohr—both took his ideas and general approach most seriously. But for Kramers, progress in physics was almost synonymous with extending, refining, and applying Bohr's ideas. For Heisenberg, Bohr's ideas were suggestions, directions to pursue; they should be scrutinized, investigated, and, if needed, rejected and replaced. Heisenberg did not have the a priori deference for Bohr's ideas that Kramers did. In the arguments and disputes which inevitably arose as new problems were encountered, Kramers was kind, reasonable, and accommodating, whereas Heisenberg was stubborn, tough, and unyielding. It is not surprising that Bohr's influence on Kramers was more pronounced and more permanent than his influence on Heisenberg. As the difficulties of the Bohr theory became more serious, Kramers' reaction was to refine and modify that description; he was most hesitant to make radical breaks. Heisenberg, on the other hand, was quite willing to contemplate any and all revolutionary changes. Heisenberg was not (as Kramers was) hampered by strong emotional and intellectual attachments to the Bohr concepts. It was *this* distinction that was ultimately to prove of decisive importance.

IV. The Gospel According to Kramers and Holst

Kramers was well aware of the special advantages he enjoyed by having daily contacts with Bohr. In keeping with the Dutch tradition, it was natural that he felt that such a privilege conferred corresponding obligations. He therefore considered it his duty to present the Bohr approach to physics as clearly as possible and as frequently as needed. He already handled a great deal of the institute's and Bohr's correspondence, but he still felt that he should do more. About the time the Bohr Institute was opened, it occurred to Kramers that the time had come to make Bohr's work available to a much larger audience.[89] This idea eventually resulted in a popular book by Kramers and Helge Holst. Holst was a librarian at the Royal Technical College of Copenhagen. There was no guile about the purpose of this book; it was written in Danish and entitled *Bohr's Atomteori.* Its explicit purpose was to acquaint the Danish public with the achievements of its great compatriot. The book appeared in 1922, exactly the same year that Bohr received the Nobel prize for physics.

It was uncommon at the time to present current scientific research in popular books. The only other example was a book published by Einstein in Germany in 1917 on the special and general theory of relativity. It is, of course, extremely difficult to present theoretical ideas without a considerable amount of mathematical background, but Kramers and Holst made a valiant try to do so and by and large they were very successful.† Their main purpose was to make the Danish public aware of the importance of Bohr's work; this aspect could be managed with a minimum of technical background. The central role they attributed to Bohr can be gleaned from the preface to the English translation (published in 1923): "The ideas of Bohr founded as they were on the quantum theory were startling and revolutionary but their immense

† It would appear that Kramers and Holst were considerably more realistic about the level of understanding of physics and chemistry by the public, or a typical high school student, than Einstein was, but even they were at times overly optimistic.

success in explaining the facts of experience after a time won for them wide recognition of the scientific world.... The past decade has witnessed an enormous development at the hand of scientists in all parts of the world of Bohr's original conceptions; but through it all Bohr has remained the leading spirit and the theory... therefore must properly bear the name of Bohr."[90] It is hard to imagine greater acclaim or a more complimentary testimonial.

Although no doubt Holst was a faithful collaborator and a great help, there cannot be any question that this book acquired its scientific authenticity through Kramers' cooperation. Indeed, in the English edition, there is a foreword written by Rutherford in which he makes this explicit: "Dr. Kramers is in an especially fortunate position to give a first hand account of this subject, for he has been a valued assistant to Professor Bohr in developing his theories and has himself made important original contributions to our knowledge in this branch of inquiry."[90] Without Kramers, the book would have been limited to a local Danish audience; with his active participation it became of significance to the worldwide physics community. In fact, the style and type of examples used unmistakingly show Kramers' influence.

As might be anticipated, the book gives a careful, almost meticulous description of the conflicts and problems in physics before Bohr's entry into atomic physics. Bohr's assumptions and postulates are enunciated with great precision, always using simple language and interesting examples. It is also not surprising that the many successes and the new insights of the Bohr approach are reported, with a sense of satisfaction and even pride. There is almost a sense of triumph when Kramers and Holst recount the argument between the Bohr theory and the Stark effect experiments.† "Not until we think of the extraordinary accuracy of the measurements... can we thoroughly appreciate the importance of the quantitative agreement between theory and observation.... Moreover we must remember how completely helpless we previously were in the strange puzzles offered even by the simplest of all spectra, that of hydrogen."[91] The authors clearly were overjoyed with the successes they could report. As a special feature, Kramers and Holst produced colored pictures of the various orbits of electrons in atoms. These diagrams gave a (perhaps unwarranted) degree of reality to the electrons, their paths, and their motions in the Bohr picture.

Even though the avowed intent of this book was to familiarize many nonscientists with the results and achievements of Bohr, serious attention was paid to the shortcomings, limitations, and downright bizarre characteristics of the theory. Kramers and Holst make it quite clear that Bohr's studies lead to hitherto uninvestigated realms, and no one has any a priori knowledge or understanding of the type of laws operating in those domains. They are also honest in admitting that Bohr had not as yet succeeded in formulating a set of quantitative laws to replace the older, no longer applicable laws. They are equally frank in expressing their confusion about what and how to think about

† This is the same subject discussed so extensively in Chapter 11, Section III.

an electron when it makes a quantum jump. They stress one rather weird feature in particular. If an electron jumps from level 6 to level 1, a characteristic frequency is emitted. If the electron jumps from level 6 to level 2, and subsequently from 2 to 1, two quite different frequencies are emitted. All this is standard in the Bohr theory. The remarkable point is that these two electrons starting out under identical circumstances, nevertheless behave quite differently. It even appears that the electrons "know" from the start where they are going and what they are going to do; all this is certainly curious: "The electron seems to arrange its conduct according to the goal of its motion and also according to future events. But such a gift is wont to be the privilege of thinking beings that can anticipate certain future occurrences. The inanimate objects of physics *should* observe causal laws in a more direct manner."[92] Such candor about matters not understood is unusual and refreshing. No doubt part of the willingness of Kramers and Holst to be so open and frank was based on the belief that these difficulties would soon be solved, surely along the lines of Bohr's thinking. The difficulties were seen as challenges to the theory, not crippling handicaps. The overall tenor of the book is one of great optimism: With Bohr's ideas a new era of great promise had started; the theory might be incomplete and need modifications, alterations, and interpretations, but certainly not a major revolution. Bohr had shown the way.

The book was a great success. It spread the fame of Bohr wide and far and was appreciated by physicists and lay persons alike; it was used by many physicists as a first readable introduction to atomic physics. Von Laue, a brilliant but quite critical German physicist, gave the book an absolutely glowing review in *Naturwissenschaften* in 1924. He mentions that "the Bohr theory bears the stamp of genius," thereby fulfilling all the hopes and expectations the authors had for their book. It is not inappropriate to describe the book by Kramers and Holst as a truly missionary venture to spread the gospel according to Bohr.

V. The Great X-ray Paper: The Last Success

Kramers' generally positive attitude toward the "old" quantum theory of Bohr was not just the result of personal devotion to a revered teacher. The theory did reorient the thinking of a whole generation of physicists; Kramers was aware of the impressive successes in spite of some irritating failures. In the spring and summer of 1923, while Pauli was visiting Copenhagen, Kramers wrote a major paper[93] which established the effectiveness of Bohr's version of quantum theory in an altogether new area. This was a new venture for Kramers. He had already written papers in a number of different fields, but this paper was a substantial extension of the Bohr formalism: it went beyond a purely technical application of known theories. It was certainly Kramers' most imaginative paper (up to that time). The success of that paper was instrumental in bolstering Kramers' confidence in the validity of Bohr's theory.

The purpose of this study was to give a theoretical account of certain experimental results in x rays. Specifically, Kramers wanted to explain the observed x-ray absorption coefficient d and the measured energy distribution of the x-ray radiation. This required, first of all, a careful scrutiny of the experimental results in this area. A glance at Kramers' paper shows that he had become thoroughly conversant with the literature on that topic. It should be stressed that at that time there was very little theoretical understanding of these results, and the quantum ideas had never been invoked to understand these data. It should also be emphasized that the x-ray spectrum, apart from occasional lines, was a *continuous* spectrum. Quantum theory had not been used in that domain. This was the problem Kramers tackled.

The basic processes envisaged in this treatment are an absorption process of a quantum by an atom with the subsequent emission of an electron. There is also an inverse process where an electron is captured in an atomic orbit with the subsequent emission of radiation.† The first process is Einstein's photoelectric effect: A quantum of energy $h\nu$ interacts with an atom and an electron of energy $\frac{1}{2}mv^2$ is liberated. If W is the energy needed to free the electron from the atom, $h\nu = W + \frac{1}{2}mv^2$. This relation connects the optical frequency (ν) to the mechanical velocity (v). Kramers introduces a probability P for this process to occur, and the whole point of the paper is the calculation of that probability. The introduction of P is reminiscent of the introduction of the Einstein coefficients; the whole treatment is strongly influenced by Einstein's methods. Kramers next relates the probability P for the photoelectric process to the probability Q for the inverse process—the process whereby an electron is captured and monochromatic radiation is emitted. The connection between P and Q is obtained by assuming thermodynamic equilibrium, closely paralleling the manner in which Einstein obtained the relation between the coefficients for spontaneous and stimulated emission.

To make further progress, the probability Q for the electrons captured with emission of radiation must be estimated. Kramers first decides to analyze the continuous spectrum. The basic idea is to calculate the classical radiation emitted when an electron approaches and moves in the field of a positive nucleus. The orbit of such an electron is approximately a hyperbola. From the properties of the orbit and the known laws of classical electrodynamics (the rate of radiation emission is proportional to the square of the acceleration), the *classical* radiation emitted (per second) can be calculated. It is then possible to make a Fourier decomposition of the radiation emitted. The total classical radiation emitted then becomes an integral over the frequencies of a spectral density function, $S(\nu)$. This function $S(\nu)$ is proportional to the energy emitted in the frequency range ν. This classical density depends on other parameters characterizing a specific orbit; for example, $S(\nu, v, L)$ is the radiation density of frequency ν resulting from an electron orbit with initial velocity v and angular momentum L. Classically, $S(\nu, v, L)$ is a definite computable

† This clearly means that the atom must be missing at least one electron.

function for given classical orbits. From S, the energy radiated per second can be found directly. Kramers' basic idea is to connect the classical energy with the quantum energy. His fundamental assumption is that the quantum probability Q for the emission of a quantum of energy $h\nu$ per second is proportional to the *classical* function S: $h\nu Q(\nu) \approx S(\nu, v, L)$. This is the basic innovation. With this assumption and the known relation between Q and P, all aspects of the continuous x-ray spectrum became computable. By further combining the proposed connection with the photoelectric equation and using for W the energy of a stationary state, the discrete features of the x-ray spectrum can also be obtained. The basic novelty of Kramers' treatment is the conjectured relations between the quantum probability and the classical spectral density. It is in fact this relation which makes the calculation possible. This is certainly in the spirit of Bohr's correspondence principle; it is also very much in line with Kramers' earlier intensity calculations, but it is a significant and novel extension.

What is particularly impressive about Kramers' paper is that it not only provided the physical basis for the phenomena and performed the complicated analysis with superlative skill, but the calculations were carried out so far that the results could be compared directly with experiment. This is an ideal most theorists strive for but rarely achieve. For example, in the actual experimental situation, there will not be just a single electron with a given velocity which is captured by an atom and emits a quantum. Instead, there will be many electrons, all with different velocities which undergo the same type of process. The net emitted radiation will consist of all these quanta; thus,the observed x-ray spectrum must be calculated by averaging the process described over the velocity distribution of the electrons. After carrying out all these computations, Kramers ends up with a numerical expression for the x-ray absorption coefficient and a precise numerical answer for the total intensity distribution. The x-ray absorption coefficients α were found to be proportional to $Z^4\lambda^3$, where Z is the atomic number λ is the x-ray wavelength. The intensity came out as proportional to $Z(\nu_0 - \nu)$. This is exactly the same form as the empirical relation used to describe the experiments. Even the observed and calculated proportionality constant agreed. (In this formula, ν_0 is the so-called quantum limit of the continuous spectrum, related to the experimentally applied voltage V of the x-ray tube by $h\nu_0 = eV$.) The numerical value of the computed absorption coefficient also agrees well with experiment. Kramers' theory gave an excellent account of the x-ray data then available.

It is perhaps unusual to describe a scientific paper as strong, mature, or confident, but this is exactly the impression created by Kramers' paper. The general approach is well defined, the physical ideas are clear, and the author is a complete master of the technical apparatus necessary to carry out his ideas. The new concepts are introduced with verve and confidence, causing the treatment to appear definite and persuasive. (The agreement with experiment also helps!) It was rapidly recognized that this paper was a major contribution; Pauli wrote on June 6, 1923 (well before the paper was published): "Kramers has a very beautiful theory of the continuous X ray spectrum."[94] The paper was published November 1, 1923. On December 12, 1923, Eddington[95] wrote to Kramers congratulating him on this major achievement. But perhaps the most telling assessment comes from Kramers himself when he writes to Romeyn: "With my own work and in my new surroundings I now have a lot of success. The paper mentioned has all the beautiful drawings from Groet* it appeared Nov. 1 and makes a good impression on those who understand these things."[96]

The successful applications of Bohr's philosophy and what certainly would seem to be a natural generalization of Bohr's ideas to the x-ray problem must have given Kramers increased confidence in the general approach. These successes more than made up for the failures and difficulties discussed earlier. The assurance so acquired was evidently great enough that Kramers felt safe in criticizing de Broglie's treatment of x-ray absorption. He adds in a footnote to the paper that de Broglie has treated the same problem and even gets similar results but "this author arrives at a result which [just as Compton's formula] differs from our formula (9) only by the value of the numerical factor, however his treatment of the problem, which *at all points differs* essentially from that followed in this paper, seems not to be *consistent* with the way quantum theory is at present applied to atomic problems."[93] Civilized and polite as this might appear, there is no mistaking the withering criticism these gentle words imply. Divested from politeness, he asserts that de Broglie either does not understand quantum theory or does not know how to use it.

The paper on x rays was written during one of the happier periods in Kramers' life. He had success in his work, was recognized as one of the masters of quantum theory, and was a trusted collaborator of Bohr. (He was often referred to as "Bohr's lieutenant."†) Discussions with Bohr were often carried out via Kramers. As tangible recognition he was made a lecturer (roughly an associate professor) at the Bohr Institute in the fall of 1923, with a corresponding increase in salary. He had growing confidence in his own ability, and the turbulence in his personal relations had apparently subsided. Kramers of course recognized the criticisms and reservations about quantum theory which many physicists, especially Pauli and Heisenberg, expressed so forcefully and so frequently. At times, he might even have had his own doubts and misgivings,‡ but in the daily contact with Bohr these concerns gradually disappeared. His confidence and faith in the basic tenets of Bohr's theory,

* Groet is a small seaside resort town in Holland. It appears that Romeyn and Kramers spent some time there together and that some of the extensive calculations in the paper were performed at that time.

† In certain circles, Kramers was referred to as the "Paladine of Bohr."[97] As the paladines were the heroic knights defending Charlemagne, Kramers was seen as the "defender of the faith."

‡ See especially Chapter 14.

already strong, were greatly enhanced by his own successful treatment of x rays. In one of these remarkable ironic twists which abound in history and in life, the continued reliance of these same ideas led him simultaneously to some of his greatest achievements and curiously enough to some of his most dismal failures.

CHAPTER 13

The Multifarious Consequences of a Desperation Revolution

I. The Struggle About the Photon

The main preoccupations of the physicists in and around the Copenhagen Institute were no doubt with the structure of atoms and the analysis of atomic spectra. Kramers' study of x rays veered in a slightly different direction. It necessarily involved the nature of the interaction of radiation and matter. This subject was customarily treated in a rather gingerly fashion. Since no exact laws or precise description was available, a detailed analysis of that interaction was usually avoided; alternatively, the *results* of the interaction of matter and radiation were summarized in terms of probability laws. In such interactions, the dichotomy between the discrete character of stationary states of atoms and the continuous character of the emitted classical radiation became especially pronounced. In the Bohr theory, the discontinuous transition, in principle, could not be described; it was only possible to describe the physical results of such transitions. The electromagnetic radiation once produced was believed to be completely and totally classical. That left the highly nontrivial problem of providing a framework that contained the discrete and continuous features in one harmonious whole. This problem was especially urgent in view of the Einstein photoelectric effect in which the free electromagnetic field—light—exhibited (or seemed to exhibit) quantum features. Indeed, in Kramers' treatment of x rays, the quantum character of the radiation is precisely introduced by the Einstein equation $h\nu = \frac{1}{2}mv^2 + W$ (see Chapter 12, Section V), where $h\nu$ is the energy quantum of the radiation of frequency ν. This in fact was the *only* way in which quantum ideas made their appearance in his paper. Kramers, however, was extremely careful not to imply or infer anything about the nature of that light quantum. This is in marked contrast to Einstein's interpretation; he associated specific particlelike properties with the photon.† (For a lucid discussion, see Pais.[98])

Although Kramers in this paper was noncommittal about the particle aspect of the energy quantum, this in no way should be interpreted as tacit or implied acceptance of the photon idea. To the contrary, there is a great deal of evidence to indicate that the nature of radiation was extensively discussed in Copenhagen and especially by Bohr and Kramers.[99] In spite of this interest in and attention to the nature of radiation, Bohr for some time did not publish any papers on this problem.[100] His first public comment on the topic was actually a side remark made in a lecture in Berlin in 1920 about what he was not going to discuss: "I shall not ... discuss the familiar difficulties to which the hypothesis of 'light quanta' [real photons] leads in connection with the phenomena of interference, for the explanation of which the classical theory of radiation has shown itself to be so remarkably suited. Above all I shall not consider the problem of the nature of radiation."[101] This quote does not indicate a great enthusiasm for photons; perhaps this somewhat tentative assessment was influenced by Einstein's presence in the audience.[102] But there was nothing tentative about Bohr's criticism of the photon idea in subsequent publications and lectures.

In a long paper written in 1922, Bohr writes: "We can even maintain that the picture which lies at the foundation of the hypothesis of light quanta [photons] excludes in principle the possibility of a rational definition of the frequency ν, which plays a principal part in this theory. The hypothesis of light quanta, therefore, is not suitable for giving a picture of the processes in which the whole of the phenomena can be arranged, which are considered in the applications of the quantum theory."[103] This is a remarkably categorical dismissal of the photon notion. (Note that Bohr usually writes "light quanta" for what is here called photons in quantum theory.) That the photon frequency ν is defined and can only be defined operationally through interference experiments is a recurring theme in all the critiques of Bohr (and his collaborators) of the photon idea. In the paper (to be discussed in more detail in Section III of this chapter) with Kramers and Slater in which a new theory is advocated, the same deep dissatisfaction with light quanta is expressed once again. "Although the great heuristic value of this hypothesis is shown by the confirmation of Einstein's prediction concerning the photoelectric phenomena, still the theory of light quanta can *obviously* not be considered as a satisfactory solution of the problem of light propagation. This is even clear from the fact that the radiation frequency ν appearing in the theory is defined by experiments on interference phenomena which apparently demand for their interpretation a wave constitution of light."[104]

About the same time that Bohr expressed his tentative doubts about the photon in his lecture in Berlin, he expressed himself much more forcefully in

† In the sequel, whenever "photon" is used it is intended to denote typical particlelike properties, such as localized energy or momentum.

a lecture in Copenhagen: "We must admit that at the present time we are entirely without any real understanding of the interaction between light and matter; in fact in the opinion of many physicists it is hardly possible to propose any picture which accounts at the same time for the interference phenomena and the photoelectric effect without introducing profound changes in the viewpoints on the basis of which we have hitherto attempted to describe the natural phenomena."[105]

These quotations give an accurate picture of Bohr's thinking on photons. He clearly felt that a literal acceptance of Einstein's photon was completely out of the question. In the paper[103] in *Zeitschrift für Physik*, he insisted on the necessity of the wave picture for the explanation of dispersion and absorption. To make some sense or to function at all in such a tangled situation with so many contradictory elements, it is necessary to decide what principles or laws are fundamental for further progress and which ones are dispensable. Bohr left no doubt what his own order of priorities was. He had unquestioned confidence in the existence of stationary states; he certainly trusted the adiabatic theorems and was profoundly convinced of the correspondence principle. In effect, he was convinced that quantum theory had to be a "natural," smooth generalization of classical electrodynamics. This was precisely the reason that Bohr could not accept the photon; it is necessarily outside the correspondence principle limit. Bohr was aware quite early that a possible way to reconcile the *discrete* stationary states with the *continuous* electromagnetic radiation was to give up the strict conservation of energy (and momentum) in the individual interactions. A number of other physicists had considered similar possibilities. Darwin wrote to Bohr[106] expressing severe doubts about the validity of the conservation of energy. Bohr composed a lengthy answer[107] which contained almost "ruminations" on the nonconservation of energy. Bohr was certainly not ready to reject that possibility out of hand. The answer to Darwin was never sent. Later, Bohr and Darwin met and Darwin's unquestioned confidence in the wave theory was certainly sympathetic to Bohr. Darwin, in turn, encouraged by Bohr's reception of his ideas, published a note[108] in which nonconservation of energy is explicitly invoked. Darwin also expresses a variant of Bohr's favorite objection: "The frequency is not at all the same thing as it is in mechanics and is not susceptible to any clear definitions."

The precise relation between the photon idea and the possible nonconservation of energy and momentum was made explicit in Bohr's Solvay report (published as late as 1923): "Such a concept [photons] ... seems on the one hand to offer the only possibility of accounting for the photoelectric effect if we stick to the unrestricted applicability of the ideas of energy and momentum conservation."[109] It is abundantly evident that if a choice had to be made between accepting the photon concept and retaining the conservation laws or rejecting the photon and with it the unrestricted validity of the conservation laws, Bohr's sympathy was altogether with the second alternative. Bohr's aversion to the photon notion was probably matched only by his attachment to the stationary states and the correspondence principle. Thus, any further development of quantum theory, according to Bohr, would have to be based on the rejection of the photon ideas, the acceptance of quantum states, and the insistence that quantum theory was to be a natural generalization of classical electrodynamics. The remainder of physics would just have to adapt to these requirements, whatever the cost and however curious and unintuitive the resulting picture would turn out.

It is of course to be expected that Kramers' views on photons closely paralleled those of Bohr. Kramers was in daily contact with Bohr and Bohr's opinion on photons was a strongly held opinion. Given the nature of the personalities, it would be inconceivable that Kramers' views would be substantially different from Bohr's. Actually, Kramers in his own writings expresses a substantial dislike for Einstein's photons. This is most pronounced in the rather sharp and definitely unflattering comments made about photons in the popular book by Kramers and Holst.[90] In the discussion of the photoelectric effect, Kramers and Holst[110] first emphasize the curious fact that although classically the electromagnetic energy should be distributed evenly over the field of radiation, the energy is given to some localized atoms and not to others. Furthermore, the atoms which are "selected" to receive the radiation acquire a definite energy quantum, independent of the intensity. All this is hard to understand in the classical wave picture. After mentioning Einstein's photon interpretation of the photoelectric effect, they remark: "But even if in this [Einstein's] theory the difficulties mentioned are overcome, far *greater* difficulties are introduced; indeed it may be said that the whole wave theory becomes shrouded in darkness. The very number ν which characterizes the different kind of rays loses its significance as a frequency and the phenomena of interference—reflection dispersion, dispersion diffraction and so on—which are so fundamental in the wave theory of the propagation of light, and on which for example the mechanism of the human eye is based,† receive no explanation in the theory of light quanta." This is a sweeping and unequivocal judgment on the role of photons, in perfect agreement with Bohr's views. It is interesting that the dubious role of ν, the frequency, is again made a major objection. To dispense with light quanta once and for all,‡ Kramers and Holst appeal to the operation of a diffraction grating. As is well known, for such a grating to operate at all, the incoming light must hit the different slits in exactly the same phase. Only in that case can the path differences of the diffracted rays in different directions yield interferences. The phase notion is, of course, typical for wave phenomena; it is crucial for the interference produced by the grating. Kramers and Holst argue that, on the photon picture, the slits in the grating would be hit by photons which are emitted by atoms in the light source. Since it is a matter of pure chance when atoms emit

† It is interesting and perhaps not altogether coincidental that Lorentz was the first one to point out how the photon picture led to difficulties, in the explanation of the operation of the human eye.

‡ This example will play an important role in Chapter 14.

such photons, these photons would arrive at the slits at random times. It is difficult if not impossible to see how interference could result from photons arriving at unrelated instants. They conclude: "An understanding of the observed effect of a grating on light seems then out of the question."[110] This again is in agreement with Bohr's ideas but the language used is less tentative than Bohr's language. The overall assessment of Einstein's photons and his efforts to construct a theoretical structure based on those ideas are summarized in an unmistakingly negative evaluation. "The theory of quanta may thus be compared with medicine which will cause the disease to vanish but kills the patient. When Einstein, who has made so many essential contributions in the field of the quantum theory, advocated these remarkable representations about the propagation of radiant energy, he was naturally not blind to the great difficulties just indicated. His apprehension of the mysterious light in which the phenomena of interference appear in his theory is shown in the fact that in his considerations he introduces something which he calls a 'ghost' field of radiation to help to account for the observed facts. But he has evidently wished to follow the paradoxical in the phenomena of radiation out to the end in the hope of making some advance in our knowledge."[110] This rather lengthy quote is remarkable on several accounts. The language used is surprisingly strong. Kramers usually was controlled and quite guarded in his pronouncements. The condemnation of photons and Einstein's efforts to construct a scheme in which they fit seems unusually severe especially if it is compared with Kramers' rather mild criticism of de Broglie's work (see Chapter 12, Section V) for which he had a rather low regard. It is also somewhat surprising that in his implied critique Kramers would refer to Einstein's "ghost field." It is true that Einstein at one time seriously contemplated the introduction of "guiding fields" which would direct the photon along its path, however, this work was never published.[111,112] It was probably known to the physics community only through rumors and possibly through letters and discussion remarks. Einstein for many years was deeply preoccupied with the photon problem and at one time even tried to eliminate photons. He writes to Jakob Johann Laub: "At present I have great hope of solving the radiation problem, that is without light quanta. I am tremendously curious how the matter will turn out. One would have to renounce the energy principle in its present form."[113] Einstein failed in this effort. He writes a week later: "There is again nothing on the solution of the radiation problem. The devil has been playing a dirty trick on me."[114] Apparently, the idea of the ghost field remained of interest to Einstein as the recollections of Born[111] and Wigner[112] demonstrate, but nothing ever appeared in the published literature. It seems a little harsh and even out of character for Kramers to refer to these personal, speculative, and unpublished efforts in a popular, public denunciation of the photon ideas. Conceivably, Kramers intended no more by these comments than to indicate to what unusual constructs the incorporation of the photon notion would lead; they certainly left no doubt about Kramers' profound misgivings with photons. As a somewhat ironic postscript to this particular criticism of Kramers, it should be pointed out that about 2 years later Bohr, Kramers, and Slater, in their concerted and even desperate effort to *eliminate* photons altogether (see Section III of this chapter), arrived at a theoretical notion, which was surprisingly similar to the previously disparaged ghost field of Einstein. As might be anticipated, Einstein by that time had very little use for that theoretical proposal. That development was still in the future when Kramers and Holst wrote their book (1923), but the discussion in that book (necessarily representing the prevailing opinion in Copenhagen) leaves no doubt how Einstein's photon was regarded at that time. Their discussion ends with a remarkable disclaimer: "This matter [the matter of photons] is introduced here because the Einstein light quanta have played an important part in discussions about quantum theory and some readers may have heard about them without being clear as to the *real* standing of the theory of light quanta. The fact must be emphasized that this theory *in no way* has *sprung* from the Bohr theory, to say nothing of its being a necessary consequence of it."[115] The intent of this paragraph is clear. It is a warning to the reader not to confuse Einstein's light quanta with the Bohr quantum theory of the atom. According to this quote, they have little or nothing to do with each other. It is in particular a warning that the Bohr theory, which to be sure uses quantum ideas, cannot be held responsible for the Einstein quantum (the photon) and whatever paradoxes or conflicts this concept generates. By this statement, the Bohr theory disavows any *connection* with Einstein photons; by the previous quotes, Kramers repudiates the photon altogether. This is an expression of the almost emotional and intense antagonism that the Einstein photon generated in Copenhagen.

II. The Kramers Dispersion Theory: Not a Consequence but a Success

A. The Unreliability of Chronology

It is too bad that in writing a book it is impossible to arrange it in such a way that two sections both precede each other. This is particularly pertinent for Sections II and III of this chapter. As should be clear from the discussion on the photon (Section I), it was inevitable that in Copenhagen, sooner or later, an effort would be made to produce a theory that would eliminate the objectionable photon notion. This became especially urgent in late 1922 (December to be precise), when Compton reported on an experiment which could convincingly be explained (both qualitatively and quantitatively) using Einstein's photons. The construction of a theory eliminating photons, in the face of such strong evidence to the contrary, could only be considered by physicists who had an uncompromising aversion to the photon notion, while recognizing the subtleties of quantum theory at the same time. There were very few such persons around; very likely Bohr and Kramers were the only ones. Using some ideas by Slater, Bohr and Kramers in January 1924 succeeded in formulating a theory of radiation—without photons. This theory, which will be called the BKS theory, is discussed in Section III. As with all theories, this one led to a variety of consequences. One of the most important consequences was a set of relations, describing the dispersion of light by atoms, named the Kramers' dispersion relations. These were published by Kramers on March 25, 1924, a little after the BKS theory was published. Most books and discussions follow the order of publication in their presentations (e.g. Jammer,[116a] and van der Waerden[116b]). To appreciate Kramers' role and the decisive influence that these events had on Kramers' subsequent scientific career, it must be realized that the actual situation was much more tangled than a mere reference to the publication dates would indicate. There is now substantial evidence that chronologically Kramers' results were actually obtained well before the Bohr, Kramers, Slater paper was composed. (See Ref. 119, Slater,[133] N. Wasserman,[139] and Stolzenburg[138].) Furthermore, Kramers' results are logically independent of the assumptions made in the BKS theory. In particular, they do not depend on the philosophical attitudes expressed in that theory. Finally, the derivation of these relations follows the pattern of Kramers' investigations so closely that an examination of Kramers' work on dispersion theory shows it as a direct and natural continuation of his previous work. The procedure actually followed in the derivation of the dispersion relation is a combination of methods of Kramers' work on the Stark effect with ideas and interpretations of the x-ray paper.

For these reasons, Kramers' dispersion theory will be presented first, not as a success or consequence of the BKS theory, but as an independent result. Although this is not difficult to do, as the following should demonstrate, the resulting situation is again not as simple or unambiguous as one might wish. There is no doubt that Kramers' dispersion relations are logically completely independent of the BKS theory. But it is also true that very soon (a few months at the most) after obtaining the form of the dispersion relations (in the fall of 1923), Kramers became heavily involved in the BKS theory. It is possible to describe the dispersion theoretic results in terms of the BKS ideas. Evidently, Kramers found this a natural language to use. In his first published report on dispersion he refers to the forthcoming BKS paper as providing the general framework presupposed, and he uses some of the terminology of that paper. Succeeding papers, comments, and letters leave no doubt that to Kramers the dispersion theory and the ideology of the BKS theory had become totally intertwined,† even though initially they were obtained independently. This close link eventually led to a number of serious conflicts which will be examined in Section IV. To get a clear picture of the remarkable series of twists and turns, the curious alternation of successes and failures, it is best

† Some speculations on possible reasons for this tight connection will be presented in Chapter 14.

to present Kramers' results on dispersion theory—as they were obtained—unencumbered by the BKS philosophy. The dispersion theory results then emerge as a stage in a series of investigations in which Kramers tried to extend the range of the correspondence principle to broader and broader areas. Apart from his usual technical mastery, these papers also exhibit Kramers' increasingly sophisticated approach to physics.

B. Dispersion Theory: Background and Basic Ideas†

The classical theory of the dispersion of light was one of the signal successes of the classical theory of Lorentz. This theory combined a well-defined, simple physical picture with a straightforward mathematical treatment to yield results that can be compared directly with experiment. Physically, monochromatic light of frequency ν_0 is incident on an atom. As a consequence, the atom then becomes a source of light, and the emitted scattered (or dispersed) radiation can be described as the radiation from an oscillating dipole of dipole moment P. P is proportional to the electric field of the incoming radiation E: $P = \alpha E$. It is important that α, the polarizability, can be measured directly as a function of ν_0; for example, by measuring the index of refraction n, α is related to n via another famous Lorentz (Lorenz) formula:

$$\frac{n^2 - 1}{n^2 + 2} = \frac{4\pi}{3}\alpha$$

Thus, if a model or theory allows a *calculation* of P, and hence of α, the theory and experiment can be compared directly.

In the classical Lorentz–Drude model, the atom is schematically represented by a set of (charged) harmonic oscillators of mechanical frequencies ω_i.‡ Under the influence of an oscillating external field E of frequency ν_0, they will execute a forced harmonic motion described by the equation for the displacement x:

$$m(\ddot{x} + 4\pi^2\omega_i^2 x) = eE\exp(i\nu_0 t)$$

Since the dipole moment P is given by ex, the stationary state (nontransient) solution of that equation yields immediately that $P = \alpha E$, with

$$\alpha = \frac{e^2}{4\pi^2 m(\omega_i^2 - \nu_0^2)}$$

In general, there will be a number of oscillators, say f_i, of different frequencies ω_i, all contributing to the dipole moment. This leads to the important result

† This section is reasonably technical.

‡ In the sequel (as already done before), the mechanical frequencies are generically denoted by ω. Optical frequencies or Bohr transition frequencies are called ν. This gives the formula a slightly awkward appearance (2π factors may appear wrong), but the distinction between the two types of frequencies is so basic that it is well to denote them by different symbols.

that the polarizability is given by

$$\alpha = \sum_i f_i \frac{e^2}{4\pi^2 m} \frac{1}{(\omega_i^2 - \nu_0^2)}$$

This is the classical result of Lorentz (among others) and it is referred to in Kramers' first paper on dispersion.[116] The importance of this relation is that it can be tested experimentally. The frequencies ω_i in the test are the frequencies of the absorption lines, and they are given by the experiment. ν_0 is the frequency of the incident light, and it can be varied. Experiments provide strong support for this relation, since it is possible to fix the f_i as constants, so that this expression for α represents the data with considerable accuracy, over a wide range of values of ν_0. The harmonic oscillator picture also provides an immediate explanation of the fact that the scattered and incident radiation have the same frequency, with (for a given ν_0) a fixed phase difference; that is, they are coherent. This follows from the well-known fact that a harmonic oscillation of natural frequency ω_i, forced by a harmonic force of frequency ν_0, will lead to a stationary (nontransient) motion again of frequency ν_0. (Parenthetically, this coherence is a strong argument for the wave picture as opposed to the photon notion.) These successes of the classical dispersion treatment were a source of constant embarrassment to Bohr's quantum theory of the atom. According to that theory, the absorption frequencies ν_{ij} are related directly to the energy differences between the levels, $h\nu_{ij} = E_i - E_j$. These ν_{ij} have nothing to do with the mechanical frequencies ω_i of the mechanical motion, yet these are the frequencies that occur in the expression for α. (The ω_i also occur in the mechanical equations of motion.) Thus, the successful dispersion theory in which only the mechanical frequencies played a role seems to contradict Bohr's claim that only frequencies ν_{ij} determined by energy differences should occur. Immediately after the very first talk Bohr ever gave on atomic structure (Birmingham, September 1913), Lorentz asked how Bohr's concepts of stationary states and quantum transitions could possibly be compatible with Lorentz's results in dispersion theory. Bohr at that time and for some time to come had no answer. Lorentz remained deeply concerned about the possibility of combining classical dispersion theory and stationary states. By 1921, Lorentz was persuaded of the reality of stationary states, but he remained convinced that the dispersion phenomena required that oscillators of some kind played a significant role in the emission of light. He speculated to that effect in the discussions of the Solvay Conference of 1921: "One could imagine for example, that in addition to the 'Bohr atoms', the gas would contain 'true' oscillators and one of these might provisionally *store* the energy lost by an atom in its passage from one stationary state to another—for this to happen the energy would have to be exactly the energy quantum corresponding to the proper frequency of the oscillator.... Perhaps atoms even change into oscillators, temporarily."[117] These remarks show quite clearly to what lengths Lorentz was prepared to go just to incorporate oscillators in the description of dispersion phenomena. The radical character of these proposals already indicates how difficult it was to construct a comprehensive scheme which would accommodate both the results of classical dispersion theory and Bohr's stationary states. As late as 1922, no solution was available and one was left with what Bohr called a "paradoxical contrast" between classical and quantum demands in dispersion theory. "On the one hand—the phenomena of dispersion in gases show that the process of dispersion can be described on the basis of a comparison with a system of harmonic oscillators, according to the classical electron theory with very close approximation, if the characteristic frequencies of these oscillators just equal to the frequencies of the lines in the observed absorption spectrum.... On the other hand the frequencies of these absorption lines according to the postulates of quantum theory are not connected in any simple way with the motion of the electrons in the normal state of the atom, since of course they are determined according to the condition for the frequency by the difference in energy of the atom in this state and in another [excited] state essentially different from this one."[118] This was the state of affairs in 1922. The ultimate clarification of the relationship between the classical dispersion theory and the Bohr view of the quantum origin of atomic spectra was given by Kramers in 1924. An important intermediate insight was contained in the studies of Ladenburg.

It would appear natural (although this may only be so in retrospect) to inquire about the relation between the oscillators in the Lorentz–Drude model and the electrons (and their motions) as constituents of atoms. If the oscillator picture is merely a calculation scheme, the precise relation does not matter much. If, however, the spectrum of an atom is viewed as a consequence of the motions and arrangement of electrons in atoms, this effective oscillator picture, which presents the optical data so well, must be related to the mechanics of these orbits and to Bohr's frequency conditions. As actually used, the ω_i are the actual oscillator frequencies, and there are f_i such oscillators. Neither f_i nor ω_i have anything to do with the electrons; in particular, the number f_i is *not* the number of electrons. It *is* assumed that the oscillators are charged, with the electronic charge e, and that as oscillating charges they follow the usual laws of electrodynamics. The power they radiate is

$$\frac{2}{3}\frac{e^2}{c^3}\dot{v}^2$$

They experience a radiation reaction

$$\frac{2}{3}\frac{e^2}{c^3}\ddot{v}$$

(This last term is used as a friction term in the equations of motion.) However, there is no obvious connection between the objects called oscillators in the dispersion approach and the electrons in Bohr's theory of the atom. The relation between f_i and the number of electrons in the atom is of course not fixed by these considerations.

Ladenburg in his papers[119] did *not* address himself to these questions, nor did he seem particularly concerned about the logical conflict between the classical dispersion theory and the quantum theory of the atom. His main, and quite legitimate, concern was that in the basic formula

$$\alpha = \sum_i f_i \frac{e^2}{m} \frac{1}{4\pi^2(\omega_i^2 - \nu_0^2)}$$

the coefficients f_i were adjustable parameters. To be sure, it was possible to determine the coefficients from experiment, and this was satisfactory. It would be considerably more satisfactory if an independent method of determining the f_i were available which would allow a tighter check of the theory. In Ladenburg's paper, he suggested a method that would allow the independent determination of the numbers f_i. (The f_i were called the number of "dispersion" electrons, presumably indicating that only these objects participate in the dispersion process.) Ladenburg's suggestion was simply to equate the energy emitted per second by the system, calculated classically according to the oscillator picture, with that calculated theoretically according to the quantum picture. The classical energy in a given frequency range involves the number of oscillators of that frequency (the f_i numbers); the quantum energy depends on the quantum of energy $h\nu_{ji}$ and the Einstein A coefficient for that transition. Exactly the same type of equality was assumed by Kramers in his paper on x rays. There Kramers used this connection to calculate the A coefficients from classical considerations; Ladenburg used the connection to obtain the f coefficients from the A coefficients. The justification is provided by the correspondence principle. For long waves the classical and quantum calculations of the power emitted should give the same result; it is assumed that the same relation holds generally. The result of equating the classical oscillator energy to the quantum energy yields a simple connection between the number f_i and the Einstein A coefficients. For simplicity, consider atoms capable of just two states 1 and 2. Let f_2 be the fraction of the dispersion electrons in state 2. Then the relation in question is given by

$$f_2 = \frac{mc^2}{8\pi^2 e^2 \nu_{21}^2} A_{21}$$

A_{21} is the Einstein coefficient for spontaneous emission and ν_{21} is the Bohr frequency. It is possible to obtain experimental information about the A coefficient from the measurements of lifetimes of atomic states. With this information and the relation just obtained, the coefficients f are now determined. Thus, the polarizability α can now be calculated without any adjustable constants. To do this, one has to use the connection between f and A in general (not for atoms with just two states). Substituting the result in the expression for α in the basic formula leads to

$$\alpha = \sum_i \frac{A_{i1}}{\nu_{i1}^2(\omega_i^2 - \nu_0^2)}$$

(numerical constants have been omitted). As written, this is a funny formula; that is, it is "mixed." It contains Bohr frequencies ν_{i1} (frequencies for transition from i to 1) and ω_i, which are classical oscillator frequencies. Actually, writing α as a sum over i, with terms such as ν_{i1}, already implies that one has replaced a sum over classical oscillators by a sum over quantum states. (The sum over i is over all states above the ground state.) In the same spirit, ω_i is now replaced by ν_{i1}; that is, the mechanical oscillator frequency is just replaced by the Bohr frequency (to the ground state). The final expression for the polarizability is (constants are again omitted)

$$\alpha = \sum_i \frac{A_{ji}}{\nu_{ji}^2(\nu_{ji}^2 - \nu_0^2)}$$

This relation, which implies a connection between lifetime experiments (to obtain A) and absorption experiments (to determine α) is experimentally well satisfied. This is the main result; the arguments presented in the derivation of this formula are of course suggestive rather than deductive. But the experimental agreement so obtained must nevertheless be taken seriously. In a letter to Bohr,[120] Ladenburg reports that he has come to the conclusion that in dispersion phenomena the Einstein coefficients play an essential role. He especially emphasizes the satisfactory experimental agreement obtained. This is as far as the Ladenburg analysis went.

C. The Kramers Derivation

It is possible to obtain Kramers' dispersion relations as a rather formal but still direct generalization of the result of Ladenburg. This is not the way Kramers obtained these relations: His derivation, motivation, and conclusions were very different from Ladenburg's. Actually, Kramers in his own way added to the subsequent confusion when he published his results very slowly (almost grudgingly) and in a curious order. The first published paper (March 25, 1924)[116] is little more than an announcement of the dispersion formulas; no hint of a derivation is given. The paper concludes that the reaction of an atom to incident radiation can formally be compared to the reaction of a set of oscillators to that radiation. Kramers also shows how Ladenburg's earlier results are a special case of his formulas. The second note, submitted July 22, was just a comment on a letter by Breit.[121] It contains a brief, terse outline of Kramers' derivation. In an earlier letter to Ladenburg[122] (June 8, 1924), Kramers notes that he plans to write a lengthy paper on dispersion theory with all derivations given in detail. Such a paper was eventually written; it was a collaboration with Heisenberg. The paper was submitted January 5, 1925 and contained proofs of the results previously announced, as well as a number of further details and elaborations.[123] Only one of these extensions—the formula for the incoherent scattering—is due to Heisenberg (see v.d. Waerden[124]). The main results and arguments in the paper are all Kramers'. There are even some intimations that Kramers put Heisenberg's name on that paper as a courtesy to a younger colleague who had shown such great enthusiasm for the general topic.[125] Some time later, Bohr expressed doubts about the wisdom of that particular action.[126] It is certainly true that Kramers did not guard the best results he had ever obtained in a jealous manner, nor did he in any way rush into print with his important findings.

Kramers, in contrast to Ladenburg, was deeply upset by the conflict between the classical dispersion theory and the quantum theory of atomic structure. The paradoxical contrast between dispersion ideas and quantum theory, explicitly emphasized by Bohr,[118] was bound to cause Kramers deep concern. But there were other, perhaps almost personal, reasons for Kramers' preoccupation with this general subject. The conflict between the classical dispersion theory and quantum theory was tied directly to differences between two imposing personalities: It could easily be seen as a conflict between Lorentz and Bohr. Although at that time Kramers was totally committed to Bohr's views in physics, it would be remarkable if there did not remain a substantial residue of the youthful veneration with which Kramers earlier had regarded Lorentz. There are numerous indications—spread throughout Kramers' life—that Lorentz's influence on Kramers persisted and even seemed to increase as time went on.[127] Certainly, Kramers must have known and reflected on Lorentz's comments[117] about the oscillator features associated with dispersion. As a total believer in the correspondence principle, Kramers recognized that the quantum theory of dispersion had, at least in some limit, to resemble the classical theory of forced oscillators. Kramers was continually thinking about this set of questions and problems, during and after the writing of the x-ray paper in 1923. During that period, he eventually developed a dispersion theory which was, as mentioned earlier, published in bits and pieces. As a consequence, its direct connection with Kramers' earlier studies was not apparent. Perhaps more seriously, because of the somewhat erratic schedule of publication, Kramers' dispersion theory became entangled with the Bohr–Kramers–Slater theory, which tended to obscure both its independence and its own logical structure.

To exhibit this structure it is best to enumerate the individual steps in Kramers' derivation.

(1) Kramers starts out by considering the atom as an s-dimensional multiple periodic mechanical system with frequencies $\omega_1, \ldots, \omega_s$.
(2) This system is perturbed by a monochromatic electromagnetic wave of frequency ν_0. Next, the resulting electric dipole moment of this system is calculated and expressed as a Fourier series.
(3) This dipole moment can be reexpressed in terms of derivatives. In the next major step the derivatives are replaced by difference quotients. This is suggested by the correspondence principle.

Before discussing these points, it should be remarked that the correspondence principle is used twice; once to connect the mechanical frequencies ω to the Bohr frequencies ν, and twice to relate the Fourier coefficients to the Einstein A coefficients. The resulting expressions for the dipole moment or α are Kramer's formulas.

Discussion of step (1). There is nothing unusual about Kramers' starting point; he used the identical description of atoms in studying the Stark effect in his thesis. Classically, the (unperturbed) dipole moment P^0 of the atom can be developed in a Fourier series:

$$P^0 = \sum_{l_1,\ldots,l_s} C_{l_1,\ldots,l_s} e^{i(\omega_1 l_1 + \cdots + \omega_s l_s)t + \gamma_i}$$

The ω_i are the mechanical frequencies; the Fourier coefficients C depend on the action variables $I_1, \ldots, I_s$. One has the usual relations of classical mechanics $\omega_i = \partial H/\partial I_i$, where H is the Hamiltonian. In the Bohr–Sommerfeld method, quantum numbers were introduced as usual via the quantization conditions $I_i = n_i h$. The method, language, and formalism Kramers had used time and time again.

Discussion of step (2). This system is subject to a monochromatic wave; its electric field is given by $E = E_0 e^{i\nu_0 t}$. The problem is now to calculate the dipole moment of the system as a function of time. This is a problem in classical Hamiltonian mechanics and can be handled by performing an infinitesimal canonical transformation. The resulting dipole moment is a fairly complicated expression; its time dependence contains a number of frequencies, including ν_0. In the earlier papers, Kramers concentrates on the coherent dispersed light, that is, light of frequency ν_0, the same frequency as the incoming light. Thus, one has to separate those particular terms from the general expression for P. The result for the coherent part of the dipole moment can be written (constants are omitted)

$$P = \mathrm{Re} \sum_{l_1,\ldots,l_s} \frac{\partial}{\partial I}\left(\frac{C_l^2}{\omega + \nu_0} + \frac{C_l^2}{\omega - \nu_0}\right) E_0 e^{i\nu_0 t}$$

Here Re is the real part, ω is an abbreviation for

$$\omega = l_1\omega_1 + l_2\omega_2 + \cdots + l_s\omega_s$$

$\partial/\partial I$ stands for

$$l_1 \frac{\partial}{\partial I_1} + \cdots + l_s \frac{\partial}{\partial I_s}$$

and C_l is an abbreviation for $C_{l_1,\ldots,l_s}$.

The implementation of this step takes some skill in classical mechanics, but as stressed numerous times, Kramers was an expert in that field. The method used (an infinitesimal canonical transformation) is actually a generalization to a time-dependent problem of a method Kramers had used previously for time-independent problems. For an unperturbed problem—an atom not in an external field—the Fourier coefficients C are independent of the time. This is *no longer* so for an atom in a varying field. The time dependence of the C coefficients must then be calculated. This is not easy, but Kramers did it with his usual skill. Again this calculation was totally consonant with Kramers' ideas and methods.

Discussion of step (3). The form obtained for the classical expression for the dipole moment P contains derivatives explicitly. To obtain a quantum transcription of this, the correspondence principle is invoked. Recall that the Bohr frequency $\nu_{\rm qm}$ between quantum states is $\nu_{\rm qm} = \Delta H/h$. Here ΔH is the energy difference between two stationary states E and E' whose actions differ by $l_1 h_1$, ..., $l_s h$, so that

$$I_1 - I_1' = l_1 h \cdots I_s - I_s' = l_s h$$

while

$$I_1 = n_1 h, \quad I_1' = n_1' h, \ldots, \text{etc.}$$

If n_1 and n_1' are large compared to l_1, these quantum frequencies should approximate the classical frequencies $\omega_i = \partial H/\partial I_i$. This suggests that in the quantum transcription an I derivative generally should be replaced by a difference. Formally, a function $F(I)$ is replaced by

$$\frac{\partial F}{\partial I} \leftrightarrow \frac{1}{h}[F(n_1) - F(n_1')]$$

This replacement must be applied to the expression for P (the definition of $\partial/\partial I$ must be kept in mind) in this replacement. This substitution will clearly cause each term in P to become a difference of two terms. The correspondence principle in the same form also relates the mechanical frequencies ω_s occurring in P to the quantum frequencies. The precise relation is

$$\omega(l_1,\ldots,l_s) \leftrightarrow \nu_{qm} = \frac{1}{h}[(E(n_1,\ldots,n_s) - E(n_1',\ldots,n_s')]$$

where $l_1 = n_1 - n_1', \ldots, l_s = n_s - n_s'$. Note that the classical frequency ω,

$$\omega = l_1\omega_1 + \cdots + l_s\omega_s = (n_1 - n_1')\omega_1 + \cdots + (n_s - n_s')\omega_s$$

refers to a specific orbit (as is classically to be expected) while the quantum frequency involves two states $E(n_1,\ldots,n_s)$ and $E(n_1',\ldots,n_s')$. Call these *states i* and *j*, respectively, and the corresponding quantum frequency ν_{ij}, so that $\nu_{ij} = (1/h)(E_i - E_j)$. This replacement of ω by ν completely eliminates the mechanical frequencies from P. The only remaining mechanical elements in P are the Fourier coefficients of the classical motion:

$$C_{l_1,\ldots,l_s} \quad \text{or} \quad C_{n_1-n_1',\ldots,n_s-n_s'}$$

It will be recalled that in Kramers' thesis the intensity of the spectral lines of the Stark effect were obtained by conjecturing a relation between the Fourier coefficients and the Einstein A coefficients for spontaneous emission. This was first suggested by Bohr on the basis of the correspondence principle; it led to remarkable agreement with experiment. It is therefore not surprising that Kramers invokes the identical connection once again. Formally, this amounts to the replacement

$$\frac{(2\pi\omega)^4}{3C^3}|C_{l_1,\ldots,l_s}|^2 \leftrightarrow A_{i\to j}h\nu_{ij}$$

If these three replacements are substituted in the expression for P, a slightly messy calculation leads to the final result (again all constants are omitted):

$$P_k = E_0\left(\sum_i \frac{A_{i\to k}}{\nu_{ik}^2(\nu_{ik}^2 - \nu_0^2)} - \sum_j \frac{A_{k\to j}}{\nu_{kj}^2(\nu_{kj}^2 - \nu_0^2)}\right)$$

In this formula, k denotes a particular state of the atom. In the first sum, only terms where E_i is larger than E_k occur; in the second term, the energy in the state j is less than E_k. (Or in the first sum ν_{ik} is positive, while in the second sum ν_{kj} is positive.)

This rather detailed description of Kramers' path toward the dispersion relation shows that his derivation is a natural continuation, and in a sense a culmination, of his earlier investigations. The description of the atom as a multiple periodic system, the exploitation of Hamiltonian mechanics, and the use of the correspondence principle occur in almost the same form in his previous work. The expressions relating ω and ν and the relation between C and A can all be found in Kramers' thesis. The replacement of classical derivations by difference quotients occurs first in his thesis and later in a number of places. The dispersion relations emerge as a consequence of the systematic application of these physical principles Kramers trusted most: namely, classical mechanics, the existence of stationary states, the correspondence principle, and classical electrodynamics for free fields. It is very important to stress that this *derivation* involved *no* new physical assumptions beyond those Kramers trusted implicitly.

The result obtained was not only an impressive demonstration of Kramers' intellectual power but also a deeply satisfactory result. As such, it is not surprising that these results gave Kramers additional confidence in the principles on which his results were based. It was (pretty) easy to show (as this derivation would immediately suggest) that this result did satisfy the Bohr correspondence principle: For high quantum numbers the classical results are recovered. This removes one of the paradoxical features noted earlier by Bohr. It is also satisfactory that the Kramers' result contains the result of Ladenburg as a special case. For if the state k, in the Kramers' formula, is the *ground* state of the atom, there are *no* states j whose energy is less than E_k. Consequently, the second term in the expression for P_k is absent, and the Ladenburg result is reobtained (this does not yield the correspondence theory limit—which for a ground state quantity is not really all that surprising). The consistency of the approach can be checked further by computing the classical energy radiated per second by a system of dipole moment P as computed by Kramers. If in this classical expression for the power emitted, the classical quantities ω, C, and $\partial/\partial I$ are replaced by their quantum counterparts, as is done in the Kramers' derivation, the quantum energy is obtained. One can then check that the power radiated in a given frequency range in the quantum language contains (apart from constants) the terms $A_{ij}(h\nu_{ij})$. In this manner, one obtains the equality of the energy calculated by classical and quantum methods, which was exactly the starting postulate of Ladenburg's calculation. This exhibits the internal consistency of the distinct approaches very clearly.

It is well known that classically a *single* oscillator of frequency $\bar{\nu}$, mass m, and charge e, in response to an external electromagnetic field $E_0 e^{i\nu_0 t}$, will produce coherent scattered radiation, described by a dipole moment

$$P = \frac{e^2}{4\pi^2 m} \frac{E_0}{\bar{\nu}^2 - \nu_0^2}$$

Comparing this result with Kramers' formula shows that the dispersion of light by an atom can be described in terms of a *number* of classical oscillators of frequencies ν_{ik} with ∂^2/m replaced by $A_{i\to k}/\nu_{ik}^2$ (constants are again omitted). This indeed fulfills Lorentz's expectation that something is oscillating or acts as if it were oscillating in the atom. As a formal description there is nothing against this; however, it should be stressed that these pseudoclassical oscillators are rather peculiar entities. The formal replacement of e^2/m by $A_{i\to k}/\nu_{ik}^2$ which gives the oscillator picture does little to elucidate the physical nature of these oscillators. Furthermore, it is most important that the second term in Kramers' formula also demands the possibility of negative dispersion. To maintain the oscillator picture this requires that e^2/m be replaced by $-A_{k\to j}/\nu_{kj}^2$; formally, there is no objection against this but it diminishes the intuitive physical understanding of these oscillators even further. The dispersion of light, according to Kramers' formula, can be described by two processes (corresponding to the two terms in Kramers' formula): absorption and emission. In the first term $E_i > E_k$, and the atom goes from a higher to a lower state as it emits energy. In the second term $E_j < E_k$, and the atom absorbs energy. Instead of describing these processes in terms of the atom, they can also be redescribed in terms of the oscillators themselves.

Relative to this oscillator picture, there are two extreme positions which can be adopted; it appears that at different times (perhaps even at the same time) Kramers held both. One extreme position would be to ascribe direct physical reality to the oscillators. It is then incumbent on future studies to investigate their properties and analyze their relevance for other physical phenomena. There are indications from discussions with Slater that early in January 1924 Kramers thought along those lines. The other extreme position is that no reality of any kind should be ascribed to these oscillators. It is just a "classical manner of speaking" to describe a quantum formula. The formula as it stands contains only observable quantities. All references to the multiple periodic mechanical system have disappeared. So it might even seem inappropriate to reintroduce a fictitious, pseudoclassical oscillator language to describe the atom. Kramers was quite emphatic about the importance of the occurrence of observables referring just to the atom. In his answer to Breit's criticisms, Kramers writes: "The advantage of [our formula] is that it contains only such quantities as allow of a direct interpretation on the basis of the fundamental postulates of the quantum theory of spectra and atomic constitutions and exhibits no further reminiscence of the mathematical theory of multiple periodic systems."[121] This would seem to indicate that Kramers, at that time (July 1924), was willing to ascribe considerable autonomy to his results, while the oscillator picture was primarily considered as a means to provide a suitable terminology to characterize the connection between optical and atomic phenomena. Indeed, Kramers in his papers uses the terminology of positive and negative virtual oscillators (or absorption and emission oscillators). But he is also quick to point out that this terminology does not imply any new hypothetical mechanism; it is merely a suitable way to give a classical oscillator description of the two terms in the dispersion formula. It is still pertinent to emphasize that the two terms with the two opposite signs originate purely and exclusively from the replacement of the derivative by a difference quotient. Thus, as far as the *derivation* is concerned, there is no need (nor is it especially desirable) to introduce two types of oscillator as physical entities. This becomes necessary only if a classically physical visualization of the mathematical dispersion formulas is required.

The derivation of Kramers' results sketched here should make it clear that his derivation fits naturally into the pattern of his previous investigations. It is not true that Kramers obtained his formulas merely by formally adding the second term corresponding to the negative dispersion to the Ladenburg formula. To pull such a term out of thin air would run counter to Kramers' methods and philosophy. Also, Kramers expresses considerable trepidation about these negative terms; he writes as partial justification—almost as a plea for tolerance: "However unfamiliar this 'negative dispersion' might appear from the point of view of the classical physics, it may be noted that it exhibits a close analogy with the 'negative absorption' which was introduced by Einstein, in order to account for the law of temperature radiation."[128] Furthermore, without Kramers' derivation it is not easy to show this dispersion relation satisfies the correspondence principle and it is quite impossible to show that it is the *only* (or most sensible) such extension of the Ladenburg formula. Kramers' confidence in his result was based on his own derivations.

Further scrutiny of the Kramers' derivation shows its independence of the Bohr–Kramers–Slater approach. In that approach, the oscillator picture plays a fundamental role. In Kramers' procedure, the oscillators emerge as a convenient description of the mathematical results, but they are not necessary for the derivation. To emphasize this point without repeating the whole derivation, recall that the dipole moment of a classical oscillator of natural frequency $\bar{\nu}$ in an external field $E_0 e^{i\nu_0 t}$ is given by

$$P \approx \frac{e^2}{4\pi^2 m} \frac{E_0}{\bar{\nu}^2 - \nu_0^2}$$

Kramers' result for the dipole moment of a multiple periodic system of frequencies $\omega_1, \ldots, \omega_s$ in the same field $E_0 e^{i\nu_0 t}$ is a *sum* of terms. In the quantum language these terms have the form

$$P \approx \frac{A_{i\to k} E_0}{\nu_{ik}^2(\nu_{ik}^2 - \nu_0^2)}$$

In classical mechanics they have a slightly different form†

$$P \approx E_0 \frac{\partial}{\partial I}\left(\frac{C_l \omega}{\omega^2 - \nu_0^2}\right)$$

It is striking that the quantum expression is more immediately interpretable in terms of classical oscillators than the classical expression (because of the derivative $\partial/\partial I$ in the classical expression). The denominators

$$\frac{1}{(\nu_{ik}^2 - \nu_0^2)}$$

are of course reminiscent of oscillators; but in neither the classical nor quantum expression are oscillators presupposed. For example, the C coefficients in the classical expression describe the classical orbit which will not be oscillator orbits. To describe the quantum P in a classical oscillator picture, one has to provide these new oscillators with some rather contrived properties; for example, Kramers gives them an effective charge and mass, e^* and m^*, such that

$$\frac{(e^*)^2}{m^*} = A_{ij}\frac{e^2}{m}\cdot\frac{3mC^3}{8\pi^2 e^2 \nu_{ij}^2}$$

In that case, the quantum expression for the dipole P is a sum of terms:

$$\frac{(e^*)^2}{4\pi^2 m^*}\frac{E_0}{\nu_{ij}^2 - \nu_0^2}$$

Each term has exactly the classical form. However, the oscillators themselves are different for each atomic transition (e^* and m^* depend on i and j). Furthermore, in the derivation the oscillator properties are not used at all, rather their properties are obtained from the derivation. These observations and comments demonstrate the logical independence of Kramers' dispersion relations from the later BKS theory. As mentioned earlier, Kramers very soon used the terminology and ideology from the Bohr–Kramers–Slater theory to describe his dispersive theory results; although as stressed numerous times, it is not logically necessary to do so.

That Kramers worked along the lines outlined here, largely independent of the BKS theory, can be learned from the correspondence and the autobiography of J. C. Slater. Slater spent the fall of 1923 in the Cavendish Laboratory pondering the dual character of light. He writes to his parents that he has a hopeful idea on this topic.[129] Some of the details of this idea are discussed (in Section III) in connection with the BKS theory, but for now it is sufficient to mention that in his speculations about quantum theory and radiation, Slater associates a set of classical oscillators with each atom. He mentions his ideas in a general way in a letter to Kramers.[130] Slater had

† The difference in these terms results from the frequency factors, which relate the Fourier coefficient C to the Einstein coefficient A.

planned to spend the fall of 1923 in Copenhagen. Because of Bohr's absence at that time, he did not arrive in Copenhagen until December 21, 1923. Bohr and Kramers were both intrigued by Slater's oscillators, giving them the name "virtual oscillators."

It is quite important that by the time Slater arrived, Kramers already had the dispersion relations. On July 27, 1924 Slater writes in a letter to van Vleck: "You perhaps noticed his [Kramers] letter to *Nature* on dispersion; the formulas in that he had before I came, although he didn't see the exact applications."[131] This indicates that Kramers had obtained his dispersion relations before Slater's arrival, probably as early as November or December of 1923.

Slater, in an interview† in October 1963, was again quite explicit that Kramers' dispersion relations were completed before Slater had arrived in Copenhagen.[132] But the most detailed description of the status of Kramers' work on dispersion theory, prior to the BKS theory, can be found in Slaters' autobiography.[133] Slater starts Chapter 3 on matrix mechanics this way: "On or about Jan. 17, 1924, while the BKS paper was still being finished, Kramers showed me a theorem which he had proved shortly before, and which later proved to be practically as important in the development of matrix mechanics as the Compton effect was in the development of wave mechanics.... I believe I was one of the few people who were told this theorem at this early date and it is not always realized what an important role Kramers played in these early discoveries."[133a] The formula as I copy it from this memorandum is

$$\frac{\partial}{\partial I_l}\frac{C_l^2\omega_l}{(\omega_l^2-\nu_0^2)}\rightarrow\frac{C_1^2\omega_1}{(\omega_1^2-\nu_0^2)}-\frac{C_1^2\omega_2}{(\omega_2^2-\nu_0^2)}$$

This formula, which Kramers showed Slater, is of course just the replacement of the derivative by the difference quotient, step (1) in Kramers' derivation. As written, the Fourier coefficients C had not yet been replaced by the A coefficients.

Slater also reports that Kramers had been working on the problem of the forced oscillations of a multiple periodic system in the presence of an external electromagnetic field of frequency ν_0. Kramers had solved this problem, which is exactly that treated in steps (1) and (2) of the dispersion relation derivation, before Slater had arrived in Copenhagen. Referring to this work, Slater writes: "This was an involved problem in the classical mechanics of multiple periodic systems, but Kramers had solved it before I arrived in Copenhagen and I found it easy to reproduce his proof from the knowledge I already had of these systems."[133b] This quote leaves no doubt that Kramers had completed the derivation of the dispersion relations before Slater arrived in Copenhagen.

Slater, before coming to Copenhagen, had introduced the virtual oscillators for a specific purpose—to assign physical reality to the oscillators. By contrast, the oscillators in Kramers' derivation arose from the mathematical

† This is perhaps even more persuasive if it is realized that in the remainder of the interview, Slater is quite unkind to Kramers. He refers to Kramers as "Bohr's Yes man."

formalism; the oscillator language was especially suitable to express these results. It is natural to inquire what relation exists between these two notions. Slater's virtual oscillators were immediately incorporated into the BKS theory and became an essential ingredient of that theory. This happened in January 1924, during the discussions between Bohr, Kramers, and Slater. At about the same time, Kramers identified his mathematical oscillators with the virtual physical oscillators. This is the reason for the intimate link between the BKS theory and Kramers' dispersion relations. This interlocking of the two schemes became even closer because Kramers in his delayed publications used the language of the BKS theory to describe his dispersion theoretic results. With the BKS terminology superimposed on Kramers' formal results, it is no longer easy to disentangle the two sets of ideas. In addition, Kramers did not seem anxious to distinguish the BKS ideas from the dispersion theory. To the contrary, all evidence points to the fact that Kramers considered Slater's virtual oscillators as the physical realization (as far as possible) of the mathematical oscillators he had introduced.

With the combination of Kramers' formalism and the BKS ideas, it became possible to make a number of specific calculations. Evidently, Kramers asked Slater to do that, but Slater was not all that enthusiastic;[133c] he preferred to do his own work. Slater's proposed work was based explicitly on Kramers' formalism. He writes: "I was anxious to start with the techniques Kramers had used, but to use it to carry through the absorption probability calculation."[133c] Slater carried out these calculations, and they were published in early 1925. It is typical that Kramers, in early 1924, had not yet published the results, which Slater needed to use. So the publication of the brief notes on the dispersion relations was in part an accommodation to Slater, who then could use these results in his own work.[134] This was a remarkably unselfish thing for Kramers to do, but it was quite typical because Kramers' ambition was tempered and controlled by kindness. It is also yet another illustration of the independence and importance of Kramers' pre-BKS work.

III. The Bohr–Kramers–Slater Theory

A. Background: Slater's Ideas

The BKS theory was a combination of a number of cross currents and conflicting, puzzling results in the physics of 1923. One important element in this conflict situation was the uncompromising antagonism of Bohr and Kramers toward the Einstein photon. This antagonism was always uncomfortable in view of the Einstein photoelectric effect, but during the year 1923, the discoveries of Compton created a practically desperate situation for the opponents of the photon description of radiation. Compton's experiments, demonstrating the change in wavelength of x rays upon scattering, created enormous excitement (and corresponding scrutiny) in the world of physics.[134] In January 1923, Sommerfeld while on a lecture tour in California wrote

Bohr: "The most interesting thing that I have experienced scientifically in America ... is the work of A. H. Compton in St. Louis .... After it the wave theory of x rays will become invalid."[135] It is easy to imagine Bohr's reaction to this categorical dismissal of the wave theory! Sommerfeld goes on to say that he is not totally sure that Compton is right, but nevertheless he lectured on the Compton effect wherever he went. In the same letter to Bohr, Sommerfeld wrote: "I only want to call your attention to the fact that eventually we may expect a completely fundamental and new insight." By the fall of 1923 not only Sommerfeld but many other continental physicists accepted the Compton effect and its photon interpretation.[134] Instrumental in this acceptance was an independent paper by Debye,[136] a theoretical study that arrived at many of the same conclusions as Compton had obtained earlier. By the end of 1923, Sommerfeld writes to Compton, reporting on the reaction of the German physicists to Compton's discovery: "There can be no doubt that your observation and theory are completely accurate."[137] He also mentions that an understanding of the crystal interferences of x rays is now completely absent. How important Sommerfeld considered the Compton effect can be inferred from the way Sommerfeld refers to that effect in the new edition (1924) of his book (*Atombau und Spektrallinien*): "It is probably the most important discovery which could have been made in the current state of physics."

Although there is little direct† documentation to this effect, these developments, which appeared to indicate the necessity or even inevitability of the photon notion, must have been deeply disturbing to Bohr and Kramers. It was during this unsettled period that Slater came up with a number of unconventional and daring ideas concerning the interaction of matter and radiation.[138,139] The unconventional features introduced by Slater were *not* intended to eliminate photons, but rather to reconcile the apparently contradictory mechanisms that governed the interaction of matter and radiation. However, as soon as Bohr and Kramers became aware of Slater's suggestions, they recognized immediately that Slater's proposals provided a possible means to obtain a radiation theory without photons, and therein lay, for them, the enormous—in fact irresistible—appeal of Slater's ideas. Not surprisingly, Slater's ideas were immediately preempted by Bohr and Kramers, the combination resulting in the BKS paper, while the photon which occurred (rather prominently in fact) in Slater's original version was unceremoniously ousted.

Slater's‡ main purpose in developing his theory was to obtain a scheme including both the continuous wave ideas and the discontinuous photons. He evidently thought a good deal about this problem and felt that in November he was getting somewhere. He writes to his mother: "You know those difficul-

† The indirect documentation consists precisely of the BKS theory.

‡ Only those parts of Slater's work needed for the appreciation of the later BKS theory are outlined here. More details can be found in Notes 128 and 139. Actually, Slater's physical ideas as reported in his autobiography, his published paper,[140] and a more detailed unpublished manuscript[141] are not precisely equivalent. As could be expected, his ideas were in a state of flux. Consequently, a schematic presentation of the basic ideas and purposes is most instructive for this discussion.

ties about not knowing whether light is old fashioned waves or Mr. Einstein's light particles or Silberstein's light darts or what. Well that is one of the topics on which I perpetually puzzle my head; and about a week and a half ago I had a really hopeful idea on the subject.... It is really very simple. I have *both* the waves and the particles and the particles are sort of carried along by the waves, so that the particles go where the waves taken them, instead of shooting in straight lines as other people assume."[142] What enabled Slater to include both contradictory features in his description was a radical reinterpretation of the processes of emission and absorption. Slater was struck (and not impressed) by Bohr's requirement that monochromatic light of frequency ν is emitted *instantaneously* when an atom makes a quantum transition. He argued that a wave of frequency ν is a periodic phenomenon of period $T = 1/\nu$. Thus, the *duration* of the emission process T_e should be at least as long as T in order to produce this periodicity in time. One would actually expect $T_e \gg T$, but for instantaneous processes $T_e = 0$, so this is impossible. From this Slater concluded that the Bohr instantaneous emission of light is incorrect. Instead, he proposed that when an atom is in an excited state, say k, it emits electromagnetic waves of all frequencies ν_{kj}, where j is a state of lower energy than k. In fact, Slater envisaged the atom in state k as a set of *classical* oscillators of frequencies ν_{kj}. He then *treated* the radiation emitted as ordinary classical electromagnetic radiation. In particular, it is possible to consider the Poynting vector of the electromagnetic field produced by the oscillator of frequency ν. Call this $\mathbf{P}_\nu$. Then by ordinary electromagnetic theory, $(\mathbf{P}_\nu \cdot d\mathbf{S})\,dt$ is the energy outflow in time dt through the element of area $d\mathbf{S}$. The next new assumption Slater makes is that the *probability* that the atom will emit a photon of energy $h\nu$ per second is given by

$$p_\nu = \frac{1}{h\nu} \oint \mathbf{P}_\nu \cdot d\mathbf{S}$$

This assumption is not all that different in spirit (apart from the explicit use of the photon concepts) from that made earlier by Kramers and Ladenburg, where the radiated energy computed classically was equated to the energy contained in the emitted quanta. Here it is assumed that the probability of emission of a photon per second is just the ratio of the classical power emitted (by the oscillators) to the photon energy. But it should be emphasized immediately that, with this assumption, Slater goes far beyond the domain of classical physics. He totally reinterprets the electromagnetic field of the oscillators. Although he treats it as a classical electromagnetic field, he interprets it as a probability field. This is made explicit by defining p_ν as the probability that an atom emits a photon and also by identifying the probability to localize a photon at a point with the local energy density of

$$\frac{1}{8\pi}(\mathbf{E}^2 + \mathbf{B}^2)$$

of that same field, at that point. The ultimate nonelectromagnetic character of the field produced by the oscillators is contained in the claim that this field does not carry or transport energy. For this reason, the field is referred to as the virtual field from now on; this name does not occur in Slater's work (it was coined by Bohr and Kramers), but it is convenient to use it to distinguish this field from the usual classical electromagnetic field.

Thus, the picture of the emission of radiation according to Slater, from an atom in state k, proceeds by the establishment of a classical electromagnetic (virtual) field as produced on classical grounds by oscillators of frequency ν_{ki} ($i < k$). The probability of emission of a photon of frequency i is determined by the Poynting vector. Once a photon is actually emitted, the atom jumps instantaneously from state k to state l; in this new state, a new virtual field of frequencies ν_{li} is produced ($i < l$). The photon's subsequent path is, according to Slater, determined in some manner by the virtual electromagnetic field. Slater does not produce detailed formulas that would determine just how these photon paths should be obtained, but he makes plausible that the tangent to the photon path at a point is given by the direction of the Poynting vector at that point. He also hints that the probabilities of photons actually moving along these paths might account for interference phenomena.[143] Consequently, the (virtual) field guides the photons.† Slater was also able to describe the absorption and scattering of light via an oscillator picture. When external radiation falls on an atom, this would produce induced oscillations which, via interference with the incident radiation, would yield absorption. In pure oscillator language, the absorption oscillators would contribute to the virtual field only in the presence of an external field. Scattering could be described by a similar appeal to the properties of classical oscillators. A harmonic external field that is out of phase by exactly 90° with a harmonic oscillator does not produce a change in the amplitude in that oscillator; the (virtual) oscillator under those circumstances does not absorb energy from the external field. The net effect of this process is a change in the *direction* of the Poynting vector of the virtual field, leading to scattering but not absorption of the incident energy. These ideas, in rather broad outline, defined Slater's approach. Slater had great hopes for future applications of his theory; he was optimistic about the possibility of detailed calculations of optical phenomena. Slater was more interested in specific results than general philosophic viewpoints; he certainly did not share Bohr's and Kramers' a priori dislike of photons.

Slater's great hopes were not realized. When Slater explained the overall ideas to Bohr and Kramers, they were immediately interested. They were very enthusiastic about some features, but quite critical about others. In the intense discussions that followed, Slater altered his ideas a number of times. It is rather remarkable that even the short note[140] in which Slater outlines his own theory contains an explicit reference in the text detailing Kramers' criticism of some of the underlying ideas. To get a clear understanding of the questions involved, it is instructive to enumerate Slater's assumptions and to analyze the way in which these were modified in the BKS theory:

† The general flavor of this approach is very reminiscent of Einstein's earlier efforts of the guiding field.

(1) An atom in a stationary state emits, produces, or otherwise generates a continuous classical radiation-type field. This field is *not* the usual classical electromagnetic field; it neither carries nor transports energy or momentum. In a sense, Bohr's stationary state assumption is maintained: An electron in a stationary state does not emit the classical electromagnetic radiation. In another sense, it violates Bohr's ideas, because "some" field is produced. This field will be called the virtual field.
(2) The virtual field can be described *formally* as the usual electromagnetic field produced by a set of harmonic oscillators of frequency ν_{kj} $(j < k)$. These frequencies are the Bohr frequencies of the atom.
(3) The virtual field is interpreted as a probability field† in the following sense: The local energy density of the virtual field at a point is interpreted as the probability to find a photon at that point, while the *probability* that an atom emits a photon is determined by the surface integral of the Poynting vector around the atom. If a real photon is emitted, the atom changes discontinuously to a new stationary state. Absorptive processes are described by a similar mechanism. If a number of atoms are present, each one finds itself both in its own virtual field and the virtual fields generated by all others. The absorption of a real photon would again be determined by a surface integral over the Poynting vector (this time the direction of the normal is opposite to that for emission).
(4) Once a photon is emitted, it behaves very much like a classical particle: In particular, it is possible to ascribe a path to a photon. The virtual field guides the photon along its appointed path—again determined by the virtual field. (Although Slater called the objects photons, they were not all that similar to Einstein's photons. Slater's particles did not move in straight lines nor was their speed the speed of light.[130])

These comments contain the essential features of Slater's proposals. It is evident that he accepted photons; the real novelty was his introduction of the virtual field. Although that field was not as such observable, it regulated and monitored the probabilities of real processes. To Slater, the use of the probability notions resulted from the provisional character of the theory; he did not view it as an intrinsic feature and he hoped that further investigations would eliminate the need for probability concepts.[139] This in broad outline was the status of Slater's ideas when he arrived in Copenhagen.

B. Bohr–Kramers–Slater: The Interplay of Personalities

Slater arrived in Copenhagen on December 21, 1923. The few weeks following Slater's arrival must have been a period of feverish activity for Bohr, Kramers, and Slater. Not only did Slater have to explain his ideas in detail, to perhaps

† The term "probability field" as such is not used in either Slater's work or the BKS paper. It is used here because it gives an accurate description of the way the virtual field is interpreted and used. See also the comment on Slater's assumptions on the next page.

two of the worst listeners in the world (Kramers and Bohr), but the three of them managed to write and complete a most significant paper in about 3 weeks. The paper was submitted (that is, typed and proofread) to the *Philosophical Magazine* on January 21, 1924, exactly 1 month after Slater's arrival. This is extremely fast for anyone; but for a paper in which Bohr was associated, the speed was phenomenal.† If nothing else, this demonstrated the extreme importance that Bohr attributed to the topic.

The speed of writing is even more remarkable if it is remembered that the collaboration between Slater, on the one hand, and Bohr and Kramers on the other was far from smooth and harmonious. The content of the Bohr–Kramers–Slater paper differs in significant respects from the original suggestion of Slater. The obvious and expected difference is that anything having to do with the photon notion, so explicitly contained in Slater's original theory [points (3) and (4)], was ruthlessly and unequivocally eliminated. Slater was not altogether happy about that, but he certainly acquiesced and at times he even appeared to agree. It is not all that easy to reconstruct precisely what happened during that month, in part because Slater's retrospective long-term memory as recorded in later interviews and letters (1963–1964) is rather different from his immediate reactions as written around 1923.‡ It is clear that this was a period of controversy, but also a time of excitement, promise, and incessant discussions! Slater describes the general scene in a letter to his parents: "I got started a couple of days after Christmas telling them about this theory, and that has got them decidedly excited, I think. Of course they don't agree with it all yet. But they do agree with a good deal and have no particular argument except their preconceived opinions against the rest of it and seem prepared to give those up if they have to .... I spent most of Thursday and a while Friday and almost all Saturday afternoon talking with one or the other of them. And Prof. Bohr wanted me to write it down for him to see (which I just finished today) and he told Dr. Kramers not to talk to me about it until that was done, and then went and spent all the time talking to me himself." [144] Slater's assessment that Bohr and Kramers appeared willing to give up their preconceptions (about photons) was as wrong as could be. There were lengthy and detailed discussions about that point, but Slater was unable to convince Bohr and Kramers. Not quite a week later Slater again writes to his parents: "Part of it they believe and part they don't. But I have been thinking about it and have come to the conclusion that the part they believe is the only part that leads to any results anyway, and the rest is more a matter of taste than anything else, and you have about the same thing whether you keep it or not. So I am willing to let them have their way, and shall probably agree soon that

† The speed of producing this paper was so great that in spite of the Bohrlike language, it might well be that Kramers wrote major portions. In any case, Slater did not write any of it.

‡ Apart from a few letters by Bohr, neither Bohr nor Kramers seem to have commented on this period. Kramers' diaries contain occasional oblique references to these discussions. These are incorporated in the text without specific attribution.

it is better. I haven't anything against it now. But they continue quite excited about it."[145] Thus, in not quite a week, Slater had changed his attitude to the extent that instead of trying to persuade Bohr and Kramers to consider photons, he instead was willing to compromise on leaving them out, arguing that it did not matter much. In about another 2 weeks, Slater writes his parents again: "I have finally become convinced that the way they [Bohr and Kramers] want things, without the little lumps carried along on the waves [the photons guided by the waves], but merely the waves which carry them, is better."[146] It is impossible to see from this phraseology whether Slater was indeed convinced, or whether he was merely "agreeing to help the peace" as he said later. There are several hints, however, that Slater was—at least at that time—converted to the thinking of Bohr and Kramers. It seems unlikely that Slater, in a personal letter to his parents, would not admit that he agreed with Bohr and Kramers for convenience, just to keep the peace. Also, one day after the BKS paper had been submitted, Slater once again writes his parents: "Dr. Kramers and I are working on the more exact theory and are making rapid progress."[147] That does not sound like a person who is coerced into an approach he does not believe. Finally, in a paper published well after the BKS theory had been shown to lead to experimentally unacceptable results, Slater still writes: "The essential feature was the emission of the field before the ejection of the corpuscle, that is during the stationary state before the transition .... When this view was presented to Prof. Bohr and Dr. Kramers they pointed out that the advantages of this essential feature would be kept, although rejecting the corpuscular theory, by using the field to induce a probability of transition rather than guiding corpuscular quanta. On reflection it appeared that no phenomena at that time demanded the existence of corpuscles. Under their suggestion, I became persuaded that the simplicity of mechanism obtained by rejecting a corpuscular theory more than made up for the loss involved in discarding conservation of energy and rational causation."[148] These words, written in 1925, seem to show unequivocally that under the influence of Kramers and Bohr, Slater had given up the photon idea, abandoned the idea that the virtual field guided the photon, and willingly and with good grace accepted the limitations of the validity of the law of conservation of energy.

In spite of this apparent unanimity (at least in 1925) of the authors of the BKS theory, the cooperative effort caused strains and difficulties. Thus, Slater writes to van Vleck: "The paper with Bohr and Kramers was got out of the way the first six weeks or so, written entirely by Bohr and Kramers."[149] It is hard to decide from this letter whether Slater was pleased or displeased with the way the BKS paper was composed. But just a little later Slater, in a letter to Bohr, refers to some differences of opinion: "And please let me thank you again for your great kindness and attention to me while I was in Copenhagen. Even if we did have some disagreements, I felt very well repaid for my time there and I look back to it very pleasantly."[150] In spite of these mild, almost kind comments, there was an increasing intellectual separation between Slater and Bohr and Kramers. Slater, in a letter to Bohr (dated January 6, 1925) about a year after the BKS theory, comments on his own efforts, which are now decidedly different from those of Bohr and Kramers in Copenhagen: "Just before the holidays I sent you a copy of the paper which I have been writing, but I didn't have time to write you then. I suppose by this time you have received it and have found from it, that I haven't changed my opinions in any important way .... I don't expect you to believe anything in the paper, but I hope perhaps you won't think it is all completely foolish. Unless Dr. Kramers has changed his opinions considerably in the last few months, I suppose he won't even go that far." This is still quite civil, but there is more than an undertone of disenchantment. In the very few later communications between Bohr, Kramers, and Slater, this correct relation was scrupulously maintained. When Bohr wrote to Slater on January 29, 1926 that he had a bad conscience because he persuaded him (Slater) to give up the photon idea, Slater answered (May 27, 1926) that Bohr need not have a bad conscience. After all, the BKS suggestion led to new experiments, which in turn led to new information. Consequently, Slater concludes it was all worthwhile.

All this appears highly proper; everything is as it should be. Scientific differences should be unrelated to personalities, to personal likes and dislikes. Almost 40 years afterward, Slater comments, probably for the first time, on these events. He gave an interview,[132] answered a letter to van der Waerden,[124] and later wrote an autobiography.[133] The propriety and restraint of the earlier reactions are completely gone. His comments are angry, and it is evident that he harbored great resentment against Kramers and Bohr—a resentment that had not diminished in any way over all the many years. In the interview, he bitterly complained that the paper which went out under his name[132] was dictated to him "very much against" his wishes. Slater's rather noncommittal letter to van Vleck, in which he reports that just Bohr and Kramers composed the BKS paper, is actually an expression of his outrage that "they kept me in the back room or they kept me in the front room while they went in the back room ... they kept on changing it." In the interview and also in the later letter, Slater adamantly denies that he had anything to do with, or even agreed to, a theory where the law of conservation of energy is merely a statistical law. "I was all in favor of letting the energy be the energy of the photons, saying that the wave did not carry energy, that the energy was conserved and it was conserved because photons carried when they went from one .... No I was in favor of exact conservation. Now we will come to the point which is that those papers were dictated by Bohr and Kramers very much against my wishes."[132] Slater expresses much the same recollection (and feelings) in answer to a question put to him by van der Waerden: "When you first had the idea of a virtual field and explained it to Kramers, was the idea of a statistical conservation of energy and momentum already in your mind or was it the result of your discussion with Kramers?"[124] (This to be sure is a somewhat tendentious phraseology for a question intended to elicit factual information.) As was to be expected, Slater reacted vigorously and unequivocally. Slater answered: "As you suspected, the idea of statistical conservation of energy and momentum was put into the theory by Bohr and Kramers, quite against my better judgment. I had gone to Copenhagen with the idea that the field of the oscillators would be used to determine the behavior of the photons, which, I preferred to regard as real entities, satisfying conservation as we now know that they did and I wished to introduce probability only insofar as the waves determined the probability of the photons being at a given place at a given time. Bohr and Kramers opposed this view so vigorously that I saw that the only way to keep peace and get the main part of the suggestion published was to go along with them, with the statistical idea." This passage, written on November 4, 1964,[124] shows the ferocious pressure brought on Slater to change his ideas. It clearly still rankled, perhaps even more so because Slater *did* alter his views. In spite of Slater's later denials, he wrote and published a paper on optical phenomena under his own name, in which he uses all aspects of the BKS theory.[151] He expresses none of the misgivings or doubts contained in the letter to van der Waerden. Quite the contrary, he writes: "Although the atom is radiating or absorbing during stationary states, its own energy does not vary, but changes only discontinuously at transitions, as has always been supposed. It is quite obvious that the mechanism becomes possible only by discarding conservation."[151]

Clearly, the views expressed by Slater in 1925 are strikingly different from those expressed in the interview in 1963, in the letter in 1964, and in the autobiography in 1974. There is, of course, nothing unusual about a radical change in viewpoint over an interval of 40 years, one in which the scientific ambience itself has changed so markedly. It is not surprising either that the memories of events long ago are selectively different from the reactions to those same events at that time. But the vehemence which the recollections of the events in 1924 evokes in Slater in 1964 strongly suggest that Slater felt coerced, misled, used, and exploited by Bohr and Kramers. They had (in Slater's view) taken his ideas and browbeaten him into changing them in a direction which turned out to be incorrect, thereby denying him the proper recognition and credit. The resentment lingered and the anger remained—to be released only by a public denunciation of Kramers and Bohr in the interviews and letters. This public, or semipublic, forum also gave Slater a chance to clarify his role and his insights, as he saw them in 1964.

Although it might appear that Slater's resentment was equally directed toward Kramers and Bohr (if anything more toward Kramers), the actual strangely triangular relation between Bohr, Kramers, and Slater was much more subtle. The initial conversations leading eventually to the BKS paper were carried out primarily between Kramers and Slater. Bohr did not enter the discussions until later. In the fall of 1924, probably as early as September, Kramers and Slater met at the Liverpool Meeting of the British Association (of Physicists). As can be inferred from a letter by Slater to Kramers,[152] they discussed the problem of radiation at that meeting. In that same letter, Slater writes to Kramers about his ideas on oscillators and virtual fields. It is understandable that Kramers, who by that time had pretty much finished his papers on dispersion theory, would be most intrigued by Slater's oscillators. He certainly talked to Bohr about it. The early contacts between Slater and Kramers appear to have been quite congenial and pleasant. In several early drafts of the letter that Slater eventually sent to *Nature*, Slater thanks Kramers in a very cordial manner for his help and support. So at this stage Kramers and Slater got along well. Slater acknowledged Kramers' help and interest in his work, and as his autobiography and later interviews show, he fully appreciated Kramers' contributions to dispersion theory. Slater, as much as anyone (see Section II.C in this chapter), stressed the importance of Kramers' contributions to matrix mechanics. In the succeeding discussions between Bohr, Kramers, and Slater, that situation changed rather rapidly. In those discussions, as had almost become traditional by that time, Kramers was an intermediary, representing, explaining, and justifying Bohr's views (especially on photons) to Slater. But because of Kramers' own interest in the oscillator picture of an atom, he was extremely sympathetic to Slater's approach. Thus, Kramers from a more formal vantage point defended and explained some of Slater's ideas to Bohr. However, as the discussion in Section I should have made abundantly clear, Kramers had no more use for photons than did Bohr, while they were essential to Slater. In the ensuing conflict—whose intensity can be inferred from the violent reactions it still provoked 40 years later—Kramers and Bohr prevailed. Obviously stung and angered by what Slater interpreted as a lack of respect for his scientific insight and status, he withdrew from cooperation with both Bohr and Kramers. Slater comments later in his autobiography: "Even in that short time, I realized that I was not temperamentally suited to be an assistant to the professors in the European tradition of carrying out his calculation as he requested .... The result was that proceeding as amicably as possible, I let Bohr and Kramers know that I preferred to work on my own."[133c] Although this declaration of independence sounds reasonable and controlled, Slater in his autobiography, as well as in later interviews, shows a lasting and intense antagonism against both Bohr and Kramers. While this antagonism appeared initially directed toward both Kramers and Bohr, as time went on, the anger seemed more and more pointed at Bohr. Thus, Slater writes in his autobiography: "The conflict, in which I acquiesced to their point of view but by no means was convinced by any arguments they tried to bring up, led to a great coolness between me and Bohr, which was never completely removed."[133d] It is impossible to decide from this sentence whether Kramers was pointedly omitted, or whether he was just not considered important enough for inclusion. Later, after Kramers already had left Copenhagen, Slater comments that he was never invited back to Copenhagen. "I had made myself sufficiently unpopular with them, so that apparently they did not care to see me again." However, after World War II, Slater wrote a number of cordial letters[153] to Kramers with no trace of resentment or anger. Quite the opposite is true. On November 3, 1949, Slater, in his capacity as Chairman of the Physics Department at M.I.T., invites Kramers to be a visiting professor as a replacement for Weisskopf—who had taken a leave of absence to accept an important position at CERN. It is, of course, a great honor to be invited to a world renowned institution such as M.I.T., but Slater in his letter stresses the scientific prestige Kramers' presence would bring to M.I.T. It seems evident that by that time (1949–1950) little if any resentment toward Kramers remained. Although Kramers for personal reasons was unable to come to M.I.T., the cordial relations between Kramers and Slater persisted. Kramers and Slater met for the last time in Copenhagen (July 1951) during a meeting of IUPAP (International Union of Pure and Applied Physics), of which Kramers was President. By the time the next IUPAP meeting took place (May 29–June 7, 1952), Kramers had died—and Slater writes: "I was sorry at this IUPAP meeting not to have a final chance to see my friend Kramers."[134]

There was no comparable softening of Slater's attitude toward Bohr. Slater attended a conference in Copenhagen from July 5 to 14, 1951. This was the first time Slater had been back in Copenhagen since 1924. It was a special occasion with many guests and reunions, and everything appeared to be going well; yet as always, the past was present. Slater reports his impressions: "If it had not been for one thing I would have thought that Bohr mellowed completely in his old age and I would have forgotten about my feelings concerning him and Copenhagen more than 25 years earlier."[133e] This "one thing" was Bohr's reaction to a talk given by Brillouin on the relation between thermodynamics and information theory. Slater reports "when he [Brillouin] finished, Bohr got up and attacked him with a ferocity that was positively inhuman. I have never heard one grown person castigate another in public, emotionally and without any reason whatever, as far as I could judge, the way Bohr mistreated Brillouin. After that exhibition, I decided that my distrust of Bohr dating from 1924, was well justified."[133e] Whether indeed Bohr was as harsh as Slater remembers or whether perhaps Brillouin was wrong are matters that cannot be decided and it probably does not matter much now. But it is completely clear by 1952 that Slater's sentiments toward Bohr were very different from those toward Kramers.

It is perhaps not too surprising that in Slater's first public recounting of the events of 1924, during the interview on October 3, 1963, the hurt, anger, and resentment of those early days would dominate the discussions. It is also not surprising that Bohr and Kramers would share about equally in the rancor expressed at that time. What is surprising is the vehement, belittling, almost vicious manner in which Slater refers to Kramers. On several occasions he characterizes Kramers as "Bohr's Yes Man." This is a deeply insulting remark. Although at the time of the interview both Kramers and Bohr were no longer alive, it almost seems as if Slater is speaking to Kramers and Bohr as if they were present. It has the earmarks of a comment he had wanted to make for many years, but never dared. Slater's remark about Kramers' dependence and subservience to Bohr, thereby casting serious doubts about Kramers' scientific independence and integrity, is puzzling in view of the fact that the same Slater, 14 years earlier, offered Kramers a prestigious position at M.I.T. It may, of course, be that the interview in 1963 just rekindled the old animosities to such an extent that Slater expressed his anger in this violent outburst; but it is of some interest to speculate on other possible reasons for this vehemence.

Slater's specific description of Kramers as "Bohr's Yes Man" expresses more than frustration and anger about scientific disagreements, no matter how intensely felt. Rather, it has elements of deep disappointment, of being "let down," of a betrayal of confidence. It must be remembered that Slater first presented and discussed his ideas with Kramers. (It is important in this connection that van der Waerden, in his letter of inquiry, refers to the early discussion in which Kramers but not Bohr was involved.) Kramers liked the idea of virtual oscillators; he explained and described them to Bohr. At that time, Kramers was an accomplished scientist with a growing reputation. Slater was just a year beyond his Ph.D. degree. It seems very likely that Slater felt that in the conflict with Bohr, Kramers should have taken his side against Bohr. When Kramers, with his deep understanding and formal power, instead sided with Bohr and did so vigorously, Slater was deeply disappointed. Slater's outburst, even so many years later, is the outcry of a person who feels abandoned by one he had entrusted with a precious idea. In this context Slater's intemperate comment becomes quite understandable. It is even possible to understand why, in the course of time, Slater's relation to Kramers improved, while no rapprochement with Bohr ever took place. Slater was hurt and angry because Bohr pushed him in an incorrect scientific direction. But Slater, of all people, knew that Kramers' adherence to Bohr's ideas and the BKS theory was ultimately more detrimental to Kramers than it was to Slater. With the recognition of a shared failure, presumably due to Bohr, a reconciliation gradually set in. What Slater did not know, and indeed could not know, was that at that time it was psychologically impossible for Kramers to challenge Bohr on photons. Slater was not aware—as indeed most people were not—that by one of those ironic twists so common in the development of physics, sometime earlier (1920, see Chapter 14) Kramers had had his own fight with Bohr regarding photons. This conflict was as intense as the later conflict between Kramers, Bohr, and Slater. In the argument with Bohr, Kramers had assumed a position similar to the one Slater was to adopt later. The discussions between Bohr and Kramers took place in the privacy of the Bohr Institute, and little of it ever came to public attention. But the arguments were relentless, fierce, even heated, leading to a considerable strain in the otherwise almost idyllic relation between Kramers and Bohr. Kramers eventually was persuaded by the relentless and insistent power of Bohr's logic. But unlike Slater, his was not a grudging acquiescence; instead Kramers was totally convinced of the cogency of Bohr's position. His resulting antagonism toward photons, recorded in Section I, has all the passion of a repentant convert. Kramers' conversion also reestablished, perhaps even strengthened, the close relation between Kramers and Bohr.

C. The Bohr–Kramers–Slater Theory: Basic Ideas

The result of the many acrimonious discussions, the painstaking and searching investigations, was the celebrated Bohr–Kramers–Slater (BKS) paper, written in 3 weeks by Bohr in English and translated during the next equally hectic $3\frac{1}{2}$ weeks by Kramers into German. It was submitted and published both in the *Philosophical Magazine* and in the *Zeitschrift für Physik*. This double publication is already an indication of the significance the authors attributed to this paper. In spite of the fact that this paper has been extensively discussed, analyzed, and interpreted (just recently two important theses and one summary article were completed, all devoted to the BKS theory[154–156]), it remains an unusual even enigmatic paper. It contains a thorough but highly personal critique of the radiation theory at that time, proposes a rather vague general program—more precisely it suggests a general approach toward the problems of radiation theory, and is an unabashed advocacy for a quantum theory of radiation *without* photons. The paper does not contain any formula (except the Bohr relation $E_2 - E_1 = h\nu$) or numerical discussion of either experiments or theory. (In this respect, the paper is totally different from any paper Kramers ever wrote, either before or after.) The paper must therefore be seen as the delineation of a *framework* for the description and interpretation of radiation phenomena without ever using photons. In this sense, the paper is like a mathematical existence proof. It demonstrates that it is *possible* to obtain a framework for a quantum theory of radiation without photons. In the design of this framework certain heuristic principles of quantum physics are now considered as basic. The principles so considered by Bohr and Kramers were the following:

(1) The existence and stability of the stationary state of atoms.
(2) The unrestricted validity of the correspondence principle; the quantum theory must be a natural (smooth, continuous) generalization of classical electrodynamics.
(3) The unrestricted validity of classical electromagnetic theory for free fields: In other words, there are *no* photons.

Bohr in many papers and Kramers in his popular lectures often used the notion of a *formal* theory. (In Kramers' inaugural lecture in Utrecht this is used extensively.) A formal theory is a theory which contains rules, relations, and connections between physical quantities, but the theory does *not* and sometimes *cannot* provide a physical mechanism or a physical justification for such relations. The original Bohr theory is an example of such a formal theory. It contains quantum transitions as an essential ingredient and contains rules describing the features and results of quantum transitions. However, the transitions themselves as discontinuous events cannot be described further in terms of either classical electrodynamics or the Bohr theory. Bohr, in his many descriptions of his theoretical constructs, was always careful to stress that the mechanism of the transitions as such is *not* describable within his theory. In that sense, the Bohr theory is a *formal* theory. The proposed BKS theory similarly has formal features; it contains relations for which no mechanism is provided.

There are several distinct ideas, elements, and suggestions which together comprise the BKS framework.† Fundamental in that framework is the definition and interpretation of the virtual field.

(1) It is assumed that every atom in a stationary state (say k) emits, generates, or otherwise produces a purely classical field. This field, which will be called the *virtual* radiation field, is mathematically *identical* with that of an electromagnetic field produced by a classical set of harmonic oscillators of frequencies ν_{kj}. Here the ν_{kj} are the usual Bohr frequencies of the atom. It should be noted that this virtual field always accompanies the atom when it is in the stationary state k; it is not implied or intended that the atom actually performs a transition to state j. The total virtual field is the field generated by a set of classical oscillators of frequencies ν_{kj}, where k and j run over all the possible states of the atoms. A typical atom will necessarily find itself in a virtual field which is a sum of the virtual fields generated by other atoms and its own virtual field. (It will be a sum because as classical Maxwell-type fields, the field equations are linear.)

(2) Rule (1) defines the virtual field. BKS proceed next to give it a physical interpretation. The virtual field is not directly observable; it carries neither momentum nor energy (a real classical electromagnetic field carries both). Instead, the physical significance attributed to the virtual field is that its intensity (computed classically of course) at the location of an atom determines the *probability* that the atom will make a *real* quantum transition. Thus, in the BKS theory, the virtual field plays exactly the same role in producing real transitions as does the electromagnetic field in Einstein's derivation of the Planck formula. If an atom makes a *real* transition from state k to state l, the virtual radiation field produced by that atom is changed. It will again be a classical field, but this time it is generated by oscillators of frequency ν_{lj}. Thus, an actual transition alters the virtual radiation field and with that the probabilities for subsequent real transitions.

Assumptions (1) and (2) are basic for the BKS theory. Before proceeding with further assumptions it is instructive to discuss (1) and (2) and contrast

† The formulation and especially the phraseology of the BKS theory given here is somewhat different from what is used by BKS and the standard references. The connection with these versions will be given later in the text. This method of presentation is intended to separate the issues involved.

them with Slater's ideas. It is evident that the virtual oscillators [in (1)] which are the sources of the virtual field are taken over directly from Slater's original theory. Kramers in his dispersion formula also notes that atoms respond to incident radiation as if they consisted of a set of harmonic oscillators of frequencies ν_{kj}. Thus, assumption (1) is a straight adaptation of the earlier work of Kramers and Slater. However, whereas in Slater's theory the virtual field guided the photons, which in turn transmitted energy and momentum, in the BKS theory the photons have disappeared and they or their functions are replaced by a virtual field. The only physical role of the virtual field is the determination of the transition probabilities of real transitions in atoms. In particular, the virtual field should *not* be considered as a medium that transmits energy and momentum. It is most appropriate to describe the virtual field as a probability field. To see that this designation closely expresses the physical circumstances, consider the following example:

Suppose an atom A performs a real transition from state k to state l. This will change the virtual field everywhere, but in particular at the location of atom B. Thus, the probability that atom B will make a transition from its initial state to another state is now changed. The net result is that a real transition at A changes the transition probability at B. The virtual field thus acts as an intermediary agent, transmitting physical influences from one atom A (at one location) to another atom B (at another location). It is important to emphasize again that the virtual field transmits *probabilities* for physical events. By this mechanism subsequent events are related to initiating events by probability rules, not by causal rules. The virtual field has a physical reality by what it *does* (it determines probabilities for real processes), not by what it *is* (it is in fact unobservable). It will be clear from this description that the BKS theory has nothing to say about the actual mechanism of this transmission process; it is a typical formal element in the theory. (The *mathematics* of the transmission is precisely that of classical electromagnetism.) A further important consequence of this picture is that the transmission of probability by the virtual field (hence the relation between real processes at distinct locations) is separated and in principle independent of the transport of real energy and momentum. (The virtual field does not carry either.)

The interpretation of the virtual field as a probability field led to further differences between the original Slater theory and the BKS theory. Slater envisioned the virtual field as ultimately caused by the motions of the electrons in the atom. He thought of this relation as totally classical and causal, with all conservation laws rigorously satisfied. A given set of electronic motions would produce a unique, specified virtual field. Probability entered Slater's theory only through the relation of this causal virtual field with photons. The probability to find a photon at a given location and the probability that an atom would emit a photon could both be expressed in terms of that virtual field. However, to Slater the connections in the following chain were all causal with the specific exception of the last one: electronic motion → Fourier components → oscillators → virtual fields → emission of photons. Furthermore, at each step including the last one in that chain, the conservation laws were rigorously satisfied. Slater actually expected (or in any case hoped) that the last step would also be deterministic in an eventual more detailed theory. However, both Bohr and Kramers felt (much influenced by Einstein's derivation of the Planck law) that a strictly causal description of atoms and radiation was impossible. Since an atom in an excited state can make transitions to a number of lower-lying states with just their relative probabilities prescribed (a point so striking to Kramers that he signaled it as "most unusual" in his popular book), it is hard to see how such probability features can be achieved by causal interactions with causal virtual oscillators. Kramers objected strenuously to the mixing of the deterministic and probability procedures in Slater's treatment of photons. According to Slater, the virtual field is the agent governing the emission of a photon. However, the result is *not* deterministic; it is only possible to assign a *probability* to the event. But once the photon exists, this same virtual field guides the photon by perfectly deterministic laws along a well-defined path. Thus, the coupling between the photon and the virtual field is treated sometimes as deterministic and sometimes not. This, Kramers felt, was inconsistent.

In the BKS version, the photon is eliminated altogether, the virtual field remains purely as a probability field, and the mixing between deterministic and probabilistic features is eliminated. All the aspects of the dynamics of atomic transitions, emission, and absorptions are determined by the virtual field via probability laws. This was formalized by BKS in principle (3):

(3) The transition probability of an atom is determined *exclusively* by the state of the atom and by the virtual field at the location of the atom.

This assumption asserts that the only way an atom is aware of or can physically respond to the presence of other atoms is via the virtual fields that these atoms produce. If there are just two atoms A and B—atom A in state k and atom B in state l—then the *net* virtual field is composed of oscillators with frequencies ν_{ki} (from atom A) and ν_{lj} (from atom B), where i and j run over the possible states of atom A and atom B, respectively. To study the consequences to which the BKS interpretation leads, consider an initial situation where atom A is in state 1 and atom B is in state 2. The virtual field then contains the frequencies $\{\nu_{1i}\}$ and $\{\nu_{2j}\}$. Now a large variety of *real* processes can take place, all stimulated by these frequencies. For example, atom B can emit a real electromagnetic wave or atom A can stay in its ground state. This is allowed by the usual theory; but it is possible (not equally likely) that under the influence of the virtual field A makes a real transition from state 1 to state 3 ($j = 3$), while atom B remains unchanged. This process is not allowed by the usual theory. Many other processes are also possible; for example, B could make a transition to its ground state, while A is excited to state 17. The point of these comments is to stress that the *virtual fields* at

A and B, respectively, determine the transition probabilities. Thus, the actual emission or absorption processes at A are distinct and separate from the corresponding processes at B. This feature, which is the extreme logical consequence of the probability interpretation of the virtual field, leads to the *independence* of distant transitions. The word distant and the corresponding notion are not as harmless or as simple as one might hope. It was made into a major feature of the BKS framework. Formally, assumption (2) in the case of two atoms (A in state k, B in state l) allows the transition probability for A from k to k' to be a function of P_A of $(k, k', \{\nu_{ki}\} \{\nu_{lj}\}, l)$. Assumption (3) requires instead that P_A is a function of $(k, k', \{\nu_{kj}\}, \{\nu_{lj}\})$ *not* of l. This assumption therefore disconnects the events at A from the actual state at B. In particular, no conserved matter, or equivalently no conserved energy and momentum, is transmitted from B to A. Such transfers involving a conserved quantity would necessarily couple the physical conditions at B and A. This is precisely what the BKS theory wanted to avoid. In Slater's earlier description, and certainly in Einstein's radiation theory, it was precisely the function of the photon to provide such a connection, via conservation laws. In the BKS theory, where no photons were allowed, a definite coupling between such distant events would be too close to maintaining the photon notion (or something like it). Hence, BKS opted for a theory in which no coupling remained between distant events to remove all vestiges of a photon even if it were merely a carrier of a conserved quantity.

It is now quite evident that a theory based on assumptions (1), (2), and (3) cannot maintain the validity of conservation laws in individual events. In the example given earlier of the two atoms A and B, starting with A in state 1 and B in state 2, there was a finite probability for A to make a real transition to state 2 while B remained in state 2. This will, of course, change the virtual field, but since it does not carry energy the net result of this individual process is that energy is not conserved. This nonconservation was made into a principle:

(4) "We abandon on the other hand any attempt at a causal connection between the transition in distant atoms and especially a direct application of the principles of energy and momentum."[157]

Returning once again to the two atoms—A in state 1 and B in state 2—we see that the traditional (Einstein) view of the excitation of A would be that B makes a transition from state 2 to 1, emitting a photon. The photon is then absorbed by A, exciting it from state 1 to state 2. The BKS description of this same process proceeds along quite different lines. Because of the virtual field at B (in state 2) there is a finite chance that B makes a real transition to state 1. Suppose such a transition takes place. This will have as a consequence that the virtual field changes everywhere, in particular at the location of A. It is assumed that this change in the virtual field causes an *increase* in the probability that A (in state 1) will make a real transition to state 2. The increase in the probability of excitation of A produced by the changing composition of the virtual fields replaces the excitation via photon transport. Thus, all the processes which the photon picture allows can be accommodated in the BKS scheme; but there are real processes that the BKS theory allows which are forbidden in the photon picture. For example, in the present case, BKS gives a finite probability for the process where A goes from state 1 to 2, while B is unchanged (remains in state 2). This shows once again the radical difference between the BKS and photon theories.

The basic BKS framework is contained in principles (1), (2), (3), and (4). It is clear that these principles do not define a precise scheme or a deductive theory. The principles as stated are not specific, quantitative, or logically independent. No basic equations can be deduced from these principles. However, they do provide a setting within which problems in radiation theory can be formulated. It is of course necessary to provide further explanations and elaborations before these ideas can be explored, developed, and tested. However, even in this embryonic stage the BKS proposal had, to some physicists, many appealing features, and if nothing else it provided a suggestive alternative in their search for a systematic quantum theory. The originators of the theory were all enormously intrigued by the potential of the BKS theory; so much so that, to them, the qualitative advantages far outweighed the vagueness and incompleteness of the theory.

One point that particularly pleased Bohr was the disappearance of the (Einstein) distinction between spontaneous and stimulated emission. In the Einstein scheme, an atom in *free* space has yet a finite probability to decay from a higher to a lower state. That probability is determined by the A coefficient. This process must be distinguished from the stimulated emission which occurs when the atom is acted on by an external field. That probability is proportional to the B coefficient and to the local energy density. In the BKS theory, where *all* transitions are caused by the local virtual field, this difference disappears. In spontaneous emission, the total virtual field is generated by the atom itself; in stimulated emission the total virtual field is the sum of that of the atom itself and the other atoms. Thus, all that is different in stimulated and spontaneous emission is the makeup of the virtual field. The mechanism (although unknown) of the two types of transition is the same. To have a single agency, the virtual field, responsible for both types of transition was especially satisfactory to Bohr; he needed to invoke the correspondence principle only once, for the virtual field.

Another point which needed further explanation was the nonconservation of energy and momentum. As formulated so far, the theory does not conserve energy and momentum in individual processes. On the other hand, for macroscopic processes and presumably for any combination of a large number of individual microscopic processes, the conservation laws should certainly possess an average or approximate validity. It is necessary to show that this feature can be incorporated in the BKS theory in a natural manner.

The only dynamics occurring in the theory is the classical interaction between the virtual oscillators (which replaces or symbolizes the atoms) and the various fields. BKS suggest a way in which the interaction between the virtual field and the virtual oscillators can be specified so that conservation laws will be reestablished in an average sense. Exploiting the classical wave nature of the virtual field, BKS assume that Huygens' principle is valid for the virtual radiation field. Consequently, an atom under the influence of virtual radiation would itself become a source of new (secondary) virtual radiation of the same frequency. Suppose that the frequency of the incident virtual radiation is ν_0, while the set of Bohr frequencies of the irradiated atom is $\{\nu_{ki}\}$, where k is the state of the atom. Now several possibilities arise: the numerical values of ν_0† and $\{\nu_{ki}\}$ could be very different. In that case, it is assumed that the amplitudes of the secondary virtual waves are quite small. However, if one of the virtual frequencies is very near the incident frequency, that is, $\{\nu_{ki}\} \simeq \nu_0$, it is assumed that the amplitude of the secondary wave emitted by the atom is quite large. In the spirit of a classical wave theory, there will then be an interference between the incident virtual wave and the secondary virtual wave of (about) the same frequency. Whether this interference is constructive or destructive depends, of course, on the relative phases of the incident and secondary virtual waves. It is by stipulating these phase relations that BKS introduce interactions. Constructive interference will lead to a larger virtual field. Such a larger field in turn will be more effective in producing transition of a certain type. Thus, the interference determined by the phase relations will alter the virtual field, thereby changing the transition probabilities for real processes. The verbal rule proposed (it is not formalized) by BKS is that if a real transition has occurred whereby energy is created, the virtual field will be so changed that the probability that a subsequent transition will *destroy* energy is increased. Correspondingly, the probability that a subsequent transition would again create energy is decreased. This stipulation makes it increasingly unlikely that there will be a succession of transitions all increasing the energy. Consequently, for a large number of transitions, the probability of a unidirectional change of energy becomes particularly small. It is to be expected and it is very easy to show that this mechanism provides for energy conservation on the average. The statistical conservation of energy is therefore achieved by invoking a particular set of phase relations between incident and secondary virtual radiation. It would be hard to argue that this insight is a special success of the BKS theory, but the explanation agrees with the basic ideas. For later reference it should be noted that this proposed mechanism necessarily yields fluctuations in the total energy.

Another aspect of the classical wave description of the virtual field is the use of spherical waves in analyzing the physical effects of this field. That is of course in harmony with Huygens' principle, which requires the secondary waves to be spherical waves. The use of spherical waves is totally at variance with the Einstein description of radiation processes. Einstein in discussing spontaneous emission states totally without equivocation: "If a molecule undergoes a loss in energy of magnitude $h\nu$ but without external excitation,

† In an actual physical situation, there would be a *set* of incident virtual frequencies $\{\nu_0\}$.

by emitting this energy in the form of radiation (outgoing radiation) then this process too is directional. Outgoing radiation in the form of spherical waves does not exist!"[158] This is yet another instance of the sharp conflict between the photon and the BKS theory.

Summarizing, the BKS theory succeeded in outlining a framework from which photons were eliminated. The cost of the assumptions necessary to achieve this long sought goal were:

(1) An atom in a stationary state k is accompanied by a virtual field, a classical field generated by harmonic oscillators of frequencies ν_{kj}.
(2) The interaction between an atom and a radiation field possesses an intrinsic probability character; the classical virtual field is interpreted as defining the probability of a real transition in an atom.
(3) The only link between distant atoms is provided by the virtual field. The virtual field carries no energy or momentum and is unobservable; it is a genuine ghost field.[†] Its composition in terms of virtual oscillators determines the probabilities of real transitions. Distant atomic transitions are independent of one another.
(4) There is no conservation of energy and momentum in individual events. By assuming the validity of Huygens' principle for the virtual field, it is possible to adjust the phase relations between incident and secondary virtual radiation so that energy and momentum will be conserved on the average.

Thus, the total cost of the elimination of the photons is the intrinsic probability nature of the interaction between atoms and radiation; the absence of a spatial, temporal, or causal connection between distant events; the statistical character of the conservation laws; and an unspecified transmission mechanism of probability via a virtual (ghost) field.

D. The BKS Theory: Reflections and Second Thoughts

Whether this cost is high or not, worth it or not, is ultimately determined by the confrontation of the predictions of the theory with experimental results. But even before this decisive stage is reached, theories can be (and certainly are) judged. It is important whether the explanations the theory provides are compelling and unambiguous; whether they are natural within the structure or merely possible; whether the theory has a tight, well-defined logical character; whether it provides a new viewpoint which although incomplete is yet promising: All these are significant factors in the assessment of a new theory. It is evident that at this stage the personal predilections, hopes, expectations of individual physicists play a decisive role in that assessment.

[†] It will not escape the reader that this field is embarrassingly similar to Einstein's ghost field or *Führungsfeld*.[113,114]

As a theory in which photons were systematically eliminated, the BKS theory was an unquestioned success. In providing a well-defined, unambiguous scheme, the theory did not do all that well. There was a groping, tentative quality about the theory which is perhaps not surprising for such a first radical attempt. However, such qualities are not conducive to either a systematic implementation or an understanding of the theory. One difficulty or ambiguity that caused severe conceptual problems is the somewhat vague, tenuous relation between the virtual field and the real electromagnetic field. In the original Kramers–Slater scheme, the real electromagnetic field interacted according to classical laws with the virtual (mechanical) oscillators, which were a replacement for, or a symbolization of, the actual motions of electrons in an atom. It was stressed in the discussion of Kramers' and Slater's work on virtual oscillators (see Section II) that the kind and degree of physical reality which should be attributed to these oscillators was a point of confusion and controversy. This issue was not really settled in the later BKS formulation; there it reappeared in two different guises; (1) in the ambiguous properties associated with the virtual radiation field (2) and in the equivocal description of the interaction between the virtual field and a real electromagnetic field. Let us take a look at both points.

According to the BKS prescription, the virtual field is formally identical with the radiation field produced by a set of (charged) oscillators of the Bohr frequencies ν_{ij} of the atom. That is a completely classical, methematically unambiguous, causal prescription. However, the interpretation of this causal, classical field is neither causal nor classical. The field itself defines probabilities; this removes the causal nature. The probabilities are for quantum transitions, removing the theory from the classical realm. The only reason BKS manage to get away with assigning contradictory properties to a single agent is by a *strict* separation of the physical functions of the virtual field. It is produced classically from quantum frequencies but it propagates classically, again it stimulates quantum processes via probabilities. To avoid ambiguities, these distinct, contradictory features must not be mixed in any way. It is exactly the assignment of this double set of properties which makes the virtual field so interesting and powerful, but it can also lead to problems. For example, in the correspondence theory limit by Bohr's fundamental rule, this virtual field should become a purely classical field.

In the limit of high quantum numbers (of the atom), the frequencies $\{\nu_{ij}\}$ will approximate the mechanical frequencies of the classical motion ω. Consequently, in that limit the virtual field is formally identical with the electromagnetic field produced by the moving charges. By the correspondence principle, the virtual field *is* then the electromagnetic field. But it is now no longer obvious that the probability interpretation of the virtual field can be maintained. The virtual field defines the probability for quantum transitions. In the correspondence theory limit this would transcribe to assigning a *probability* for the electromagnetic field to produce a change in a classical orbit. For in that limit the virtual field becomes the electromagnetic field, while a quantum transition becomes merely a change in the classical orbit. Since in classical electrodynamics the effect of a given electromagnetic field on the motion of charges is completely deterministic, the probability for a given change in a classical orbit through the action of a given field is either one or zero. Thus, unless it can be proved from the BKS theory (and since there is no formalism, this is almost impossible to do; certainly BKS do not do it) that the transition probabilities defined by the virtual field approach zero and one in the appropriate limit, the systematic applications of the BKS approach would confer probability features on a classical electromagnetic field. This is an example of the peculiar and worrisome results of the BKS theory referred to above. It is interesting to pause and see how this ambiguous feature of the BKS theory was later instrumental in providing a stimulus for the interpretation of quantum mechanics.

It will be recalled that Einstein (and later Slater) attributed probability characteristics to a classical electromagnetic field, but these features were introduced specifically to accommodate the photon properties. The field defined the location or emission probability of photons. The purpose of the BKS theory was, of course, exactly the opposite; they wanted to eliminate the photons. It is most interesting that in spite of this opposite purpose, BKS were also led to a probability interpretation—this time of the virtual field. For the development of physics, both were quite important because both had a pronounced influence on Born in preparing the way for his eventual probability interpretation of Schrödinger's waves.[159] Born comments on the great influence Einstein's ghost field[159] had on his thinking; he was certainly impressed by the BKS theory. (See Section IV of this chapter for Born's enthusiastic reaction to that theory.) According to an interview with Jordan,[160] Born's attempts to construct a probability interpretation of the Schrödinger waves was in part stimulated by the probability aspects of the BKS theory. This is but one of many instances where approximate or ambiguous concepts in an earlier theoretical structure become the exact and rigorous notions of a later theory.

The BKS theory stipulates explicitly, even emphatically, that the transition probabilities of an atom are determined exclusively by the local virtual field at the position of an atom. However, the paper is less explicit about the determination of the virtual field itself if external, real fields are present in addition to the virtual fields. The general question is how the nature and composition of the virtual field of an atomic system changes if that system is subject to a real external electromagnetic field. BKS do not address this general question as such, but they touch on it indirectly in connection with their remarks on the Stark effect. BKS were interested in lifetime effects in inhomogeneous electric fields. In particular, they wished to consider situations in which the atoms moved so fast in the electric field that the lifetime of the states considered was large compared to the time it takes for the atom to move from one field strength to a quite different field strength. (During its lifetime the atom "samples" many different field strengths.) According to Stark's experiments, the data could be represented by a radiating atom in a *single* fixed field, with a Doppler shift superimposed. For the present purposes, the way in which BKS try to deduce the effect of the field on the virtual oscillators is of more interest than the actual results. They claim that the *motion* of the electrons in the stationary states of the atoms changes continuously as the atoms traverse the field. In turn, the virtual oscillators associated with the atomic transitions will also change continuously. From this BKS conclude that the virtual radiation field of the moving atoms will be the same as if the atoms throughout moved in a constant field.† In this way BKS obtained agreement with Stark's data. The main reason to record this somewhat incomplete argument is to stress that, according to BKS, the effect of a real field on the virtual oscillators is obtained from an analysis of the influence of the external field on the *actual* electronic motion. Subsequently, this altered electronic motion is redescribed in terms of altered harmonic oscillators, yielding an altered virtual radiation field. The oscillators play an auxiliary role; they do not possess an actual physical existence. In particular, there is *no* direct coupling between the real, external field and the virtual oscillators.

In more complicated circumstances, this same relation is presumably maintained; the effect of the external field on the virtual oscillators is always indirect and can be summarized by the following scheme:

Stationary states	→ Bohr frequencies $\{\nu_k\}$	→ electronic motion in stationary states	→ virtual oscillators	virtual fields
External fields	→ changed stationary states $\{\nu'_{kl}\}$	→ changed electronic motion	→ new virtual oscillators	new virtual fields

To the extent that BKS specify the coupling between the real field and the virtual field at all, it is according to the scheme outlined here. It must be reiterated that in this scheme the virtual oscillators do not possess an independent existence; the oscillator picture provides just a language to express the results of the mathematical analysis of the atoms considered as dynamical systems. (This in particular implies that *other* dynamical systems might not be describable in the same oscillator terms as atomic systems.)

Even though this is a perfectly possible and sensible approach, it appears that BKS at other times refer to the virtual oscillators and the virtual field they generate as autonomous objects, whose properties and behavior can be discussed independent from their origin as a mere redescription of the motion of electrons in atoms. The interpretation given to the virtual field as a probability field for atomic transitions is logically independent from the mathematical origin of the oscillator picture. Even if the virtual oscillators do not possess an independent existence in the BKS theory, they *do* possess an inde-

† This conclusion is at best an approximation.

pendent interpretation. The distinction between the real field and the virtual field becomes especially ambiguous if Kramers' result in dispersion theory is contrasted with the assumption of Huygens' principle for virtual radiation. Kramers' result was that the response of an atom to *real* electromagnetic radiation can be described as the response of a suitably chosen set of oscillators, subject to the same radiation. On the other hand, the response of an atom to virtual radiation is that the atom itself becomes a source of secondary *virtual* radiation in the form of spherical waves. It is reasonable to expect that there is a close connection between these two behaviors of an atom; however, the precise nature of this connection is not spelled out in the BKS theory. As will be seen later in the BKS discussion of the Compton effect, the virtual oscillators there are treated as quite independent objects, even possessing a velocity different from that of the emitting electrons. Thus, the treatment of the virtual oscillators is ambiguous within the BKS theory.

Questions of this general type abound in the BKS theory. They have an almost exasperatingly vague character. They are hard, almost impossible to settle because of the incomplete and tentative manner in which the theory is formulated. The absence of a clear-cut distinction between "actual" field and "virtual" field and the lack of a rule that specifies the respective domains of applicability of these notions were confusing and upsetting to many physicists. In this connection it is interesting that Schrödinger, who was enthusiastic about the BKS approach, still did not understand the need for both the real and virtual fields. He wrote to Bohr:† "I can not go quite along with your calling this [radiation] virtual. (Perhaps you have reasons for that, which I do not yet appreciate) but which is then the 'real' radiation, if not *that* which causes the transitions, i.e. (the radiation) which creates transition probabilities.... From a purely philosophical viewpoint one could even dare to doubt which of two electron systems possess the greatest reality, the 'real' which describes the stationary states or the 'virtual', which emits the virtual radiation and scatters incident virtual radiation."[161] This quote shows that Schrödinger was unhappy and unclear about the distinction BKS made between the real and virtual fields. In a later paper,[104] Schrödinger gives a summary and some applications of the BKS theory without ever mentioning the virtual fields. He stressed the probability (acausal) aspects and especially the statistical nature of the conservation laws. He sidesteps the question of the relation between the real and virtual BKS fields by eliminating the virtual field altogether. (His discussion contained just *one* field; he never said whether it was real or virtual.)

Apart from the indefinite relation between real fields and virtual fields, the BKS theory contained other equivocal features. The tentative nature of some of the suggestions can be appreciated best by referring directly to the original language of the paper.‡ The key concept of the virtual field is introduced by the following sentence: "We will assume that a given atom in a certain

† A rather free translation by M. Dresden.

‡ It is at this point that occasional comparisons with the more traditional treatments are made.

stationary state will communicate continually through a space time mechanism which is virtually equivalent with the field of radiation which on the classical theory would originate from the virtual harmonic oscillators corresponding with the various possible transitions to other stationary states."[162] The probability interpretation of the virtual field (in the language used here) is expressed in this manner: "Further we will assume that the occurrence of transition processes for the given atom itself, as well as for the other atoms with which it is in mutual communication, is connected with this mechanism by probability laws which are analogous to those which in Einstein's theory hold for the induced transition between stationary states when illuminated by radiation." This last formulation shows that the transition mechanism proposed is patterned after Einstein's stimulated emission processes. That is reasonably well defined, but the crucial points in the quotes are clearly the definitions of "communicate" and "virtually equivalent," while a further specification of "space time mechanism" would also help. Actually, the BKS paper does not contain a further clarification of these concepts, beyond the use which is made of them. The authors, however, were well aware of the somewhat nebulous character of those concepts. Bohr, in a letter to Pauli, requested Pauli's help in preparing the German version; he also pleads for understanding and sympathy:† "The three authors would be very grateful, if you would make suggestions for the improvement of the language in the presentation. You know how meticulous I am as far as details are concerned and I would therefore propose that if possible you would have mercy on the words 'communicate' and 'virtual', for after lengthy cogitation we have agreed to these as the foundation of our presentation."[163] Pauli answered in a somewhat lighthearted fashion.[164] He promises not to tamper with the words "communicate" and "virtual." He adds an interesting aside, that although he knows the meaning of these words he was still unable to figure out what the paper was all about. He was anxiously awaiting it. It was certainly refreshing to see a slightly frivolous reaction to these otherwise ponderous matters, but it also shows that some of the most important concepts in the BKS theory remained obscure. In the early spring of 1924 Kramers wrote to his good friend Oscar Klein in Ann Arbor about his great excitement and great expectations of the "new theory of radiation."[165] He sent Klein a copy of the completed BKS paper at the same time. That paper led to many discussions, great interest, and great confusion in Ann Arbor. Klein wrote to Bohr asking, almost plaintively, to please explain what was meant by "virtual radiation."[166] Another participant in the Ann Arbor discussions, Walter Colby (a good friend of Kramers from Copenhagen), was most anxious to know, what in the world BKS meant by "communicate." These are just a few indications of the problems and bewilderment these basic ideas of the BKS theory caused many physicists. Unfortunately, neither in the paper itself nor in later amplifications did any crisp answers appear. The concepts were used in an intuitive, almost handwaving manner. "Communi-

† The translation by M. Dresden is again rather free.

cate" clearly implies some kind of transmission process, but what was transferred and how this was accomplished was left open to debate. Under these circumstances, the language used in the present discussion, characterizing the virtual field as a probability field, is perhaps the easiest to accept.

Another somewhat ambiguous element enters the BKS theory through the statistical version of the law of conservation of energy. This formulation of that law contains the idea of "distant" atoms in an essential manner. But just what "distant" atoms are is defined only in a qualitative manner. The precise words used are: "As regards the occurrence of transitions which is the essential feature of quantum theory, we abandon on the other hand any attempt at a causal connection between the transitions in distant atoms and especially a direct application of the principles of conservation of energy and momentum, so characteristic for the classical theories. The application of these principles to the interaction between individual atomic systems is in our view limited to interactions which take place when the atoms are so close that the forces which would be connected with the radiation field in the classical theory are small compared with the conservative parts of the fields of force originating from the electric charges in the atom."[167] Thus, BKS were not proposing an indiscriminate renunciation of the law of conservation of energy, but they did suggest that its applicability be restricted to particular classes of interactions. These energy-conserving interactions are presumably singled out by the condition that the radiation reaction forces [$\frac{2}{3}(e^2/c^3)\dddot{x}$ for a single electron] are very small compared to the electrostatic, Coulomb forces (e^2/r^2). Thus, for close collisions, say electron–atom collisions, where the electrostatic forces clearly dominate, energy conservation continues to hold. This is also necessary to maintain the usual interpretation of the Franck–Hertz experiment which provides such a convincing demonstration of the existence of stationary states.

However, BKS propose that in circumstances and configurations where the radiation reaction forces are comparable to the electrostatic forces, the situation vis-à-vis the conservation laws is different. It is possible (although BKS never mention it) that the authors were thinking of the special character of the electromagnetic field produced by an accelerated charge. As is well known (at that time it was certainly known to Lorentz and undoubtedly to Bohr, Kramers, and Slater), this field splits in a sum of two distinct terms: a distorted Coulomb field, which behaves as $1/r^2$, and a radiation field, which behaves roughly as $1/r$ (r is the distance from the charge). Thus, far from the charge, the radiation field dominates. Possibly this separation is the origin of the notion of "distant" in the BKS theory. From the BKS discussion, it is apparent that they took the separation of the total field into Coulomb-like fields and the radiation field extremely seriously. They certainly did *not* consider charges and radiation as a single dynamical system to be treated by one single set of principles. The interaction between atoms and the field assumed in the "near" region is quite different from that assumed in the "far" (radiation) region. In the near region it is assumed, even required, that the conservation laws hold; in the far region these laws only hold in an average sense. Furthermore, in these far regions the virtual field plays an essential role, while its role in the near region is not important. (It is never mentioned!) "The cause of the observed statistical conservation of energy and momentum, we shall not seek in any departure from the electrodynamic theory of light as regards the laws of propagation of radiation in free space, but in the peculiarities of the interactions between the virtual field and the illuminating atoms."[168] It is precisely at this point that BKS introduce Huygens' principle for the virtual radiation, with the suitably adjusted phase differences to reestablish the conservation of energy on the average.

These selected quotes demonstrate the equivocal nature of the BKS scheme. For near events, energy conservation holds; there is no virtual field, the field is an ordinary electrostatic field, and there is strict causality. For distant events, energy conservation holds only on the average, achieved via phase relations of a virtual field, which is a probability field; there is no causal relation between distant events. It is unclear whether a single framework exists which can encompass the required description of both types of event. Bohr, Kramers, and Slater were well aware of the tentative character of their suggestions, the mixing of various concepts. Nothing could express the guarded, groping character of the BKS theory better than the conclusion of Section 2 of the BKS paper: "In spite of the time–spatial separation of the processes of absorption and emission of radiation characteristic for the quantum theory, *we may nevertheless expect*, in our view, a far reaching analogy with the classical theory of electrodynamics as regards the interaction of the virtual radiation field and the virtual harmonic oscillators conjugated with the motion of the atom. *It seems actually possible*, guided by this analogy, to establish a *consistent* and *fairly complete* description of the general optical phenomena accompanying the propagation of light through a material medium, which accounts at the same time for the close connection of those phenomena with the spectra of the atoms."[169]

E. The Compton Effect in the BKS Theory

As detailed in Section III.A, by the end of 1923 the Compton effect was widely believed to be an unequivocal demonstration of the existence of photons. After producing a radiation theory without photons in January 1924, it was incumbent on Bohr, Kramers, and Slater to analyze Compton's result from their new viewpoint. It is therefore somewhat surprising, even disappointing, that in the BKS paper only a scant $1\frac{1}{2}$ pages out of 18 are devoted to the Compton effect. The discussion given is in the same vein as the rest of the paper; it is quite general, occasionally vague, and without a single formula. As the BKS paper itself, the discussion purports to show that it is *possible* to give a description and interpretation of the Compton effect based on the ideas of the virtual field (as summarized in points (1)–(4) on page 178). However, in so doing, the ambiguities discussed at length in Section III.C became especially acute, leading to possible but quite awkward constructions. It was also necessary, in order to maintain the statistical conservation of momentum, to introduce discontinuous elements in their otherwise continuous wave theory. BKS were willing to consider these extreme versions of their theory because without a compelling explanation of the Compton effect, it would be impossible to take their theory seriously. In addition, the interaction between an electromagnetic field and an electron is the *simplest* possible interaction between matter and radiation. Surely the theory should describe this in succinct simple terms. It is perhaps surprising that, in spite of its vagueness, the BKS analysis of the Compton effect still leads to strikingly different predictions from the photon theory.

The classical mechanism of the scattering of electromagnetic waves by charges was straightforward and simple. If a plane monochromatic electromagnetic wave of frequency ν is incident on an electron at rest, the electron will experience forces; that is, it will be accelerated. As an accelerated charge it will radiate in all directions. The frequency of the emitted radiation is again ν. It is also possible to visualize the motion of the electron under the influence of the monochromatic wave as a *forced* oscillation of frequency ν. This forced oscillator then emits classical radiation of the same frequency ν. Both the frequency and the angular distribution of the scattered radiation computed on this classical basis agreed with experiment, provided ν is sufficiently small. In this picture, the scattered radiation is really secondary reradiated radiation from the moving electrons. This classical picture was of course totally continuous; it provided a satisfactory description of the phenomena. It is therefore understandable that Compton's discovery—that for x rays the scattered radiation had a frequency ν', different from ν, that furthermore depends on the scattering angle—created great excitement. Although the realization that this was a quantum effect, an extreme manifestation of the photon character of radiation, came pretty fast, it inevitably caused much soul searching, especially on the part of Compton. (For an extensive and incisive discussion of Compton's ideas see Stuewer.[170])

The photon picture indeed provides a simple and direct description of Compton's results. Once the photon idea is accepted, all of Compton's findings can be understood qualitatively and quantitatively as due to a mechanical-type collision between the photon and the electron. The quantitative results emerge straightforwardly from an application of the laws of conservation of energy and momentum to that collision. A photon of frequency ν has absolute value of momentum $h\nu/c$. Call the frequency, wavelength, energy, and momentum of the scattered photon ν', λ', $\varepsilon' = h\nu'$, and $h\nu'/c$, respectively. Finally, call the wavevectors of the incident and scattered photon $\mathbf{k}$ and $\mathbf{k}'$, respectively. The momentum acquired by the electron is $\mathbf{p}$, and its energy is $E = \sqrt{p^2c^2 + m^2c^4}$. The process can be represented by the following diagram:

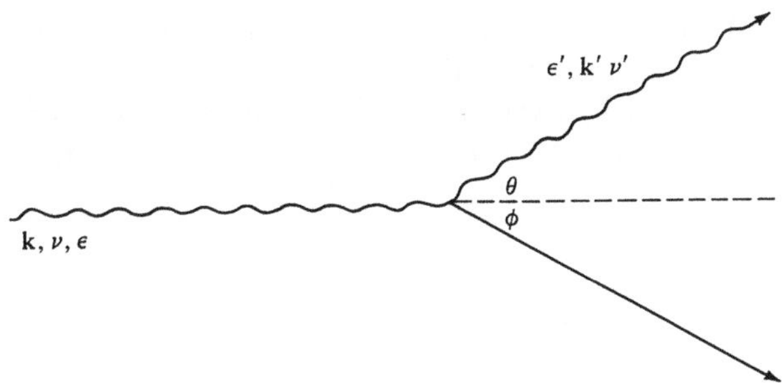

θ is the photon scattering angle, and ϕ is the electron recoil angle. In terms of these quantities, the conservation laws simply read:

$$\hbar\mathbf{k} = \hbar\mathbf{k}' + \mathbf{p}$$

$$E + mc^2 = E' + \sqrt{p^2c^2 + m^2c^4}$$

(As written $\hbar$ is the Dirac $\hbar$ equal to $h/2\pi$. Thus, the absolute value of $\hbar\mathbf{k}$ is $(h/2\pi)(2\pi/\lambda) = h\nu/c$ as required.) All the physics is contained in these conservation equations; the rest of the analysis is just manipulation of these equations and interpretation of the results. Of the greatest significance are the wavelength shift, $\Delta\lambda \equiv \lambda' - \lambda$, and the relation between the photon scattering angle θ and the electron recoil angle ϕ:

$$\Delta\lambda \equiv \lambda' - \lambda = \frac{h}{mc}(1 - \cos\theta) = \lambda_C(1 - \cos\theta)$$

$$\tan\phi = \frac{1}{1 + \lambda_C/\lambda \tan(\theta/2)}$$

$\lambda_C = h/mc$ is the Compton wavelength. Compton derived the expression for $\Delta\lambda$; extensive comparisons with his data showed that it agreed well with the experimental results. The second formula, first obtained by Debye,[136] shows that there is a unique, definite relation between the scattering angle θ and the electron recoil angle ϕ. In the early Compton experiments, this relation was not tested, but it played an important role in later developments.

These results were all known before the BKS theory was written. Thus, BKS were obliged to provide a possible explanation of the wavelength shift in their theory without photons. They attempted to accomplish this by combining their ideas with an earlier suggestion of Compton. It must be stressed that Compton was as aware as anyone of the dilemma the photon idea introduced in physics. As mentioned by Stuewer,[170] Compton on December 6, 1922 submitted an extensive report to the *Philosophical Magazine* on the internal reflection of x rays. This report, forcefully demonstrating the wave character of x rays, was sandwiched between Compton's presentation of the "photon paper" on December 1 and its submission for publication on December 13. It is only possible to speculate about the conflicts and confusion these contradictory properties of x rays produced for a scrupulously honest thinker, but there can be no doubt that these questions preoccupied Compton for a long time. The idea to which BKS attached their theory was Compton's notion that the observed wavelength shift was basically a Doppler effect. It is well known that a source at rest, emitting waves of frequency ν, will, when moving with velocity $v = \beta c$ relative to the laboratory system, give rise to waves of frequency ν':

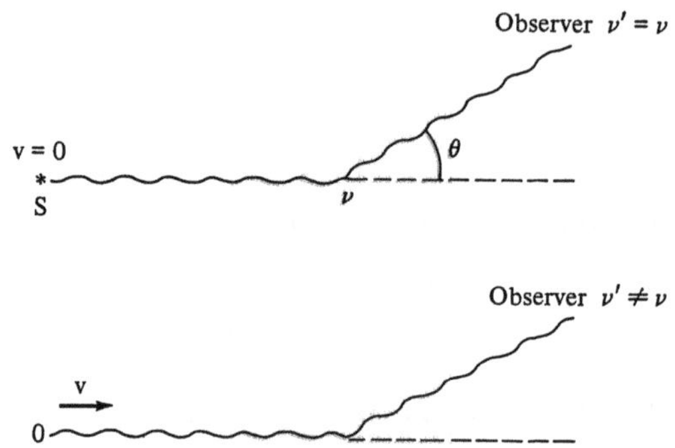

$$\nu' = \nu \frac{1 - \beta}{1 - \beta \cos\theta}$$

Thus, Compton envisaged his effect as the result of a two-step process. In the first step, the x rays would cause the electron to move and it would acquire a velocity. This moving electron would then reradiate x rays, which because of the electron's velocity would have undergone a Doppler shift. This is an interesting and suggestive idea. From the Doppler effect formula, it follows, using $\nu\lambda = c$, that there will be a wavelength shift given by

$$\lambda' - \lambda \equiv \Delta\lambda = \lambda \frac{\beta}{1 - \beta}(1 - \cos\theta)$$

Inspection shows that this Doppler effect explanation gives exactly the same angular dependence $(1 - \cos\theta)$ of the wavelength shift as the photon picture and Compton's experiments. That is satisfactory. However, to obtain numerical agreement with the data, it is necessary to identify $\lambda\beta/(1 - \beta)$ with $\lambda_C = h/mc$. This gives an expression for the velocity of the source emitting the radiation as $\beta_{eff} = c\lambda_C/(\lambda_C + \lambda)$. This is a perfectly well-defined velocity; Compton called it the "effective electron" velocity. However, this velocity cannot be related in a very direct way to the velocity of the electron. Thus, in his early investigations, Compton ascribed the wavelength shift to the Doppler effect of an "imaginary recoiling" source that most definitely was not the electron. Although logically possible, Compton later abandoned this explanation in favor of the photon interpretation described earlier. Bohr, Kramers, and Slater incorporated the Doppler effect mechanism of the wavelength shift in their explanation of the Compton effect. Specifically, they assumed that corresponding to an atom making a discontinuous transition is a "Compton event," where the electron discontinuously acquires a momentum and an electromagnetic wave is emitted. Just as the virtual field determines the probability of an atomic transition, in the BKS theory it is suggested that in the Compton effect the virtual field determines the *probability* of a given momentum transfer to the electron. The essential discontinuous element of a quantum transition is transcribed to a discontinuous momentum transfer. The scattering of the waves, by contrast, was assumed to be continuous, and it was envisaged as the emission of secondary coherent waves by the virtual oscillators. These oscillators were moving, hence a Doppler shift would result. By assigning to the virtual oscillators Compton's effective electron velocity β_{eff}, this accounted precisely as in Compton's early theory for the wavelength shift. BKS tacitly assume that just as the virtual field necessarily accompanies an atom in an excited state, a moving electron too is necessarily accompanied by such a virtual field.

With these stipulations, BKS manage to describe the features of the Compton effect. They could maintain the completely classical character of the scattered radiation via the Doppler effect, which further retained its usual causal and continuous character. The only quantum (discontinuous) element introduced was the probability of a finite momentum transfer to the electron. The BKS explanation, like the original Compton explanation, again involves two distinct aspects. The emission of the secondary (scattered) radiation is separated from the momentum transfer: the former is classical, causal, and continuous; the latter is discrete and noncausal. It must be admitted that the mechanism suggested is possible and even in the general spirit of the BKS theory. However, the resulting picture is far from intuitive. The most conspicuous feature is no doubt that the *source* of the emitted radiation moves with a velocity different from—and unrelated to—the electron's velocity. This unusual feature was not lost on BKS; they write: "That in this case the virtual oscillator moves with a velocity different from that of the illuminated electrons themselves is certainly a feature strikingly unfamiliar to the classical conceptions. In view of the fundamental departures from the classical space time description, involved in the very idea of virtual oscillators, it seems at the present state of science, hardly justifiable to reject a formal interpretation as that under consideration as inadequate."[171] The formal character of the theory, which merely requires the formulation of a set of unambiguous rules without necessarily relating these rules to specific physical mechanisms, is invoked here in an essential manner. Although the BKS position is logically unassailable, many physicists expressed severe misgivings about the BKS constructs. In his description of the BKS theory, Pauli, although not devastatingly critical, is extremely circumspect; he takes great pains to emphasize the curious consequences of the BKS theory.[†] "To take the Compton effect into account, it must be admitted that the velocity (and hence also the position) of the center of the emitted spherical wave are in general distinct from the velocity (and position) of the corresponding atom. It must further be admitted that the velocity changes discontinuously everytime the atom—or the free electron—

[†] The translation by M. Dresden is quite free.

experiences a recoil, which accompanies the scattering of radiation."[172] In this and other publications, Pauli does not criticize the theory further; he is satisfied with a painfully clear exposition of its curious consequences. In private letters, Pauli was a good deal more negative; likely his respect and admiration for Bohr kept him from a public expression of his unhappiness. Many other physicists were also baffled by the distinction between the electron's location and velocity and the location and velocity of the virtual oscillator. After describing the BKS ideas in his textbook,† Andrade adds an almost plaintive footnote, which expresses his deep uneasiness: "Presumably if incident radiation from more than one direction falls on a free electron the virtual oscillator can also be in several places at once, none of them that of the electron."[173] It is clear that the BKS explanation of the Compton effect, although logically possible, leads to unusual, unexpected, and somewhat confusing situations. The ambiguities and tenuous nature of the theory discussed earlier become especially pronounced. The unclear distinction between actual fields and virtual oscillators, stressed in Section III.C, becomes especially important here. There can be no doubt that the incident x rays are "real," yet BKS continually refer to "the incoming virtual radiation." Presumably, this virtual field is a redescription of the interaction between the x rays and the electron. The electrons in turn scatter this virtual field in a classical manner, emitting secondary waves, which however are real and observed in Compton's experiments. On the other hand, to maintain the Doppler effect explanation, the sources emitting the observed radiation must be moving with speed β_{eff}; they cannot be the electrons. Consequently, the virtual oscillators are totally autonomous objects with their own velocity and position. They and the virtual field are no longer just the catalysts which induce atomic transitions, but they are genuine sources of real radiation. The virtual field retains the property that it determines the probability of a momentum charge (which is a discontinuous quantum event). Thus, the virtual field is given an intermediate type of reality; it is definitely more than a mathematical redescription of a given physical configuration. This is not wrong; it is unusual and certainly unclassical.

These summarizing remarks also point the way toward experimental tests (or disproof) of the BKS theory. One of the main characteristics of the theory is that the emission of the scattered wave and the acquisition of momentum by the electron are related by probability laws only. In particular, the emission of the wave ν' and the recoil of the electron would be uncorrelated. This has several striking consequences which can be experimentally checked.

(1) Because of the independence of recoil and emission there could be (and in general should be) a *time delay* between the emission of a scattered wave and a recoiling electron. One could attempt to measure this time delay.
(2) For the same reason, there would not be a one-to-one correspondence between a scattering event and a recoil event. There could be scattering

† It is an indication of Bohr's reputation in physics, that a new (1924) and clearly speculative theory would already be included in a textbook written about a year later!

without recoil. Thus, by measuring the coincidences between Compton electrons and scattered x rays it would in principle be possible to check whether there is a one-to-one correspondence or a statistical relation between these events. Bothe and Geiger proposed and eventually carried out this experiment.

(3) The independence of emission and recoil also implies a statistical relation between the scattering angle θ and the recoil angle ϕ, while the photon theory (the Debye formula) predicts a unique relation between θ and ϕ. Compton and Simon set out to measure the experimental relation between these angles. Both the Bothe–Geiger experiment and the Compton–Simon experiment were difficult; it took more than a year to arrive at definitive results.

The results showed unequivocally that the predictions of the photon theory were correct in all details. The experimental findings were incompatible with the BKS predictions; they specifically excluded the kind of theory envisaged by Bohr, Kramers, and Slater. But this did not happen until April 25, 1925. In the interim period, from January 1924 until the spring of 1925, the BKS theory with its new and bold ideas created enormous excitement in the world of physics. Great as that influence was, it was not greater than the influence the theory was to have on one of its originators, Kramers.

F. The Intellectual Framework of Kramers' Physics

It is not at all unusual for researchers to develop a special fondness for their results. It also often happens that they develop a strong commitment to particular methods and research directions. But it is not always recognized that the attachment to a particular set of ideas often involves powerful emotional elements. These emotional aspects can become so strong that they assume an almost obsessive character. Perhaps the frankest, most candid description of the deep feelings engendered by research ideas is contained in Feynman's Nobel Prize lecture. In this lecture, Feynman describes the evolution of his brilliant and incisive ideas on quantum electrodynamics. He records his initial reaction: "The idea seemed so obvious to me and so elegant, that, I fell deeply in love with it. And like falling in love with a woman, it is only possible if you do not know much about her, so you can not see her faults ... so I was held to this theory, in spite of all difficulties by my youthful enthusiasm."[174] After relating the further development of the space–time approach to quantum electrodynamics and the now fundamental path integral formulation, he concludes: "So what happened to the old theory that I fell in love with as a youth? Well I would say its become an old lady who has very little attraction left in her and the young today will not have their hearts pound when they look at her anymore. But we can say the best we can for any old woman and she has been a very good mother and has given birth to some very good children." Feynman's profound emotional involvement with his theory is explicit and manifest.

It would be preposterous to equate the extroverted, rakish Feynman with the controlled, careful Kramers. Also, the scientific style in 1924 was more rigid and circumscribed than in 1966. Nevertheless, the intensity of Kramers' emotional reaction to the BKS theory was quite comparable to Feynman's reaction to his space–time approach. To Kramers, the ideas, general approach, and philosophy—if not the specific details—of the BKS theory signaled a new beginning which not only would bring solutions to the outstanding problems of radiation theory but also promised the realization of his deeply felt scientific aspirations. It appeared to Kramers that with the advent of the BKS theory many of his scientific goals had come within reach.

Actually, it is not all that easy to discern just what his scientific aspirations and goals were. Like the vast majority of physicists, Kramers in his publications was primarily preoccupied with specific technical questions; matters of general philosophy were touched only rarely. However, in his popular writings, starting with the joint book with Holst[175] on Bohr's theory of the atom, Kramers reveals his personal feelings regarding the nature and goals of physics to a surprising extent. This is not done in a particularly systematic or organized manner, but scattered throughout the book there are a number of observations on the purposes, limitations, and scope of physical theories. Taken together, these almost inadvertent comments give illuminating insights into Kramers' personal philosophy of physics.

Kramers comes very close to a confession of his scientific faith, in a passage which follows a spirited enumeration of the problems and paradoxes of the Bohr theory: "It must not be forgotten that in science we must always be patient with the question 'Why'? *We can never get to the bottom of things.* On account of the nature of the problem, answers can not be given to the questions why the smallest material particles (for the time being hydrogen nuclei and electrons)—the elementary physical individuals—exist, or why the fundamental laws for their mutual relationships—the most elementary relationships existing between them—are of this or that nature, a satisfactory answer would necessarily refer to something even more elementary. We can not claim more than a complete description of the relative positions and motions of the fundamental particles and of the laws governing their mutual action and their interplay with the ether.† If we examine our knowledge of the atomic processes in the light of this ideal, we are tempted to consider it as boundless ignorance."[175a] This was written in 1923, about 2 years before the BKS theory. In this passage, in fact throughout this section, the incomplete, unintuitive, and paradoxical features of the Bohr theory are continuously emphasized. It is impossible to escape the feeling that the authors get some wry pleasure out of recounting these features: "During the transition from one stationary state to another, we have no knowledge at all of the existence of the electron, we do not even know whether it exists at that time,

† In the German edition[176a] an important phrase is added at this point: "To get any further is impossible anyhow."

or whether perhaps it is dissolved in the ether to be reformed in a new stationary state."[175b] But more important than the fact that the Bohr theory is very far from an ideal theory is the recognition that any ideal theory itself, according to Kramers, is necessarily incomplete and of limited scope. This same theme—the intrinsic and unavoidable limitations of scientific theories and their necessary tentative and provisional character—recurs time and time again in Kramers' letters and lectures. It is the key element in Kramers' philosophy. A telling example is the letter Kramers writes to Romeyn about the time Kramers was completing the derivation of the dispersion relations: "In my field I struggle for the broadest points of view, will I ever be so fortunate as to reach a new mountain top? Perhaps, just perhaps, but one thing is certain; never shall I be able to fathom, just what the struggles and triumphs are in the life of my best friend. His thinking, his perception and understanding of the world, I shall never know. Here the intellect can accomplish nothing, the instinct a little, but love everything .... I think you will surely understand what I mean. It is insane to try to understand the world in the manner of the rationalists. This becomes evident once one has really tried and even succeeded to understand a little on that basis .... That we believe in what we do makes life possible and occasionally happy, but we still believe lies; for everything understood as true, (by us), can be understood as false by Mephistopheles and God doesn't consider eliminating him .... I would not like to pass for an intellectual."[177]

This letter is remarkable for many reasons. It was evidently written with great intensity, even passion. That was unusual for Kramers who was usually guarded and controlled even in his personal letters. The letter clearly expresses very personal and deeply held beliefs. As if to stress the depth of his convictions, Kramers repeats in this same letter that: "I very seriously mean everything that I have written."[177] It is perhaps not so surprising that Kramers, who was acutely sensitive to literature and music, would recognize that science does not encompass all facts of human life. But the harsh judgment that "it is insane† to try to understand" as quoted in the letter, is quite surprising. If Kramers merely wanted to say that science as a human activity is circumscribed, that the discipline itself imposes boundaries on its scope, this could be done without such a vehement outburst. It appears that Kramers struggled long and hard with the proper delineation of science and its relation to other human activities. The restriction to scientific activities was at the core of the conflicts that Kramers, as a student, suffered in Leiden. Although by this time he had long since settled these questions, it is understandable that, when writing to Romeyn, some of the old feelings would reappear and give rise to an emotional insistence on the limitations of science itself. It is striking that Kramers kept on reiterating throughout his life that just those who attempted to produce rational explanations of the natural world are most cognizant of the limitations of their efforts. Although phrased in general terms,

† The Dutch word Kramers uses here is extremely pejorative.

he clearly means himself. The allusion to Mephistopheles in the quote refers to the first encounter of Mephistopheles and Faust in Goethe's play, *Faust.* It is in these initial discussions that Mephistopheles ("the spirit who denies") suggests, even taunts, that all truths, values, and purposes, which Faust seeks so avidly, can be interpreted as lies, deceits, and falsehoods. Kramers transcribes this to the laws and concepts of physics which, although useful and important, *might* with equal effectiveness be replaced by totally different rules and notions that could contradict the traditional ones. The impossibility of logically excluding this contingency caused Kramers to view natural laws as a mixture of definition, convenience, and convention. He never considered scientific laws as eternal and immutable truths. This of course was very much in line with his general antidogmatic attitude, but it is important that he carried this same attitude over to the scientific sphere.

The last comment in the letter, that God isn't about to eliminate the devil,† does not apply directly to physics, but it is a significant residue of the Calvinist theology that shaped his thinking. To a true Calvinist, a world without the devil is unthinkable; to Kramers, this implied that a world without evil, human weaknesses, ambiguities, and contradictions was inconceivable.[178] This attitude, although not science, was bound to influence his approach to physics. Thus, at this stage in Kramers' career, there were two guiding themes underlying his approach to physics. They were only occasionally formalized, but there can be no doubt that they exerted a great and continual influence. One was the intrinsic incompleteness of scientific knowledge, best summarized by his warning "We can never get to the bottom of things." The other was the pervasive belief that laws of nature are primarily conventions, ultimately justified only by their efficiency in correlating experimental data. They need not and indeed cannot express ultimate reality, nor need they be eternal or universal truths. There is more than a hint in Kramers' writings that *this* level of comprehension is accessible only through different and nonscientific means, possibly through music, love, religion, or theology. The exclamation in Kramers' letter that "he would not like to pass for an intellectual" must be understood in this context.

These two themes remained the basis for Kramers' general, philosophical orientation toward physics. As time went on, he expressed these ideas in more detail and occasionally elaborated on them in public lectures. It is remarkable that even though physics itself changed radically during his career, Kramers' basic attitudes changed hardly at all.

The extraordinary appeal that the BKS theory had for Kramers can now be understood; it realized and indeed gave a concrete form to the general, and up to that point only partially formulated, elements of Kramers' deeply felt basic philosophy. No wonder that to Kramers the BKS theory was not just a useful calculational or descriptive scheme; instead, it was the formal realization of his deeply felt physical aspirations. At a Congress on Physics

† Although not literal, this is actually a better translation of the sense of this remark than what was given earlier.

and Medicine,[179] Kramers gave a lecture in which he discusses the situation in basic physics generated by the BKS theory. Not surprisingly, this "new situation" is compatible with—in fact an expression of—the general philosophical principles described in Section III.F. These ideas pervade the German edition of the popular book with Holst, from the foreword to the detailed examples of the BKS theory, in a treatment which is without much doubt the most understandable exposition of the BKS ideas. This inclusion of a brand new, still incomplete theory testifies to the importance Kramers attributed to the philosophy and the approach. Kramers was well aware of the limited and partial results BKS had so far achieved. He writes: "The new conceptions of the postulates, which requires the independence of the atom processes, is in no way a completed theory, it is only an attempt to throw a little light in the great darkness of our ignorance about atomic processes; for the time being it should be considered merely a working program for the theorists."[176,179] Even so, the ideology of BKS is evident all through the book. In his presentation, Kramers does not shy away from profound and controversial questions.† In connection with the probability character of the BKS theory, he raises (perhaps for the first time) the possibility that the laws of physics are *intrinsically* probability laws. This is different from their use in classical statistical mechanics. Kramers envisages the possibility that the strict law of causality does not hold and that there is no more fundamental level on which it would be reestablished. This is an explicit instance of Kramers' general attitude that the laws of nature are conventions useful for the ordering of phenomena, but a breakdown or limitation of such laws is neither a reason for despair nor surprise. Even so, it is still startling when Kramers intimates that: "For the moment it seems to me rather a matter of taste which alternative (causality or probability) is selected and perhaps this will remain so forever. The actual choice affects the methods of physics research far less than one would at first be inclined to believe."[176b] He stresses this same point again and again: "One may never forget that the principle of causality ... was never a logical necessity but rather a fact of experience ... one could easily imagine that it breaks down for atomic processes."[176b] All features of the BKS program fit in an unforced way into Kramers' philosophical framework. The sentence, "to get any further is impossible anyhow" (see p. 192), added in the German edition after an enumeration of the limitations of atomic theory, has an almost triumphant ring, as if to celebrate the official recognition of the limitations of physics.

G. Kramers' Personal Elaborations

Since the BKS ideas dominated Kramers' thoughts to such an extent, it is hardly surprising that all his scientific work would be closely related to that theory. Several different activities all spring from the same source: the develop-

† This itself is out of character for Kramers, who tended to avoid confrontations. Here he went out of his way to seek them out, most likely because he had great confidence in the theory.

ment of the BKS ideas. The dispersion relations, discussed in Section II, which preceded the BKS theory were extended in a study with Heisenberg. Because of their importance in connection with the development of matrix mechanics, they are discussed independently in Section IV. In this extension, the ideas of the dispersion relations were cast in the framework of the BKS theory, so that it is not easy to disentangle the dispersion theory results from the BKS formalism. Kramers certainly wished them to be organically connected. As described in the previous section, Kramers' revision of his popular book on the Bohr theory was strongly influenced by the BKS theory. But in addition there are a number of private manuscripts (unpublished), sketches of theoretical ideas, written in 1924–1925, which demonstrate how the BKS ideas permeated Kramers' thinking at that time. Among Kramers' papers is a typewritten draft of a note entitled "On the Possible Meaning of the Breakdown of Energy Conservation in Atomic Processes."[180] This is a comment on an earlier paper by Schrödinger[181] on the BKS theory. Although as explained earlier, Schrödinger was not altogether happy about the distinction between real and virtual radiation (see Section III.D), the general approach of BKS was completely in Schrödinger's line. He was especially excited by the violation of the conservation laws in radiative processes, as he writes to Bohr: "I long ago 'made friends' with the thought that the basis of our statistics is likely not microscopic regularity, but rather 'absolute chance' and that perhaps the energy and momentum conservation laws have only statistical validity."[182] Schrödinger was comfortable and pleased with the statistical conservation laws. Pursuing these ideas, Schrödinger in his paper[181] pointed out some unusual consequences of the BKS theory. Systems for which energy is only conserved statistically will inevitably exhibit fluctuations in the energy; call these energy fluctuations ΔE. Schrödinger showed that the average of the square of the fluctuations $\langle(\Delta E)^2\rangle$ is proportional to the time; for large enough time this will become important. Schrödinger draws some far reaching conclusions from that observation: "An *isolated* system exhibits a definite *average* behavior only for relatively short time periods. In the limit as time goes to infinity, its behavior becomes completely undetermined, since its energy content disperses according to a $\sqrt{t}$ law from its initial value. We can only depress the fluctuations if the size of the system is increased, or by considering the system as a subsystem of a larger system. (A heat bath.) The exact validity of thermodynamics can only be maintained for a system in a heat bath, and then only in the double limit (size $\to \infty, t \to \infty$)."[183] The result of the Schrödinger analysis shows that a strictly isolated system will not maintain a fixed energy "without connections with a continuously regulator (heat bath), it would, as far as energy is concerned, exhibit aimless fluctuations."[183]

Kramers arrives at much the same conclusions, but he stresses that for realistic systems, with the experimentally available duration of observations, the effects discovered by Schrödinger are totally negligible. Thus, the energy fluctuations, required by the BKS theory, do not lead to any conflict with experiment. Once again the BKS ideas, although perhaps unusual and unexpected, did not lead to a clear-cut disagreement with experiment. Shortly afterward (September 21–27, 1924), at a meeting of German scientists and doctors, Einstein, Kramers, and Pauli discussed the BKS theory at great length. Einstein, who more than any other physicist had obtained surprising and profound results from the analysis of fluctuations, discussed the energy fluctuations of the BKS theory with Kramers and Pauli. Einstein showed that the BKS theory would lead to systematic deviations from the first law of thermodynamics. For example, the total kinetic energy of a cavity filled with radiation could become arbitrarily large. In fact, the average of the fluctuations $\langle \Delta E \rangle$ [Schrödinger had calculated $\langle (\Delta E)^2 \rangle$] is not equal to zero as might naively be expected, but it is positive and proportional to the time. Consequently, the average energy keeps on growing and can become arbitrarily large. Similar results (the calculations are also quite similar) occur in the analysis of Brownian motion. However, in Brownian motion there is a dissipative mechanism (a friction force) that keeps the energy finite; there is no corresponding dissipative mechanism in the BKS theory, so the result that $\langle \Delta E \rangle$ increases with time must be faced. Kramers, in the second part of his manuscript, shows that for systems actually studied in the laboratory the effect discovered by Einstein is so small as to be unobservable. In fact, for all *terrestrial* experiments carried out over times realistically available, the effect would be totally negligible. Nevertheless, Kramers recognized that the BKS theory in the form as given would lead to the spontaneous increase of the average energy of a closed (isolated) system. He speculates that this energy increase is real and might be observable for large stars. Kramers adds immediately that further careful thought and detailed calculations are necessary to support this idea. In a postscript to a letter to Kramers, Pauli, who did not care for the BKS theory anyhow, admonishes Kramers to "leave the stars alone."[184] Kramers in no way considered the Einstein fluctuation objection as decisive; quite the contrary, he thought that the energy increase might well be the key to the understanding of the energy production by stars. Even if it turned out that this did not work and that such an increase in average energy contradicted experiments, Kramers still would not admit the end of the BKS theory. In discussion with Bohr, Kramers hinted that it might very well be possible to modify the laws regulating the relation between radiation and quantum jumps (the probability laws whereby the virtual field stimulates the real transitions) in such a way that the increase in average energy would be avoided. He had great confidence in the theory; his hopes that it would lead to new results in different areas were undiminished.

How high Kramers' expectations and hopes for the BKS ideas were can be inferred from a note Kramers wrote in 1945 upon finding a calculation he had made years ago (in 1925). There exists an unpublished manuscript written in February 1925, "On the Properties of Atoms in a Radiation Field,"[185] which Kramers evidently planned to published in *Naturwissenschaften.* This manuscript contains a summary of the BKS theory and a number of other proposals and suggestions. But as in the BKS papers, there are few formulas and almost no calculations. In the Kramers papers,[186] there is a small folder containing handwritten calculations, complementing the qualitative arguments in this manuscript. These very rough calculations are dated January 29, 1925. It appears that Kramers had forgotten that he had made these rough calculations. More interesting than the calculations themselves is Kramers' assessment of this work, 22 years later. On the cover of the file containing these notes he writes (in a mixture of three languages):

> This was the best idea
> that—I—in the old days—
> produced in Copenhagen.
> It just wasn't published
> and of course it was totally
> superceded by the later developments

This remark was written March 13, 1945, in a mixture of Dutch, Danish, and German.† The manuscript intended for *Naturwissenschaften* and the handwritten notes refer to the same subject matter: speculations on the thermal equilibrium between matter and radiation. This is an extension or application of the BKS ideas that specify the interaction between matter and radiation. The basic new idea is that if an atom is in a certain energy state, it can *under certain circumstances* behave as if it were in another energy state. Suppose an atom is capable of two states A and B, with A is being the ground state. Then according to Kramers if the atom is in state A, there is a finite probability k that it acts as if it were in state B. For example, if an electron is incident on this atom in state A, the ground state, there is a finite probability that in the collision the electron will pick up energy from this atom (for the atom can act as if it were in state B).

In somewhat more contemporary terms, Kramers appears to say that an atom in a radiation field is never in a single stationary state, but necessarily in a mixed state, a superposition of stationary states. It is clear that this idea is an extension of the BKS idea that the virtual field can induce emission or absorption, without energy conservation in the individual events. In this study, Kramers tries to calculate the statistical effects of such processes.

Kramers speculates further that the impossibility of defining an energy in a state introduced by this assumption will make it impossible to maintain the classical orbit notion. The further development of these ideas is extremely sketchy.‡ Kramers' notes have all the earmarks of a first attempt; there are changes, numerical errors, and personal comments: "How can I be so stupied," "this is not the way," "leave out these terms for now." About all that can be concluded from these papers is that Kramers tried to push the BKS ideology as far as possible. In these notes he insists that the energy balance between

† The translation is a literal one by M. Dresden.

‡ The notes are so rough that any exposition of Kramers' ideas is bound to be very incomplete. Since his ideas were in the process of formation, only the general flavor can be trusted.

atoms and radiation is of *statistical* character, in complete agreement with the BKS approach. That statistical character, Kramers seemed to feel, necessitated a statistical treatment of the type attempted. It is hard and probably impossible to reconstruct exactly what Kramers had in mind in these investigations. One could, with hindsight, see elements of the superposition of states or the linear combination of eigenstates in his basic idea. But the same hindsight offers no help in understanding the separation of optical and thermal properties which seems forced and artificial. But more important than these results is the recognition that Kramers' thinking was firmly rooted in the BKS framework. It was the basis for his published works, unpublished manuscripts, and private speculations. His attachment to those ideas was scientific, philosophic, and personal. Nothing could express Kramers' deep emotional attachment to the BKS ideas better than his nostalgic wistful reference to a speculative contrived elaboration of that theory 20 years after its irretrievable failure: "The best idea that I produced in the old days in Copenhagen."

H. The Reception of the Theory

It was inevitable that the BKS theory would create enormous excitement in physics. Bohr's reputation was at its zenith; his ideas pervaded and dominated atomic physics. Results emerging from Bohr's Institute usually signaled new directions in, or novel approaches to, atomic physics. They were eagerly awaited by the physics community. Furthermore, the BKS theory dealt with radiation theory, a topic that since the discovery of the Compton effect had become of paramount interest. Indeed, so intense was the interest in the conflicting approaches to the quantum theory of radiation that even the public at large showed a lively interest in the problems facing physics. Thus, Einstein wrote in a German newspaper: "The positive result of the Compton experiment proves that radiation behaves as if it consisted of discrete energy projectiles, not only in regard to energy transfer (energy conservation) but also in regard to momentum transfer (momentum conservation)."[187] To stress the unusual, even paradoxical situation produced by the Compton effect, Einstein writes in this same article: "There are therefore now two theories of light, both indispensable and—as one must admit today in spite of twenty years of tremendous effort on the part of theoretical physicists—without any logical connection."

Einstein had personally carried out a major part of this tremendous effort; that his comments were solicited by and published in a daily newspaper shows just how widespread the general interest was. It is therefore not surprising that a theoretical effort such as the BKS theory which attempted to formulate a single unified theory of light would be of immediate concern to the whole physics community. Finally, the daring character of the BKS theory, its unusual features such as the statistical conservation laws, and the absence of strict causality were so striking that they by themselves would send shock waves through the world of physics. Since the BKS theory was a program with, at best, an incomplete mathematical structure without decisive experiments which would confirm or reject the theory, the reactions to the BKS theory became a very personal matter. It remained so for a little more than a year (February 1924 until April 1925). Such individual responses naturally depend on the personal hopes and expectations of just what constitutes an acceptable theoretical scheme in physics and, although this ideally should not be the case, they also depend to a considerable extent on the status, reputation, and prestige of the authors of the theory. The combination of these factors guaranteed that the BKS theory would create an uproar; it would be scrutinized, analyzed, and discussed by practically all theoretical physicists. It might appear that because of its almost provocative character, the BKS theory would stimulate serious debates throughout the physics community. This is not untrue, but it is not exactly true either. Since the formulation of the theory was incomplete and the paper itself was difficult to understand, many physicists were reluctant to adopt a strong position either for or against the theory unless of course they had personal or philosophical reasons for doing so. Thus, Schrödinger liked the BKS theory because it formalized one of his cherished unspoken principles, the statistical nature of conservation laws. Sommerfeld, who had been so impressed with the Compton effect, was clearly uneasy with the unorthodox features of the theory and tried to maintain an open-minded neutrality. It became clear very soon that Einstein had a variety of serious objections against the BKS theory. In spite of this, Bohr and Einstein never discussed the BKS theory either privately or publicly, nor are there written records of such discussions. Kramers did discuss the BKS theory with Einstein, but the most substantive debates were between Pauli and Bohr and Pauli and Kramers. In the debates with Bohr, Pauli presented Einstein's objections as well as his own. In the discussions with Kramers, Pauli repeatedly emphasized that the BKS approach was on the wrong track altogether. Heisenberg took an intermediate position; he did not criticize the particulars of the BKS theory to the same extent that Pauli did. He was generally sympathetic to the BKS approach and accepted some (but not all) of its features and used them in his own highly personal manner to construct matrix mechanics. The ensuing polemics marked a turning point in physics. Bohr's ideas up to that time had always been a guide, perhaps the only guide in deciphering atomic structure and atomic spectra. In the discussions about the BKS theory, Pauli simply refused to follow Bohr along this road. Heisenberg was not so explicit,but he did not exactly follow Bohr's lead either. Thus, a definite separation resulted; it was not exactly a rift but it was unmistakingly a parting of the ways. The direction advocated by Pauli led to the exclusion principle, Heisenberg's approach led to matrix mechanics, and the BKS theory as formulated led nowhere. These developments, which took only about a year, put Kramers in a difficult and ultimately detrimental position. He totally believed the BKS theory: He spent a great deal of time thinking about it and therefore did not contemplate related, alternate possibilities (as Heisenberg did). Thus, after the demise of the BKS theory he needed some time; but more importantly, he needed energy and motivation to reorient his thinking. Because of his close relation to Bohr, Kramers would certainly side with Bohr in the conflict with Einstein.

As in all human relations, imponderable elements played an important role. Pauli and later Heisenberg came to the Bohr Institute at a time when Kramers was already an accomplished physicist and the acknowledged spokesman of Bohr. In 1924–1925, at the time of the BKS debate, Pauli and Heisenberg both achieved scientific autonomy, which certainly recognized Bohr's eminence but which nevertheless enabled them to differ sharply from Bohr and go their own way. Kramers did not desire this level of independence from Bohr. This separation in turn caused subtle changes in the personal relations between Kramers on the one hand and Pauli and Heisenberg on the other. Pauli especially made numerous derisive remarks about Kramers' approach to physics. That Pauli made critical sneering comments about physicists is of course nothing new, but up until the time of the BKS theory his criticism of Kramers had been somewhat muted. At times, Pauli even showed respect for Kramers' physics; he occasionally showed some slight if irreverent deference. All this changed perceptibly with the advent of the BKS controversy. Pauli's critique of Kramers' physics became sharper and the irreverence turned into sarcasm. To get a clear understanding of these subtle matters, it is best to analyze these changing relations after the reactions to the BKS theory and its experimental failure have been described (see Section IV).

The scientific ferment produced by the BKS theory is vividly expressed by Born in a letter to Bohr. Born had heard about the BKS theory from Heisenberg who had visited Copenhagen in March 1924. Born writes: "I would like to tell you, how happy I am about the new orientation you have given the problem of radiation ... naturally we have had similar thoughts here, but it was you who introduced a decisive new point ... the association of two oscillators [one for emission and one for absorption] ... with this association I am almost convinced all contradictions disappear. Transitions are instantaneous and absorption and emission occur in the stationary states. I am totally convinced that your new theory is essentially correct and also that in a certain sense it is the last word that can be said about these questions, it is the rational extension of the classical radiation phenomena to the discontinuous elementary processes."[188] This letter reads more like a personal glowing tribute than a considered judgment of a new, radical, and speculative proposal. Born does not comment at all on the breakdown of either the principle of causality or the conservation laws. It is interesting that in spite of his enthusiasm Born made little use of the BKS theory. It is likely that the probability ideas in the BKS theory predisposed Born to his later probability interpretations of wave mechanics, but the connection is, of course, not very direct. It is more than likely that Born was strongly influenced by Bohr's renown. The last sentence in the letter, that the BKS theory is "the rational extension of the classical radiation phenomena," is actually a *code* often used by the Copenhagen school to denote the future, more complete theory, incorporating quantum and classical ideas in a coherent scheme. It is evident that Born was preoccupied with the search for such a theory and when writing the letter to Bohr he must have believed (as of course did Kramers) that the BKS scheme provided a promising framework for that future theory. That might account for Born's uncritical acceptance of the BKS ideas. Born later was reluctant to accept the failure of the BKS theory. However, only a few months after the demise of the BKS theory, Born recognized the basis for such a new comprehensive scheme in Heisenberg's matrix mechanics. He became an enthusiastic supporter of that theory and very soon made fundamental contributions to matrix mechanics.

Heisenberg, who informed Born about the BKS theory, was initially quite skeptical. In a postcard to Pauli he writes: "Bohr's paper on radiation is very interesting, but I don't see that it represents a fundamental advance."[189] Especially through his association with Kramers, Heisenberg became more tolerant of the BKS theory. With the incisive insight to focus on essentials so characteristic of extraordinary genius, Heisenberg used just the formal virtual oscillator picture of the BKS theory to develop matrix mechanics and ruthlessly discarded everything else. He did not get embroiled in philosophical arguments about the scope and purposes of physical theories; but felt that experiments would decide between competing schemes. He states this explicitly in a letter to Sommerfeld: "For the rest I believe more and more that the question 'photons or correspondence principle' is a question of semantics. All effects in quantum theory must after all have a classical counterpart, for the classical theory is almost correct; thus all effects must have two names, a classical and a quantum [name]. Which one prefers is really a matter of taste. Perhaps the Bohr radiation theory is a very happy description of this dualism; I am anxiously awaiting the results of the Bothe–Geiger experiment"[190] Heisenberg was certainly influenced by the BKS theory, but he never became a partisan and he adapted selected pieces of the theory to his own purposes.

Several other physicists, such as Ladenburg and Becker, took the theory at face value. Becker used the BKS ideas pretty much literally to obtain a *unified* (this was the basic point) description of absorption and dispersion. Ladenburg, who after all had an early version of the virtual oscillators, wrote to Kramers expressing his interest in the BKS theory; he planned to use the theory for his further studies in dispersion. There were other such reactions but none of them led to any important consequences. Sommerfeld's reaction was quite different. He was in the process of preparing the fourth edition of his celebrated treatise *Atombau and Spektrallinien*. As a brilliant expositor, he was naturally anxious to give an organized, systematic treatment of radiation theory. But accepting the photon, which Sommerfeld tended to do on the strength of the Compton experiment, he was, as Einstein before him, compelled to say that there were two logically independent, even contradictory, theories of light. Sommerfeld was uneasy about the BKS theory but he was not willing to assert publicly that Bohr was wrong. Sommerfeld was one of the invited speakers at the aforementioned Meeting of German Scientists and Physicians in Innsbrück (September 21–27, 1924). It was at that same meeting that Kramers and Einstein discussed the fluctuation problem. Sommerfeld's topic was the foundations of quantum theory so there was no way in which he could avoid a discussion of the BKS ideas. But he was unsure and apprehensive about how that should be done so he asked Kramers to help him with the presentation of the BKS ideas. One of the reasons for Sommerfeld's discomfort was no doubt that he did not wish to slight or offend Bohr in any way. In the letter Sommerfeld wrote to Kramers there occurs this remarkable passage: "In Innsbrück, I would like to contradict Bohr as little as is possible, without compromising my own ideas."[191] Nothing could illustrate the intricate interplay between personalities more clearly than this request, where Sommerfeld asks Kramers' help to avoid the appearance of insulting Bohr. Kramers must have considered Sommerfeld's request as a golden opportunity to present, if only by proxy, the BKS ideas to a very distinguished and influential audience. Consequently, in his answer to Sommerfeld, Kramers gives a careful description of the methods and procedures of BKS. He stresses that one of Bohr's main goals is to develop viewpoints that will allow a *unified* treatment of all the phenomena. He reiterates that the correspondence principle is the basis of all arguments in quantum theory and that it, or a natural extension of it, must be an integral part of the theory of the future. Kramers' valiant effort to convince Sommerfeld ends with a flourish: "It is just an attempt to represent the different observations as simple and as correct as possible; it is of special importance to us that this is possible without attributing an enigmatic dualistic nature to radiation [waves and quanta]; the radiation is described in a continuous manner as all optical experiments require, while the changes in the atom–again in harmony with our experience and experiments—are discontinuous. That we thereby have to renounce the strict validity of the conservation of energy is merely the expression of the desire not to make assumptions about the phenomena which are not directly based on experiments."[192] This is a most persuasive, almost seductive, formulation of the BKS philosophy. Kramers is quite correct in stressing that at that time, and indeed up until the time of the Compton–Simon experiment in 1925, there was no evidence that energy and momentum were conserved in the individual processes. In spite of Kramers' beautiful letter, with its impeccable logic, Sommerfeld remained unconvinced that the BKS framework would provide the desired unifying viewpoint. For a person like Sommerfeld who liked specific detailed models, and who was comfortable with well-defined mathematical structures, a virtual field that determines the probability for atomic transition must have been difficult to accept. In fact, he did not accept it, although he did not reject it either. In the Innsbrück lecture, Sommerfeld referred to the BKS theory as a compromise which allowed the simultaneous use of wave and quantum concepts, but only by giving up the possibility of a single unified viewpoint. He ended his lecture by expressing the hope that experiments would soon decide whether the BKS proposal was possible or not.

The reception of the BKS theory at the Innsbrück Conference was mixed. Many persons like Sommerfeld were cautious but unsure, others had great sympathy, but Einstein and Pauli were totally opposed. Einstein, on May 28, 1924, gave a colloquium in Berlin on the BKS theory. Although he did not totally reject the theory, he had both specific and general objections. In his letters Einstein left no doubt† that he found the BKS ideology objectionable.[193,194] Einstein had considered similar ideas much earlier and even had a name for the BKS virtual field; he called it the "ghost field" (as reported by E. Wigner[112–114]). However, Einstein abandoned this approach after he realized that it would be impossible to maintain the conservation laws in individual processes. This was a step Einstein was not prepared to take. Understandably, he objected to a theory which did exactly what he (Einstein) was unwilling to do. Furthermore, all experiences with quantum processes seemed to indicate to Einstein that the conservation laws were rigorously valid. Another objection of Einstein to the BKS theory (as reported in a letter from Pauli to Bohr[195]) concerns the photon, the issue which started the whole controversy. Einstein felt that the distinction madé by BKS between conservative and nonconservative forces was artificial and unnecessary; he strongly felt that energy and momentum conservation should make sense for *all* processes. He therefore concluded that the photon as a carrier of momentum and energy is a fundamental object. In fact, he attributed a greater reality to the photon than to the wave field. (It is hard to imagine a greater heresy to Bohr.) Since the frequency of the emitted light is different from the orbital frequency of electrons (this was Bohr's great discovery), Einstein felt that the production of light with its wave properties was a secondary effect. But Einstein's main objection was the almost casual abandonment of the principle of causality. Einstein had struggled with this issue for some time (and he continued to struggle with it later); he had thought long and hard about it. To abandon it for a number of speculations of vague and doubtful character was just impossible. It is interesting that Ladenburg who attended Einstein's colloquium on the BKS theory seemed to feel that Einstein's reaction to the BKS theory was not all that negative. It must be remembered that Ladenburg was very partial to the BKS theory; he no doubt mainly heard what he wished to hear, for there can be little doubt that Einstein (as written in his letters at that time) objected strongly to the BKS theory. Nevertheless, Ladenburg writes to Kramers 2 weeks after Einstein's colloquium: "His [Einstein's] opinion was decidedly not negative; he asserted that the new view was complete and consistent and not in any direct conflict with experiment.... He did stress the conceptual logical difficulties of the theory"[196]

Many of Einstein's objections to the BKS theory were reflections of his personal philosophy of physics. The frequently used phrases—"Einstein feels,"

† A detailed, incisive analysis of Einstein's objections to the BKS theory can be found in Martin Klein's discussion.[102] Here only a few of Einstein's comments, pertinent to Kramers' role, are discussed.

"I expect," "Einstein believes—all indicate personal preferences and preconceptions about the nature of physical theories. For that reason, Kramers was not particularly upset about Einstein's criticisms; he discussed them in a rather cavalier fashion as "emotional arguments" (feelings).† Kramers answers Ladenburg: "It was very interesting to hear Einstein's considerations; as he says himself they are all just feelings [emotional arguments]." [197] Whether it was wise to dismiss Einstein's arguments is an open question, but there is no doubt that the emotional attachment of Einstein to the principle of causality played an important role in Einstein's rejection of the BKS theory; he writes in a letter to Hedwig Born: "But I shouldn't let myself be pushed into renouncing strict causality before it had been defended altogether differently from anything done up to now. The idea that an electron ejected by a light ray can choose of *its own free will* the moment and direction in which it will fly off is intolerable to me. If it comes to that, I would rather be a shoemaker or even an employee in a gambling casino than a physicist" [194] This passage certainly shows a deep emotional involvement.

Kramers was not wrong in characterizing Einstein's objections against the theory as emotional arguments, but since Kramers had an equally strong emotional attachment in favor of the theory, such exchanges led nowhere. The only substantive Einstein objection—the increasing average energy of an isolated system—Kramers turned around as a possible argument in favor of the theory by relating this energy increase to the energy production in stars. Einstein's arguments against the BKS theory certainly deserved to be taken seriously, but although cogent and perceptive, Einstein's arguments were not decisive and the standoff between Kramers and Einstein remained.

At the time of the Innsbrück meeting (September 1924), Pauli was strongly opposed to the BKS theory. His attitude toward the theory underwent several changes. He was never particularly enthralled by the virtual fields but for a while he was willing to consider the ideas involved. It is clear that his admiration and deference for Bohr conflicted with his strongly developed physical sense. Pauli's ambivalence and his vacillating attitude are summarized in a letter he wrote to Bohr: "We discussed at that time [during Pauli's earlier visit to Copenhagen] many problems in physics, especially the ideas on radiation theory as contained in the paper by you, Kramers, Slater. At that time you, by your various arguments, managed to quiet my scientific conscience, which strongly objected to the ideas. This lasted only for a short time and as Kramers will have told you already I now—as a physicist—am totally opposed to the basic notions which underlay this approach to radiation theory. I am encouraged in this viewpoint, since many other physicist, perhaps most, also reject the approach, although their reasons differ in part from mine." [198] This is an unequivocal beginning; it reads a little like a declaration of independence. Pauli then proceeds to inform Bohr about Einstein's arguments; evidently Bohr was very anxious to hear Einstein's objections even if Kramers was not.

† The German term used is *Gefühls argumente.*

Although it was evident that Pauli had his own emotional objections against the theory, he, like everyone else, was anxiously awaiting the outcome of the experimental tests of the BKS theory, which Geiger started about that time. Somewhat surprisingly, Pauli writes in this same letter that he does not much care what the outcome will be. This is surprising because only a little later Pauli, in a letter to Sommerfeld[199] gives a most sweeping condemnation of the complete BKS approach, whatever the outcome of the Geiger experiment might be. Sommerfeld never overcame his worries about the photon–wave duality of light. In the fourth edition of his book he was quite open about it: "Modern physics is confronted here with two incompatible features and must frankly confess non lignet."[†200] Pauli in his letter to Sommerfeld strongly agreed with Sommerfeld; he goes further than Sommerfeld when he writes: "Your frank 'non lignet' is a thousand times more sympathetic to me, than the ad-hoc, artificial pseudo solution of Bohr, Kramers, Slater, even if the experiment of Bothe and Geiger would turn out to support that theory."[199] This is an unequivocal statement; it captures Pauli's annoyance and irritation with the BKS approach. It is important that in his communications with Bohr, Pauli expressed himself in a more controlled manner. He certainly made no secret of his opposition to the BKS ideas, but he is somewhat guarded in his criticism, and at times it is even possible to discern glimpses of his deference for Bohr. In a letter dated October 1924, Pauli writes to Bohr: "You see even if it were psychologically possible for me to base my scientific opinions on some kind of belief in authority (which however as you know is not the case) this would [in this case] be logically impossible, since the opinions of the two authorities, are in such conflict." It is a compliment for anyone, even Bohr, to be compared with Einstein, and Pauli undoubtedly intended it as such. Pauli's ambivalent reaction to Bohr is especially clear in the way he ends his thoughtful, serious, professional, and critical letter to Bohr: "Now I just hope that things go well with you and that you soon will solve the problems of atoms with more than one electron. I know further that it only pleases you when I express my opinions about questions in physics, honestly and warmly—even when you consider them quite stupid."[198] This sounds more like an adolescent boy who is at once proud and contrite at his own audacity, contradicting his father so forcefully, than an exchange between coequal scientists. There was nothing guarded, deferential, or contrite in Pauli's communications with Kramers about the BKS theory. In letters he was brutally frank, and in discussions he employed his formidable critical powers to point out weaknesses in the theory. It appears that Kramers, although obviously not pleased, was also not particularly upset. Everyone, including Pauli and Einstein, agreed that the BKS theory was a *possible* theory; they just did not care for the general approach. Kramers could and did take solace from one of Pauli's observations: "Of course all arguments are 'emotional arguments', one can not prove anything

† This Latin expression is a fancy but colloquial way of saying that the situation is unclear, or not evident, or unresolved. In legal parlance the expression means "not settled."

logically and the existing data are not sufficient to decide for or against the (BKS) ideas."[198] In spite of the ferocious criticism of Pauli, Kramers remained confident in the BKS approach.

The different expectations between Einstein and Bohr, well known in the world of physics, eventually reached the public press. On October 25, 1924, the Danish newspaper *Politiken* carried a story on the conflict between Einstein and Bohr. The story was not all that clear but it had something to do with the nature of light. An editor (K. Joel) of the German newspaper *Vossiche Zeitung* in Berlin wrote to Einstein asking for clarification, information, and Einstein's opinion.[201] Einstein, in a rather curt note, acknowledged that there was a difference of opinion between him and Bohr but offered no further explanation.[202] The original story in *Politiken* was far from transparent: Joel's reporting did not improve its clarity, and Einstein was not about to enter into lengthy polemics in the public press. Similar stories were published in a number of European newspapers. One of these was published in the Dutch newspaper *De Telegraaf* in October 1924. Evidently, the newspaper received a telegram from the Bohr Institute about the BKS theory and the impending struggle between Einstein and Bohr. The report in the Dutch newspaper presumably based on this telegram was just as unclear as the earlier reports. This article came to the attention of Jan Romeyn, Kramers' lifelong friend. Romeyn was very interested, in part because he was interested in anything that had to do with Kramers and Bohr, in part because, as explained earlier, the BKS philosophy seemed in close harmony with his political philosophy. Romeyn was a very intelligent person and although he was not a physicist he had a remarkable grasp of scientific arguments. But he was totally confused by the newspaper article. He asks Kramers for enlightenment: "A curious newspaper article came to my attention about a new theory of light of Bohr (and you?), which I really can't make heads or tails of."[203] Kramers did not answer this letter directly. However, he did write a letter to the newspaper *De Telegraaf*, in which he requested, insisted, that a telegram published on the BKS theory by that newspaper be withdrawn.[204] This letter was rather forceful; it is clear that Kramers felt embarrassed and somewhat irritated by the misrepresentation of his cherished ideas in an irresponsible and superficial manner. There is no evidence that the newspaper ever published such a retraction. Likely the issue was not important enough. But for Kramers and the world of physics, the validation or rejection of the BKS ideas was of overriding importance. The outcome of the experiments were anxiously awaited by everyone.

I. The Crash

The explanation of the Compton effect according to the BKS theory was described in detail in Section III.E. It was emphasized there that this explanation assigned different mechanisms to the causes for the emission of radiation and the momentum transferred to the electron. The Compton effect was considered as the result of two unrelated processes. The momentum transfer to the electron was a discontinuous quantum event; just its probability was determined. The radiation, by contrast, was emitted by the moving virtual oscillators; it was a purely classical effect. The two processes were unrelated. Consequently, the BKS theory allowed for a time delay between the emission of the scattered radiation and the recoil of the electron. Furthermore, there need not be, and in general would not be, a one to one matchup between the scattered radiation and the recoiling electron. Such a coincidence would, in the BKS picture, occur only as a pure chance event. This remark was the basis for the experimental check of the BKS theory by Bothe and Geiger. (The suggestion for such a check was made shortly after the BKS theory appeared.[205]) In principle, the experiment is simple enough to perform: One has to set up a counter for light waves and a second counter for the scattered electron beam, and then one has to measure the coincidences between the two counters. In practice, Bothe and Geiger had to develop the coincidence technique from scratch.† This was not easy; it took close to a year before they had refined their methods sufficiently to yield reliable results. Bothe and Geiger were well aware of the importance of their experiment and they took great pains to be as careful as possible. They had understandable troubles with the efficiency and stability of their counters, but after about a year they obtained significant reproducible results. On April 18, 1925, they reported a preliminary result;[206] a week later they submitted a detailed study.[207] (Rumors about their experiments made the rounds as early as January 1925, but Bothe and Geiger did not wish to publish anything until they were quite sure.) The statement of their results reflects the care they took in this experiment: "Under ideal experimental conditions, according to Debye and Compton, with each (*hν*) count [in counter 2], there should simultaneously occur an (*e*) count [in counter 1]. In reality, due to the unavoidable imperfect experimental conditions, one could expect only one coincidence for every 10 *hν* counts, as extra experiments actually showed. On the other hand, coincidences should practically be excluded, according to Bohr's views (the BKS) theory. The experiments yielded one coincidence per 11 counts approximately, after deducting the counts expected purely accidentally.... Our experiments therefore, decide in favor of the older conception [i.e. the light quantum theory]."[206] The experiments showed that the light quantum and the recoiling electron were produced within a time interval smaller than 10^{-3} s. The statistical independence of the two processes, so essential for the BKS approach, was decisively contradicted by the correlation found in the experiments. As an aside it is amusing to note that some 10 years later, Shankland[208] reported an experiment (of the Compton–Simon type) that gave rise to a brief resurgence of the BKS possibility, only to be refuted again by a better experiment of the Bothe–Geiger class by Hofstadter

† It can legitimately be claimed that the development of the coincidence counter techniques is one of the most important consequences of the BKS attempt.

and McIntyre.[209] They established simultaneity up until a time interval less than 1.5×10^{-8} s. Still more recent experiments (1955) put limits on the time interval of less than 10^{-11} s.[210] The results of Bothe and Geiger were thus the first results (to be confirmed later with greater and greater accuracy) showing that the BKS theory was untenable.

About the same time that Bothe and Geiger had started their coincidence experiments, Simon and Compton initiated a series of cloud chamber experiments designed to test another feature of the BKS theory. They developed techniques enabling them to distinguish two different types of track; one type was a recoil electron track, the other the track of a photon electron (which allowed the identification of the Compton scattered quantum). In their early communication (November 28, 1924) they reported that on the average one recoil electron is produced for every scattered quantum. This is in complete harmony with the Bothe–Geiger result and again in stark conflict with the BKS theory. In a later study, Compton and Simon[211] verified the numerical relation between the scattering angle θ and the electron recoil angle ϕ:

$$\tan\phi = \frac{1}{1 + (\lambda_C/\lambda)\tan(\theta/2)}$$

It is important that a pure wave theory can never lead to a unique relation of this type. The measurements of Compton and Simon thus excluded the possibility that a purely classical wave phenomenon, even one produced by virtual oscillators, would be the mechanism responsible for the scattered waves. Yet BKS explicitly assumed just such a mechanism. Since the relation between θ and ϕ is a direct consequence of the exact conservation laws, it follows that by verifying the relation between ϕ and θ, Compton and Simon also demonstrated the strict conservation laws in the individual scattering events. According to BKS, the conservation laws had only statistical validity, so no precise relation between θ and ϕ could hold. With the Compton–Simon and Bothe–Geiger experiments, the basic ideas of the BKS theory were all shown to be in direct conflict with observation. The experimental verdict was unequivocal: The BKS theory was finished.

The reactions to the demise of the BKS theory were varied. There was a mixture of relief, disappointment, even occasional disbelief. But as intense as the reaction was, it did not last very long. By June 1925, the time that Compton and Simon announced their final decisive results, Heisenberg's first paper on matrix mechanics was just about finished. As matrix mechanics and wave mechanics, some 6 months later, came onto the scene, physicists turned their attention from speculations about photons and waves to the implementation of the new formalism. So rapidly did physics evolve that when Compton reported on his effect at the Fifth Solvay Conference in Paris (October 24–29, 1927) and noted that the observations excluded the Bohr–Kramers–Slater theory, it seemed to many of the participants that Compton was merely rehashing ancient history. Through the work of Pauli, Heisenberg, Schrödinger, and Dirac, physics had passed well beyond the concerns of the BKS considerations. Although the physical problems remained, the BKS theory itself was no longer the center of interest.

It would be wrong to say that the BKS theory did not exert a lasting influence on physics; many of its ideas—and some of its programmatic points—remained of importance well after the details of the theory were forgotten. The comet-like appearance and disappearance of the theory also influenced several individual physicists. There can be little doubt that Kramers suffered the most profound and lasting effects, but many others, in their efforts to take a position, were strongly affected as well. Born's ambivalence and vacillations relative to the BKS fortunes are noteworthy, especially in view of Born's later relation to Kramers (Section IV.C). It was noted earlier that Born's first reaction to the BKS theory was one of unqualified enthusiasm. As late as January 1925, Born was still a strong proponent of the BKS theory. This in spite of the fact that by that time the preliminary results of Bothe and Geiger strongly suggested that the theory could not be maintained. Born wrote to Bohr, saying that everyone in Berlin believed that the issue was settled in favor of photons: "Einstein Triumphs; I however am in no way concerned."[212] Born proposes an alternate interpretation of the observed coincidences by stipulating that the primary x rays consisted of short pulses followed by long pauses. Thus, the electron could only be emitted during the period of the emission of a pulse. Bohr's reaction was guarded; he professed great interest in Born's suggestion (just what Bohr actually meant by "great interest" was not always clear), but he insisted that the experiments had to be completed and analyzed before any conclusions should be drawn. Evidently, Bohr was immediately convinced that the Bothe–Geiger experiments spelled the end of the BKS theory. He received advance information of the Bothe–Geiger results on April 17, 1925. Four days later he answers: "I was completely prepared [for the news] that our proposed point of view on the independence of the quantum processes in separated atoms should turn out to be incorrect."[213] On April 25 Bohr adds a postscript to a long letter to Fowler (written earlier): "Just in this moment I have received a letter from Geiger, in which he tells that his experiment has given strong evidence for the existence of coupling in the case of the Compton effect. It seems therefore that there is nothing else to do than to give our revolutionary efforts as honorable a funeral as possible."[214] Bohr repeats much the same sentiments in a long letter to James Franck: "There is nothing much else to do except to forget our attempt at revolution in as painless a way as possible. Our goals we will not be able to forget so easily, and in the last few days, I have been suffering through many wild speculations, to find an adequate foundation for the description of the radiation phenomena. I discussed this at length, with Pauli, who is here now, he for a long time was unsympathetic toward our 'pocket revolution'."[215] It is evident that, even though Bohr is not exactly clear what to do next, he has given up on the BKS theory and is groping for the proper next step. The reactions of Franck and Born to Bohr's relinquishing the theory were strikingly different. Franck's answer is respectful, generous, and good natured: "Your pocket revolution was really a grandiose effort, which was unusually stimulating. Even if you now hold another opinion, this episode still remains of great importance for all ideas concerned with atomic structure. Personally however I am pleased, that we again are allowed to believe in the conservation of energy, otherwise even the possibility of a description seemed to be absent."[216] By contrast, Born who had previously shown such uncritical enthusiasm for the BKS ideas reacted in a more intense, personal, even disingenuous manner. Born writes Bohr the same day as does Franck, April 24, 1925: "Today Franck showed me your letter, which interested me tremendously. I was shocked, even dumfounded, that you have given up the radiation theory without conservation laws. For about 8 days, I had started a letter to you, in which I wrote that I myself through much 'intellectual rummaging' (for which I had enough time during my Swiss trip), came to the conclusion that this theory was impossible. At the same time I wanted to inform you about a sketch for a new theory which Jordan and I worked at for some weeks. But I didn't have the courage to send you the letter, since I had the impression this theory was a 'pet idea', a very favorite idea of yours, and I hesitated to contradict you except on the most compelling grounds. But now the situation is different, I am happy that I will not have to fight with you. I also would not have dared to attack your ideas, if I could not have made a positive suggestion for a new theory. Now I believe I can do that."[217] This is an unusual letter; it certainly demonstrates Born's respect for, and even subservience to, Bohr's ideas on atomic physics. It is after all remarkable that Born would not have the courage to suggest that Bohr give up ideas which Bohr himself was quite willing to renounce. It is further curious that Born states that he derived the courage to attack Bohr's ideas only from the study he had undertaken with Jordan which would be a possible replacement for the BKS theory. Born had been and remained preoccupied with a systematic replacement of the Bohr atomic theory, but Born did not succeed in accomplishing that until after Heisenberg's fundamental advances. It is hard to see how Born's papers with Jordan could be considered as a serious alternative to the BKS theory. However, Born evidently felt that some justification was necessary to criticize Bohr's ideas, even in private.

Bohr adjusted very rapidly and with very good grace, to the failure of the BKS theory. He pointed out repeatedly, and quite forcefully, that although the BKS attempt had failed, the problems which stimulated the attempt in the first place remained unsolved. If anything, these problems were exacerbated by a number of new phenomena, such as the Ramsauer effect—the anomalously small scattering cross section of slow electrons in inert gases. It is possible that Bohr had some doubts about the BKS theory even before the Bothe–Geiger experiments. This would account for his extraordinary rapid acceptance of these results. Also, he does refer to Einstein's earlier thermodynamic objections as disturbing—a concern that Kramers never acknowledged. Bohr's main conclusion from the failure of the BKS theory was that a profound, radical revision of all the physical concepts had now become inevitable. Only by a complete renunciation of the usual space–time methods of visualization of the physical phenomena would further progress become possible. The impossibility of maintaining the usual causal space–time description was a recurring theme in Bohr's thinking; it eventually culminated in the *complementarity formulation* of quantum mechanics.[218] Thus, for Bohr, the failure of the BKS theory was merely a stage in his gradual intellectual emancipation from classical to quantum physics. To Slater, the demise of the BKS theory was not a major matter. He reacted more like a person vindicated by the turn of events than one who has been hurt in any way. Right after the Bothe–Geiger–experiments, he suggested that radiation theory could and should be saved by a return to his original idea: "The simplest solution of the radiation problem then seems to be a return to the view of a virtual field to guide corpuscular quanta."[219] There were some later exchanges between Bohr and Slater. Bohr, after discussing the radiation theory with Einstein in Leiden, writes to Slater: "Although of course we were wrong in Copenhagen as regards the question of the coupling of the quantum processes, in which respect I have a bad conscience in persuading you to our view—I believe that Einstein agrees with us in general ideas and that especially he has given up any hope of proving the correctness of the light quantum theory by establishing contradictions with the wave theory description of optical phenomena."[220] Slater's answer was similarly civil and controlled: "As far as radiation is concerned, you need not have a bad conscience for having persuaded me to think that there were no quanta. I think we did a useful service by coming on with the definite suggestions there were none; for that called people's attention to the fact that there were then no experiments which could not be perfectly well explained without assuming them. I rather imagine it may have helped in inducing Bothe and Geiger to perform their important experiments, and if so that was certainly worth while."[221] As mentioned in Section III.B, Slater might have suppressed stronger feelings when he wrote this civil letter to Bohr: Even though the BKS theory was an incident in Slater's scientific life, it did not appear to influence his later scientific style to any appreciable extent. For Bohr, the BKS episode was a necessary interlude in his development, which so sharpened the conflicts between the classical and quantum world views that he became completely convinced of the necessity of the profound alternations in the character of theoretical physics. In spite of the troubles that the BKS failure caused Bohr in his attempts to find a rational approach to atomic physics, it was part of an organic growth; he eventually arrived at a harmonious structure which incorporated all the basic elements of his philosophy.

For Kramers, it was different. The effects of the BKS debacle, although indirect, were more pronounced, longer lasting, more pervasive, and ultimately deleterious. There were a number of separate factors conspiring to make the failure of the BKS theory especially painful for Kramers. As described so extensively in Section III.F, the theory fit in beautifully with Kramers' emerging philosophy and general expectations. His most striking contributions to physics (discussed in Section II)—the dispersion relations—were, in

Kramers' thinking at that time, totally intertwined with the BKS ideology. With such an attitude and such results, Kramers would naturally be reluctant to abandon thoroughly congenial ideas which also had produced his most spectacular successes. Complicating a touchy situation even further was the coincidence that Heisenberg's matrix mechanics, firmly based on a joint Kramers–Heisenberg paper, was conceived at about the same time that Kramers tried to accommodate his thinking to the failure of the BKS theory. Thus, Kramers had to come to terms with the recognition that the BKS theory was false, while a joint effort with Heisenberg based on closely related ideas pointed the way to a correct theory (Section IV). So it is not surprising that Kramers' immediate reactions were guarded and noncommittal. He writes to Born on May 13, 1925, about 3 weeks after the Bothe–Geiger results had become public: "I have recently discussed the question of the coupling of the fundamental processes in the radiation theory at some length with Bohr. I am convinced that a systematic coupling of the individual processes is incompatible with wave optics, however we should attempt to explore the contradiction in a less naive fashion than Einstein often does. Precise experiments on the coupling are certainly of the greatest interest. Unfortunately I can not survey how convincing the Geiger–Bothe experiment is, for the case of the Compton effect."[222] It is clear that, at that time, Kramers saw no way in which photons could be a part of a theory in which wave optics retained its customary form. Kramers appears to have no doubts that the experiments eliminated the statistical conservation laws, but he was not willing to grant anymore than that. He certainly did not believe that the existence of a photon had been demonstrated. Even after the great matrix mechanics breakthrough by Heisenberg, Kramers remained doubtful about the role of the photon. He writes to Fowler: "Although the new quantum mechanics certainly mean[s] a most important progress, it is of course not yet the complete solution of all quantum problems. It does for instance not touch on the problem of light quanta and the problem of the time at which the transitions take place."[223] This letter shows that even 8 months after the Bothe–Geiger experiment Kramers was still struggling with the photon concept. In a letter to Born (dated May 15, 1925[222]) Kramers thanks Born for the invitation to visit Gottingen; he is not anxious to give a talk in Gottingen because "All the essentials done in Copenhagen, Heisenberg knows at least as well as I." Kramers describes Born's invitation as *Grossmutige Wohltatigkeit* ("generous charity"). This indicates that Kramers did not wish to talk about the BKS theory at that time. The joint work on dispersion theory with Heisenberg to which Kramers refers was finished in December 1924; that material Heisenberg knew well. Since December 1924, Kramers had single-mindedly pursued the BKS theory; that work Heisenberg did not know, but Kramers evidently felt it would be inadvisable to discuss it after the Bothe–Geiger experiments and after Bohr had publicly rescinded the theory.

Since Kramers did not wish to speak on what concerned him most, nor about work done earlier with Heisenberg, he did not give any lectures at all during his Göttingen visit. Actually, this was a curious visit. Kramers describes it in a letter to Urey; he mentions pleasant social occasions, playing the cello, meeting Born and Franck—but he says nothing about discussions with Heisenberg.[224] Born comments on Kramers' visit in a letter to Einstein; he describes Kramers as an "extremely gifted and likeable person." Born's wife writes Kramers a postcard expressing her pleasure at his visit; she says: "It isn't all that often that physicists are worth while human beings." There is no reason to doubt that Kramers had a good time in Göttingen: He always enjoyed social occasions in cultural and intellectual surroundings. The reports of both Max Born and Hedwig Born all show that the visit to Göttingen was certainly a personal and social success. But there can be no doubt either that his disappointment about the failure of the BKS theory must have weighed heavily on him. In his letter to Urey he describes the disproof of the BKS theory as an accomplished fact. The description is precise, careful, and detached as behooves an objective physicist. But the language is brittle, in marked contrast to Kramers' earlier enthusiastic description of the BKS theory only a few months earlier.

Perhaps the most compelling evidence about Kramers' state of mind during and after his visit to Göttingen (June 8–20, 1925) is the total silence Kramers maintained about his meeting and discussions with Heisenberg. There is no doubt that Kramers and Heisenberg met and discussed Heisenberg's paper on matrix mechanics at that time. In a postcard to Pauli,[225] Heisenberg refers to these discussions; he even comments that Kramers accuses him of "optimism." About 2 months later, in a letter to Bohr, Heisenberg refers again to his discussion with Kramers on quantum mechanics: "In the last month I have actually not thought about physics, and I don't know whether I understand anything about it. Earlier I had, as Kramers has perhaps told you, produced a paper on quantum mechanics; I would like to hear what you think of it."[226] But Kramers in his letters and discussions mentioned neither the meeting nor Heisenberg's paper for at least 2 months. In the open, congenial atmosphere of Copenhagen where the discussions, seminars, and informal talks were the rule—where minor ideas and minute details were routinely discussed in agonizing detail—it is impossible to believe that Kramers would have just forgotten to mention Heisenberg's paper and results or that they would merely have slipped his mind. Heisenberg's methods and results were just too striking and impressive for that. Thus, for some reason, Kramers was unwilling or unable to communicate Heisenberg's ideas to the Copenhagen circle. It is not unreasonable to speculate that as soon as Kramers learned of Heisenberg's ideas in Göttingen, which were in fact a brilliant reorientation of the programmatic ideas of the joint Kramers–Heisenberg paper, he recognized immediately that a new era had started in atomic physics. This recognition must have come as a sudden shock to Kramers who had anticipated a prolonged period of searching for suitable transcriptions of classical ideas into quantum ideas, guided by the Bohr correspondence principle. This shock could easily have been the reason for a semiderisive comment on Heisenberg's expectations as "too optimistic." In spite of this remark, Kramers must have recognized the far-reaching implications of Heisenberg's approach. As a profoundly decent human being and a committed physicist, Kramers could not but rejoice in that development; as an extremely sensitive person with hopes, passions, and ambitions, the combination of his own failures and Heisenberg's successes, almost grafted onto those failures, must have caused deep conflicts and profound turmoil. As a very controlled and private person, Kramers would not say anything about either. Since it is overwhelmingly likely that the failures of the BKS theory and the promises of the Heisenberg matrix mechanics were uppermost in Kramers' mind—topics he was unable to discuss—Kramers' silence about Heisenberg's results becomes quite understandable. It took several months before Kramers had accommodated to the new reality of matrix mechanics and he could discuss it; he even made a number of (relatively minor) contributions to its further development. However, after the summer of 1925, he rarely referred to the BKS theory. In fact, he commented on it only once more, as the result of what Kramers felt were a number of provocations by R. de L. Kronig. Early in 1926, Kramers and Kronig exchanged a number of letters that dealt with the Kramers–Kronig dispersion relations. In one of these letters Kronig gratuitously attacked the BKS theory. Even at this late date when the issue of the BKS theory was effectively settled, Kronig aggressively attacked the theory, referring to the prevalence of "Copenhagen myths" and the suffocating influence of "Copenhagen superstitions." (This—on Kronig's part—might be a delayed echo of Pauli's criticism.) Kramers was evidently quite irritated by Kronig's attack, but in his usual controlled manner he proceeded to give a thoughtful, careful answer: "After the demise of Bohr–Kramers–Slater, I no longer believe in a simple space time description of resonance radiation."[227] Kramers took great pains to explain to Kronig (and possibly to himself) the rationale of the BKS approach; it was not a capricious attempt to be different or to create myths. Kramers considered Kronig's comments as a cheap, superficial dismissal of a profound idea, but he was sensitive to the criticism anyhow. Kramers by this time knew that the BKS ideas could not be maintained; atomic physics had taken a different and enormously successful direction. After this episode, Kramers never again—not in lectures, papers, or letters—referred to the BKS theory nor to the agony its demise caused him.

IV. The Road to Quantum Mechanics: The Changing of the Guard

A. The Kramers–Heisenberg Dispersion Theory Paper: Background and Chronology

The joint paper by Kramers and Heisenberg, "On the Dispersion of Radiation by Atoms",[228] was completed in Copenhagen in December 1924. It was a remarkable paper on several counts. It was the last, and certainly as far as physics was concerned, the most important paper in which the BKS philosophy was expressed explicitly. Although the detailed dependence of the Kramers–Heisenberg results on the BKS approach was rather slight, there are frequent allusions to the BKS papers, and at least in one important instance, specific use is made of the BKS ideas. Furthermore, the Kramers–Heisenberg paper contains the first (and only) organized, systematic exposition of Kramers' ideas on dispersion theory. Up until that time ideas were hinted at in short notices and abbreviated comments; Kramers had long planned to write a detailed paper on this material but it took about a year (from December 1923 until December 1924) before such a paper was completed. Thus, the Kramers–Heisenberg paper includes—apart from a number of new results such as the Smekal–Raman effect—an overview of Kramers' thinking during that period. But perhaps most important is that the Kramers–Heisenberg paper contained the main elements and basic method from which Heisenberg later developed his matrix mechanics. The notation, general approach, and some specific results of the Kramers–Heisenberg paper were all essential ingredients in Heisenberg's fundamental paper on matrix mechanics.[229] None of this detracts in any way from Heisenberg's monumental contribution, but the recognition of these particular circumstances is necessary to put Kramers' contributions in their proper perspective. The influence of the Kramers–Heisenberg paper on the later Heisenberg paper was certainly considerable, and Kramers, without any doubt, was the senior and major author of the Kramers–Heisenberg paper. To appreciate Kramers' contribution, it is necessary to analyze the evolution of his thinking on dispersion theory during this period. Unfortunately, this is not made any easier by Kramers' erratic publishing schedule and his simultaneous preoccupation with the BKS theory. Nevertheless, published papers, available letters, and recent interviews allow the reconstruction of a coherent, plausible picture of this development. The innovations in the Kramers–Heisenberg paper can be understood most simply if its contents are presented against the background of Kramers' earlier work in dispersion theory. To make the discussion transparent and self-contained, it is best to summarize† some of these earlier results. (The details are contained in Sections II.B and II.C.)

The basic phenomenon to be described is the dispersion of light by atoms. Monochromatic radiation of frequency ν_0 is incident on an atom, of presumably given constitution. The atom then emits radiation, and the problem is to give a qualitative and quantitative description of that radiation. This radiation can be characterized as the radiation produced by an induced dipole moment $\mathbf{P}$. The whole question is the determination and calculation of this vector $\mathbf{P}$. For weak fields, the dipole moment $\mathbf{P}$ and the incident field $\mathbf{E}_0$ are linearly related. In simple isotropic cases $\mathbf{P}$ and $\mathbf{E}_0$ are proportional: $\mathbf{P} = \alpha\mathbf{E}_0$,

† One of the important purposes of this compilation of results is to fix the notation. Throughout this book, the notation has been uniform, but it differs from the original papers. As will become evident, the proper notation is important as a means to suggest new directions.

where α is the polarizability. In general, such a linear relation is described by a polarizability tensor α_{ij}. All discussions are restricted to the linear domain, causing α to be constant (independent of $\mathbf{E}_0$). The comparison of theory and experiment almost always proceeds via an analysis of the data for α. If an atom is represented as a set of classical harmonic oscillators of frequencies ω_1 (as it is in the classical Lorentz theory), classical dynamics yields an unambiguous answer for α:

$$\alpha = \sum_i \frac{e^2}{4\pi^2 m(\omega_i^2 - \nu_0^2)} f_i$$

Here f_i is the number of electrons (better as the number of charged oscillators) in the atom. This formula agrees rather well with experiment—and this was precisely the problem. According to the Bohr theory, the classical frequencies of motion ω are in no way relevant for the emission and absorption of radiation. The relevant frequencies are the Bohr frequencies ν_{ij}, determined by $h\nu_{ij} = E_i - E_j$. They are distinct from the mechanical frequencies ω_i. The central question was to combine these conflicting ideas in a coherent framework. The atom behaved as if it contained classical oscillators, but the frequencies of these classical oscillators were not the classical mechanical frequencies; instead they were the quantum Bohr frequencies.

In Kramers' approach, this general set of problems was always split into two distinct questions, denoted here by (1) and (2). The first set of questions (1) was the approximate calculation of the dipole moment $\mathbf{P}$. This was always a purely classical calculation, based on classical Hamiltonian mechanics; the motion of the electrons in the atom was described as an s-dimensional multiple periodic motion with frequencies $\omega_1, \ldots, \omega_s$. Although the calculations were approximate, the procedure was nonetheless well defined and unambiguous. The second set of questions (2) was the quantum transcription of the results so obtained. Several distinct elements are involved in this quantum transcription:

(i) The selection of the stationary orbits. According to the Bohr–Sommerfeld theory these are the states characterized by $I_i = n_i h$ $(i = 1, \ldots, s)$.
(ii) The connection and eventual replacement of the mechanical frequencies ω_i by Bohr frequencies ν.
(iii) The replacement of derivatives by finite differences.
(iv) The relation and eventual elimination of classical Fourier coefficients in terms of Einstein A coefficients.

It is evident that the second set of questions (2) is not treated in as systematic a fashion as the first set. This was of course impossible; Kramers was very conscious of the fact that he tried to *infer* quantum relations from classical results. So there can be no question of a precise derivation: The results were obtained as much by sensible inferences or inspired guesses as by deductive logic. It is important to emphasize that this was exactly the approach adopted in the Kramers–Heisenberg paper. This element of intelligent guessing was characteristic of Kramers' earlier work and played an important role in the Kramers–Heisenberg paper. In his major book on quantum mechanics, written many years later, Kramers still refers to this guessing feature:† "The classical perturbation calculations were the basis, and by a correspondence type reinterpretation of the derivatives as differences, the exact formula for the dipole moment of the atom was guessed."[230] This mixture of classical analysis, correspondence arguments, reinterpretations, and guesses was evidently very congenial to Heisenberg. For upon his return to Göttingen (from Copenhagen) in the spring of 1925, he decided to attack the problem of the intensities of the hydrogen lines in a similar manner. In his autobiographical essays, Heisenberg writes: "I made a first attempt to guess what formulae would enable me to express the line intensities of the hydrogen spectrum, using more or less the same methods that had proved so fruitful in my work with Kramers in Copenhagen."[231] In conversations with Born, as reported by Mehra,[232] Heisenberg told Born: "One has such an impression that one could now almost guess all the intensities, if only one did it well enough." According to Heisenberg, Born encouraged him and said: "That is a good idea, you try it. Then you probably want to know something about classical Fourier components, there you need Bessel functions." Although this particular attempt of Heisenberg was not successful, it shows that the procedure advocated by Kramers—to guess at the quantum formulation at a certain stage of the discussion—had a long lasting and pervasive effect. It is interesting to recall further that the relation between the Fourier components of the classical motion and the intensities of spectral lines played a basic role in Kramers' thesis. Thus, several features of Heisenberg's investigations, culminating in matrix mechanics, were based on earlier ideas of Kramers.

The procedure reviewed on p. 217, problems (1) and (2), led to the original Kramers dispersion formula. Since practically the identical pattern is followed in the Kramers–Heisenberg paper, it is convenient to recall‡ the notation and pertinent results. The mechanical frequencies of the atomic system are $\omega_1, \ldots, \omega_s$, the Hamiltonian of the system is H, and the action variables are $I_1, \ldots, I_s$. The canonical equations yield $\omega_i = \partial H/\partial I_i$ $(i = 1, \ldots, s)$. The dipole moment P_0 of this system can be expanded in a Fourier series:

$$P_0 = \sum_{l_1, \ldots, l_s} C_{l_1, \ldots, l_s} e^{i(\omega_1 l_1 + \omega_s l_s)t + i\gamma} \tag{a}$$

The $C_{l_1, \ldots, l_s}$ are the Fourier coefficients; they depend on the values of I. The sums over $l_1, \ldots, l_s$ are over *all* positive and negative integers. It must be remembered that the classical orbit is fixed by the numerical values of the actions I_i. As written, the Fourier coefficients refer to a specific, single orbit and P_0 is the dipole moment of that orbit. If this system is influenced by a monochromatic light wave of frequency ν_0, an induced dipole moment will be produced. The result of the approximate calculation mentioned earlier yields

† Reading Kramers' frequent references to this guessing aspect of the analysis, one gets the distinct impression that Kramers rather enjoyed this (nondeductive) feature of the method.

‡ In the formulas collected here, numerical factors m, c, and π are often omitted. The general structure is the main concern.

for the coherent part of the dipole moment

$$P = \mathrm{Re} \sum_{l_1,\ldots,l_s} \frac{\partial}{\partial I}\left(\frac{|C_l|^2}{\omega + \nu_0} + \frac{|C_l|^2}{\omega - \nu_0}\right) E_0 e^{i\nu_0 t} \tag{b}$$

where

$$\omega = l_1\omega_1 + \cdots + l_s\omega_s$$

$$\frac{\partial}{\partial I} = l_1\frac{\partial}{\partial I} + \cdots + l_s\frac{\partial}{\partial I_s}$$

The *coherent* part of P is that part which has the same time dependence as the external field $e^{i\nu_0 t}$. There is also an incoherent part with a different time dependence but this was *not* considered in Kramers' early work. Also in the early papers, Kramers considered only situations where the polarizability α was a scalar (not a tensor), so that $\mathbf{P} = \alpha\mathbf{E}_0$. Formula (b) ends the classical perturbation discussions; it is this formula that must be translated into quantum terms. Part of that is obvious; a specific orbit, with prescribed values of I_i, becomes a specific state with quantum numbers n_i, where $I_i = n_i h$. This in turn makes the dipole moment P a function of the particular stationary state, characterized by $n_1, \ldots, n_s$. The transcription of the frequencies proceeds as always via the correspondence principle. Let i be a state where the action variables have the values $n_1, \ldots, n_s$, so that $I_1 = n_1 h \ldots I_s = n_s h$. Let j be a state where $I_1 = n_1' h \ldots I_s = n_s' h$. Also, call $n_1 - n_1' = l_1 \ldots n_s - n_s' = l_s$. Then the quantum or Bohr frequencies are

$$\nu_q = \nu_{i\to j} = \frac{1}{h}(E_{n_1,\ldots,n_s} - E_{n_1',\ldots,n_s'}) \tag{c}$$

The E's are the energies of the respective states. For $l_1 = n_1 - n_1', \ldots, l_s = n_s - n_s'$ small compared to the values of n, the classical and quantum frequencies should agree; thus, the correspondence should be

$$\omega_{l_1,\ldots,l_s} \equiv \omega_{n_1-n_1',\ldots,n_s-n_s'} \leftrightarrow \frac{1}{h}(E_{n_1,\ldots,n_s} - E_{n_1',\ldots,n_s'}) \tag{d}$$

This correspondence shows that a given classical Fourier component (i.e., $l_1, \ldots, l_s$ are given) corresponds to an infinite set of states, $n_1, \ldots, n_s, n_1', \ldots, n_s'$, where $n_1 - n_1' = l_1 \ldots n_s - n_s' = l_s$. Consequently, a *sum* over all Fourier components, as occurs in (c), will lead to a sum over a double infinite set of states. The replacement (d), which for one variable can be written

$$\frac{E - E'}{h} = \frac{\Delta H}{h} = \nu_q \leftrightarrow \omega = \frac{\partial H}{\partial I_i}$$

can be used to suggest a general connection between derivatives and differences of the type

$$l_i\frac{\partial F}{\partial I} \leftrightarrow \frac{1}{h}[F(n) - F(n')] \tag{e}$$

where F is any classical function of I. This was indeed the connection used by Kramers in his first derivation of the dispersion relations. Finally, it is necessary to replace the Fourier coefficients of the classical motion by their quantum counterparts. For this, Kramers again refers back to his thesis where such a relation is obtained by equating the classical and quantum energies radiated per second. If i is the quantum state $n_1, \ldots, n_s$ and J is the quantum state $n_1', \ldots, n_s'$, with $n_1 - n_1' = l_1, \ldots, n_s - n_s' = l_s$ and $C_{l_1, \ldots, l_s}$ the classical Fourier component of the motion, the connection is

$$\frac{(2\pi\omega)^4}{3c^3} |C_{l_1, \ldots, l_s}|^2 \leftrightarrow A_{i \to j} h\nu_{ij} \tag{f}$$

Equations (c), (d), and (f) applied to (b) yield the original Kramers' formula for the (induced) moment of an atom in state k:

$$P_k = E_0 \left(\sum_i \frac{A_{i \to k}}{\nu_{ik}^2(\nu_{ik}^2 - \nu_0^2)} - \sum_j \frac{A_{k \to j}}{\nu_{kj}^2(\nu_{kj}^2 - \nu_0^2)} \right) \tag{g}$$

Kramers, in his first notes, wrote (g) in a slightly different notation: He used ν_q, the quantum or Bohr frequency for ν_{ij}, and wrote $A_j^{(a)}$ for the Einstein coefficient of the absorption processes (while the index k was suppressed). This obviously does not change anything, but it does not bring out the essential dependence on *pairs* of states as explicitly as does the notation used in (g). It is this dependence on pairs of states which Heisenberg exploited so brilliantly in his development of matrix mechanics.

The repetition of the basic ideas leading to Kramers' dispersion relations presented here was deliberate. It emphasized once again that the dispersion relations were logically totally independent of the BKS theory. In the arguments reviewed so far, no hint of the BKS ideas was introduced. It also showed that in spite of the obvious similarity between Ladenburg's and Kramers' results, their motivations, status, and logic were altogether different. This becomes most evident if it is realized that Ladenburg's paper[119] contains nothing remotely resembling relation (f)—the relation between the Fourier components of the motion and the radiation emitted. Ladenburg equated the classically and quantum theoretically calculated energy. He did not attempt to relate the dynamics of the electrons in the atom to the radiation emitted. In Ladenburg's paper, no electronic orbit is ever mentioned, nor do Fourier coefficients occur. The transcription from the mechanical description of the motion of the electrons, as provided by classical mechanics, to a quantum description of emission and absorption of radiation was of paramount importance to Kramers; this issue was neither raised not treated by Ladenburg.

A final reason for repeating Kramers' ideas at this juncture is to stress how important and pervasive Kramers' influence was in the development which eventually led to Heisenberg's breakthrough. Although other persons, especially Born, made signal contributions, it is undoubtedly correct that from the early announcement of the dispersion relations, up to and including the Kramers–Heisenberg paper, Kramers' ideas guided that development. To appreciate the various stages through which this development actually progressed, it is instructive to give a selected chronology of the events that culminated in Heisenberg's fundamental advance. That way, Kramers' contributions as well as those he missed can clearly be exhibited.

Chronology

December 21, 1923	Slater arrives in Copenhagen.
January 17, 1924	Slater reports that Kramers showed him the replacement of derivatives by finite differences.[133]
January 21, 1924	BKS paper submitted to *Philosophical Magazine.*
March 25, 1924	Kramers submits his note on dispersion relations to *Nature.*
June 8, 1924	Letter from Kramers to Ladenburg[122] promising a detailed description of his derivation.
June 13, 1924	Born's paper,[233] referring extensively to Kramers' procedure, submitted to *Zeitschrift für Phusik.*
June 21, 1924	Kramers' talk for the German Physical Society, Hamburg. In this talk, Kramers announces that he had obtained extensions of his dispersion relations.[234]
July 22, 1924	Kramers' second note, submitted to *Nature*, outlines the steps needed to arrive at the formula already quoted in the March 25,1924 paper.
September 1924	Heisenberg arrives in Copenhagen during Kramers' absence.
November 1, 1924	Bohr refers to a forthcoming Kramers–Heisenberg paper.[235]
December 22, 1924	Heisenberg leaves Copenhagen for vacation. Kramers finishes writing the paper on December 24, 1924.
January 8, 1925	Heisenberg writes to Kramers, agreeing with the final draft.
April 21, 1925	Heisenberg writes to Bohr from Göttingen about his general plans.
May 16, 1925	Heisenberg writes to Bohr that he is investigating (guessing) the intensities of the hydrogen lines.[236]
June 5, 1925	Heisenberg, in a letter to Kronig, has shifted his interest from the calculation of the hydrogen intensities to those of the anharmonic oscillator. The basic ideas of matrix multiplication are already evident.[237]
June 24, 1925	Heisenberg outlines the basic ideas of matrix mechanics in a letter to Pauli.
June 29, 1925	In a postcard to Pauli, Heisenberg refers to a meeting with Kramers. Heisenberg believes that matrix mechanics is basically complete, but Kramers accuses him of optimism.
July 4–5, 1925	Heisenberg completes the manuscript of the matrix mechanics paper and gives it to Born to read—about July 10.

The information in this chronology, especially if it is combined with letters and interviews with Slater, shows clearly that Kramers had the dispersion relations before Slater had arrived in Copenhagen in January 1924.[131] In Slater's autobiography, he also writes that Kramers had completed the problem of the forced oscillations of a multiple periodic system in an external electromagnetic field.

These contributions of Kramers have been generally recognized. In an investigation devoted to interactions *between* electrons, Born was quite candid in attributing the new approach to Kramers: "In this situation one might consider whether it would not be possible to extend Kramers' ideas, which he applied so successfully to the interaction between radiation field and radiating electron, to the case of the interaction between several electrons of an atom." [233] In his autobiography, Born is even more explicit: "One can say that Kramers, guided by Bohr's principle of correspondence, guessed the correct expression for the interactions between the electrons in the atom and the electromagnetic field of the light wave. That is at least the way I regarded his results. It was the first step from the bright realm of classical mechanics into the still dark and unexplored underworld of the new quantum mechanics. I made the next step." [238] It is evident that Born openly acknowledges that he follows the direction initiated by Kramers. There is, however, a certain amount of confusion about the replacement of the derivatives by difference quotients, a most essential step in obtaining the form of the dispersion relations. This confusion was undoubtedly caused by Kramers' delay in publishing his result. Thus, Jammer[239] and Mehra and Rechenberg[240] refer to Born's version of the replacement of derivatives by differences in the form:

$$l\frac{dF(n)}{dn} \leftrightarrow \frac{i}{h}[F(n) - F(n-l)] \tag{h}$$

This is not incorrect; Born's paper[233] certainly contains this formula. It is evident, however, from Slater's comment (see the second item in the chronology) that Kramers had the explicit relation between derivatives and differences as early as January 1924. No hint of this relation was given in Kramers' first paper, so that Born, who wished to use Kramers' method, was left to his own devices to obtain Kramers' results.† In so doing, he obtained relation (h). It is amusing that in his second note Kramers gives some indications about the derivation of the dispersion relations; he mentions, almost in passing, the replacement of derivatives by differences. By that time, Born had already obtained relation (h) on his own.

A persuasive, albeit indirect, argument that Kramers considered the replacement of the derivatives by finite differences as his own contribution is the remarkable circumstance that the Kramers–Heisenberg paper, where this replacement plays a crucial role, does not contain a single reference to Born's results which were published before that paper was completed. Kramers—who was meticulous as well as generous in giving credit—just organized and formalized his earlier considerations, and he clearly did not feel it necessary to acknowledge Born's similar but later contributions. If Kramers had felt that his results were in any way dependent on Born's paper, he undoubtedly would have given the appropriate acknowledgment.

† Whittaker[241] explicitly states that Born gave a general procedure for translating classical formulas into quantum mechanical formulas, based on Kramers' original paper.

B. General Approach Between Virtual Oscillators and Matrices

Kramers was well aware that his early derivations of the dispersion relations made rather restrictive assumptions.[121] He specifically assumed that the induced dipole moment **P**(t), which is the source of the secondary waves emitted by the atoms, has the identical harmonic time dependence as the incident radiation **E**(t). Furthermore, the phase difference ϕ between these waves was assumed to be small; this is equivalent to the requirement that the external frequency ν_0 is "sensibly" different from the classical oscillator frequencies ω_1. Only under these circumstances can the classical formula for the polarizability be expected to give reasonable results. It was the "reasonable" agreement of these classical results with experiment which was such a powerful impetus for Kramers to develop a theory that combined the predictive power of the classical dispersion theory with Bohr's quantum ideas. On the basis of the classical ideas of Lorentz, it was totally reasonable to expect that the secondary radiation would have the same frequency ν_0 as the incident radiation and experiments supported the claim that ϕ is small. However reasonable these assumptions might appear, Kramers recognized early that the quantum transcription of classical results was a delicate and tricky process in which classically unimportant or negligible features might well lead to unanticipated quantum effects. (The knowledge that the scattered frequency ν_1 in the Compton effect was *not* equal to ν_0 was a strong hint in that direction.) Thus, Kramers embarked on a systematization and extension of his earlier studies. Both assumptions, that the phase difference ϕ between the incident and emitted wave was small and that the time dependence of **P**(t) is harmonic with frequency ν_0, were subjected to a serious reexamination. As early as June 1924, Kramers announced that he had obtained an expression for the incoherent part of the dipole moment **P**(t); that is what he called the part of the dipole moment **P**(t) whose time dependence is *not* harmonic with frequency ν_0.

There were several motivations for Kramers to undertake these extensions. He saw in these dispersion relations the starting point of a new, complete, and autonomous quantum theory, so a detailed and systematic development was certainly required. He was at this same time deeply immersed in the BKS ideology, and he undoubtedly viewed the dispersion relations and their contemplated extensions as significant support of those ideas. In fact, the abstract of the Kramers–Heisenberg paper contains this interesting passage: "The considerations [presented here] are based throughout on the viewpoint proposed in a new paper by Bohr, Kramers, Slater, which relates the emission of light by an atom to its stationary states; and the conclusions [reached here] if they would be confirmed might provide an interesting support for this viewpoint."[228] Although, as the sequel will show, the BKS ideas were used only in a rather indirect manner, it is clear from the passage just quoted that Kramers considered the results obtained as concrete confirmation of the whole BKS scheme. This by itself would have made him impatient to pursue this line of research. But probably the most compelling reason for this study, was Kramers' wish to provide a wave theoretical (continuum) interpretation of a number of results obtained earlier by Smekal.[242] Smekal's considerations were based firmly and explicitly on the photon theory of light. By using the photon picture of light, much in the same way Compton had,† Smekal arrived at the interesting and surprising conclusion that monochromatic radiation incident on an atom should yield scattered radiation, containing a number of frequencies, among which would be $\nu_0 + \nu_k$ and $\nu_0 - \nu_l$. Here, ν_0 is the incident frequency of the radiation; ν_k and ν_l are both possible Bohr frequencies of the atom. Thus, according to Smekal, there should exist a *discrete* frequency displacement of the scattered light. It is remarkable that although these effects were only found several years later by Raman[243] and Landsberg and Mandelstam,[244] Kramers nevertheless was convinced that a pure wave theory without photons, such as the BKS theory, should also be capable of explaining such discrete frequency shifts. It is a testimony to Kramers' deep physical insight that instead of ignoring Smekal's predictions, based as they were on the unpalatable photon notions (remember the whole BKS theory was constructed to avoid that notion), he recognized the physical plausibility of the predicted phenomena. He also recognized that the effects predicted by Smekal, which so far were not included in his dispersion relations, could be obtained in a natural manner from the incoherent part of the induced dipole moment. Referring to Smekal's result, the Kramers–Heisenberg paper reiterates the BKS view: "The calculation in terms of light quanta is above all of importance because it enables us to connect the macroscopic energy and momentum conservation laws with the concepts of quantum theory in a simple and instructive way. *But* by the nature of the case, such considerations *do not allow* us to draw any conclusions regarding the *corpuscular nature* of light, we always have the requirement that we must be able to get the results so obtained, to agree with the wave theoretical description of the optical phenomena in a way which is free from contradictions."[228] Thus, one of the most important motivations (and probably the least important result) of the Kramers–Heisenberg paper was the explanation of Smekal's results on the basis of a pure wave theory.

It might appear curious that the ideas expressed here are all attributed to Kramers; after all, they are contained in a joint paper. But what is even stranger is that this joint paper actually acknowledges Kramers' priority: "The idea of relating the scattering action of atoms in the presence of external radiation according to Smekal, to the scattering action of atomic systems expected from the classical theory, in accordance with the correspondence principle first occurred to Kramers in connection with his work on dispersion theory."[228] It appears more than a little unusual that a joint paper would so explicitly assign the priority of its guiding idea to a single author. Since the final redaction of the paper was entirely due to Kramers,[245] this sentence sug-

† The main technique used by Smekal was the judicious application of the energy and momentum conservation laws to the atom–photon system.

gests a lingering resentment that, at Bohr's urging,[246] these results appeared as a joint publication by Kramers and Heisenberg rather than as a paper by Kramers alone. (See Section 13.C for further elaborations of this point.)

The systematic exposition of the dispersion theory begins by considering a plane monochromatic polarized wave incident on the atom. The electric field is written throughout as the real part of a complex vector:†

$$\mathbf{E}(t) = \mathrm{Re}(\mathbf{E}_0 e^{i\nu_0 t})$$

It is *assumed* that the radiation emitted by the atom can be represented by an oscillating dipole of dipole moment $\mathbf{P}(t)$. This is again written as the real part of a complex vector:‡

$$\mathbf{P}(t) = \mathrm{Re}(\mathbf{P}_0 e^{i\nu_0 t}) \tag{a}$$

$\mathbf{E}_0$ and $\mathbf{P}_0$ are time-independent, complex vectors; both could depend on ν_0. In general, $\mathbf{P}_0$ and $\mathbf{E}_0$ are in different directions, but $|\mathbf{P}(t)| = |\mathbf{E}(t)|$. If assumption (a) is combined with the (classical) expression for the intensity of the scattered light

$$\text{Intensity} \approx \nu_0^4(\mathbf{PP}^*) \tag{b}$$

one can account for the dispersion, scattering, and absorption of light. The somewhat tenuous status of the formula for the dipole moment (a) is made explicit in the present paper by references such as: "We *assume* expression (a) to be valid as long as the atom remains in the stationary state under consideration."[247a] "Without considering the question of the limits within which the initial assumption (a) was valid one of us [HK] started...."[247b]

Strictly speaking, in a complete theory it should not be necessary to make assumptions about the form of the dipole moment $\mathbf{P}(t)$. In principle, the dipole moment for a given system can be calculated; thus its form should be deduced rather than assumed. For the damped harmonic oscillator, which is the prototype of all such calculations, form (a) is precisely correct and no assumption is necessary. This leads one to hope that for more complicated systems, where the calculation is more difficult, the same general *form* might still be correct, at least in some approximation. That was the basis of Kramers' original assumption. Recognizing the approximate character of that assumption, Kramers was quite willing to contemplate generalizations and modifications of assumption (a). The BKS framework suggested a very natural generalization for the formula for the induced dipole moment. According to BKS, an atom in a stationary state responds to an external field as if it consisted of a set of virtual oscillators possessing all the Bohr frequencies ν_{ik}. Consequently, if an atom in a stationary state k is irradiated by a wave of

† The use of the complex representation for $P(t)$ and $E(t)$ allows a convenient analysis of the general (linear) configuration, where P and E are no longer parallel.

‡ The notation used is the same as used throughout the book. ν_0 is always the impressed external frequency, ω always stands for mechanical frequencies, and ν_g or ν_{ij} are atomic transition frequencies. Constants and the factor π are usually omitted. A star (*) denotes complex conjugate.

frequency ν_0, there is a finite probability that via the intervention of the virtual field the atom makes a real transition to a higher state i. The net effect of the interaction would be the emission of a secondary wave of frequency $\nu' = \nu_0 - \nu_{ik}$. Schematically,

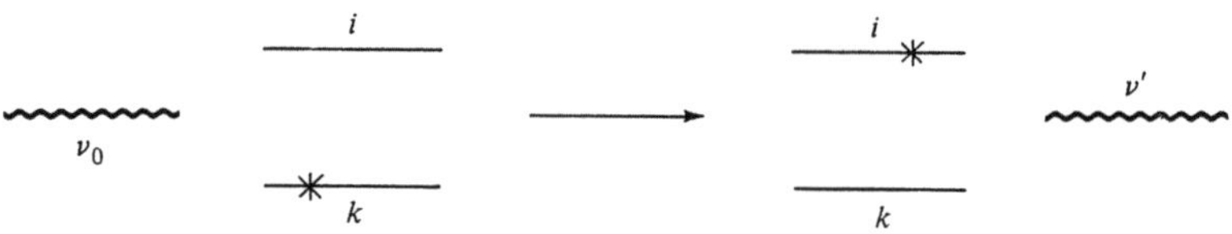

Similarly, there is a finite probability for the process that the atom will make a transition to a state j, lower than k. In that case, the emitted light has a frequency $\nu'' = \nu_0 + \nu_{kj}$. The BKS theory provides for, or indeed requires, such processes to occur. If one still wants to represent the secondary radiation under these circumstances in terms of a quantum dipole moment $\mathbf{P}(t)$, one would expect that instead of (a) it would have the form

$$\mathbf{P}_k(t) = \mathrm{Re}\left(\mathbf{P}_k e^{i\nu_0 t} + \sum_j \mathbf{P}_j e^{i(\nu_0+\nu_{kj})t} + \sum_i \mathbf{P}_i e^{i(\nu_0-\nu_{ik})t}\right) \qquad \text{(c)}$$

Here $\mathbf{P}_i$ and $\mathbf{P}_j$ are linear functions of $\mathbf{E}_0$. It was just noted that for a given system the dipole moment can be calculated; its form cannot arbitrarily be assumed. It must therefore have been deeply satisfying to Kramers (and unfortunately profoundly misleading) that a detailed analysis of the atom as a multiple periodic system led via the same methods used earlier to an expression for $\mathbf{P}(t)$ of the form (c)—precisely that suggested by the physical ideas of the BKS theory. It is easy to anticipate and easy to show that this particular form, with the frequencies $\nu'' = \nu_0 + \nu_{kj}$ and $\nu' = \nu_0 - \nu_k$, will indeed reproduce the results of Smekal obtained in the photon picture.

Thus, Kramers' belief in the validity of the BKS theory was considerably strengthened by his successful incorporation of the Smekal processes within that general framework. The natural and smooth manner in which the dispersion theoretic formalism leads to an expression of the form (c) must have increased his already considerable confidence in the BKS ideology even further. Actually, the BKS ideas do not enter into the derivation of $\mathbf{P}(t)$, so successes or insights based on the expression for $\mathbf{P}(t)$ in no way support the physical ideas of the BKS theory.

It is most ironic that another *formal* element in the Kramers–Heisenberg paper also suggested by the BKS theory led Heisenberg to his brilliant development of matrix mechanics while Kramers' confidence in the *physical* basis of the BKS theory, based on the identical paper, turned out to be totally misplaced and led nowhere.

The idea in question is the introduction of amplitudes a, related to—but not identical with—the Fourier coefficients C. According to the BKS picture, an atom in a stationary state acts as a source of spherical waves of frequencies

ν_{ki}, the Bohr frequencies of the atom. These spontaneous transitions occur even in the absence of an external field. The BKS theory suggests strongly that this spontaneous radiation can again be described in terms of a classical dipole whose dipole moment can be represented as

$$\mathbf{P}_{\text{spont}}(t) = \text{Re}\left(\sum_i \mathbf{a}_{ki} e^{i\nu_{ki}t}\right) \equiv \text{Re} \sum_q \mathbf{a}_q e^{i\nu_q t} \tag{d}$$

The **a**'s are the amplitudes in question. The notation used here is consistent with that used throughout this book; it deviates considerably from the original notation. In the original notation (also adopted by most of the analysts and historians), the *state* dependence of the dipole moment is usually suppressed. Just as classically the dipole moment depends on the orbit, specified by the values of the actions I_i, in quantum theory the dipole moment depends on the stationary state considered, here usually called k. In the notation used in this book, P_k depends on symbols with two indices, such as $A_{i\to k}$ in Equation (g) on p. 220, or $\mathbf{a}_{ik}$ in (d) on p. 227. As specified, the quantities $\mathbf{a}_{ik}$ are complex vectors depending on two states. In the earlier discussion, certainly in the Kramers–Heisenberg paper, the frequencies ν are written with a single index q and ν_q, to stress their quantum origin. Similarly, the amplitudes are written as $\mathfrak{A}_q$ and the Einstein coefficient for spontaneous emission is written as a_q. Usually, the notation is a matter of history, taste, or convention and not of any importance. The changes made here are done with the help of hindsight; however, the recognition that the basic quantities ν_q, $\mathfrak{A}_q$, and a_q all depend in an essential manner on *two* states—so that they are called ν_{ki}, a_{ki}, and A_{ki}, respectively—was one of the insights that played a role in Heisenberg's matrix mechanics. To call the amplitudes a_{ki}, rather than $\mathfrak{A}_q$, further stresses that these quantities were the precursors of the probability amplitudes, or matrix elements. It is helpful that the notation employed should stress the similarity. Occasionally, a_q will be written for a_{ki}, but the notation used from here on adheres to the conventions noted here: They are; $\mathbf{a}_{ik}$ are the amplitudes, A_{ik} the Einstein coefficients, and C_l, the Fourier coefficients.

It is clear that development (d) is based on the BKS picture, where the atom is envisaged as a set of oscillators of frequency ν_{ki}. Energy considerations suggest that the relation between the Einstein A_{ki} and the amplitudes is given by

$$A_{ki} h\nu_{ki} \leftrightarrow \frac{(2\pi\nu_{ki})^4}{3c^3}(\mathbf{a}_{ki}\mathbf{a}_{ki}^*) \tag{e}$$

[In the Kramers–Heisenberg paper, Equation (e) is the "operational definition" of the amplitudes a. It is described there as the "simplest description of the radiation emitted."]

It is important to distinguish the quantum development (d) from the multiple Fourier series development of the classical dipole moment [formula (a) in the summary]

$$\mathbf{P} = \sum_{l_1,\ldots,l_s} \mathbf{C}_{l_1,\ldots,l_s} e^{i(\omega_1 l_1 + \cdots + \omega_s l_s)} \tag{f}$$

$\mathbf{C}_l$ is a complex vector; it possesses the important symmetry that $\mathbf{C}_l = \mathbf{C}^*_{-l}$. Development (f) is totally classical; it refers to an atom in a single classical orbit (characterized by specific values of I), the $l_1, \ldots, l_s$ assume positive and negative integer values, and the C coefficients are ordinary Fourier coefficients. By contrast, development (d) is partially quantum theoretical; the ν_{ki} values are determined by the stationary states of the atom, and the coefficients a_q (or a_{ki}) are not simply Fourier coefficients. Development (d) is suggested by the *physical* picture of the BKS theory where an atom behaves as if it contained (classical) oscillators of frequencies ν_{ki}, while development (f) is a purely mathematical expansion that merely requires the multiple periodicity of the system. It is evident that in spite of these differences the Fourier coefficients C and the amplitudes a_{ki} are not unrelated. It was, of course, well known to Kramers (he used it numerous times in his thesis) that the square of a Fourier component $|C|^2$ is proportional to the intensity of the radiation emitted. The precise relation was used in the derivation as (f) on p. 220:

$$\frac{(2\pi\nu_{ij})^4}{3c^3}|C(l_1,..,l_s)|^2 \leftrightarrow A_{i\to j}h\nu_{ij} \tag{g}$$

Here the quantum state i has quantum numbers $n_1, \ldots, n_s$, the quantum state j has quantum numbers $n'_1, \ldots, n'_s$, while $l_1 = n'_1, \ldots, l_s \sim n_s - n'_s$. C is the Fourier component of a single orbit, depending on action variables $I_i = l_i h$. Equation (g) is, of course, a version of the correspondence principle, valid for high quantum numbers n_i, n'_i. On the other hand, Equation (e) as written is presumably always valid; the amplitudes are square roots of the A coefficients. The identity in form of Equations (e) and (g) suggests that the amplitudes a_{ki} must asymptotically approach C_l:

$$\lim \mathbf{a}_{ki} = \lim \mathbf{a}_{n_1,\ldots,n_s;n'_1,\ldots,n'_s} \to \mathbf{C}_{n_1-n'_1,\ldots,n_s-n'_s} \tag{h}$$

The nature of the limit is not all that precisely defined, but it is important to stress that in that limit a *two*-index object a_{ki} becomes a *one*-index entity C_l. Kramers and Heisenberg "attempt to further define or describe the a_{ki} as a symbolic average over the Fourier coefficients C, taken over a region between two stationary states." In the notation used here, they appear to hint at a relation of the type

$$a_{n_1,\ldots,n_s;n'_1,\ldots,n'_s} = \int_{n_1,\ldots,n_s}^{n'_1,\ldots,n'_s}\!\!\cdots\!\int dl_1 \cdots dl_s C(l_1,\ldots,l_s)G(l_1,\ldots,l_s) \tag{i}$$

Here $G(l_1,\ldots,l_s)$ is an appropriate weight function. Kramers and Heisenberg neither write nor use this relation; but they are emphatic in commenting that, as far as radiation is concerned, no significance can be attributed to the Fourier coefficients of the individual orbits (states) $C_{n_1,\ldots,n_s}$, $C_{n'_1,\ldots,n'_s}$. Only the average of these coefficients over a region [of which Equation (i) is an example] can be expected to have a meaning. Thus, in this paper a marked shift occurs from the Fourier coefficients C, which depend on a single state (or orbit), to the characteristic amplitudes a, which necessarily depend on two states. The reason for this shift is the validity of expansion (d), which just formalizes the basic tenet of the BKS theory, that an atom responds to radiation as does a set of classical oscillators of frequencies ν_{ki}. The characteristic amplitudes a_{ki} are *not* defined for a single state, while the Fourier coefficients can be defined for a single state. It might appear that the shift from the Fourier coefficients C to the characteristic amplitudes would not be of great consequence, because both $|C|^2$ and $|a|^2$ are proportional to the Einstein coefficients A. Indeed, the early versions of the dispersion relations by Kramers and Ladenburg contain just the Einstein coefficients. Since only squares $|C|^2$ and $|a|^2$ seem to occur, both directly related to A, it might seem a matter of indifference whether one deals with the $|C|^2$ or the $|a|^2$; either is proportional to A. In fact, Born writes: "In this case it would *not* seem reasonable to look for [quantum] quantities, corresponding to Fourier components C_l themselves. Evidently it is only the quadratic expressions $|C_l|^2$ which have a quantum theoretical meaning."[248] Born was misled by the formalism; he mistakenly believed that the earlier Kramers' formulas, which were valid for the case that $P(t)$ and $E(t)$ were parallel, would retain their form in more general circumstances where P and E are no longer parallel. This is not the case.

In the Kramers–Heisenberg paper, which treats the general linear connection between P and E, it is demonstrated that the relation is of the type†

$$\mathbf{P}_k(t) \approx \sum_i \frac{\mathbf{a}_{ki} \cdot (\mathbf{E} \cdot \mathbf{a}_{ki}^*)}{\nu_{ki} - \nu_0} \tag{j}$$

It is clear the P does *not* just depend on $|a_{ik}|^2$. If one calls P_k^s the s component of P_k, Equation (j) leads to terms of the type

$$P_k^s = \sum_i \frac{a_{ki}^s (a_{ik}^*)^{s'} E_{s'}}{\nu_{ik} - \nu_0}$$

which shows explicitly that terms of the form $a_{ki}^s (a_{ki}^*)^{s'}$ with $s \neq s'$ occur. Thus, Born's guess was not right; Kramers and Heisenberg correctly inferred that it was necessary to attribute individual physical significance to both a_{ik} and a_{ik}^*. The intrinsically complex character of the characteristic amplitudes a_{ik} implies that both the magnitudes and the phases will play a significant role in physical processes. This dependence on the phases leads to a number of unexpected interferences and, at first sight, puzzling cancellations. These were analyzed in great detail in the joint Kramers–Heisenberg paper. The experience with the phases so obtained was instrumental in focusing Heisenberg's attention on the importance of the characteristic amplitudes. These amplitudes, introduced in an almost cavalier manner in the Kramers–Heisenberg paper, became in Heisenberg's hands the fundamental ingredients from which matrix mechanics would be constructed. The physical idea for these amplitudes originated directly from the BKS picture; that information is contained in the Kramers–

† The precise formulas are given in Section IV.C; here just the directional features are important.

Heisenberg paper. It was also suggested that the amplitudes, both their magnitude and phases, are significant objects. But the decisive step whereby the amplitudes become autonomous, and are in fact the basic entities from which a theory should be constructed, was taken by Heisenberg alone in his seminal paper on matrix mechanics.

C. Technical Matters: The Old Method and New Puzzles

The Kramers–Heisenberg analysis of the response of an atom to monochromatic light leads to *two* important physical results:

(1) Under the influence of monochromatic light of frequency ν_0, the atom *emits* coherent spherical waves of the same frequency ν_0 as the incident waves.
(2) In addition, the system emits *incoherent* spherical waves of frequencies $\nu_0 + \nu_{kl}$, where ν_{kl} represents the possible Bohr frequencies of the atom.

The dispersion and absorption of the incident light is described by the coherent secondary waves of frequency ν_0; the incoherent waves of frequencies $\nu_0 \pm \nu_{kl}$ appear as scattered light and do *not* contribute to either absorption or dispersion. The principal *formal* result is an explicit quantum theoretical expression for the dipole moment of the atom as a function of the time. From this expression, the intensity of the various frequencies could in principle be obtained. This requires a Fourier decomposition of the power the atom emits; this power W is related to the dipole moment P by a well-known formula of the type $W \approx (PP^*)$. The expression for P, which is the central result of the Kramers–Heisenberg paper, is a purely quantum theoretical expression; as in Kramers' original formula, all references to the classical mechanical system have disappeared. With an almost obvious interpretation of the entities, this expression is identical with the quantum mechanical form; it was truly the first quantum mechanical expression. The method used to obtain these results is a straightforward adaptation of the procedure used earlier which has been summarized twice before (see Sections II.B, II.C, and IV.A).

Although the methods employed are the same as used previously, the physical results are not. A number of new situations arise which require further discussion. It is most illuminating to list the logical outline of the derivation and comment only on those aspects not previously encountered. The essential steps involved are:

(I) The classical (perturbation) calculation of $P(t)$.
(II) The selection of the orbits according to the Bohr–Sommerfeld quantum rules.
(III) The quantum transcription $\omega \to \nu$.
(IV) The quantum transcription from derivatives to differences.
(V) The quantum transcription from Fourier coefficients to characteristic amplitudes.

All these have been analyzed on previous occasions; it is unnecessary to do so again. Apart from more detailed explanations of these matters, the Kramers–Heisenberg paper contains methodologically little that is new. (It would also get rather boring to repeat this analysis.) But it *is* of interest to point out that the pattern of the derivation presented in the Kramers–Heisenberg paper is *identical* with that in Kramers' earlier papers. The only difference is (V) in the list above, where the Fourier coefficients are transcribed to the characteristic amplitudes. In previous derivations, the squares of the Fourier coefficients were replaced by the Einstein A coefficients.

In the notation† which has been used throughout, the system considered has s degrees of freedom, the action variables are $I_1, \ldots, I_s$, the frequencies are $\omega_1, \ldots, \omega_s$, $\omega_l = l_1\omega_1 + \cdots + l_s\omega_s$, $C_l = C_{l_1,\ldots,l_s}$ are the Fourier coefficients,

$$\frac{\partial}{\partial I} = l_1\frac{\partial}{\partial I_1} + l_s\frac{\partial}{\partial I_s}$$

$\sum_l$ is an s-fold sum over positive and negative integers $l_1, \ldots, l_s$, and ν_0 is the incoming frequency. The electric moment P is

$$\mathbf{P}(t) = \mathbf{P}_0(t) + \mathbf{P}_1(t) \tag{1}$$

Here $\mathbf{P}_0(t)$ is the dipole moment of the unperturbed system [written before as Equation (f), p. 220]

$$\mathbf{P}_0(t) = \sum_l \mathbf{C}_l e^{i\omega_l t} \equiv \sum_{l_1}\cdots\sum_{l_s} \mathbf{C}_{l_1,\ldots,l_s} e^{i(l_1\omega_1+\cdots+l_s\omega_s)t} \tag{2}$$

The abbreviation $\sum_l \equiv \sum_{l_1}\cdots\sum_{l_s}$ is used from here on. Thus, $\sum_l\sum_{l'}$ implies a $2s$-fold sum: $\sum_{l_1}\cdots\sum_{l_s}\sum_{l'_1}\cdots\sum_{l'_s}$.

The classical result for P_1 is

$$\mathbf{P}_1(t) = \mathrm{Re}\sum_l\sum_{l'}\left[\frac{\partial \mathbf{C}_l}{\partial I'}\frac{(\mathbf{E}_0\cdot\mathbf{C}_{l'})}{\omega'_l + \nu_0} - \mathbf{C}_l\cdot\frac{\partial}{\partial I}\frac{(\mathbf{E}_0\cdot\mathbf{C}'_l)}{(\omega'_l + \nu_0)}\right]e^{i(\nu_0+\omega_l+\omega'_l)t} \tag{3}$$

A glance at this classical result shows that $P(t)$ contains not just the frequency ν_0, but all kinds of other frequencies as well. (Recall that $\omega'_l = l'_1\omega_1 + \cdots + l'_s\omega_s$.) In the earlier discussions, only that part of $P_1(t)$ was considered which had the time dependence $e^{i\nu_0 t}$. The corresponding terms in the double sum in Equation (3) reduces to a single sum where $\omega + \omega' = 0$. Call that part of $P_1(t)$ the coherent part, $P_1^{\mathrm{coh}}(t)$. Some simple, but rather lengthy, manipulations then show that

$$\mathbf{P}_1^{\mathrm{coh}} = \mathrm{Re}\sum_l\frac{\partial}{\partial I}\left(\frac{\mathbf{C}_l(\mathbf{E}_0\cdot\mathbf{C}_l^*)}{\omega_l - \nu_0} + \frac{\mathbf{C}_l^*(\mathbf{E}_0\cdot\mathbf{C}_l)}{\omega_l + \nu_0}\right)e^{i\nu_0 t} \tag{4}$$

This is still a purely classical result. The replacement of the orbit parameters $I_1, \ldots, I_l$ and the other quantum transcriptions, steps (III)–(V), can be carried

† It should be stressed again that factors m, c, and especially 2π are just omitted from the formulas. On the other hand, the vector notation is maintained because it does affect the structure, the factor 2π does not either.

out exactly as before, and the result is the quantum form of the coherent part of the dipole moment when the atom is in state k:

$$
\begin{aligned}
P_k^{\text{coh}} &= \operatorname{Re} \sum_i{}' \frac{1}{h}\left(\frac{\mathbf{a}_{ik}\cdot(\mathbf{E}_0\mathbf{a}_{ik}^*)}{\nu_{ik}-\nu_0} + \frac{\mathbf{a}_{ik}^*(\mathbf{E}_0\cdot\mathbf{a}_{ik})}{\nu_{ik}+\nu_0}\right)e^{i\nu_0 t} \\
&\quad - \operatorname{Re} \sum_j{}'' \frac{1}{h}\left(\frac{\mathbf{a}_{kj}\cdot(\mathbf{E}_0\mathbf{a}_{kj}^*)}{\nu_{kj}-\nu_0} + \frac{\mathbf{a}_{kj}^*\cdot(\mathbf{E}_0\cdot\mathbf{a}_{kj})}{\nu_{kj}+\nu_0}\right)e^{i\nu_0 t}
\end{aligned}
\tag{5}
$$

In the summation over i, ν_{ik} is positive, so the energy $E_i > E_k$. In the sum over j, ν_{kj} is positive, so the energy $E_j < E_k$. Expression (5) thus presupposes a particular ordering of the energy levels denoted by i, j, k as follows:

Expression (5) is not exactly the notation Kramers and Heisenberg used; they used $\mathfrak{A}$ for the amplitudes and $\mathfrak{E}$ for E, which makes the formulas more difficult to decipher. More important is that they did not use a sum over states, but instead a sum over *types* of oscillator. Kramers in his first paper on dispersion, where E and P were assumed parallel, wrote the counterpart of Equation (5) not in the form expected from (5),

$$
P = E\left(\sum_{i>k}\frac{A_{ik}}{\nu_{ik}^2(\nu_{ik}^2-\nu_0^2)} - \sum_{j<k}\frac{A_{kj}}{\nu_{kj}^2(\nu_{kj}^2-\nu_0^2)}\right) \tag{5a}
$$

(here A_{ik} and A_{kj} are the Einstein coefficients for spontaneous transition from i to k and k to j, respectively) but instead in the form

$$
P = E\left(\sum_a \frac{A_a}{\nu_a^2(\nu_a^2-\nu_0^2)} - \sum \frac{A_e}{\nu_e^2(\nu_e^2-\nu_0^2)}\right) \tag{5b}
$$

The physical meaning of (5a) and (5b) is identical; the sum over j is over all states j, to which k can decay by spontaneous emission. The corresponding sum over e in (5b) is over all emission processes. Similarly, the sum over i in (5b) is over all states that can spontaneously go to state k. The sum over a in Kramers' original language is "over all states a which via spontaneous emission can end up in the prescribed state as final state." This same terminology is used again in the Kramers–Heisenberg paper, where Equation (5) is written in the form

$$
\begin{aligned}
P_1^{\text{coh}} &= \operatorname{Re}\frac{1}{h}\sum_a\left(\frac{\mathfrak{A}_a\cdot(\mathbf{E}\mathfrak{A}_a^*)}{\nu_a-\nu_0} + \frac{\mathfrak{A}_a^*\cdot(\mathbf{E}\mathfrak{A}_a)}{\nu_a+\nu_0}\right)e^{i\nu_0 t} \\
&\quad - \operatorname{Re}\frac{1}{h}\sum_l\left(\frac{\mathfrak{A}_e\cdot(\mathbf{E}\mathfrak{A}_e^*)}{\nu_e-\nu_0} + \frac{\mathfrak{A}_e^*(\mathbf{E}\cdot\mathfrak{A}_e)}{\nu_e+\nu_0}\right)e^{i\nu_0 t}
\end{aligned}
\tag{5c}
$$

The similarity (identity) of Equations (5) and (5c) is no doubt clear; the sum over a is over all absorption oscillators [in direct analogy to the first term in

Equation (5b)] or as Kramers and Heisenberg call it "over those frequencies ν_a where the system exhibits selective absorption." The sum over e is as before over all frequencies of spontaneous emission. It is probably also clear that to obtain form (5) or (5c), the Fourier coefficients were replaced by the characteristic amplitudes $\mathfrak{A}$. In Kramers' original derivation, the product $|C_l|^2$ was replaced by an Einstein A. Form (5) appears as a factorization in linear terms of the resonance denominator for $\nu_{ik}^2 - \nu_0^2$. This also shows that the amplitude combinations occurring in the i sum (and also the j sum) in Equation (5) refer to the same type processes.

The different notations actually reflect different underlying physical pictures. Kramers interpreted the expression P^{coh} in Equation (5b) for the dipole moment of the atom as a sum of the dipole moments of individual oscillators. These oscillators, the BKS virtual oscillators, were considered as having a certain physical reality, as was described so extensively in the BKS theory. There were *two* classes of such oscillators—the *absorption oscillators* which yield the first term in (5b) and the *emission operators* which are responsible for the second (negative) term. They even possessed rather different properties, one exhibiting *positive dispersion* and the other *negative dispersion.* This same distinction is maintained in the more general formula (5c) of Kramers and Heisenberg. Both expressions are consistent with the picture that the induced dipole moment of the atom can be described as a combination of the dipole moments of individual virtual oscillators. All this is in tune with the BKS ideas.

However, expression (5), which is numerically identical to (5c), describes this same dipole moment in rather different terms as due to transitions within the atom. Instead of attributing the dipoles to individual (virtual) oscillators, the dipoles are properties of pairs of atomic states, as expressed, for example, by the a_{ik}. The individual terms in the sum thus refer *not* to individual (virtual) oscillators, but to particular transitions. It is very interesting that this new point treats *all* transitions uniformly. The two sums in Equation (5) $\sum_i, i > k$, and $\sum_j, j < k$, corresponding to the two classes of oscillators, can be written as a single sum over *all* i.[†] To show this, recall that the Fourier coefficients C_l possess the symmetry $C_l = C_{-l}^*$. Since the Fourier coefficients transcribe to amplitude from state $i \to k$, this imples that $a_{ik} = a_{ki}^*$ (Kramers and Heisenberg would write $\mathfrak{A}_q = \mathfrak{A}_{-q}^*$.) One has always that $\nu_{ik} = -\nu_{ki}$. If these symmetries are used in the second (negative) term of Equation (5), it is easy to see that the sign changes and two terms result which are identical in form to the first two (positive) terms, allowing the dipole moment to be written (the common time factor is omitted)

$$P_k^{\text{coh}} = \operatorname{Re} \sum_i \frac{1}{h}\left(\frac{\mathbf{a}_{ik}\cdot(\mathbf{E}_0\cdot\mathbf{a}_{ik}^*)}{\nu_{ik}-\nu_0} + \frac{\mathbf{a}_{ik}^*(\mathbf{E}_0\cdot\mathbf{a}_{ik})}{\nu_{ik}+\nu_0}\right) \tag{6}$$

The sum is now over *all* states i. To check this, consider the first negative terms in Equation (5), say

[†] It is not clear whether this was known to Kramers and Heisenberg. It is pretty clear that this observation goes against the BKS philosophy, so it is unlikely that they would have pursued it.

$$-\mathrm{Re}\sum_j{}' \frac{1}{h}\frac{\mathbf{a}_{kj}(\mathbf{E}_0\cdot\mathbf{a}^*_{kj})}{\nu_{kj}-\nu_0} = -\mathrm{Re}\sum_j{}' \frac{1}{h}\frac{\mathbf{a}^*_{jk}(\mathbf{E}_0\cdot\mathbf{a}_{jk})}{-\nu_{jk}-\nu_0}$$

This has the identical form as the second positive term in (5). The same applies to the second negative term, so that form (6) results. In this version the two types of oscillator of the earlier formulation have disappeared; all the frequencies refer to transitions between states i and k, where now E_i could be smaller or larger than E_k.

It is possible to carry the interpretation of this expression for the induced dipole moment in terms of stationary states and the transitions between them somewhat further. In a classical description, the external field $E_0 e^{i\nu_0 t}$ alters the orbits of the electron, and the classical expression (4) is just an indication of how these distorted orbits produce a new net electric dipole moment. In the BKS description, this same dipole moment is influenced and affected by the external field. In the quantum description as given by (6), just properties referring to the stationary states occur, such as ν_{ik} and a_{ik}. It is now very suggestive to associate a possible set of processes, or a possible sequence of atomic transitions, with the individual terms of formulas (5) and (6). If it is recalled that the characteristic amplitudes a_{ij} were introduced from the Einstein A coefficients, it becomes clear that in some way a_{ij} measures the likelihood of a transition from i to j. Since $a^*_{ij} = a_{ji}$, it measures the likelihood of a transition from j to i. Thus, a term of the form $a_{ik}(Ea^*_{ik})$ [as occurs in Equation (5) for $i > k$] can be described as a process where an atom in state k under the influence of the field E makes a transition to i (as indicated by $Ea^*_{ik} = Ea_{ki}$) and subsequently the state i decays to k (as suggested by a_{ik}). The process may be symbolized by the diagram

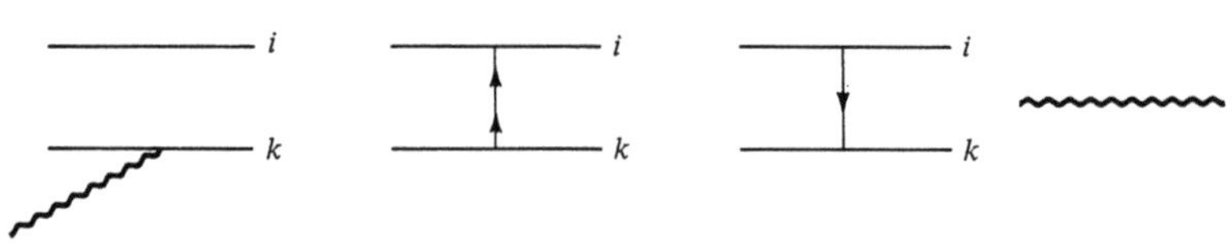

More precisely, there is a finite probability that the sequence of processes described takes place. There is also a finite probability that in this process light of frequency $h\nu_0 - h\nu_{ik}$ is emitted. This is in harmony with the resonance denominator $\nu_0 - \nu_{ik}$ in (5).† The other terms in Equation (5), $(a^*_{ik})(Ea_{ik})$, can be similarly described. It should suffice to just exhibit the diagram

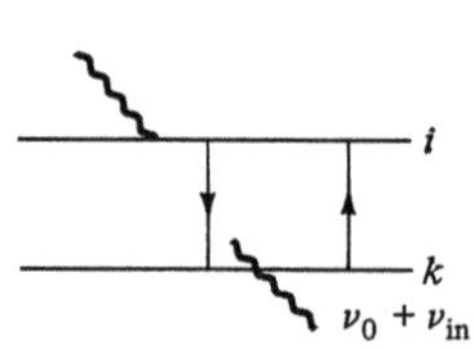

† There are hints in Kramers' diaries that he thought along the pictorial lines indicated here. The language and the type of description employed in the Kramers–Heisenberg paper certainly suggest such an interpretation.

This representation of the induced moment in terms of the transition frequencies ν_{ik} and the characteristic amplitudes a_{ik} must be understood as a means of visualizing the response of the atom to the incident radiation. The description in terms of the possible transitions is the quantum counterpart of the distortion of the classical orbit by the field. But the use of the transition amplitude language no more implies the occurrence of actual physical observable processes than the earlier use of the oscillator language implies the existence of physical observable oscillators. It marks the beginnings of a shift from virtual oscillators to virtual processes.

It is interesting to point out that a full-fledged quantum mechanical treatment (which of course did *not* exist at the time the Kramers–Heisenberg paper was written) gives for the expectation value of the electric dipole in state k (the kk matrix element of P) exactly the value (6) guessed by Kramers and Heisenberg. The only thing quantum mechanics supplies, which Kramers–Heisenberg did not (could not) have, is an explicit expression of the characteristic amplitudes a_{ik} in terms of the wavefunctions of the stationary state. The a_{ik} conjectured by Kramers and Heisenberg to be important are indeed of fundamental significance; they are exactly the matrix elements or transition amplitudes of the electric dipole

$$\mathbf{a}_{ik} = \int u_i^*(e\mathbf{x})u_k\, d^3x \tag{7}$$

The u_i, u_k are eigenfunctions of the atom in the i and k states. As noted earlier, the a_{ik} do depend exclusively on the unperturbed system. The property $a_{ik} = a_{ki}^*$, derived by Kramers and Heisenberg from the symmetry $C_l = C_{-l}^*$, expresses just the Hermitian character of the dipole matrix. But none of this was known or even fathomed by Kramers and Heisenberg when they introduced the "characteristic amplitudes."

It was a reasonably simple exercise to reobtain the original Kramers' formula (5b) from the more general expression (5). It is only necessary to assume that all vectors $\mathbf{a}_{ik}$, $\mathbf{a}_{ik}^*$, $\mathbf{a}_{kj}$, $\mathbf{a}_{kj}^*$ are parallel to $\mathbf{E}$. Then substitution in (5) and use of Equation (e) on page 227 leads directly to (5b). The main importance of this observation is that the particular form (5b) is *not* universally valid; for general orientations, Equation (5) must be used.

It can further be seen from Equation (5) [or (5b) for that matter] that as ν_0 approaches an absorption frequency [ν_{ik} in our notation for a transition from k (lower) to i (higher), ν_a in the old notation], the absorption will increase sharply as was indeed observed. For values of ν_0 too near ν_a, the formula will cease to be valid. (This is also true for the classical result.) If ν_0 approaches an emission frequency ν_e [or ν_{kj} in our notation for a transition from k (higher) to j (lower)], formula (5) again indicates that the spherical wave of that frequency will be reinforced. But this is *not* the total effect. In addition to this stimulated emission by the external field ν_0 from k to j, there will be the *spontaneous* emission from the higher to the lower state of exactly the same frequency ν_{kj}. (The BKS picture, where the *virtual* field induces these transitions, *requires* spontaneous emission.) The net effect is a combination of the stimulated emission [as computed via Equation (5)]† and the spontaneous emission. Consequently, to obtain the result, the precise phase relations between these two waves are essential. Without detailed knowledge of these phases, it cannot be concluded that the waves of frequency ν_{kj} will be reinforced.

The results of the Kramers–Heisenberg paper so far discussed were extensions and refinements of earlier studies. The main novelty of the paper was the treatment of the incoherent part of $\mathbf{P}_1$, corresponding to the terms in Equation (3) where $\omega + \omega' \neq 0$. The principal technical difficulty in handling the expression is the proper grouping of the terms so that they can be interpreted physically. Once a convenient grouping has been found, the quantum transcription is again straightforward. The principle on which the organization of the terms in the sum is based is to isolate the terms, with the identical time dependence. As might be anticipated the incoherent part of the dipole moment in state k, from now on called P_{2k}, can be written as a *double* rather than a single sum over states:

$$P_{2k} = \mathrm{Re}\left(\sum_i \sum_j M(i,k|j) + \sum_i \sum_j M(k,i|j)\right) \tag{8}$$

where‡

$$M(k,i|j) = \frac{1}{4h} e^{i(\nu_{ik}+\nu_0)t}\left(\frac{a_{ij}(E\cdot a_{jk})}{\nu_{jk}+\nu_0} - \frac{a_{jk}(E\cdot a_{ij})}{\nu_{ij}+\nu_0}\right) \tag{9}$$

This form [Equations (8) and (9)] is obtained by combining all terms in the sum (3) with the time dependence $e^{i(\nu_{ik}+\nu_0)t}$, then transcribing these terms to quantum theory by the procedure used earlier. It is evident that, apart from possible signs and complex conjugates, the form of (9) and (8) is very similar to that of the coherent part. Expressions (8) and (9) are somewhat abbreviated; some further explanation is needed to provide all details necessary for a complete understanding. (In the Kramers–Heisenberg paper some of these details are implied, while others are omitted.) It is clear that the expression for the incoherent part P_{2k} involves double sums over two sets of states (i and j), where the coherent part P_{1k} only involves a single sum (over i). As before, it is possible to give a pictorial interpretation of the individual terms in (8) and (9). The complete specification of expressions (8) and (9) is most easily given in terms of the pictorial representation. Consider, for example, a term $a_{ij}\cdot(Ea_{jk})$ in the expression for the dipole moment. This can be described by a sequence of transitions where the atom in state k under the influence of E makes a transition to state j (term $E\cdot a_{kj}^*$); subsequently, it makes a transition to state i. Equations (8) and (9) state that the response of the atom is expressed

† It is perhaps not superfluous to stress that although the expression for P^{coh} is expressed in terms of the Einstein coefficient A, this is for mathematical convenience; the process itself is a stimulated emission process.

‡ To not encumber the notation more than is strictly necessary, the vector signs are omitted *from here on.* However, $(E\cdot a_{jk})$ is a scalar product throughout.

as a transition, that is, over i and j. It is necessary to specify the limits of those sums. In these sums, as direct inspection of P shows, there are terms involving ν_{jk} and ν_{ik}. It is possible that in the sums over i and j, $E_k < E_j$ or $E_i < E_k$. The corresponding frequencies then will be negative. It is asserted that in such circumstances—say $E_i < E_k$, so $\nu_{ik} < 0$—the corresponding amplitude a_{ik} must be replaced by a_{ik}^*. (This is perhaps not so surprising since classically $C_l = C_{-l}^*$, but its demonstration takes some care.†) The precise limits in the i and j summations can be determined either from a detailed scrutiny of the transcription process‡ or from physical arguments based on the pictorial representation of the response in terms of transition processes. The limits on the summations are connected with just what states are involved in the transition process. The words used in the argument sound strikingly similar to those used in the BKS theory. After observing that a spontaneous emission consists of harmonic components, linked to the combination of two stationary states, with the amplitudes determined by the correspondence principle, they continue: "This principle does not immediately allow us to decide from which of these two states, it is that the atom emits this spontaneous radiation, this can only be settled by an argument linked to the energy principle.... This argument states that emission always takes place from a state whose energy content is the greater."[249] This is the basic idea which will be used to find the limits on the i and j summations in Equation (9). The argument used is an interesting mixture of classical and quantum ideas, of formal manipulations and intuitive pictures.

Classically, an oscillator or an atom when acted on by an external field can absorb energy from that field or give up energy to the field. What actually happens depends on the phase relations. In a quantum version all the corresponding processes and the radiation emitted must be expressible in terms of transitions between the stationary states. The two classical possibilities, just noted, transcribe to two distinct quantum processes. Let the atom be in a state k and irradiated with a frequency ν_0. Consider the following sequence of possible events denoted by (1), (2), and (3):

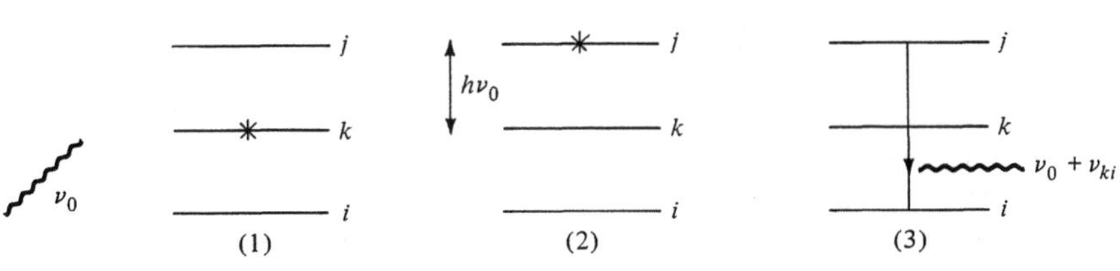

There is a finite probability that the atom will pass from configuration (1) to configuration (2); that is, there is a finite probability that the radiation is

† This is one of the results used but not proved in detail in the paper.

‡ This is not done either.

absorbed, while the atom is excited to state j.[†] In turn, the atom now has a finite probability of making a *spontaneous* transition to any state i, when $E_i < h\nu_0 + E_k$. The net result is the emission of light of frequency $\nu_0 + \nu_{ki}$. This was exactly Smekal's original result. It can be verified that processes of this type are all represented by the first sum in Equation (8). The limit on the i summation is exactly that $E_i < h\nu_0 + E_k$. The j summation is over all states j, where a_{ji} and a_{jk} are not zero. Kramers and Heisenberg warn that processes (2) and (3) are really occurring simultaneously in the atom; it is pushing the picture too far to insist on a strict temporal order. The process just described crudely corresponds to the classical resonance absorption process where an oscillator absorbs energy from the field. The classical (resonance emission) process where the oscillator releases energy to the field also has a quantum counterpart. To illustrate this process, consider a different set of possible events:

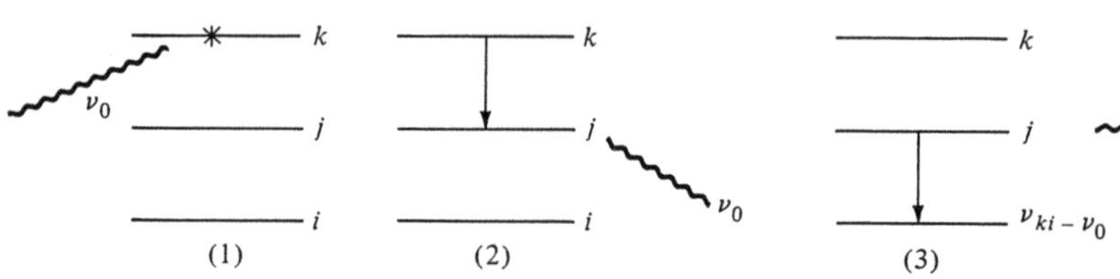

There is always a finite probability that the atom under the influence of ν_0 emits light of frequency ν_0, making a transition to state j (so that $E_k - E_j = h\nu_0$). The atom can now make a spontaneous transition from state j to i, emitting light of frequency $\nu_{ji} = \nu_{ki} - \nu_{kj} \simeq \nu_{ki} - \nu_0$. Thus, the net result is the emission of two frequencies ν_0 and $\nu_{ki} - \nu_0$. States i, to which the spontaneous transition can take place, all satisfy $E_i < E_k - h\nu_0$. These processes correspond exactly to the second terms in the sum (8). The complete result of the analysis is therefore Equation (8), which can now be written

$$P_{2k} = \mathrm{Re}\left(\sum_i{}' \sum_j M(i,k|j) + \sum_i{}'' \sum_j M(k,i|j)\right) \tag{10}$$

M is given by Equation (9) as before. The first sum, $\sum'$, runs over all states i, such that $E_k < E_k + h\nu_0$ while the second sum, $\sum''$, is over all i when $E_i < E_k - h\nu_0$. It is of interest to observe that the derivation of these results can be obtained in two quite distinct ways. The most straightforward method is by a detailed analysis of the incoherent terms in the sum ($\omega + \omega' \neq 0$), the organization of these terms, and the quantum transcription. Although in principle straightforward, this is technically not all that simple, but for two powerful formalists such as Kramers and Heisenberg, it was nearly child's play (they almost present it as such). This method requires no physical assumptions

[†] One would surely expect this process to be most efficient if $\nu_0 \simeq \nu_{jk}$.

and there is no need for a physical visualization; all that is needed and used is an organizational procedure. This is quite different from a discussion that attempts to describe the physical significance of the formulas in physical terms, as representing the effects of possible transitions within the atom. In such a description the physical picture envisaged becomes most important. The discussion presented in the last few pages employs these same intuitive, pictorial arguments, as does the paper as a whole. This is no doubt due to Kramers, who saw the Kramers–Heisenberg analysis as an amplification of the BKS ideology. Heisenberg, who was less or not committed to these ideas, tended to put more faith in the purely mathematical analysis. For example, that an atom in state k, under the influence of light of frequency ν_0, would lead to the emission of spherical waves of frequency $\nu_0 + \nu_{ki}$, as described on page 237, could be considered as a result of the mathematical analysis. The induced dipole moment contains such terms. That would be near to Heisenberg's approach. Or one could attempt to present this result in terms of transitions between atomic states. Kramers went even further; he wanted to describe these transitions as due to – or stimulated by—the virtual field. Thus, in the example given before, where an atom initially in state k absorbs light and then proceeds to state j, to later decay spontaneously to state i with emission of light of frequency $\nu_0 + \nu_{ki}$, Kramers would ascribe these transitions as mediated by the virtual field. This is also the reason that these descriptions appeal to the probability of transitions, while energy appears not necessarily conserved in the individual processes. It was mentioned that an atom in state k has a finite *probability* of absorbing light of frequency ν_0 and going to state j. If energy were to be strictly conserved, this would require $\nu_{jk} = \nu_0$ (for a heavy atom). Since j, k, and ν_0 are given, this means that the sum over j contains terms l, where $\nu_{kl} \neq \nu_0$, so that the identical visualization would lead to processes where energy is not conserved. Since the characteristic amplitudes a only supply probabilities for transitions, there was nothing contradictory about this interpretation. Rather, to Kramers these features were natural outgrowths of the BKS framework, where individual events did not require rigid conservation laws. The only remnant of the conservation of energy in this transition process description of the formal result was that spontaneous decay could take place only if $E_i > E_k$. Kramers was clearly pleased that this pictorial description not only reproduced Smekal's result, where an atom could emit the frequency $\nu_0 + \nu_{ki}$ [as in process (1)], but also provided for additional processes, where an atom under the influence of radiation of frequency ν_0 would emit two light waves of different frequencies: ν_0 and $\nu_{ki} - \nu_0$ [see process (2), p. 238]. Referring to these processes and to Smekal's work, Kramers writes: "Transitions of the latter type [here called the second type], which hardly present themselves in such a natural way from the point of view of light quanta, were not discussed." Kramers was still doing battle against the photon!

The redescription of the formal results (8) and (9) in terms of physical pictures and transition processes leads to an interesting puzzle. Consider a situation where state i is such that $E_i > E_k$. Call $E_i - E_k \equiv h\nu^*$; then $\nu^* = \nu_{ik}$. One can ask for the contribution of state i to the scattered radiation, when a frequency ν_0 is incident on the atom in state k. The scattered frequency is $\nu_s \equiv \nu_0 - \nu^*$. This is a special case of formulas (8) and (9). The result can be written

$$\begin{aligned} M_{ki}(\nu_0 - \nu^*) = \mathrm{Re}\Bigg[&\sum_{j>i} \frac{a_{ji}(E \cdot a_{kj}^*)}{\nu_1 - \nu_0} + \frac{a_{kj}^*(E \cdot a_{ji})}{\nu_2 + \nu_0} \\ &+ \sum_{k<j<i} \frac{a_{ji}^*(E \cdot a_{kj}^*)}{\nu_3 - \nu_0} - \frac{a_{kj}^*(E \cdot a_{ji}^*)}{\nu_4 - \nu_0} \\ &+ \sum_{j<k} - \frac{a_{ji}^*(E \cdot a_{kj})}{(\nu_5 + \nu_0)} - \frac{a_{kj}(E \cdot a_{ji}^*)}{(\nu_0 - \nu_0)} \Bigg] e^{i(\nu_0 - \nu^*)t} \end{aligned} \tag{11}$$

In Equation (11) ν_1, ν_3, and ν_5 are the absolute values of ν_{jk} in their respective regimes; ν_2, ν_4, and ν_6 are the absolute values of ν_{ij}. All these terms can be associated with a sequence of transitions. (A double arrow, accompanied by E, indicates a transition influenced or stimulated by the field.)

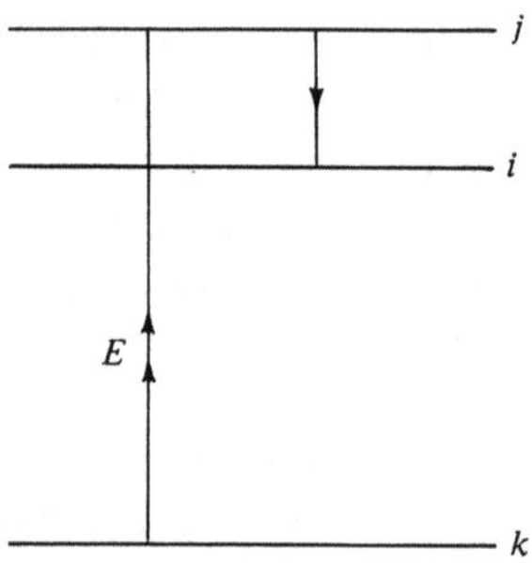

(Ia)

$$\frac{(E \cdot a_{kj}^*)(a_{ji})}{\nu_{jk} - \nu_0} \qquad (\nu_{jk} \equiv \nu_1)$$

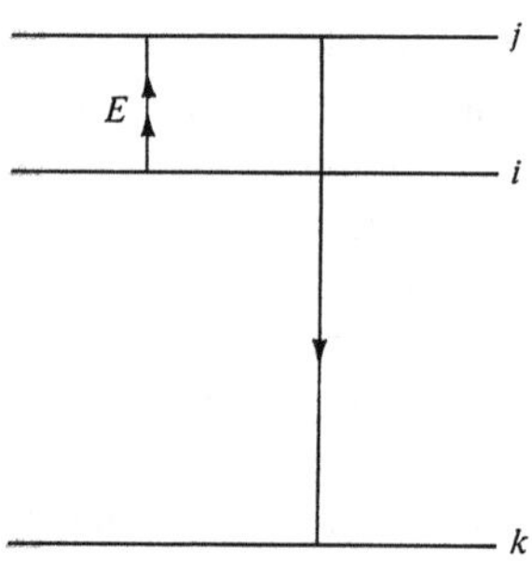

(Ib)

$$\frac{(E \cdot a_{ij}^*)a_{jk}}{\nu_{ji} + \nu_0} \qquad (\nu_{ji} \equiv \nu_2)$$

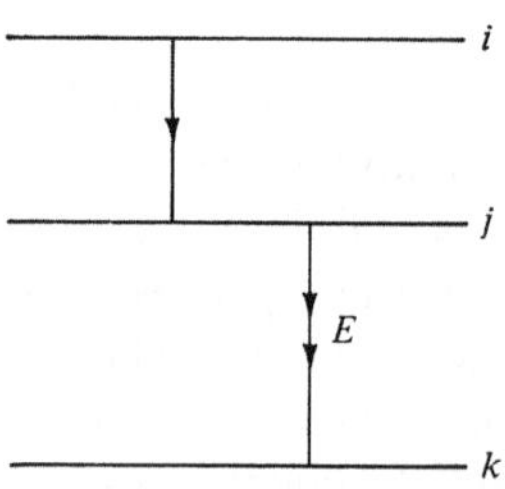

(IIa)

$$\frac{a_{ij}(Ea_{jk})}{\nu_{jk} - \nu_0} \qquad (\nu_{jk} \equiv \nu_3)$$

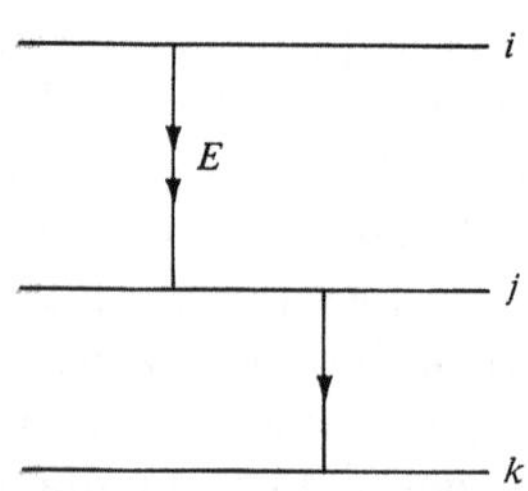

(IIb)

$$\frac{(Ea_{ij})a_{jk}}{\nu_{ij} - \nu_0} \qquad (\nu_{ij} \equiv \nu_4)$$

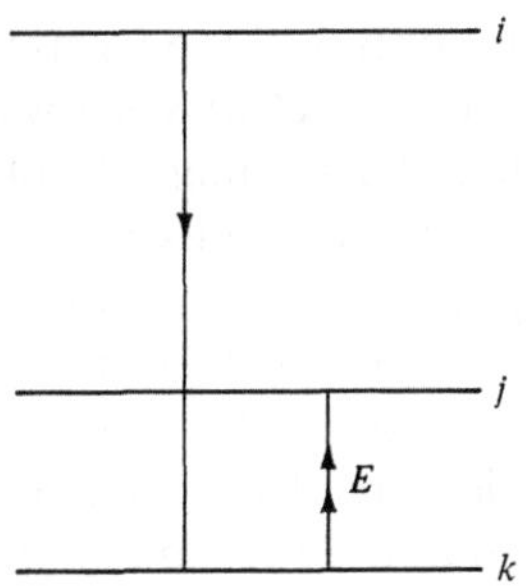

(IIIa)

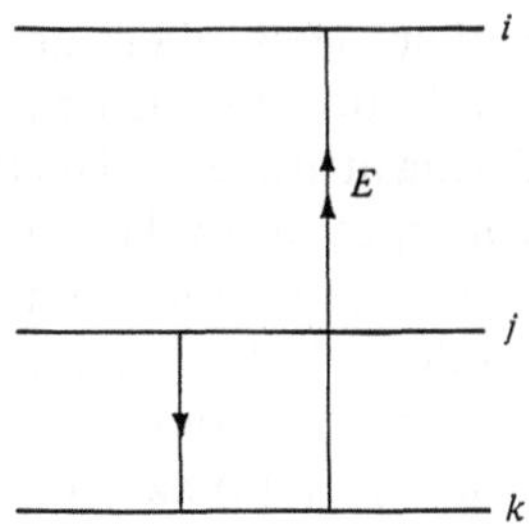

(IIIb)

$$\frac{a_{ij}(Ea_{jk}^*)}{\nu_{kj} + \nu_0} \quad (\nu_{kj} \equiv \nu_5) \qquad \frac{a_{kj}(Ea_{ji}^*)}{\nu_{ij} - \nu_0} \quad (\nu_{ij} \equiv \nu_6)$$

In obtaining these pictures, frequent use has been made of the symmetries $a_{ij} = a_{ji}^*$. As always, a_{ij} is considered as the amplitude for a spontaneous transition from E_i to E_j with $E_i > E_j$. Similarly, a_{kj}^* is a measure of the amplitude for a stimulated emission for a transition in which $E_k < E_j$. (In the figures all single arrows corresponding to unassisted transitions point down.) The anticipated physical behavior can now be read, either by examining the figures or from the formulas. It is evident, for example, by looking at (Ia) that as ν_0, the incident frequency, approaches $\nu_1 \equiv \nu_{jk}$, the dispersive moment will increase sharply. It is also clear from (Ia) that as ν_0 approaches ν_{jk}, the incoming radiation will be very effective in exciting the atom to state j, thereby increasing the absorption sharply. Under those circumstances $\nu_0 \simeq \nu_{jk}$, and one has $\nu_0 \simeq \nu_{jk} = \nu_{ji} + \nu_{ik} = \nu_{ji} + \nu^*$. Consequently, the scattered frequency ν_s approaches ν_2 for $\nu_s = \nu_0 - \nu^* = \nu_{ji} = \nu_2$. One can now expect that, if the atom is irradiated with a frequency $\nu_0 = \nu_1 = \nu_{jk}$, the scattered secondary radiation will exhibit a pronounced maximum around the frequency ν_2. Thus, the formulas, together with the figures provide [via Equation (11)] a purely quantum description of M_{ki}. M_{ki} is the total scattering moment of the atom in state k, which leads to definite frequencies $|\nu_0 \pm \nu^*|$ in the scattered radiation due to the presence of state i.†

Satisfactory as this may appear, there were still a number of subtle points, which gave rise to lively, intense, and sometimes heated arguments between Kramers and Heisenberg. Consider figure (IIIb). It appears from its form that there would be a resonance in the dipole moment M_{ki} if the incident frequency ν_0 approaches the value $\nu_6 = \nu_{ij}$. That is a little peculiar; intuitively one would expect that resonances in M_{ki} would only occur if ν_0 approaches a frequency corresponding to a transition to—or from—the state of the atom under consideration (state k in this case; but ν_0 has nothing to do with state k). So strong was Kramers' conviction that such terms were without physical relevance that he was not only willing but anxious to drop these unphysical

† In contemporary quantum mechanical terms, Kramers and Heisenberg guessed or derived the explicit expression for the (ki) matrix element of the dipole moment.

terms altogether.[†] Heisenberg argued that, important or not, these terms were there and could not just be dropped at will. This conflict clearly shows the sharply differing attitudes of Heisenberg and Kramers. To Kramers the physical ideas, concepts, and relation of the theoretical structure to experimental material were of paramount importance. The logical deductions could, for Kramers, reinforce but never replace the intuitive insight. For Heisenberg (at least at this stage in his development), the uncompromising logical pursuit of an idea was almost an obsession; he in no way could agree to dropping such terms. The resolution to this apparent conflict was found by Kramers. Recall that the scattering frequency is $\nu_s = \nu_0 - \nu^*$ ($\nu^* = \nu_{ik}$). Suppose ν_6 differs little, say δ, from ν_0, so that $\nu_6 - \nu_0 = \delta$. From graphs (IIIa) and (IIIb), one has $\nu_{ij} - \nu_{kj} = \nu_{ik}$, or $\nu_6 - \nu_5 = \nu^*$; combining these relations gives $\nu_5 = \delta + (\nu_0 - \nu^*) = \delta + \nu_s$. Thus, as ν_0 approaches the apparent resonance ν_0, ν_5 in turn approaches the scattering frequency ν_s. In fact, the contribution to the induced moment is proportional to $e^{i(\nu_5 - \delta)t}/\delta$. But as δ gets smaller, the resulting radiation can interfere with the spontaneously emitted radiation from state K to state j which is also of frequency $\nu_{kj} = \nu_5$. It is the interference between these two terms, as ν_0 approaches ν_6, which removes the apparent resonance in the dipole moment, because the combined effect of these two terms depends crucially on the phase relation between these two waves. Where one could calculate them, there was a destructive interference. This mechanism is general; if among the terms contributing to M, there is one that appears to have a resonance at a frequency *not* corresponding to a transition to or from the state (k) under consideration, there is always a term in the spontaneous emission that interferes with it and cancels the resonant term. With this satisfactory, and somewhat unexpected explanation of the nature of the "false," formal resonances, the paper was rapidly completed.

D. From Virtual Oscillators to Matrices: Heisenberg Goes It Alone

The Kramers–Heisenberg paper was well received, and its influence was immediate and widespread. That is a little bit surprising because the paper is not easy to read even now. It has unusual sign conventions and rather terse mathematical arguments that make the equations difficult to follow and hard to check. It was not an extension or application of the Bohr theory or Bohr–Sommerfeld quantum rules; instead, it was a mixture of Bohr's quantum theory, the correspondence principle, guesses, and transcriptions, all loosely based on the BKS philosophy. But perhaps most unusual, for a paper that was accepted so rapidly, is the fact that there was no experimental evidence whatsoever to support the Kramers–Heisenberg formula. The original Ladenburg formula fit the experimental data quite well, but the Kramers

[†] Almost as if to show his contempt for these terms, he called them "false resonances."

formula, which theoretically was considered superior, did not receive any empirical support until several years later. Kopferman and Ladenburg[250] observed the negative dispersion, required by the second term in Kramers' expression, only in 1928. The Smekal and Raman lines included in the Kramers–Heisenberg formula were not known either (they were found in 1928). Thus, there was literally no direct empirical evidence for the Kramers–Heisenberg results. Even so, there was a belief that relations of that general type constituted a step in the right direction, toward the elusive goal of constructing a more complete and autonomous theory of atomic phenomena. Heisenberg caught the general mood very well: "One felt that one had now come a step further in getting into the spirit of the new mechanics. Everybody knew that there must be some new kind of mechanics behind it. Nobody had a clear idea of it, but still one felt that the dispersion formula was a good step in the right direction."[251] Great as the influence of the Kramers–Heisenberg paper was in physics, its greatest impact was undoubtedly on the authors themselves. As might be expected they reacted in characteristically different ways.

There was little disagreement about the major results of the paper. It contained the exposition of Kramers' method, leading to the Kramers–Heisenberg formulas. That was generally regarded as a spectacular result. Even now—60 years later—after the systematic and explosive development of quantum mechanics, the dispersion formula stands, correct, unaltered, and valid in the domain of its derivation.† To guess such a result on such an incomplete and shifting basis was a remarkable feat. The characteristic amplitudes ($\mathfrak{A}_q$ or a_{ik}) were originally introduced in a somewhat off-hand manner, but as the paper progresses, there are increasing hints that the authors feel that these amplitudes are particularly important objects, whose full significance still remains to be grasped, like the tentative announcement of a musical theme, whose full development is yet to come. It was especially surprising that the phases of these complex amplitudes played such an important role. They after all were the essential ingredients in Kramers' resolution of the paradox of the "false" resonances. Although there was general agreement about these matters, Kramers and Heisenberg adopted distinctly different attitudes toward the meaning and significance of those results. For Kramers, the important insights obtained were all a confirmation of the basic soundness and relevance of the BKS philosophy. The description of the formulas in terms of transitions stimulated by oscillators was a vivid if not direct illustration of the importance of the virtual oscillator picture. The paper ends by relating its results directly to the BKS notions. After commenting that spontaneous emission must be described as the radiation in a definite stationary state and not as an action between states, Kramers concludes: "But this is just the hypothesis introduced by Slater which led to the views that Bohr–Kramer–Slater developed in more

† It is one of the very few results that retained its physical and formal validity after quantum mechanics replaced the classical description.

detail.... These can be formulated in such a way that the role played by the atom in optical phenomena always can be reduced to interactions between the radiation field and the atom in some arbitrary stationary state."[252] This is how the paper ends; the beginning abstract is even more explicit in acknowledging its connection with BKS. "Our conclusions, should they be justified, can be expected to constitute an interesting confirmation of these concepts" (i.e., of the BKS concepts).† Thus, all the formalism is sandwiched between tenets of the BKS philosophy. There can be little doubt that Kramers considered the Kramers–Heisenberg relations as a triumphant consequence of the BKS approach. It was quite literally its last success: This clarifies further (if such clarification were still needed) how profoundly shattering the destruction of the BKS framework must have been to Kramers—the very framework that had, as he saw it, led him to his most brilliant, most conspicuous success. That he was loath to give up the ideas of BKS and did so only reluctantly and slowly is completely understandable.

To Heisenberg, the results of the Kramers–Heisenberg paper had little—if anything—to do with the BKS philosophy. He was interested, if not intrigued, by the methodology of the paper: the curious mixture of classical physics and the judicious guessing that led to such interesting results. But he was particularly impressed by the nature of the results obtained—formulas that contained only quantum concepts. As Kramers had done before him, Heisenberg also emphasized that the dispersion formulas related observable quantities. No trace of the features of the underlying atomic model such as electron orbits remained. He was also struck by the important role played by the amplitudes and their phases. "That made a strong impression on me at that time. Then I saw that the analogy of the quantum theoretical quantities to the Fourier components in classical physics was really very close, because not only the absolute value (of the characteristic amplitudes) but even the phases must be well defined.... The important thing was that one could actually have interference. Therefore all these interference arguments of Kramers were important, one could see that there the phases come in and are important."[253,254]

Thus, for Heisenberg the main importance of the Kramers–Heisenberg dispersion paper were the formal relations—the mathematical structures—obtained. The method—although interesting—was of less significance, and the possible physical picture underlying the method was even less important. For Kramers, on the other hand, the results obtained were important and interesting, but the implied validity of the underlying physical ideas—of the intuitive almost philosophical ideas of the BKS theory—was by far of greater importance. So impressed was Heisenberg by the formal structure of the dispersion relations and by their exclusive dependence on observables, that he felt that relations of this type, and the entities occurring in them, should be the basic ingredients on which a theoretical scheme should be constructed.

† See page 223 for the complete quote.[228]

Thus, the dispersion-type relations involving amplitudes and frequencies, in Heisenberg's view, should not be considered as consequences of an underlying theory but instead as the direct expression of the physical laws governing the behavior of atoms. This general idea became Heisenberg's guiding principle which eventually led to the development of matrix mechanics. It was precisely at this juncture that Kramers and Heisenberg parted company. The inversion that Heisenberg suggested, whereby dispersion-type relations expressed in amplitudes would become the primary structures instead of derivatives of a classical mechanical picture, was—if not unacceptable—certainly foreign to Kramers. Since the underlying physical ideas of the Kramers–Heisenberg paper were so fundamental to Kramers, he would be most reluctant to replace these ideas by a selected set of consequences, no matter how important.

To Heisenberg, the dispersion results were so striking that *they*, rather than the more doubtful and somewhat vague principles on which they were based, should form the foundation for a coherent theoretical scheme. To Kramers, these same principles, with all their vagueness, *were* the theoretical scheme. Although it sounds perhaps a bit pretentious, it is nonetheless true that to Kramers the *physical reality* was to be described by classical pictures with the BKS ideas of the virtual fields superimposed. To Heisenberg, the dispersion relations themselves were the physical reality. They connected in principle observable features to each other; they did not contain *any* of the features associated with the classical mechanical characteristics of atoms, such as orbits, positions, and velocities. This basic difference in view makes it understandable why Heisenberg, after the Kramers–Heisenberg dispersion relations were completed, would avidly pursue a formulation of atomic phenomena suggested by the dispersion relations, using just amplitudes and frequencies and eliminating everything else, while Kramers would instead explore new and different consequences of the BKS conceptual scheme. With the collapse of the BKS theory, Kramers' efforts naturally came to an abrupt halt, while the collapse had no effect on Heisenberg.

Although it is undoubtedly true that the experience and knowledge Heisenberg acquired in his collaboration with Kramers were of unestimable importance in Heisenberg's formulation of matrix mechanics, it is equally true that the actual construction of matrix mechanics was a step of extraordinary boldness, brilliantly executed and fully justifying the acclaim and recognition that Heisenberg received. The initial, necessary somewhat tentative, approach of Heisenberg toward matrix mechanics was strongly influenced by the methods of the joint Kramers–Heisenberg paper. In fact, upon his return to Göttingen, Heisenberg started these explorations by an almost literal imitation of the procedures of the Kramers–Heisenberg paper. As recorded earlier (see page 218), Heisenberg tried to *guess* the intensity of the hydrogen lines, from the Fourier coefficients of the Kepler motion. As Heisenberg recalls: "I made a first attempt to guess what formulae would enable one to express the line intensities of the hydrogen spectrum, using more or less the same methods that had proved so fruitful in my work with Kramers in Copenhagen. This attempt led to a dead end but the work helped to convince me of one thing: that one ought to ignore the problem of electron orbits inside the atom and treat the frequencies and amplitudes associated with the line intensities as perfectly good substitutes."[255] This recollection of Heisenberg shows very clearly that at this early stage he was contemplating an inversion in the traditional logical patterns. Instead of considering the dispersion relations and the entities in which they were expressed as derived from a classical model, Heisenberg wanted to consider amplitudes and frequencies as primary objects, which might or might not be expressed in terms of the characteristics of the model. In any case, the basic laws—of which the Kramers–Heisenberg relation was a first example—should be expressed directly in those terms. This shift of emphasis was due to Heisenberg; there is no evidence that Kramers ever entertained such thoughts. Indeed, from Kramers' approach to physics generally and to dispersion theory in particular, it seems most unlikely that he would consider such an inversion. But the same theme—the elimination of the orbit idea and the establishment of mathematical relations between observables—recurs frequently in Heisenberg's correspondence of that period: "The basic idea [of the paper] is that in the calculation of all quantities [energy, frequency] only relationships between 'in principle' controllable quantities may occur. (In that sense the Bohr theory of hydrogen appears much more formal to me than the Kramers dispersion theory)."[256] Later in the same letter, he returns to a related point: "I would like to understand what the meaning of equations of motion really is, when they are considered as relationships between transition probabilities." It is evident that Heisenberg was attempting to carry out his program to phrase the physical laws in terms of transition probabilities—but he was not sure at that time that he had accomplished it. Since the transition probability notion was introduced in the Kramers–Heisenberg paper, it is clear that this paper provided one of the crucial ingredients for the formulation of matrix mechanics.

There is another important element in Heisenberg's considerations which also had its origin in a somewhat indirect way in the Kramers–Heisenberg cooperation. It was observed several times that the transcription from a classical to a quantum result was a combination of guesswork and formal insight guided by physical intuition, eventually justified by an appeal to the "sensible" character of the result. But it was not an organized, systematic procedure; in fact, when Heisenberg tried a similar approach, in the case of hydrogen, it did not work. Since these quantum transcriptions were supposed to be the key, it was necessary to develop a general transcription procedure that would associate a quantum concept with every classical concept. The search for a systematic universal transcription process became one of Heisenberg's main preoccupations. In this search, the successful transcription carried out in the Kramers–Heisenberg paper could serve as a model to suggest the precise nature of this transcription process. It is not a coincidence that Heisenberg's paper contains in the title the word *Umdeutung* (Reinterpretation) because it was the recognition that there existed a universal quantum mechanical reinterpretation of classical quantities which was central to Heisenberg's procedure. Heisenberg's fundamental paper thus dealt with two basic questions, both allied to the Kramers–Heisenberg investigation: the identification of a universal transcription procedure to quantum theory and the elimination of the model-dependent notions from the basic laws. Heisenberg, with extraordinary originality and daring, set up a formalism that handled both. But for this accomplishment the Kramers–Heisenberg investigation was an indispensable preliminary, as Heisenberg wrote later: "I found in the formulae, which were the result of my collaboration with Kramers, a mathematics which in a certain sense worked automatically independently of all physical models. This mathematical scheme had for me a magical attraction, and I was fascinated by the thought that perhaps here could be seen the first threads of an enormous net of deep set relations."[257]

Heisenberg was well aware that certain fragments of a transcription procedure from classical to quantum concepts existed already. The replacement of derivatives by finite differences, the substitution of Bohr frequencies ν for mechanical frequencies ω, and in the Kramers–Heisenberg paper the replacement of Fourier coefficients C by the amplitudes a were all known. So one might well ask what else could be said about this transcription process. In this connection, Heisenberg made two absolutely fundamental observations.

(1) The method whereby the power emitted by an atom was calculated was well established. Starting from the classical expression for that power, we obtain

$$\frac{dW}{dt} = \frac{2}{3}\frac{e^2}{c^3}|\ddot{x}|^2 \tag{1}$$

The development of $x(t)$ in a Fourier series allows the calculation of the energy emitted (per second) in a given frequency interval as proportional to

$$\frac{dW}{dt} = \frac{2}{3}\frac{e^2}{c^3}\omega^4|C_\omega|^2 \tag{2}$$

(The notation is the same as always.) With the usual quantum description, this becomes

$$\frac{dW}{dt} \approx \frac{2}{3}\frac{e^2}{c^3}\nu_{ij}^4|a_{ij}|^2 \tag{3}$$

(A very similar formula was employed by Kramers in his early paper on x rays. These relations were generally known and often used.) From this procedure one could already infer something about the nature of the transcription process to quantum theory. It might crudely be represented as a succession of transcriptions:

$$\mathbf{x}(t)\,(\text{position}) \rightarrow \text{Fourier coefficients} \rightarrow \text{amplitudes} \tag{4}$$

The classical radiation field is always expressed in terms of the electric and magnetic fields E and B. For a moving electron, these can be written as a series of successive approximations, each involving different functions of the position (x), velocity (v), and acceleration (a) of the electron. The classical expressions for the fields used so far were always the lowest-order approximations (dipole approximations) and these expressions were transcribed to quantum theory. The incisive question Heisenberg raised was: How does one transcribe the higher (quadrupole) to quantum theory? It was a natural question; but no one had asked it before. In these higher terms, which of course still would depend on $x(t)$, $v(t)$, and $a(t)$ (since E and B depend on these terms), various products and powers of $x(t)$ and $v(t)$ would occur. The quantum transcription then requires the translation rules for terms such as x^2, v^2, and xv. The general question is: If a classical quantity x transcribes to a quantum entity X, while y transcribes to Y, what is the transcription procedure for $x + y$, xy, x^2, y^2, and so on? In other words, Heisenberg raised the general problem of the mathematical or algebraic structure of the transcription process. This was not only a problem of the greatest logical significance, but of practical importance as well. The classical quadrupole radiation, for example, depends on the scalar product of velocity and acceleration. Since there are many physical systems where the quadrupole radiation plays an important role, it becomes necessary to know the quantum transcription of that product. Such information was not needed for the discussion of the dipole radiation. It was Heisenberg's special insight to phrase this question in an abstract setting. The distinction between incidental features and the essential aspects of the transcription process became clear only in the abstract formulation.

(2) The second observation, perhaps less dramatic, was nonetheless of comparable significance. It concerned the quantities to which the transcription procedure should be applied. It was clear from long experience that spectral lines emitted exhibited quantum features; thus the intensities of those lines must be expressible in quantum terms. As stressed in observation (1), classically the intensities depend on the electric and magnetic fields which themselves are functions of the positions $x(t)$, velocities $v(t)$, and accelerations $a(t)$. To obtain a quantum description it is therefore enough to give a quantum transcription of the *kinematic* quantities $x(t)$, $v(t)$, and $a(t)$. Once that is done, the electric and magnetic fields and consequently the intensity of the emitted radiation are obtained in quantum terms. This is precisely the generalization of the procedure followed earlier, but the emphasis that the quantum transcription, in the first instance, should apply to the kinematical characterization was Heisenberg's.

This can be viewed (and was so considered by Bohr) as a precise formulation of the correspondence principle, because the *formal* expression of the emitted radiation in terms of the kinematical quantities is the same as in classical theory—just the entities occurring in these expressions are different. Bohr had long insisted that in some limit there would be a close correspondence between the classical and quantum radiation. In Heisenberg's formulation, a relation remains; it just assumes an unanticipated form. The scheme hinted at in Equation (4) shows that the basic formulas will be expressed in terms of amplitudes. Since the amplitudes are in principle observable from the measured intensities, this procedure expresses the basic laws in observable quantities, as Heisenberg felt should be done.

In the actual implementation of these ideas, Heisenberg leaned heavily on the Kramers–Heisenberg paper. In that study, the classical Fourier expansion of the dipole moment was transcribed to an expansion in terms of amplitudes and Bohr frequencies by

$$P^{cl} = \sum C_l e^{i\omega_l t} \leftrightarrow P^q = \Sigma\, \mathfrak{A}_q e^{i\nu_q t} \tag{5}$$

Here the notation of Kramers and Heisenberg is recalled: $\mathfrak{A}_q$ stands for the characteristic amplitudes and ν_q stands for the Bohr frequencies. The first thing Heisenberg did was to change the notation (actually similar to the one used in this book) to one which shows explicitly that the amplitudes depend on two states, so that Equation (5) reads:

$$P_n^q = \sum_m a(n, n-m) e^{i\nu(n,n-m)t} \tag{6}$$

This notation stresses that the dipole moment is calculated in state n; $\nu(n, n-m)$ is the Bohr frequency for a transition from state n to $n-m$. Equation (6) shows that, in the quantum language, the dipole moment depends on the observable amplitudes and frequencies. Both are objects that depend on two indices or pairs of states.† Formula (6) is implicitly contained in the Kramers–Heisenberg paper. The quantum information about P is summarized by the set of complex amplitudes a_{nm}. Consider the position $x(t)$ of an electron in an orbit. Classically, one can expand $x(t)$ in a Fourier series

$$x(t) = \sum_m C_m e^{im\omega t} \qquad (\omega = \text{classical frequency}) \tag{7}$$

The quantum description of the position in stationary state n (corresponding to a given orbit) would, by analogy to Equations (5) and (6), be expressed as

$$x_n = \sum_m a(n, n-m) e^{i\nu(n,n-m)t} \tag{8}$$

Heisenberg now considers the classical quantity $x(t)$ as represented in quantum mechanics by the ensemble of quantities $a(n, n-m)e^{i\nu(n,n-m)t}$. This replacement

$$x(t) \rightarrow \{a(n, n-m) e^{i\nu(n,n-m)t}\} \tag{9}$$

of a classical quantity by the assembly of quantities $\{a(n, n-m)e^{i\nu t}\}$ (now known as a matrix) was Heisenberg's basic universal transcription method. It next became imperative to settle the algebraic problem: In other words, if

† A notation employing two indices was used earlier by Born;[258] it was not used by Kramers and Heisenberg.

Equation (9) represents $x(t)$, what represents $x^2(t)$? Before indicating Heisenberg's solution to this problem it should be recalled that the Bohr frequencies $\nu(n, m)$ possess the composition property

$$\nu(n, n - m) + \nu(n - m, n - m - l) = \nu(n, n - m - l) \tag{10}$$

It is to be expected that to obtain the quantum representation for $x^2(t)$ one should try to apply the transcription procedure to the classical double Fourier series for $x^2(t)$. This result is already contained (again in a somewhat implicit form) in the Kramers–Heisenberg paper. To obtain tractable expressions, which could easily be transcribed, they had to manipulate a double series. In this process they arrived at the transcription

$$(xx)_n = \sum_m \sum_l a(n, n - m)a(n, n - m, n - m - l)e^{i\nu(n, n-m-l)t} \tag{11}$$

Although they use this expression, its transcription character is not stressed. It is interesting that Equation (11) can be represented in the pictorial manner used earlier in terms of transitions between states:

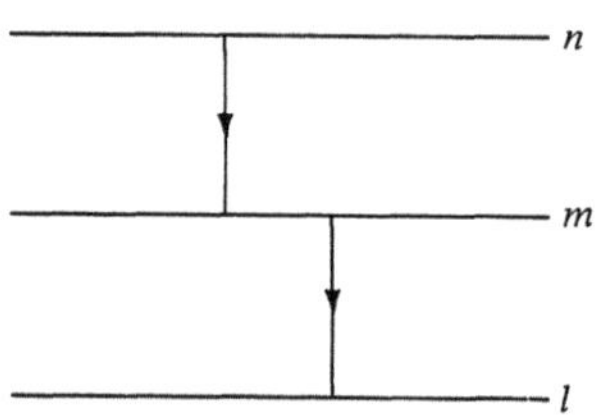

The multiplication rule of two quantities x and y, represented, respectively, by the set of amplitudes a and b, is buried in the Kramers–Heisenberg formalism. The same double Fourier series analysis yields

$$(xy)_n = \sum_m \sum_l a(n, n - m)b(n - m, n - m - l)e^{i\nu(n, n-m-l)t} \tag{12}$$

Again, in the Kramers–Heisenberg paper, no special attention is paid to this interpretation as the representative of xy; but the formula is there. A number of formulas of this type are indeed contained in the Kramers–Heisenberg paper. However, since the transcription problem was not phrased in any general way, the results contained in it do not possess a simple, discernible, and generalizable structure. By phrasing the problem abstractly, Heisenberg arrived at a tractable formulation of the transcription procedure. Associated with each classical quantity x is a quantum assembly a_{nm}, as given by Equation (8). The multiplication rule is a correspondence of the form

$$xy \to \sum a_{nk} b_{km} \tag{13}$$

Today, one would say that each classical quantity possesses a *quantum matrix representative*, and the multiplication of classical quantities is given by the matrix multiplication. But at that time Heisenberg did not know what a matrix was. It took a brilliant paper by Born and Jordan (who did know what a matrix was) to establish this connection.[259]

The transcription procedure, of course, does not yet define a complete physical framework. The dynamical problem, namely, the determination of the amplitudes, frequencies, and energies from the given forces, remains. As revolutionary as Heisenberg was in his transcription procedure from classical kinematical variables to quantum objects, he was conservative in the remainder of his program. In the usual theory there were two time honored phases in solving the dynamical problem: (1) the solution of the equation of motion and (2) the selection of the stationary states by a quantum (Bohr–Sommerfeld) condition. If $F(x)$ is the force, one has to solve the equation

$$m\ddot{x} = F(x) \tag{14}$$

while the states are determined from

$$\int p\,dq = \int m\dot{x}\,dx = nh = I \tag{15}$$

Heisenberg transcribed these equations without change to the quantum realm. In fact, the only quantum feature was the quantum interpretation of the x and $\ddot{x}$ occurring in Equation (14). Thus, Heisenberg left the dynamics unchanged but altered the nature of the kinematical variables. Using a few reformulations of (15) and the rule changing derivatives to differences, Heisenberg recast Equation (15) as

$$h = 4\pi m \sum_m [|a(n, n+m)|^2 v(n, n+m) - |a(n, n-\alpha)|^2 v(n, n-m)] \tag{16}$$

It was Heisenberg's remarkable claim (it is also true) that Equations (14) and (16) and the matrix interpretation of x and v provide a complete scheme for the description of atomic system. This then is the long sought after "future more complete theory." As Heisenberg wrote to Pauli: "I have made some progress (not much) and I am convinced in my heart that this quantum mechanics is correct as it stands, that is why Kramers accuses me of optimism."[260]

It is clear even from this highly abbreviated summary of Heisenberg's work that his paper was of penetrating originality and brilliant insight. Although the debt to the earlier Kramers–Heisenberg paper is great and noticeable throughout, the abstract formulation and implementation of his program is all Heisenberg's, and one can not help but marvel at the intellectual strength that enabled Heisenberg to obtain specific results in totally uncharted and unanticipated realms.

Original and unusual as Heisenberg's approach was, it is still inconceivable that on seeing Heisenberg's paper, Kramers would not recognize instantly that an advance of unprecedented magnitude had been made—this in spite of his derogatory comments. He knew the formalism every bit as well as Heisenberg; he had used the transcription procedure many times over. With his insight into the structure of a theory he must have seen how Heisenberg had indeed outlined the overall pattern of the "new theory." His negative reaction and his long silence can only be attributed to the combined shocks of the failure of the BKS theory and the profound disappointment of coming so near and yet missing so extraordinary an opportunity. When Kramers later commented on matrix mechanics, he always gave Heisenberg all the credit. He writes to Fowler: "Heisenberg's quantum mechanics, solves a great deal— but not yet everything, photons are still out."[261] It is remarkable how preoccupied, Kramers still was with photons. If Kramers, in September 1925, is still somewhat guarded in his praise of matrix mechanics, he becomes quite enthusiastic in the following months. He acknowledges Heisenberg's help in a most cordial way in a paper published in *Physics* in Dutch: "I want to express special thanks to Mr. Heisenberg for the great friendliness with which he has informed me about the latest (important) developments of his theory by Born, Jordan and Heisenberg."[262] At no time, in papers, lectures, speeches, or private conversations, did Kramers express anything but the greatest appreciation and admiration for Heisenberg's discovery of matrix mechanics. In private notes written after World War II, Kramers, when referring to these discoveries, characterizes Heisenberg as a "young hero."[263] If there was any difficulty or conflict, it was whether Kramers should have received more recognition and credit for his contribution to the Kramers–Heisenberg paper. As will be detailed in the next section, Kramers certainly hoped and probably expected to be the single author of the Kramers–Heisenberg paper. It is probably futile to speculate how the credit for the discovery of matrix mechanics would have been distributed in that case. There would be an indispensable preliminary paper by Kramers alone, followed by a seminal paper by Heisenberg; this might well have altered the balance of recognition. Although this kind of speculation ran counter to everything Kramers consciously admired and respected, and even though it was diametrically opposed to everything he valued in physics, there are indications that, in private talks with Klein[264] and in the total privacy of his diaries, such thoughts were not always totally suppressed. As an intensely private, guarded, and thoroughly honest person, it is doubtful that such thoughts were ever any more than fleeting moments of regrets, disappointments, and despair with opportunities gone by.

E. The Personalities of Pauli, Heisenberg, and Kramers: The Strange Path from Deference to Ridicule to Doubt

It is quite natural that in a period of rapid changes, when optimism and hope alternate so rapidly with disappointment and frustration, personal attitudes and therefore the interpersonal relations between physicists would play a dominant role. This was especially true when matrix mechanics was first formulated. The persons most directly involved in the early formulation of matrix mechanics were Heisenberg, Pauli, Kramers, and Born. Others, such as Jordan, played important scientific roles, but they were less immediately involved in the personal interactions. By contrast, Bohr, who did not contribute directly to matrix mechanics, played a pivotal role in shaping the complex personal interactions between Heisenberg, Pauli, and Kramers. In the period from 1913 to 1925, Bohr's approach and style totally dominated atomic physics. This was generally understood; it was expressed most graphically by Ehrenfest who once started a note to Bohr with: "To Bohr who searches for us all."[265] Einstein's status within physics was (and remains) unique and unchallenged, but since he did not deal directly with the problems of spectroscopy and atomic structure his day-to-day influence on the actual development of atomic physics was not as great as Bohr's. Bohr's ideas, no matter how unusual or radical, were considered as important guides that would eventually lead to a rational description of the behavior of atoms. As such, Bohr's suggestions were taken most seriously by responsible senior theorists such as Born and Sommerfeld. In Born's case, this deference to Bohr's ideas even bordered on the obsequious as the early correspondence about the BKS theory indicates. Great as Bohr's influence was in that generation of physicists, his impact on the younger physicists, starting with Kramers and followed in rapid succession by Klein, Pauli, and Heisenberg, was even greater. His influence on them was not just that of a superior intellect on his peers and pupils: He imprinted his personal approach to physics on all of them, molding their thinking, providing the language, and circumscribing the framework. All three—Kramers, Heisenberg, and Pauli—met Bohr very early in their respective careers, well before they had achieved full scientific and emotional maturity. It thus becomes understandable that Bohr's particular views and visions exerted a powerful influence on their subsequent approach to science. This was frankly acknowledged by all; Pauli wrote later: "A new phase of my scientific life began when I met Niels Bohr for the first time."[266] Heisenberg, referring to his first meeting with Bohr, recalls: "I feel it was then that I felt, I really learned what it means to work on an entirely new field in theoretical physics. The first, for me quite shocking experience was that Bohr had calculated nothing.... That was a very new way of thinking to me and it changed my whole attitude towards physics."[267] Much later, Heisenberg summarizes the major influences on his thinking: "From Sommerfeld I learned optimism, in Göttingen I learned mathematics, from Bohr I learned Physics."[268] Kramers' devotion and complete commitment to Bohr's physics and philosophy is evident throughout his work. His scientific taste was totally formed by Bohr. Generations of physicists, starting with Kramers, hoped and believed that Bohr would guide them into a promised land of a coherent atomic physics.

But Bohr's interest in and concern for his young collaborators extended much beyond the confines of science alone. He was genuinely interested in all aspects of their lives. The discussions ranged wide and far, from political institutions and social attitudes, to Danish mythology and German literature. In the Bohr Institute it was just as likely to hear philosophical arguments as an analysis of soccer strategy. No topic was excluded from this continual probing and incisive analysis. In all such discussions, Bohr was kind, gentle, and considerate, but his formidable dialectic powers were ever present so that even in the analysis of ordinary everyday events, Bohr, remaining true to his own convictions, had little patience with trivialities or superficial arguments. This particular confluence of circumstances, Bohr's concern for the personal and intellectual integrity of his collaborators, his private commitment to a rational harmonious atomic physics, and his powerful personality, all combined with the unbounded confidence in his scientific judgment, created a very intense and special atmosphere among those closely associated with Bohr. In this atmosphere, scientific endeavors, personal ambitions, and the social milieu were all strangely and strongly intertwined. The approval or sympathetic encouragement of an idea by Bohr was practically the definition of success; his disapproval or indifference would be a cause for deep concern. In that circle everyone considered physics to be of overriding importance; all felt a deeply personal stake in the development of atomic physics. It is therefore not surprising that the human relations between Bohr and the young scientists, Kramers, Klein, Heisenberg, and Pauli, would contain strong personal, even emotional, elements. Because of the very special role of Bohr, the interrelations between Kramers, Klein, Pauli, and Heisenberg were in turn profoundly affected by their actual or perceived individual relations with Bohr. As time went on, human relations necessarily change; furthermore, as time went on the younger generation, especially Pauli and Heisenberg, developed their own personal approach to the problems in atomic physics. Because of the extraordinary importance all involved ascribed to physics and to success in physics, it was almost inevitable that these interpersonal relations would be affected, perhaps subtly and indirectly, by the successes achieved in physics. Hopes for striking new results, expectations of recognition by Bohr and the physics community, secret hopes for solutions where others had failed, fears about being beaten out, and worries about ignoring new clues, all these deeply emotional elements were ever present in the physicists of that generation as they searched for a rational quantum theory. That most of the time these feelings were unspoken and often unformulated does not mean that they did not deeply influence the personal relations.

So the meandering development of quantum theory, from the Bohr theory to matrix mechanics, had its counterpart in the shifting personal relations between the main protagonists—Kramers, Pauli, and Heisenberg. In this development, Kramers played an unusual and somewhat ambiguous role. At the time Pauli and Heisenberg appeared on the scene, Kramers was already established in the Bohr Institute as Bohr's main collaborator. He was the interpreter—the spokesman—for Bohr and for Bohr's physics; he was the great proponent of the correspondence principle. Although he was only 5 years older than Pauli and 6 years older than Heisenberg, he was certainly initially considered with respect and even deference by both. On the other hand, Kramers, 9 years younger than Bohr, also had ambitions and wishes to alter, modify, and extend Bohr's ideas. As such, his perhaps unspoken aspirations were very similar to those of Pauli and Heisenberg so their relation, almost of necessity, contained elements of competition from the start.

Kramers' position at the Institute at the times when Pauli and Heisenberg appeared was a very special, favored position. Next to Bohr, he was by far the most important person in the Institute for Theoretical Physics. He had continual and direct access to Bohr, a high and unusual privilege. Not many persons had this automatic entry into Bohr's inner circle. When Pauli came to Copenhagen (September 1922), Kramers was entering a productive phase of his life, culminating in the great x-ray paper (see Chapter 12, Section V), written during the spring of 1924. When Heisenberg arrived in Copenhagen, first for a brief visit during Easter of 1924, then for an extended visit starting about September 1925, Kramers was overjoyed with the success of the dispersion relations and excited about the anticipated success of the BKS theory. Generally, Kramers was pleased with his achievements, he was totally convinced that he searched in the right directions, and he never had any doubts about his formal powers. Kramers, in contrast to Bohr, always had time to talk to visitors. Before any young physicist would get to talk to Bohr, he would, in the normal cause of events, have discussed the subject with Kramers in some detail. There can be little doubt that in these discussions in which Kramers really acted as "Bohr's representative on earth," he must at times have appeared not merely as knowledgeable and confident, but overbearing and condescending, with more than a touch of arrogance. Because of Kramers' habitually understated, gently ironic mode of expression, this was unlikely to cause serious troubles although naturally the individual reactions differed considerably. The impact of these attitudes was softened considerably by the light touch Kramers maintained thoughout. In the middle of serious discussions, he would often make facetious, irrelevant, outlandish, or mocking comments which destroyed the solemnity of the occasion. Kramers' approach to physics, in spite of the importance the field had for him, might at times appear somewhat capricious. It was this same characteristic that, in his student years, had been so objectionable to Ehrenfest. But these occasional irreverent outbursts were more an expression of Kramers' whimsical approach to life than a lack of commitment to theoretical physics (which is how Ehrenfest and Lorentz both interpreted his often disrespectful remarks).

Kramers' attitude was considerably more congenial to Pauli than it was to Heisenberg. Both Kramers and Pauli were intensely interested in advancing atomic physics, but neither was all that interested in general public recognition. This might well be for rather different reasons: For Kramers the appreciation by Bohr and his favored position in the Bohr Institute was all the recognition he needed; how others reacted was not really of any importance. Pauli was aware of his special gifts at a very early age. His monumental paper on relativity (and its immediate acceptance by the physics community) was still further confirmation that he was a scientist of world class; that was all that was important. Thus, neither Pauli nor Kramers were particularly personally ambitious; neither was especially dissatisfied with his status in physics, and neither hungered for public acclaim. Both Pauli and Kramers possessed a highly developed sense of humor; Kramers' more delicate and ironic, Pauli more biting and sarcastic. Thus, Pauli was not bothered or impressed by Kramers' occasional overconfidence and arrogance. Pauli, through his work on relativity, his studies with Sommerfeld, and his familiarity with some of the leading figures in physics (e.g., Einstein, Weyl) in no way lacked self-confidence. He expressed this in a remarkably candid, almost childish way, when Bohr invited him to be his collaborator in Copenhagen for a year: "I hardly think that the scientific demands you will make on me will cause me any difficulty, but the learning of a foreign tongue like Danish far exceeds my abilities."[269] This was not an affectation on Pauli's part, but his experiences in Munich and Göttingen had convinced him (quite correctly) that he could compete successfully with any physicist, anywhere. Pauli showed Kramers a reasonable amount of respect, due to Kramers' contributions and his close association with Bohr; but it did not seem that Kramers' initially somewhat haughty attitude bothered Pauli in the slightest. Kramers' occasional irreverent comments about physics were matched by Pauli's continual irreverence toward most physicists. Whatever the reason and whatever psychological mechanisms operated, Pauli and Kramers became close and warm friends. In the early phases of that relation, Kramers provided optimism and confidence. Pauli was given to violent fluctuations in mood; Kramers at that time was more stable, less susceptible to such large emotional swings. (At least Kramers had his moods more under control than Pauli.) Conceivably, Kramers' stability derived from his fundamental confidence in the Bohr model and in Bohr's style of physics, while Pauli with his incisive critical insight developed serious misgivings even at this early stage (1923). Pauli's close relation with Kramers did not mean that he always agreed with Kramers' approach to physics. He felt free—indeed he felt obliged—to criticize vagueness and weakness in the logical foundation with vigor and sarcasm. But in this early stage, Pauli generally showed a healthy respect for Kramers.

A remarkable example of the intense, emotional, and involved triangular relation among Pauli, Kramers, and Bohr, with its curious mixture of scientific and personal features, is contained in a letter from Pauli to Bohr. The letter was written September 12, 1923; the occasion was Bohr's sudden decision to take Kramers along on a trip to England. Bohr had planned a lecture tour of England and the United States for some time. The decision to take Kramers along on the England portion of the trip was made quite suddenly. The reason was that Kramers could help Bohr (presumably on the boat to England) with the completion of a paper for *Zeitschrift für Physik* and likely also with the preparation of the lectures that Bohr would give. Pauli was very unhappy with Bohr's sudden decision to take Kramers along:

> "I feel very hurt, that you now have induced Kramers to travel with you to England.... For I had looked forward for some time, to have a relaxing vacation with him, something I needed badly, after my recent scientific failures. You made this plan so late, that I cannot readjust my own plans accordingly. Otherwise I

> might have travelled to Bonn, that now is hardly possible. I actually talked to you this summer, to travel together with Kramers to Bonn—you told me at that time, that this would not be possible, since there would not be any one here to cover the teaching obligations. I see now from Kramers' trip to England, that this argument is completely irrelevant.... I wouldn't take this so seriously, if I could see that Kramers' presence would be of significant use to you ... but this I don't see .... The time available to you is under no circumstances enough to finish the complete paper .... If you think a little bit about it, I am convinced that you will agree with me. But it hurts me a great deal and I feel it as undeserved and unfair that you in your decision to travel with Kramers, you considered the utility and my need for Kramers so little. This decision is very important to me, since the presence of my friend, would mean a great deal to me emotionally. I have always seriously tried to help you to the best of my ability and in spite of everything it would give me great pleasure if I could do this again in the future. But now I am so depressed about the impending departure of Kramers and I feel so badly, that I must ask you to excuse me on Saturday night. I would only spoil the evening for others.
>
> Do not think badly about your very sad
> Pauli
> who up to this time has so loved and admired you
>
> P.S. I am writing you this, since I am not sure that I will feel well enough in the next few days, to come to the Institute.[270]

This is certainly an amazing letter. Kramers left for England on September 15 and returned by the end of September. Pauli recovered from his anger and depression. Hansen writes to Bohr about October 1: "Tonight Pauli gives a talk at the Physical Society, Thursday he leaves; I have the impression that he is in excellent spirits, so that his sojourn ends harmoniously in all respects."[271] Evidently Kramers, after his return from England, spent a good deal of time with Pauli, because he writes to Bohr: "I have spent the last two weeks entertaining Pauli.... Pauli left in good shape after some pleasant evenings of leave taking. He left me as Hero used to leave Leander."[272] This incident and the resulting letters demonstrate the intense nature of the relations. As time went on—especially as Pauli developed his own style—gradual changes and subtle shifts occurred. The respect and admiration Pauli had for Bohr as a person remained, but his insistence on logical clarity made him increasingly critical of Bohr's proposals in physics. But Pauli always showed a considerable deference toward Bohr, even if he did not agree with his particular suggestions. Pauli felt throughout this period that Bohr was still the person who would point the way toward correct quantum theory. Thus, after the initial discussion in the BKS theory, Pauli writes:" [I wonder] whether you have enough time to occupy yourself with full force with the scientific problems which are now so urgent in atomic physics—and of which I believe that only you are able to solve them .... I often think about your kindness and friendliness toward me—and I don't know at all how I can thank you for it."[273] These same themes—Pauli's confidence in Bohr's basic ideas, his appreciation for what Bohr has done for him, and his uncompromising criticism of Bohr's specific suggestions—recur again and again. "I hope that you are well and that you will soon solve the problems of the quantum theory of atoms with more than one electron. I know that it pleases you when I express my opinion about problems in physics, honestly and cordially—even when you consider them as quite stupid."[274]

Pauli's criticism of Kramers was considerably less restrained. The warm relations which existed between them also remained, but Pauli became quite exasperated with what he believed was Kramers' misdirected physics, especially during the BKS interlude. Pauli not only objected to Kramers' physics but also to the dogged, almost dogmatic manner in which Kramers defended his views. During the heyday of the BKS theory, Kramers was so totally convinced of the correctness of the approach that he tended to dismiss Pauli's and Einstein's criticisms as inconsequential. It is of course impossible to know to what extent Pauli's ferocious criticism of Kramers was intended as an indirect criticism of Bohr. While Bohr was in no way immune from Pauli's attacks, the tone was gentler and there were more expressions of doubts. To Kramers, Pauli was unrelentingly ferocious; he used his analytical power, sarcasm, and wit to undermine Kramers' arguments and ridicule his approach. While doing all that, Pauli would yet protest (and he certainly meant it) that he really liked Kramers. Since Kramers never deviated seriously from Bohr's ideas, where Pauli became more and more critical, even casting doubt on such sacrosanct notions as the correspondence principle, it is quite possible that only half of the criticism directed at Kramers was intended for him; the other half was likely meant for Bohr. Pauli could not bring himself to be quite as outspoken with Bohr as he was with Kramers. This, as usual, left Kramers in the middle.

Pauli had become increasingly skeptical about the use of mechanical models of atoms. This led him to analyze the meaning and reality that could be attributed to the orbits of electrons in atoms. The existence of electronic orbits had always been taken for granted in the Bohr–Sommerfeld theory; Kramers was a great expert in the analysis of these orbits—but Pauli questioned the legitimacy of the orbit notion altogether. Pauli writes to Bohr as early as February 1924: "The most important question seems to me to be the following: To what extent is one allowed at all to speak about definite orbits of the electrons in the stationary states? In my opinion Heisenberg has expressed just the right viewpoint in this regard: he doubts that one can speak about the possibility of definite orbits. Kramers has never admitted that such doubts were reasonable. Nevertheless I must insist on them, for the matter appears too important to me."[275] It is evident that Pauli took the modification or possible elimination of the orbit notion very seriously; it is also clear from the letters to Bohr that Pauli is engaged in a continual polemic with Kramers. Thus, he writes on December 12, 1924: "It seems to me that the relativistic doublet formula shows, without any doubt, that not only the dynamical concept of force but also the kinematical concept of motion will have to be profoundly modified." There is at this juncture a footnote which reads: "I am sure about this in spite of our good friend Kramers and his colored picture books.† Children, they like pictures. The desire of children for visualizability is reasonable and healthy, still such a desire can in physics never be an argument for maintaining a particular system of concepts." Pauli continues in his letter. "That is why in my paper [on the Exclusion principle] the notion of 'orbit' is avoided altogether.... I believe that energies and momenta of stationary states are much more real than orbits."[276] Pauli stresses the identical themes once again in another letter to Bohr.[303] In describing the rule for the exclusion principle[277] he writes: "For weak persons, who need the crutch of a representation by uniquely defined orbits and mechanical models, one could justify the rule simply by saying: when more than one electron in a strong field possesses the same quantum numbers, they would have the same orbit, they would therefore collide and such cases one must exclude.... I am sure that Kramers in his popular lectures could say this much more convincingly, and with such enthusiasm that among his listeners there would not remain the slightest doubt about the correctness of the rule.... If my nonsensical ideas really have the result that you would soon concern yourself with the problems of atoms with more than one electrons, I would be the happiest man on earth.... With best regards for yourself and all gentlemen in the institute, especially for our incomparable Kramers."[278] This letter contains all elements of Pauli's attitude at that time: a thorough criticism of the orbit notion—with an implied but not articulated criticism of the correspondence principle and Bohr—an explicit criticism and slight ridiculing of Kramers, and yet an expression of confidence in Bohr's ultimate sense in physics. These three themes pervaded the Pauli–Kramers–Bohr relation during that time.

These already involved relations were further complicated by the emergence of another strong personality and a major physicist—Heisenberg. Heisenberg and Pauli were both students of Sommerfeld, both of unquestioned brilliance. Their student association in Munich led to a lifelong uninterrupted friendship. As is so often the case, friendships initiated during student days have a relaxed uninhibited and permanent character. Their particular relation was special for several reasons: Both were undoubtedly in the genius class, their interests within physics were nearly identical, each recognized the greatness of the other, and both were passionately concerned with atomic physics and quantum theory. Less explicit but quite noticeable through all their correspondence is their shared conviction that they, as representatives of a young, new generation, had a significant role to play in solving the outstanding problems of theoretical physics. There were of course profound differences as well. Their methods of attacking problems were quite different: Heisenberg was more original, more daring, less systematic, and less organized than Pauli. Pauli was more deductive and considerably more critical, both of himself and

† This is a direct reference to the colored pictures of the electronic orbit in an atom, contained in the book by Kramers and Holst.

of others. He certainly expressed his criticism of Heisenberg's "haphazard methods" wherever he had a chance and whenever he felt it necessary. But this criticism was usually tempered by respect, even admiration. Thus, Pauli writes to Bohr: "Recently I saw Heisenberg at a Physical Society Meeting .... I always feel peculiar about him. When I think about his ideas—they appear gruesome to me and I grumble to myself about it. For he has 'no philosophy' and he doesn't care for a clear formulation of the basic principles .... But when I talk to him, I like him a lot and I see that he has a number of arguments—at least he feels he does. I consider him—apart from the fact that he is personally also very nice—as very important, even as a genius and I believe that some day he will produce major scientific advances."[279] These comments show Pauli's characteristic mixture of respect and impatience with Heisenberg. But Pauli clearly felt that Heisenberg under Bohr's tutelage would realize his great potential. Later in this same letter Pauli continues: "I am therefore very pleased that you have invited him [Heisenberg] to Copenhagen. Hopefully the two of you will succeed in improving atomic physics and solve a number of problems, with which I have struggled in vain and which are much too difficult for me. Hopefully Heisenberg will bring a more philosophical orientation home with him." In a later letter Pauli complains about the caliber and quality of atomic physicists in Germany; but he again makes a specific exception for Heisenberg: "He is smarter."[280]

Before describing just how productive Heisenberg's visit and subsequent interaction with Bohr were, it is interesting to stress that the basic elements of the Heisenberg–Pauli relations were persistent and did not change. Pauli's respect and admiration for Heisenberg's insight, intuition, and imagination remained. Pauli's unrelenting and uncompromising critique of vagueness and sloppiness in Heisenberg's thinking did not change either. In turn, Heisenberg's respect for Pauli's understanding of physics and his admiration for Pauli's intellectual integrity and independence also remained. In addition, they stayed close personal friends. It was for these reasons that Pauli could be an effective and trusted mediator in a serious rift which developed later between Bohr and Heisenberg in connection with the uncertainty relation. Because of Pauli's fearless honesty, extraordinary intellectual powers, and basic fairness, this conflict was kept from escalating into a major public break. Perhaps the greatest strain on the Pauli–Heisenberg relation—leading to serious public clashes—occurred in 1957 and 1958. At that time, Heisenberg was heavily involved in constructing a unified theory of elementary particles. As always, he relied heavily on Pauli for criticism, which Pauli provided in abundant doses. For a while, Pauli was swayed by Heisenberg's enthusiasm and arguments and he believed that "all elements for a major synthesis were available." Pauli's elation soon turned to bitter disappointment and he reassumed his usual hypercritical stance.[281] Pauli's letter to Heisenberg, in which he withdraw from the cooperation and the publication of the preprint, and Heisenberg's answer are typical of their contrasting scientific styles as they persisted over many years. Pauli insisted that one should only publish a paper when all consequences are clearly understood. He correctly argues that this is not the case with their preprint and concludes: "I must therefore drop the plan to publish a joint paper with you. It is the only decision which now appears to be logical and honest to me, since I no longer agree with essential parts of the preprint. But I cannot hold you back any longer."[282] Heisenberg in his answer does not agree with the principle that only papers should be published whose consequences can be clearly surveyed. "You will agree that if this principle were applied, neither Bohr's paper on the hydrogen atom, nor his paper on the periodic system, nor my paper on quantum mechanics should have been published. Your statement, that there are 'substantial uncertainties of principle in the whole approach', would be justified in all these cases. Your criteria for 'publication readiness', would have a negative effect on the progress of physics, it is therefore wrong."[283] The conflicting views of Heisenberg and Pauli are explicit and irreconcilable. But as always Pauli, in spite of his concerns and conflicts, still hopes that Heisenberg might yet be right. Heisenberg quotes Pauli as saying, even after he had given up working on their joint project and after he had assailed Heisenberg's ideas mercilessly: "I think you are doing right to continue working on these problems.... You know how much remains to be done and things will no doubt come out right one day. Perhaps all our hopes will be fulfilled and your optimism will be rewarded.... Let's hope you are still up to it ... it's all quite beyond my strength."[284] This is, if not word for word, certainly similar in spirit to Pauli's expressed hope for the advances in atomic physics that he anticipated from the impending Bohr–Heisenberg cooperation. Even Pauli's professed inability to make progress is stressed in both exclamations, which are separated by an interval of 30 years.

It is surprising that the scenario of the Heisenberg–Bohr cooperation which Pauli so optimistically outlined in his letter to Bohr[279] actually came to pass in almost the manner Pauli had envisaged. Heisenberg did come to Copenhagen (for a short visit in the spring of 1924 and for about half a year from September 1924 to spring 1925). He did come under the powerful influence of Bohr, he did alter his approach, and he did acquire a philosophy of physics. It is even true that he made signal contributions to quantum physics, but contrary to Pauli's expectations, these were in collaboration with Kramers, rather than Bohr. It was not a foregone conclusion that everything would work out this way. Prior to coming to Copenhagen, Heisenberg did have the proper respect for Bohr as a great man, but he did not automatically defer to Bohr's views and approach. As the discussion of the first encounter between Bohr and Heisenberg in Göttingen in 1922 (contained in Chapter 12, Section III) already indicates, Heisenberg's respect for Bohr did not imply that he agreed with all of Bohr's ideas, nor did it keep him from challenging Bohr publicly. Heisenberg was extraordinarily quick, unusually brilliant, and he took himself and his arguments totally seriously. Heisenberg's insistence on scientific independence and autonomy led to a number of differences with Bohr.† Their first meeting was occasioned by their first scientific difference. Another point of dispute was the use of half-integer quantum numbers which Heisenberg employed early in his career. Bohr and Kramers, and later Bohr and Coster,[286] vigorously objected to the use of half-integer quantum numbers on empirical as well as general grounds. Pauli and Kramers wrote a joint paper on band spectra of molecules, which was specifically intended to reject a previous analysis of these spectra using half-integer quantum numbers. But Heisenberg was not impressed by these arguments. (Actually, both the experimental and the theoretical situations were rather murky at that time, so firm conclusions were difficult to obtain.) In any case, Heisenberg felt strongly that half-integer quantum numbers should be considered seriously in spite of Bohr's cavalier dismissal of this possibility.‡ Thus, Heisenberg arrived in Copenhagen hoping to confront Bohr with a rather definite program of questions and challenges. This is not at all the way things worked out. Very soon after Heisenberg's arrival, Bohr took Heisenberg on a walking tour. (According to Mehra and Rechenberg,[289] this took place about March 20–23, 1924.) During this trip, Bohr started the philosophical reeducation of Heisenberg, by discussing life, culture, and civilization, as well as the grand design of science, physics, and philosophy. The breadth of the topics and Bohr's kind and compassionate view of life and persons, combined with fearless intellectual honesty, impressed the young Heisenberg profoundly. In the discussions about physics Heisenberg became aware of an altogether different approach and attitude toward that field. Physics was not a series of exercises in intellectual acrobatics to solve a set of unrelated questions, nor a set of puzzles to demonstrate mental virtuosity. Instead, it should be a fundamental search for a universal pattern. Questions of compatibility and consistency, as well as the quest for a rational design, were at the heart of the field. This view was altogether new to Heisenberg. "Bohr was more worried than anybody else about the inconsistencies of quantum theory. Neither Sommerfeld nor Born had been so much worried about things. Sommerfeld was quite happy when he could apply nice complex integrals and he did not worry too much whether his approach was consistent or not. And Born in a different way was interested mostly in mathematical problems. Inconsistencies were realized, but after all neither Born nor Sommerfeld really suffered [because of them] while Bohr couldn't talk of anything else."[290] Like Kramers and Pauli before him, Heisenberg came under the spell of Bohr's magic. Bohr's approach to physics and his grand hopes and vision provided the standards by which Heisenberg was to judge physics from there on out." Bohr's attitude was that he would

† Some of Heisenberg's comments are contained in Chapter 12, Section III; he writes about a Bohr suggestion: "This doesn't please me much, but one can't doubt that Bohr is correct."[285]

‡ It is interesting that Heisenberg appeared to become a little impatient with Pauli's deference to Bohr's ideas. Thus, Heisenberg writes to Sommerfeld[288] with some irritation that Pauli has become *sehr verbohrt* (very Bohr-like). German lends itself to constructing verbs from person's names. A student of mine (M.D.) told me he had *geschwingerd for several years.*

never stop before achieving utmost clarity. Bohr would follow things to the very end, to the point where he hit the wall. I very soon realized that there was nobody who had thought so deeply about the problem of quantum theory as Niels Bohr."[291] Heisenberg, who came to Copenhagen to argue with Bohr about specific details, left as a confirmed disciple. Bohr, who had been deeply impressed by Heisenberg during their first encounter in Göttingen, confirmed and strengthened his favorable opinion of Heisenberg in Copenhagen in 1924. So after some negotiations, it was arranged that Heisenberg would return to Copenhagen for a longer period of time in the fall of 1924. It was during that sojourn from September 1924 to the spring of 1925, that the Kramers–Heisenberg paper was written.

The organization of the Bohr Institute made it inevitable that Kramers and Heisenberg would see each other practically every day. Because of the similarity of their interests and scientific goals, it was equally inevitable that a strong personal interaction would develop. Actually, their personal relations proceeded through several quite distinct stages; neither fully realized (at least at that time) the changing and shifting character of those relations. When Heisenberg arrived in Copenhagen (Easter 1924), he was at first rather depressed by the high caliber of the young physicists in the institute. "The first few days I was deeply depressed by the superiority of the young physicists from all over the world who surrounded Bohr .... Most of them could speak several foreign languages .... they knew of the world outside .... they played various musical instruments extremely well and above all, they understood much more of modern atomic physics than I .... That I should find a place in such a circle seemed quite hopeless .... I remember with particular pleasure my first discussions with Kramers from Holland, Urey from U.S.A., and Rosseland from Norway. They all seemed to know Bohr well and respect him highly and they were full of optimism with regard to the development of Bohr's theory."[292] It is remarkable that Heisenberg's recollection in 1964 some 40 years after these events still stresses his depression and his fear of not being able to enter Bohr's inner circle. This can be understood by realizing that up until his trip to Copenhagen, Heisenberg was widely known in Germany as a young genius. He had been one of Sommerfeld's privileged students; Born thought that he was the best assistant he had ever had, so Heisenberg was used to instant recognition and immediate adulation. This was different in Copenhagen, where there were others with comparable claims—in Kramers' case with more experience and achievements—and it remained to be seen whether Heisenberg would acquire the same exalted position in the Bohr Institute that he occupied in Sommerfeld's and Born's institutes. Another significant factor leading to Heisenberg's feeling of depression is that Heisenberg was not only brilliant and ambitious but extremely competitive. He was ambitious and competitive all through his high school and college (and scientific) career; his high school teachers comment not only on his talents, but also on his strong desire to excel.[293] Heisenberg, more than Pauli—and much more than Kramers, needed the continued public recognition of his unquestioned brilliance. This need to be the best was not restricted to physics or science. He wanted to be (and was) the best skier, the best chess player. When it turned out that he was not the best ping-pong player, he went through a rigorous training regime during his trip around the world in 1929, so that he did indeed become the best table tennis player in the institute.[294] A few years later after Heisenberg was already full professor in Leipzig, a Japanese (solid state) physicist became a member of his institute. This physicist used a somewhat unusual paddle and also an unconventional grip in playing ping-pong. He could return any and all shots. Heisenberg could never beat him although he tried for a week. Heisenberg was so unhappy and upset about this failure that he did not show up in his institute for more than a week and never again played table tennis with the Japanese visitor.[295] Heisenberg's ferocious ambition and his insistence on continual admiration made it necessary that at all times and in all circumstances he be acknowledged as the best. When this turned out to be difficult, he would try his utmost to become the best; if this was impossible, he tended like so many persons to withdraw and become depressed. It is thus easy to understand Heisenberg's concern and depression upon his first visit to Copenhagen where it seemed particularly hard to achieve these goals. The confusion resulting from his ambitions and his efforts to do first rate physics and be accepted in the Bohr Institute led to a complicated vacillating relation with Kramers.

There can be little doubt that when Heisenberg returned to Copenhagen in the fall of 1924 he more than anything wanted to be accepted by Bohr as a significant coworker. Most likely he hoped to become Bohr's favorite associate as he had been of Sommerfeld and of Born. But the position of Bohr's confidente was firmly held by Kramers. Clearly, Heisenberg's perhaps unarticulated ambitions clashed directly with Kramers' position. Thus, the relations between Kramers and Heisenberg contained elements of competition from the very start. It is doubtful that Kramers perceived this very clearly. To him, Heisenberg was yet another of the energetic, brilliant, young men who routinely came to Copenhagen to be informed and acquainted with Bohr's ideas and views of atomic physics. As usual the task of familiarizing the new visitor with Bohr's approach would fall to Kramers. Apart from the observations that Heisenberg was obviously brilliant and came with the highest recommendations of Sommerfeld, Born, and Pauli, there was nothing unusual about him. Kramers never indicated in any way that he felt challenged or threatened by Heisenberg. Kramers at that time was preoccupied with the dispersion relations and the BKS theory, areas which Heisenberg up to that time had not touched, so in their early discussions Heisenberg was very much the pupil and Kramers the teacher. In their early relations, Kramers was certainly unaware of any serious competition with Heisenberg; he treated Heisenberg as he treated everyone, with a touch of friendly, ironic, facetious condescension. Heisenberg, by contrast, was competitive from the start. It is unclear whether this had a prior origin; after all, Heisenberg's first entry into the world of physics—his first contact with Bohr—was based on an incisive criticism of Kramers' work (see Chapter 12, Section III). Perhaps he believed subconsciously that he could continue to get Bohr's attention by criticizing Kramers. It is clear that Heisenberg took up the battle for Bohr's attention the moment he arrived in Copenhagen. He started this even before he came to Copenhagen. In deciding when to come to Copenhagen in the fall of 1925 (which was a somewhat arbitrary matter anyhow, since Heisenberg had a fellowship and therefore no specific obligations of any kind), Heisenberg writes the following remarkable note to Bohr: "It is not quite certain when I will be able to come. In any case, I shall make sure that I shall be there when Kramers leaves Copenhagen, so that I can help you. Therefore I would like very much to hear from you when Kramers goes away from Copenhagen† (I think this will be around Sept. 12)."[296] Although this could possibly be explained as the solicitous concern of a pupil (Heisenberg) for a revered master (Bohr), the interpretation that Kramers' absence would provide Heisenberg with a golden opportunity to have Bohr's undivided attention is by far the more reasonable one. Given Heisenberg's ambitions, Kramers' absence would be an ideal time to arrive in Copenhagen. Thus, as far as Heisenberg was concerned, the competition with Kramers started before he ever arrived in Copenhagen. On later occasions, Heisenberg would be surprisingly candid about his strong ambivalence toward Kramers.[297a] When Heisenberg first arrived in Copenhagen he was overwhelmed by Kramers' knowledge and sophistication. As Heisenberg told Mehra and Rechenberg: "He knew in general things in life so very much more than I could ever hope to know ... he spoke Dutch, Danish, German, French and English ... all these languages came out from him just as if they had been his mother tongue."[297a] Heisenberg marveled at all that Kramers knew and could do: "How can a man learn all these things?"[297a] As if this was not bad enough for a potential competitor, Kramers, who was older and married, was socially more adept and urbane than Heisenberg. He recalls that Kramers often was the center of attraction at Bohr's parties: "He could entertain the whole party when we were invited to Bohr's house."[297a] Heisenberg summarizes his reaction to Kramers: "He was the type of man who was far above my own reach and at the same time a type which was a bit strange to me."[297a] In spite of all this great admiration for Kramers, Heisenberg had a rather difficult time getting along with him. There were very few overt conflicts, but a certain uneasiness was present throughout, so that their relation never developed into a warm friendship such as the relation between Kramers and Pauli. Most of the problems between Kramers and Heisenberg were unspoken; they were also perceived very differently. The most important —and inexpressible—source of tension was Heisenberg's ambition to replace Kramers as Bohr's most favored young scientist. Heisenberg would get irritated with Kramers when Kramers did not take the problems of quan-

† It was known that Kramers would be absent from Copenhagen to attend a Congress of Scientists and Physicians in Innsbrück, September 21–27, 1924. There he talked to Sommerfeld and Einstein about the BKS theory.

tum theory as seriously as Heisenberg did: "He [Kramers] used to make jokes—when I didn't want to have any jokes made."[297a] Here Heisenberg very much adopted the attitude of Bohr:" With Bohr one got into the really desperate problems, into the discussion of those dreadful difficulties, and he would get to the bottom, while Kramers would not take these difficulties so seriously as Bohr did."[297b] Heisenberg did not realize that Kramers found it difficult to take anything very seriously; he certainly could never take himself or his concerns totally seriously. Heisenberg never had such trouble. Thus, Heisenberg, as Ehrenfest before him, mistook Kramers' irreverent manners for a lack of scientific concern. It is noteworthy that neither Bohr nor Pauli, whose total commitment to physics cannot be doubted, ever interpreted Kramers' whimsical comments as a lack of interest in science, but rather as an entertaining idiosyncrasy, possibly expressing a lack of personal ambition. There were other sources of tension. Kramers was not all that impressed with the contributions Heisenberg had made to physics—so far; Heisenberg was surely aware of this and it clearly did nothing to make their relation any smoother.

The subjectivity of personal perceptions is vividly illustrated by contrasting Heisenberg's impressions of Kramers with Kramers' own assessment. Heisenberg comments to Mehra and Rechenberg: "Kramers contributed enormously to the whole development [of atomic theory at the Bohr Institute] just because of his physical strength. He was a man of absolutely inexhaustible energy. He could work three days without sleep or anything. If Bohr wanted Kramers to calculate something, he would do it and he certainly would do it very well."[297c] This is remarkable, because Kramers did not think of himself as a man of particularly robust health, let alone of superior physical strength. Ever since Kramers' sickness in Holland in 1918 and especially after several stays in the hospitals in Copenhagen, he felt that his physical status was far from the best. He was convinced that Bohr's physical energy was greater, that Bohr possessed far greater stamina than Kramers. It is also curious to see Kramers described as a man of inexhaustible energy. Kramers in his letters often complains of lack of energy and general listlessness. Of course these conflicting views are not necessarily contradictory; it is certainly true that Kramers was capable of bursts of productive work. To Heisenberg this seemed the rule, a permanent situation; to Kramers this was the exception, usually followed by a severe letdown. Although Heisenberg initially had difficulties with Kramers, he felt (or believed) that these difficulties diminished as time went on. "[The misunderstandings] were certainly entirely my fault—because later on I liked him very much and thought he was a very nice fellow and everything went on excellently between us."[297c] It is doubtful that Kramers at that time was aware of Heisenberg's antagonism and implied challenge. Heisenberg's arrival coincided with Kramers' total immersion in the dispersion relations and the BKS theory. As such, he was convinced that he was giving a concrete form to Bohr's intuitions. He was the acknowledged master in that field; he surely did not feel threatened by Heisenberg. He treated

Heisenberg in his habitual friendly, somewhat detached, ironic, and slightly superior manner, probably unaware that this very treatment would annoy Heisenberg the most. Just as at this stage Kramers was unaware of the problems and difficulties Heisenberg experienced in his relation with Kramers, so at a later stage (after the discovery of matrix mechanics) Heisenberg was unaware of any disenchantment on Kramers' part. Interestingly enough, Pauli did perceive this—although Kramers never mentioned anything about it.

None of these tensions had a very direct operational effect; they did not keep Heisenberg and Kramers from having daily discussions or from co-operating effectively. But there were a number of indirect effects that can be traced back to the interplay of the personalities. The resulting shifting personal attitudes were in turn strongly affected by the fortunes of the BKS theory. When Heisenberg arrived in Copenhagen in the spring of 1924, the BKS theory had just been published. There was great optimism in the Bohr circles that many of the outstanding problems of quantum theory including the photon–light wave controversy could now be solved. Although Heisenberg was somewhat guarded in his detailed support of the BKS theory, he was caught up in the general optimistic atmosphere in the Bohr Institute. He certainly was in total sympathy with the general Bohr philosophy. It was Heisenberg who informed Born about the BKS theory, which led to Born's enthusiastic letter to Bohr: "I am totally convinced (although I only have Heisenberg's brief oral communication) that your theory is essentially correct—it is also in a certain sense the last thing that can be said about these questions."[298] In part because of Pauli's and Einstein's criticisms of the BKS theory, Heisenberg kept a low profile regarding the details of the BKS theory. But Bohr's influence on Heisenberg was so powerful that he gladly accepted the general BKS approach, and he used its language and its ideology in his own research. He explains his position in an informative note to Sommerfeld (who was no devotee of the BKS theory): "Besides I believe increasingly that the question 'light quanta or correspondence principle' is merely a question of words. All effects in quantum theory must of course have an analogy in classical theory, since the classical theory is 'almost' correct, hence physical effects always have two names a classical and quantum theoretical one and it is a matter of taste which one prefers. Perhaps Bohr's radiation theory is a very felicitous description of this dualism."[299] With these words, Heisenberg expresses his personal sympathy with the BKS approach. Pauli, after some initial vacillations, became more and more disenchanted with the BKS theory. This also coincided with a gradual estrangement of Pauli with the basic tenets of quantum theory. Pauli dared to call the validity of the correspondence principle into question, which obviously was heresy to Bohr—and unthinkable to Kramers. These developments had a curious effect on the Kramers–Heisenberg–Pauli interrelations. The net effect was that Heisenberg and Pauli, both as individuals and as a team, expressed increasing misgivings about Kramers' approach to physics. The criticism was sometimes muted or could merely be inferred from a special turn of phrase. But on other occasions, it could became scathing and ferocious. Often the criticism was subdued, but more often than not there was a touch of ridicule of Kramers, of his physics, of his status at the institute. Sometimes this ridiculing was merely a private joke for the amusement of Pauli and Heisenberg. Other times there would be hazing comments in letters to Bohr and Kramers himself. Although the gentle ridiculing of an older physicist (even only a little older) by a younger generation does not usually need an extensive explanation, these many negative comments appear to indicate a growing impatience and disenchantment with Kramers and his physics. This is especially interesting since Bohr at the same time was treated with respect and consideration. Although Heisenberg and Pauli often amused themselves at Kramers' expense, the reasons for their negative reactions varied widely. Pauli had always liked Kramers; he respected him as a person and a physicist. But he disliked the BKS theory and became increasingly exasperated with Kramers' dogmatic adherence to that theory. In spite of that, Pauli throughout maintained a genuine concern for Kramers' personal and intellectual well-being. (This in no way tempered his distaste for the BKS theory.) It is less clear why Heisenberg was so critical of Kramers and enjoyed making fun of him. It is unlikely that the reasons were primarily scientific. The BKS theory, so repugnant to Pauli, was quite acceptable to Heisenberg; he would not demean Kramers for participating in its development, nor would he criticize him for defending that theory. It is therefore more than likely that Heisenberg's generally derogatory attitude was determined by the competition and envy Heisenberg felt toward Kramers.

The organization of the Bohr Institute, its hierarchical structure in particular, lent itself to a burlesquing of the relations between the individuals. Bohr was referred to as the "Pope." He was first so described in a letter from Heisenberg to Pauli,[300] in which Heisenberg reports that his paper will be published with "papal blessing" (i.e., the agreement of Bohr). Pauli immediately called Kramers "His Eminence," as behooves a cardinal, while he referred to himself as an "infidel" (unbeliever),[301] with respect to the BKS dogma. From that time on whenever Kramers held his position with too much dogmatic fervor, he was always referred to as "His Eminence." Its use was definitely intended to tease; it certainly did not have a positive connotation. This period coincides exactly with Pauli's emancipation from the Bohr ideas (the correspondence principle), the collapse of the BKS theory, and the emergence of Heisenberg's matrix mechanics as the dominant theoretical scheme in atomic physics. Where referring to Kramers as the cardinal is harmless and good fun, Pauli's criticism soon became sharper. He had earlier criticized Kramers' use of the classical orbit,[278] especially the colored pictures of these orbits in the book by Kramers and Holst. He was fond of stressing that "weak persons" needed the crutch of visualizability. At the end of a letter[278] which throughout makes fun of Kramers in a somewhat sly manner, he asks Bohr to give special regards to "the incomparable Kramers." Bohr caught the irony immediately and answers that he and "the ridiculed Kramers" have been looking forward to Pauli's visit, so they could argue with him.[302] Even this criticism is still quite mild; however, in a letter from Pauli to Heisenberg, written when Heisenberg was in Copenhagen, the following passage occurs: "There are new possibilities to calculate the intensities of spectral lines .... In Copenhagen one says of course 'to sharpen the correspondence principle' ... that is the imperialism of the correspondence principle .... Could you perhaps send me page proofs of the paper of you and Kramers concerning these matters .... (Will that be allowed by his eminence, Mr. Kramers. .... I am after all an infidel) ... Further one can never tell from papers from Copenhagen if not just the main point has been missed."[303] It is possible to still interpret Pauli's comment as semijokes, but it is clear that there is an edge to all these remarks. Heisenberg as usual was more restrained but his comments also reflect unfavorably on Kramers. Thus, he writes a postcard to Pauli sending Kramers' reaction to a fundamental Einstein paper:[304] "Even 'his eminence' finds the Einstein theory beautiful and for him one should say 'there is more joy in heaven over one sinner who repenteth than ... etc."[305] This is typical for the general character; the remarks are not vicious, but certainly not admiring either. Altogether, the letters contains a succession of snide comments. However, after the BKS theory was definitely disproved and after Heisenberg had written his fundamental paper (both Pauli and Kramers were aware of this), Pauli writes a splashing, almost furious letter to Kramers.[306] Several passages are especially noteworthy. Pauli encloses with this letter a page proof for a paper he has submitted to a Danish journal,[307] for a final perusal by Bohr or Kramers. In this connection he writes: "Since it seems to me that the greatest care is necessary with respect to all people who assume the existence of some kind of true believers—and who of course count themselves among them, I would emphatically state the following. 1. You are not allowed to make any changes of any kind in my paper, without asking me beforehand whether I agree .... I want to call your special attention to the footnote on p. 5, where I quote the paper of you and Heisenberg." Pauli stresses that this footnote is necessary, so that even a possible confirmation of the consequences of his paper will never be construed as support for the BKS concepts. Then Pauli launches into a vicious criticism of the BKS theory. "Many excellent physicists (Ladenburg, Mie, Born) might have maintained [the BKS ideas] and this most unfortunate paper by Bohr, Kramers, Slater might well have become a major obstacle in the further progress of physics." Pauli continues his criticism by contrasting the BKS approach with Heisenberg's new methods. "The community of true believers would not have much success nor get much respect, if they were to oppose the present development of quantum theory, which tends to analyse the concepts of force and motion." He expresses his pleasure in Heisenberg's new efforts. "I now feel myself a little less isolated than I did for about half a year, when I felt quite alone between the Scylla of the number mysticism of the Munich school and the Charybdis of the reactionary Copenhagen pocket revolution, which you propagandized with the excessive zeal of a true fanatic. Now I hope from you that you will no longer retard the reestablishment of a healthy Copenhagen physics, which in view of Bohr's strong sense of reality cannot fail to re-emerge." Pauli then closes this letter with what might appear as a strange ending after all he has said: "Again all good wishes from your devoted Pauli."

This is a remarkably intense letter in which the personal relations of Pauli and Kramers are totally intertwined with physics. Pauli's impatience with the BKS ideas is evident; Kramers' passionate attachment to these ideas can be inferred. Pauli's uncontrollable exasperation with Kramers' dogmatic defense is evident in every sentence. Perhaps most surprising is that Pauli attributes a good share, perhaps even the major share of the repugnant BKS philosophy, to Kramers and not to Bohr. That is somewhat unexpected, for during this same period—ever since Pauli's great paper on the exclusion principle[308]—, Pauli was engaged in a running battle with Bohr about the foundations of the quantum theory generally and the correspondence principle in particular. Pauli's criticism was not only directed against the unanalyzed use of the orbit notions[309] (see page 258), but also against the correspondence principle. He explicitly states this in a letter to Kramers: "Bohr in his theory of the periodic systems has likely overestimated the role of the correspondence principle. The final outcome will likely be that the correspondence principle will be a highly important limit of quantum theory but not its foundation."[310] Kramers never answered this letter, but he evidently discussed it with Bohr and this started a lengthy, rather low-keyed discussion between Bohr and Pauli about the correspondence principle (even though the topic at times might appear to be the BKS theory).[311] Occasionally, the reference to the correspondence principle is explicit, as when Pauli writes to Bohr: "I have mentioned several times, that in my opinion the correspondence principle has in reality nothing to do with the filling up of the electronic shells in the atom. You have always answered me, that I was too critical. Now I am pretty sure about my opinion."[312] Pauli's confidence came in part from his successful and immediately appealing results on the exclusion principle.[308] There the introduction of the fourth quantum number was essential; this strengthened in Pauli the belief that the orbit notion had to be profoundly altered or modified—something Bohr did not like either. Bohr, who was anxious to save the orbit concept, tried to connect the fourth quantum number with *inner* and *outer* parts of an electronic orbit in an atom: "This would perhaps indicate that your [Pauli's] beautiful number harmonics are not so independent of our poor old classical space concepts as would be indicated by the use of 4 dimensions for the description of an electron orbit."[313] Bohr writes this in an almost apologetic manner to Pauli. Pauli's criticism really dealt with two logically independent aspects of quantum theory—the use of the correspondence principle and the manner in which the orbit notion was used. Bohr was anxious to maintain the classical orbit notion; his attempt to interpret the four quantum numbers in terms of orbital properties points in that direction. He was desperate to maintain the correspondence principle, which Bohr felt was the basis of quantum theory. He writes in the same letter: "Now I am certain that you are overstepping a dangerous boundary. When you strike up your old 'Carthaginem esse

Delendam' tune,† pronouncing the final death sentence on the use of the correspondence arguments to explain the completion of the electron groups."[313] It is clear that Bohr was deeply upset, maybe even unnerved, by Pauli's claim that the completion of the electronic shells had nothing to do with the correspondence principle. Any limitation or attack on the correspondence principle was objectionable to Bohr. That Pauli had obtained spectacular results, based on such ideas, made Bohr even more uncomfortable. This accounts for Bohr's surprisingly strong language that Pauli had overstepped a dangerous boundary. Pauli answered Bohr some 10 days later. He did not withdraw his claims; to the contrary, he was more emphatic than before: "I would like to be somewhat more complete about one point in my last letter—namely you believe that I crossed a dangerous boundary in connection with the interpretation of the completion of the electronic shells in the basis of the correspondence principle. I must nonetheless tell you honestly that I am facing this danger serenely."[314] In other words, Pauli felt that the Bohr philosophy, based as it was on the correspondence principle, was on the verge of collapse. It was exactly during this same time (December–January 1925) that Bohr, Kramers, and Slater formulated their theory. Kramers did not participate in this debate between Bohr and Pauli—but there is no doubt that he wholly supported Bohr in maintaining the primacy of the correspondence principle. The BKS theory, based on the virtual oscillators, was in Bohr's view an attempt to uphold both the orbit notions and the correspondence principle against Pauli's criticism. Pauli's dislike of the BKS theory must be viewed in the context of these conflicts. Since both Kramers and Bohr were involved, it is remarkable that Pauli's criticism of Bohr was measured and controlled, while his July 27, 1925 letter to Kramers—when the BKS theory was known to fail—was intemperate and almost cruel. It is even more remarkable that in spite of Pauli's criticism of Bohr, he, as late as May 21, 1925, a month after the experimental disproof of the BKS theory, still expects a saving idea to come from Bohr: "Physics at this moment is in a mess again, in any case it is too difficult for me and I wish I were a comedian or something like that and have never heard of physics. But I do hope nevertheless that Bohr will come up with a new idea. I implore him to do so."[315] This should be contrasted with the disparaging comments contained in the July 27th letter to Kramers. There, in addition to everything else, Pauli once again criticizes Kramers' propensity for providing actual geometrical pictures of electronic orbits. In a footnote he writes: "In a communication of Kamerlingh Onnes in superconductivity, I noticed (and it made me shudder!) in pictures [of electronic orbits] the traces of your spirit." This again refers to the pictures of these orbits in the Kramers–Holst book. It almost appears as if Pauli divides his criticism into several pieces: one against the correspondence principle, directed at Bohr and especially Kramers, while the full fury of his criticism against the BKS

† Recall that in the Roman Senate, Cicero ended every talk, every discussion, no matter on what topic, with the admonition that Carthage should be destroyed by the Romans.

theory was mainly directed at Kramers. It is not difficult to understand the controlled and moderate character of Pauli's continual critique of Bohr, who was his revered master, so that respect and admiration would always damp sarcasm and ridicule. But it is perhaps not without interest to reflect on the vehemence with which Pauli attacked Kramers. There can be no doubt that, in spite of everything he wrote, Pauli still liked and respected Kramers. The violence of his criticism is most likely a measure of his passionate concern for a rational physics. That he did not feel free to express such deep felt attitudes in such terms to Bohr is understandable; that he did so with Kramers is an expression of the geniality of their relation. But with conflicting feelings of such intensity—after all, Kramers' attachment to the BKS ideas was also deep—it is perhaps not surprising that conflicts would occur. It is also possible that Pauli was disappointed in Kramers, because Kramers did not show as much enthusiasm for Pauli's exclusion theory paper as did Heisenberg and Bohr. Kramers was preoccupied with the BKS theory at that time. Kramers was impressed with the results in the Pauli paper, but he felt that this paper, with its departure from correspondence principle ideas, pursued a wrong direction. This could have bothered Pauli. It is conceivable that Pauli had hoped for greater support from Kramers in his emancipation struggle with Bohr. Instead, Kramers became more rigid in his approach than Bohr. Whatever the reasons, at the time Pauli wrote this letter, it would be impossible to distinguish personal from scientific involvements. This letter marks a watershed in the relations between Pauli and Kramers. Kramers never answered it and Pauli never returned to the issues either. Both Pauli and Kramers followed Heisenberg's lead in developing matrix and later wave mechanics, with Pauli taking the more active part. Their personal relations after the peak of intensity expressed in Pauli's letter returned to a gentler, more even and considerate level, as exemplified by Pauli's regards for "Kramers whom he basically likes a lot"[316] and Kramers' special congratulations to Pauli on his matrix solution of the hydrogen atom.[317] It is striking that in the Pauli–Heisenberg letters written after July 1925, Kramers is never again referred to as "his Eminence." The relation between Pauli and Kramers stabilized after the summer of 1925; they remained warm friends, with Pauli maintaining a high regard for Kramers, with the concern one friend has for another. Kramers, on the other hand, always considered Pauli as the final arbiter (with the exception of Bohr) in all matters in physics. Kramers read the literature sparingly and very selectively, but he did read what Pauli had to say. Quite often he would not read Pauli's papers but instead asked him privately. But he always wanted to know what Pauli thought of it.

The relation between Kramers and Heisenberg was quite different from the Kramers–Pauli relations. It was much less intense and more formal; neither Heisenberg nor Kramers were given to the emotional outbursts of Pauli. Thus, less was said and written, more was implied or hinted. This also meant that problems or conflicts were never faced or resolved; they were circumvented or just ignored. Perhaps the single most important issue was the proper division of the credit for the Kramers–Heisenberg paper (see Sections IV.A–C). In spite of Heisenberg's initial uneasiness with Kramers as a person, he was certainly anxious to cooperate with Kramers on dispersion theoretic topics, if for no other reason than that he recognized the importance Bohr attributed to this approach. By the time Heisenberg arrived in Copenhagen (September 1924), Kramers had obtained most of the results in the paper (see Section IV.A and IV.B). Heisenberg did participate vigorously in the discussions in his usual brilliant and incisive manner. He especially stressed the role of the "false resonances." Although Heisenberg noted the formal occurrence of these terms, he did not explain why they would not lead to any physical effects. This was done a little later by Kramers. Kramers certainly originated the main questions considered in the Kramers–Heisenberg paper; he devised the methods and he carried out most (but not all) of the calculations. The interpretation was altogether due to him. So it is not surprising that Kramers always assumed that he would publish this paper by himself.† Heisenberg in his recollections reports these same events to Mehra and Rechenberg: "For some time it was not clear whether we should write a joint paper, or whether Kramers should write it alone. And actually—it was suggested—I don't know whether by Kramers himself or by somebody else‡—that Kramers should publish it alone, because most of the work was due to him."[319] It is easy to understand, given Heisenberg's ambition and attitude toward Kramers, that he would not be pleased with that suggestion. There are hints that Heisenberg mentioned his unhappiness to Bohr. Heisenberg in his conversations with Mehra and Rechenberg states: "I was a bit hurt by his idea [of Kramers' publishing the paper alone] because I felt I had made an actual contribution. But I simply said 'well I leave that entirely to Bohr (as I had left it always to Sommerfeld) Bohr shall decide. I don't care which way his decision goes. Bohr then probably thought 'After all the young man has contributed a bit, why not put his name also on it. Everybody knows that it is really Kramers' theory because Kramers had written the first notes.' This must have been Bohr's idea more or less. I think that was quite right."[326] It is difficult to deduce from this quote to whom Heisenberg was talking when he said: "I leave this entirely to Bohr," but it is clear that one way or another Heisenberg managed to invoke Bohr's authority in arriving at a decision. It is also hard to know exactly what form Bohr's involvement took. Most likely he would have mentioned the possibility of Heisenberg's coauthorship to Kramers in a tentative and oblique manner. He would certainly not have forced Kramers to do so. But there is little doubt that in the decision ultimately reached Bohr played some role. It was Bohr who as early as November 1, 1924 (not much more than a month after Kramers returned to Copenhagen) announced the imminent publication of a joint paper by Kramers and Heisenberg.[321] Bohr told Klein some years later[322] that he (Bohr) felt Kramers should have published the dispersion

† This was specifically stated by Klein.[318]

‡ The interview with Klein indicates that the suggestion was actually Kramers' own.

theory paper by himself. Although unquestionably Heisenberg was very interested in the dispersion approach to quantum theory, and although Kramers and Heisenberg had lengthy and intense discussions, Heisenberg's actual contribution (as seen by Bohr) was rather minor. This same impression is confirmed by Kramers who told Nico van Kampen (probably his best and most trusted student) that he put Heisenberg's name on the paper as a pure courtesy.[323] Kramers acknowledged to van Kampen that Heisenberg did stress the importance of the "false" resonance terms, but evidently Kramers had his doubts that this observation entitled Heisenberg to be a coauthor. Exactly how Heisenberg's name got on the paper is impossible to determine, but it is certain that once the possibility of a joint authorship was mentioned to Kramers, he would undoubtedly acquiesce. It was totally against his style to fight for priorities or insist on his due. As he once told his nephew, H. C. Kramers: "A Kramers must fight to fight."[324] This unwillingness to fight was very characteristic of Kramers. If he felt slighted or unappreciated, he was much more likely to turn inward, or become facetious than to clamor for recognition or stand up for his rights. Once it was suggested that Heisenberg might be a coauthor, Kramers would agree with melancholy good grace and would only under unusual circumstances mention it again. A remnant of the resulting coolness can perhaps be seen in the letter which Heisenberg writes to Kramers after the completion of their joint paper.[325] The letter is correct and proper, but it starts with the unusually formal (for Institute members) opening "Lieber Herr Dr. Kramers." Unreliable as personal recollections and inferred attitudes are, it is still clear from the available information that Kramers and Heisenberg differed sharply in their assessment of the credit for the Kramers–Heisenberg paper. This was in addition to (or perhaps because of) their initial dubious personal relations. Thus, Heisenberg recalls: "We did not agree to begin with and Kramers did not take the problem quite as seriously as I did."[320] This recollection of Heisenberg is in direct conflict with Kramers' total preoccupation with dispersion relations for about a year. The emotional attachment of Kramers to the BKS ideas has been mentioned repeatedly. He considered the results of the Kramers–Heisenberg paper as the outstanding success of that theory. He had gambled his total scientific prestige on the correctness of that theory over the serious objections of Pauli and Einstein. It is hard to see how anyone could be more committed. For Heisenberg to imply that Kramers did not take the problem seriously can only mean that Heisenberg was insensitive to Kramers' deep attachment to this whole set of questions. Evidently, Heisenberg confused Kramers' superficial mannerism with his genuine feelings—indicating a rather shallow personal interrelation. Heisenberg's description of some of their joint work is again hard to reconcile with Kramers' earlier known activities. Thus, Heisenberg recalls: "I did not try to discuss problems primarily from the side of Einstein's laws [of absorption and emission of radiation] and physics, but I simply said, Let us see what classical mechanics does—Let us just assume that we have a classical atom like hydrogen and we put an electric wave on it and then see what happens to the Fourier component."[326] If Heisenberg's recollection is anywhere near correct, this suggestion must have struck Kramers as especially trivial, indicating merely that Heisenberg was totally uninformed, because Kramers had settled this very problem in the fall of 1923, even before Slater arrived in Copenhagen. Kramers had informed Slater about these results; Slater was very proud that he in turn rederived these same results.[327] Kramers had used the Fourier components and their connection with the Einstein A coefficient ever since his thesis. Thus, what appeared as a brand new revelation to Heisenberg must have seemed a repetitive rephrasing of methods and results that Kramers had known for some time and that he legitimately considered his own. Because of the generally unsettled scientific situation, it is quite possible, given Heisenberg's genius, that in the passage just quoted Heisenberg was groping for the amplitude notion and its autonomy —without being able to formalize it. Naturally, Kramers would not appreciate it; as stated to him, these comments would seem a rehashing of old ideas. So there must have been a number of misunderstandings, misinterpretations, and scientific mistakes. All this superimposed on an already tenuous personal relation in which Heisenberg's fierce ambition clashed with Kramers' position and aspirations must have created substantial tension. By itself, this is not so unusual; it frequently happens that the collaboration of two brilliant minds leads to great tension and conflicts. What is special about the Kramers–Heisenberg collaboration and the resulting paper is that this paper was the direct, immediate, and exclusive precursor to the Heisenberg paper on matrix mechanics. With Heisenberg's paper, the roles of Kramers and Heisenberg were suddenly reversed. Heisenberg now was, if not the teacher, certainly the guide in the new developments; most assuredly he was not the pupil. Heisenberg's reputation rose instantly to meteoric heights. Kramers' reputation suffered a serious setback by the collapse of the BKS theory. Bohr's reputation was not helped by the BKS demise either, but Bohr's renown was so great that he could stand a little loss of face. More importantly, Bohr took the demise of the BKS theory with better grace than Kramers, so that in the end the failure of the BKS theory appeared like Kramers' failure and a mere aberration of Bohr.

Although personal likes, dislikes, sympathies, and appreciations need not be related directly to accomplishments in physics, in the Copenhagen atmosphere, so suffused by physics, success and failure in physics had a noticeable, even pervasive, effect on the subtle personal interactions. There were, of course, always the direct professional consequences of successes and failures such as job offers, recognition, and prestige (or their losses). It was inevitable that the rapid succession of the Kramers–Heisenberg paper, the failure of the BKS theory, and Heisenberg's breakthrough would affect Kramers directly and deeply. These events influenced his status in and personal approach to physics; in addition, they influenced Kramers' relations to his peer group. This last influence was subtle and implicit. Certainly no one ever said anything about it; there were no dramatic confrontations or sudden changes. But in the inner circle everyone (including Kramers) felt and knew that an important phase in Kramers' life had come to an end and that shifts and changes were bound to come. Kramers was still a major, highly respected, and significant physicist, but he no longer was Bohr's heir apparent.

F. The Changing Patterns

This rapid succession of dramatic events was bound to exert a major influence, not just on Kramers' approach to physics, but especially on his moods and general state of mind. When Kramers arrived in Copenhagen 9 years before all these events took place, he was frequently moody, given to depressions that alternated with periods of extraordinary enthusiasm and productivity. As he matured, and as his scientific reputation grew, the depressions and mood swings became less frequent, although frequently a somewhat melancholy strain is apparent in his personal letters. In any case, his public behavior showed no trace of his moods. As Heisenberg noted, Kramers was always in total command of himself and of the social situation surrounding him. Even so, with the emotional investment Kramers made in the BKS theory, with the grand hopes generated by the dispersion approach, it would be quite understandable that the unceasing sequence of disappointments would cause profound gloom and a deep depression. This is precisely what happened. But because Kramers was such an intensely private and controlled person, the depression manifested itself in unusual and barely recognizable ways. One symptom was Kramers' apparent withdrawal from physics during the summer of 1925. It is striking that after Kramers had been told about Heisenberg's matrix mechanics about June 21, 1925 in Göttingen, his only comment was that Heisenberg was too optimistic. But most striking is Kramers' silence in Copenhagen about Heisenberg's breakthrough. It is possible that Kramers informed Bohr and Kronig in Copenhagen about Heisenberg's work as Mehra and Rechenberg suggests,[320] but all evidence indicates otherwise. As was already documented in Chapter 8, page 52, as late as August 31, 1925, Heisenberg had not heard anything from Bohr regarding his matrix mechanics. In a return letter September 4, 1925, from Bohr to Heisenberg, Bohr does not mention a word about Heisenberg's work either. If Bohr had known anything about Heisenberg's work, or had any inkling at all, he would surely have commented. Clearly, Kramers did not inform Bohr about Heisenberg's results. Born, who saw Heisenberg's paper on July 11, was so impressed by it that 4 days later (July 15) he wrote to Einstein about it. Thus, Einstein was acquainted with Heisenberg's paper about 6 weeks before Bohr ever heard of it. This is remarkable because it was a breakdown of the trusted conduit; everyone assumed that Kramers would have informed Bohr about Heisenberg's work. Pauli, in his letter to Kramers on July 27, assumes automatically that Kramers and everyone in Copenhagen is familiar in detail with Heisenberg's results. Kramers' failure to inform can be understood as a part of his general withdrawal from the frontier of physics. The succession of events appears to have stunned Kramers to such an extent that he just did not want to discuss Heisenberg's breakthrough any further. In the vernacular, "he wanted to be left alone." A very similar tendency is noticeable in Kramers' letter of July 16 to Urey.† This letter is chatty, and it mentions little about Kramers' Göttingen visit, just that he met Franck and Born and "others." He does not mention Heisenberg by name or his discussions with Heisenberg—perhaps this is an inadvertent omission—perhaps not. But to say nothing whatsoever about Heisenberg's work in a letter intended to bring Urey up to date is hard to interpret as other than a deliberate omission. It should again be understood as an inability on Kramers' part to come to terms with the cumulative effect of the disappointments produced by the failure of the BKS theory and Heisenberg's successful reorientation of the dispersion theory. Because of Kramers' great personal control, and his insistence on privacy, only a few of Kramers' close associates were aware of his depression. Thus, Kramers' close friend, Oskar Klein writes: "In the summer of 1925 I had found Kramers in a state of despondence after the failure of the BKS statistical interpretation of quantum theory, although at the same time he had given an important and beautiful contribution to correspondence theory by his dispersion formula. I tried to cheer him up by pointing out in a letter that certainly I considered science a subject of great importance, just as important as the play of children. I thought that now I had to apply this to myself."[324] This shows more explicitly than most other communications that Kramers remained deeply upset by the failure of the BKS theory, and in spite of his habit to put up a front, exuding confidence, even arrogance, he continually brooded about all that transpired. The depth of Kramers' depression at that time can best be gauged from a letter Kramers wrote August 15, 1925 to his closest friend Romeyn. "That I didn't answer earlier can be blamed on this miserable mood of tiredness and overwork from which I have suffered for the last three months—and I lacked the strength to overcome that .... About July, I was for 10 days in Göttingen .... There one saw Dirk Struik and his wife." Then the letter contains an extensive description of the new house into which Kramers and his wife had just moved. (A similar description was given in the letter to Urey—evidently Kramers liked the new villa and liked to talk about it.) After these comments the letter continues: "I myself have a busy fall coming up, even if I would forget about my scientific aspirations .... In the past few months I have often had the desire to be back in Holland although I think the slogan 'changes are improvements' is not necessarily right .... This doesn't mean that I wouldn't think seriously about returning to Holland if I could get a good position there .... My friends here say that I suffered from 'Melancholia Theoretica' the last few months. There is some truth in that, no theoretician is ever entirely free from that disease."[323] This is a remarkably candid letter—for Kramers to admit that there is some truth to his suffering

† The contents of this letter are discussed in detail in Chapter 8, page 53.

from "Melancholia Theoretica" is as explicit as a public confession would be for most persons. In true Kramers fashion, this confession is immediately qualified by insisting that everyone suffers from this disease. The timing of the period of listlessness and overwork is significant. If the 3 month period Kramers mentions is about right, this mood started just about the time Kramers became convinced that the BKS theory could no longer be maintained. It makes sense that the onset of his depression coincides with that realization. It is striking that this letter, as indeed all letters of that period, again fails to mention the meeting with Heisenberg in Göttingen (although the visit to Göttingen is mentioned) or Heisenberg's new results. It appears that at that time Kramers was just unable to face up to the radical change that Heisenberg's work had produced in physics. This letter hints also that the idea to withdraw from competitive research in the frontier of physics had crossed Kramers' mind: "I have a busy fall coming up even if I would forget about my scientific aspirations." Since Kramers' deeply felt—but rarely expressed—aspirations and ambitions were to do significant original research, to forget about those aspirations would mean that he would relinquish those goals. This would constitute an actual as well as symbolic renunciation from his lifelong ambitions—a very radical change indeed! Such thoughts are not at all unusual in periods of deep depression. The generally resigned quality of Kramers' letter indicates the depth of his gloom.

It is impossible to know to what extent Kramers' state of mind was caused exclusively by the failure of the BKS theory or whether the events surrounding the publications of the Kramers–Heisenberg paper also contributed. Whatever the effects of the original Kramers–Heisenberg cooperation might have been, the sudden confrontation of Kramers by Heisenberg's completed results on June 21 must have come as a shock to Kramers. It is curious that after leaving Copenhagen Heisenberg did not communicate with Kramers at all, although he pursued a program suggested by their cooperation. He did inform Pauli and Kronig in detail; he mentioned his activities in a most cursory manner to Bohr, so that Bohr would at best have a general inkling of what Heisenberg was up to. Kramers, on the other hand, had no information of any kind. Thus, Kramers first learned about Heisenberg's approach in their discussion in Göttingen on June 21. At that time, the Heisenberg paper was already completed. Kramers had no input of any kind in the Heisenberg paper, nor did he have a chance to reflect on it as Pauli and Kronig had. It is hard to guess how Kramers would feel about this intended or possibly inadvertent exclusion. But whatever Kramers might have felt about Heisenberg's behavior, he no doubt recognized instantly that Heisenberg's paper bore the stamp of genius, and he must have either realized or sensed that with Heisenberg's work a new epoch in physics had started—an epoch in which his own role would be ambiguous, but certainly quite different from what it had been up until that time. Whatever Kramers' precise reaction to Heisenberg's revelation might have been is of course a matter of pure speculation, but it is a safe bet that it did nothing to lift his deep depression. Most likely it caused him to turn even more inward, for a painful scrutiny of his personal and scientific goals.

In such a period of reexamination, alternatives and options rejected in happier times often reemerge and are considered seriously. One such previously discarded possibility was just how long Kramers should stay at the Bohr Institute in Copenhagen where he would always be subordinate to Bohr. It appears that only Kramers' friend Romeyn felt free to bring up this somewhat touchy issue: "Come to Bergen where you can rest and reflect on the new phase in your life, for you cannot remain an appendage to Bohr forever."†[329] Kramers answers this letter some months later in a rather factual, unemotional manner. He will stay in Copenhagen, but he does not react to being "an appendage to Bohr:" "I have no time or opportunity for desires and dreams .... I will for the time being remain in Copenhagen, there was a possibility of a position in the U.S.A. (at $4000 a year) but that was not a permanent position."[330] Romeyn does not return to this issue for some time. The number of letters in the Romeyn–Kramers correspondence diminished sharply from that moment on. During the period of public interest in the BKS theory, Romeyn raises the issue once again: "I read in the newspaper a report about a new theory of light by Bohr (and you?) .... I would not only like to know about the theory of light but also about the plans for your future .... I would be sorry if you would stay in Copenhagen for good."[331] Kramers does not answer this letter; he does answer a later letter from Romeyn written March 1, 1925. Kramers' letter, written 2 weeks before the experimental disproof of the BKS theory, is again a rather matter-of-fact letter, but he does mention that there will be a professorship open in Utrecht.[332] However, Kramers gives no indication that he is at all interested in that position.

Up to this time, Kramers had not been particularly interested in considering other positions. After all, he was doing well in physics; recognition of his work and achievements had been swift—he had been appointed a lecturer (about an associate professor) in 1924. He was elected to the Danish Academy of Science and Letters in 1925, an unusual honor at such a young age. Kramers was at ease in the social and intellectual circles of the theoretical institute. Above all, the Bohr Institute and Copenhagen were the world center of atomic physics. There would appear to be no reason to contemplate a change. However, the availability of a professorship in Kramers' native Holland would be intriguing to him anytime. His ties to Holland, his friends, and his family were always strong so that the prospect of a university position would always present a temptation; but the Copenhagen location was scientifically so advantageous that this would counterbalance the attractive possibility of a position in Holland. However, during the turbulent period of the summer of 1925, this situation changed radically. With the failure of the BKS theory and the advent of matrix mechanics, Kramers' future scientific direction was less

† Actually, he brings up the length of Kramers' stay in Copenhagen as early as July 31, 1922: "How long do you plan to stay with Bohr?"[328]

clear than it had been before. He might have been concerned about his future status at the Bohr Institute. In a letter to Romeyn,[333] Kramers even hints that he considered the eventuality that his position would primarily be pedagogical. In this process of reassessment and reevaluation, with the accompanying withdrawal, the prospect of a university professorship in Holland must have appeared increasingly attractive. A departure from Copenhagen for such a position would be the ultimate withdrawal—one that is not only face saving but that because of the prestige associated with full professorships could legitimately be interpreted as a promotion. In the event of such a change, no question of failure to meet personal or other expectations need ever be brought up. It is impossible to know whether Kramers, in his depressed state, ruminated in this manner, but it is certain in his letter to Urey[334] that he is contemplating the possibility of applying for a position in Holland. This is the first reference to this possibility, and he repeats it in a letter to Romeyn a month later.[333] In that letter, Kramers mentions somewhat wistfully that in the last few months (i.e., during his depressed state) he often had the desire to be back in Holland; he would consider returning if he could get a suitable position.

Professorial vacancies in Holland at that time were filled by a well-defined, rather lengthy process. A committee was set up to consider and screen candidates. To be considered a candidate one had to express an interest (actually one almost had to agree to accept if one wished to be considered a full fledged candidate). Then the committee would collect letters of recommendation and there would be the inevitable meetings before the committee would present a rank ordered slate to the Minister of Education who would make the final choice. The chairman of this particular committee for the Utrecht chair was Ehrenfest,† Kramers' earlier mentor at Leiden. But the whole process can only be initiated by an expression of interest on the part of the candidate. All in all, this is generally a time-consuming procedure. Kramers must have made the decision to apply to Utrecht very soon after he wrote to Romeyn on August 15, because by the middle of September the selection process was already in full swing, and letters of recommendation had been solicited and some had come in. Since Bohr was one of the referees, Kramers must have informed Bohr of his decision around September 1. For Kramers, who always had a hard time making decisions, this decision was reached in record time. It seems that the choice to go to Utrecht was Kramers' alone. The move would mean uprooting his family. His wife and children were totally steeped in the Danish culture and society; they were not anxious to go but they were not seriously consulted in the decision. The speed with which Kramers reached this decision shows again how deeply the events of the summer of 1925 affected him. Kramers knew very well that after he had applied for the position in Utrecht, he had set an irreversible process in motion. But difficult as the decision might

† The rather peculiar details of the activities of this committee will be given in Chapter 15, Section III; here only the sequence of Kramers' actions is presented.

have been, after it was made, his mood improved slowly and the deep depression began to lift. He evidently decided against withdrawing from research physics; instead, he returned to the frontier by investigating some of the consequences of Heisenberg's approach to quantum theory with his usual flair and skill (for an example see Chapter 15, Section I). It is likely that Kramers knew the instant he saw Heisenberg's work that this was the proper direction to pursue: It was impossible for him to accept the changed situation right away. Even so, the decision to leave Bohr and the institute had not been easy. Although there was now little he could do about it, he still had frequent misgivings about his impending departure. Thus, he writes to Fowler "The idea of leaving Bohr and Copenhagen after almost ten years is not very agreeable, but on the other hand I feel it is a wholesome change to obtain an independent post."[335] Kramers' scientific observations in this letter to Fowler are as interesting as his personal comments: "Not all problems are solved, [Heisenberg's] quantum mechanics is of course not yet the complete solution to all quantum problems, it does not touch the problem of light quanta, nor the time at which a transition takes place."[335] The photon and the spirit of the BKS theory still haunted Kramers even after all that had transpired. After it was known definitely that Kramers would leave Copenhagen for Utrecht, in May 1926 he experienced, as have so many others in similar circumstances, that although physically present he was already considered as having left the institute. He had become, or was becoming, an outsider, a visitor; his true allegiance lay elsewhere. In a letter written to Ehrenfest in early 1926, Kramers comments nostalgically that Bohr now has many new faces around him and was doing very well. But the process Kramers had set in motion moved inexorably. On February 15, 1926, Kramers gave his inaugural address in Utrecht. He returned to Copenhagen until May 1926, then made the definitive move to Utrecht. Kramers' lengthy sojourn in Copenhagen was over. It is interesting that as soon as Kramers arrived in Utrecht he started what was probably the first organized quantum mechanics course in the world. The date of the first lecture was May 5, 1926.[336]

Heisenberg took over Kramers' position and duties in the Bohr Institute starting May 1926. Thus, at the same time that Kramers started his first quantum mechanics course for about four to six students in Utrecht, Heisenberg started his lectures on classical electrodynamics in Copenhagen. (Quantum mechanics was not taught as an organized lecture course in Copenhagen until Oskar Klein replaced Heisenberg as a lecturer in January 1928.[337])

Thus, not quite 2 years after Heisenberg arrived for his first extended visit in Copenhagen in September 1924, he had achieved everything he dreamt about and perhaps even some things he only vaguely sensed. He had learned philosophy from Bohr and certainly had made major contributions; he returned not to learn from Kramers but to shape and guide new developments. Kramers was gone and Heisenberg had taken his place. The personal relations between Heisenberg and Kramers, which initially had their tense moments, seem to have remained civil and pleasant. If Kramers was disenchanted with Heisenberg or disappointed at the ultimate outcome of the Kramers–Heisenberg collaboration, he never made this known to anyone. To the contrary, his behavior toward Heisenberg was friendly and considerate. Evidently, his admiration for Heisenberg's scientific contributions outweighed whatever personal qualms he might have had. Likely one contributing factor was that Heisenberg (having achieved all his goals) became less competitive.[297c] After Kramers had left for Holland, Heisenberg and Kramers never again cooperated, although they saw a good deal of each other at conferences, colloquia, seminars, and reunions in Copenhagen. Their relation if not especially close was still friendly and congenial.

Kramers' wife, Storm, had very mixed feelings about Heisenberg. Storm felt strongly that it was easy to take advantage of Kramers, and she was further convinced that Kramers never got the scientific credit to which his contributions entitled him. Thus it is easy to see that she was particularly weary of Heisenberg.[338,339] Storm often expressed the opinion[340] that Kramers in his relations with other persons, especially with physicists and scientists, was hopelessly naive, which made it especially easy to take advantage of Kramers. She felt that Bohr, but Heisenberg especially, had exploited Kramers unmercifully. In spite of these feelings, she was generally guarded because Kramers was totally devoted to Bohr and appreciated Heisenberg. A number of years later, a remarkably poignant recurrence of these events occurred.[341] In the middle of World War II, to be precise on November 18, 1943, Heisenberg visited Leiden; while there, he gave a seminar on his ideas of the *S* matrix.† Evidently, Kramers was very excited and he worked all through the night. What he actually did was extend Heisenberg's method to *complex* values of the parameters in the *S* matrix elements; Heisenberg in turn was very impressed by Kramers' ideas. After the seminar, Heisenberg came to dinner to the Kramers' home. After dinner, Mrs. Kramers and Heisenberg washed the dishes together. During that time, Heisenberg told Storm that he hoped that Kramers and he could work together. Storm was (according to her recollection) somewhat nervous; she knew Kramers liked and appreciated Heisenberg, and she told Heisenberg: "Why shouldn't you work together, you have worked together before." Heisenberg then answered: "That is true, but I don't want to work together in the same manner." It is anyone's guess what Heisenberg precisely meant by that remark, but it is pretty obvious that the lingering memories of the earlier cooperation contained some jarring elements.

Shortly after Kramers died (1952), Heisenberg told Mrs. Kramers that Kramers' role in the formulation of quantum mechanics was generally and vastly underrated.[342] This made Mrs. Kramers more unhappy than she had been before; she felt that Heisenberg should make this known to the world of physics. She already knew it. Whether this particular admonition had any influence is hard to know, but it is certain that in his later statements

† This important incident will be described in more scientific and political detail in later chapters. Here just the personal ramifications are mentioned.

Heisenberg appeared very anxious to give Kramers credit for all his achievements. Thus, Heisenberg in an interview[343] expresses his regrets that Kramers who did so much for the discovery of quantum physics did not get a Nobel prize. According to Heisenberg, Kramers was overshadowed by Bohr and appeared dependent on Bohr for physical ideas. In the conversations Heisenberg had with Mehra and Rechenberg, the same point comes up again: "[Kramers'] contributions to physics were very large and perhaps it had not been seen so clearly only because he had always been together with Bohr and so far was a bit dependent on him."[297c] Heisenberg's insistence to van der Waerden that the final version of the Kramers–Heisenberg paper was totally due to Kramers (except for the final formula)[344] is yet another example of Heisenberg's campaign to give Kramers appropriate if perhaps belated credit. But Heisenberg was fully aware of the disappointments that Kramers had experienced. At a dinner for outstanding scholars (summer of 1975), Heisenberg sat next to a world famous Latin scholar, Professor Wassink of the University of Leiden.[345] They naturally talked about Leiden and about Kramers. Heisenberg told Wassink that Kramers and especially Mrs. Kramers suffered badly because Kramers did not get a Nobel prize. Heisenberg also said that the influence of the Nobel prize in physics was unfortunate, since it encouraged spectacular, even sensational ventures which could give rise to quick fame—to the detriment of serious thoughtful efforts. Kramers was not given to sensationalism—in fact he abhorred it—but he was unhappy with the lack of recognition of his sober efforts by the physics community. Heisenberg was evidently aware of Kramers' feelings and attempted later to redress the rather unappreciative treatment that Kramers had received. But all this started much after the momentous summer of 1925. The relation between Kramers and Heisenberg, which started in such a stormy manner, settled down. Kramers' respect for Heisenberg's physics never wavered; Heisenberg, from his initial vacillations of Kramers as a person "far above him" to a person whom he had surpassed, settled down to a healthy respect for Kramers' powers and accomplishments. Socially, their joint interests in music, philosophy, and literature made for a pleasant, easy going if not especially deep relation. Only later in 1943, during World War II, and again shortly after the war did their relation once again become more intense. At that time, in addition to the personal and scientific elements, political considerations were interwoven in a complex pattern. Heisenberg's public recognition of Kramers' role in the development of quantum mechanics did not really start until well after both Kramers and Mrs. Kramers had died.

Kramers' planned departure from Copenhagen had much less influence on the Pauli–Kramers relation than on the Kramers–Heisenberg relation. The nadir of the Pauli–Kramers relation was certainly the Pauli letter of July 27. Although Kramers knew perfectly well that Pauli's criticism, fierce and unsparing as it was, was always directed against fuzzy thinking and pretentious claims, and not against the individual, it is doubtful that Kramers in his despondent state would derive much solace from this knowledge. In the

Pauli–Kramers relation, the personal relations and the positions adopted toward physics were coupled so strongly that a separation between personal appreciation and scientific approach is scarcely possible. Pauli was well aware that he often hurt people unintentionally. In fact, it was Kramers who told Pauli on one occasion when Pauli had been especially acerbic in his criticism. "Pauli your feelings are better than your brain."[346] But it is anyone's guess whether Kramers' understanding of Pauli's motives was very helpful in dealing with the critique that Pauli had hurled at him in this letter. As was mentioned earlier, Pauli's attitude to Kramers softened noticeably after this outburst, and their relation gradually reassumed its earlier gentler, warm character. Pauli showed more understanding of Kramers' problems, in his struggles to decide whether to leave Copenhagen, than anyone else. For example, in a letter to Heisenberg, Pauli makes a side remark concerning Kramers. From the context of this letter it can be inferred that Pauli had visited Kramers in Utrecht in the fall of 1926 where Kramers had been a professor for about 5 months. Pauli reports in a footnote:[†] "Kramers wishes to calculate the normal state of helium. So far he just worked on other things. He is in good spirits and he is capable of work (*Er ist sehr froh und arbeitsfähig*)."[347] Although a mere offhand comment by Pauli, this nevertheless shows great sensitivity. Pauli sensed clearly that it was not self-evident that Kramers would be happy in Utrecht after being in Copenhagen, nor was it obvious that he would be capable of doing physics. There is even an inference that Kramers had (presumably this last year) not always been happy and had not always been able to do physics. The comment also shows a genuine concern for Kramers' well-being. The change to Utrecht also marked a reorientation of Kramers' physics. He began to deal more with specific, technical problems (often very involved), and less with speculations about the approach to or philosophy of physics. As a consequence, the disagreements and disputes between Pauli and Kramers became less frequent and certainly less sharp. From time to time, Pauli even publicly expressed some respect for Kramers' capabilities. For example, Pauli, who was very proud of his ability in classical analysis, used to tell Jost:[348] "I am a better analyst, I can use classical analysis better than any other physicist—with the possible exception of Kramers." Whether it was so intended or not, this amounted to a rave notice for Kramers. On other occasions, Pauli would stress to the new generation of physicists in Holland (Korringa and de Groot[349]) "that Kramers was really a very fine physicist." But to dispel any thought that he was becoming a mindless admirer of Kramers, he told Casimir: "Yes Kramers is certainly very good, but I am still somewhat better."[350] When van der Waerden wanted to collect the important seminal papers which eventually led to quantum mechanics, it was Pauli who called van der Waerden's attention to Kramers' significant contributions. It was due to his insistence that this collection includes Kramers' papers on dispersion theory.

[†] For some reason, this footnote, which of course occurs in the original letter, was omitted in the transcription of this letter to the Pauli correspondence.

Sometimes rather minor events characterize a relation better than extensive descriptions. An incident at the meeting of the "h" club† illustrates the nature of the Pauli–Kramers relations very clearly.[351] Pauli gave a talk for the "h" club, discussing some new and abstract topic in 1938 or 1939. (The topic might have been the relation between spin and statistics, published a year or so later.[352]) During the talk, Pauli needed a rather technical result. When asked where that came from, Pauli tried to derive it but after some effort he was unable to do so. Then Kramers got up and gently said "Pauli this is too simple for you." Kramers obtained the needed result, sat down, and Pauli resumed his talk. Even if the details of this incident are incomplete, the general tone it conveys is certainly accurate. At that time Kramers was the "dean" of the Dutch physicists: His formal powers were legendary and in formal matters he was easily a match for Pauli. This was generally known and understood. But Pauli had become the conscience of physics; he was in constant touch with the frontier fields of physics and Kramers deferred to him. There was a mutual understanding and appreciation between Kramers and Pauli, with a full acceptance of their individual talents and limitations. Perhaps more important, the warm relation and friendship of their youth had evolved into genuine concern. Clearly, the Kramers–Pauli relation started and remained different from the Kramers–Heisenberg relation.

G. Why Didn't Kramers Discover Quantum Mechanics?

The previous discussion covered the tortuous paths that led from the analysis of the interaction between matter and radiation, to the dispersion relations, the BKS theory, the Kramers–Heisenberg formulas, and eventually to matrix and quantum mechanics. Kramers' contributions were major throughout except for the very last step which is (correctly) attributed to Heisenberg alone. It is very tempting‡ and more than a little dangerous to inquire why Kramers was unwilling or unable to participate in this last step. Not only did he not take this last step, he did not even exploit its consequences to the fullest after it had been taken. It is always intriguing to speculate about the reasons for success or failure, and it is especially pertinent in Kramers' case since this is only the first of a number of occasions that he came tantalizingly close to a major advance without making the decisive breakthrough. The pattern is so persistent that at least some attempt at understanding is required. Even so, such inquiries are not without their dangers. It is after all hard enough to explain even the simplest kind of behavior. The creation of a new scientific theory in the midst of rapid changes with very few secure guiding principles is certainly very far from being simple. To ask why a given person made one

† The "h" club was a rather informal organization of Dutch physicists. Their meetings were devoted to analyzing and scrutinizing recent results in theory. The "h" stands either for the H in H. A. Kramers or in H. A. Lorentz, or otherwise for Planck's *h*.

‡ It will be evident from what follows that the temptation is not resisted!

set of contributions and not another, or gives a partial or incorrect discussion rather than a complete and correct answer, comes very near being silly. The danger in even asking such questions is that once asked there is a tendency to give one-sentence answers or one-cause explanations of very involved, fluid, ambiguous phenomena. To argue as is sometimes done[297d] that Kramers was unable to "deviate from the canonical path of Copenhagen physics" and was therefore unable to appreciate and follow Heisenberg's radical departure from that theory contains some elements of truth. It is also true as Heisenberg himself recalled[343] that Kramers was dependent on Bohr for physical ideas. But even though both observations are not incorrect, they explain very little. When Mehra writes[297d] "the canonical path of Copenhagen physics," it is important to realize that such a path simply did not exist as an organized, formalized scheme. Copenhagen physics was more a general style, a loose collection of procedures and hopes, a curious mixture of classical ideas and quantum rules. Thus, the canonical path that Kramers was supposed to follow was not there. Everyone, including Kramers, was trying out new schemes; if specific results such as dispersion relations could be found, so much the better. If these were successful, these same methods were applied to new problems. Heisenberg's comment that Kramers tried to follow Bohr's general directive is certainly true, but it is equally true that Pauli and especially Heisenberg also followed Bohr's directives. It is interesting in this connection that Kramers and Bohr were at one time convinced that the BKS theory was the true, natural extension of the correspondence principle. On the other hand, Bohr, only a little later, characterized Heisenberg's matrix mechanics as the rigorous formalization of the correspondence principle. Thus, the general direction, precisely because of its generality and corresponding vagueness, could yield very different theories. These examples indicate what is obvious anyhow: That in producing scientific results, a large number of distinct, not easily discernible and distinguishable, factors interact and interfere in a most complex and unexpected manner. To attribute a particular result to a specific set of causes is very likely misleading, superficial, and wrong. Recognizing all these caveats and the tenuous nature of the results to be anticipated, we still inquire why certain investigations consistently achieved results, while others did so only sporadically.

As phrased, the question derives its sense from the belief that the scientific work of a great physicist exhibits a definite discernible structure. There is a pattern to the problems considered: The methods of approach and the goals pursued appear as part of a larger scheme. Within such a context, it makes sense to ask why certain questions were asked, why certain approaches were stressed while others, apparently equally pertinent, were ignored. This is the context in which the question of this postscript must be viewed. Examining this question may lead to a better appreciation of Kramers, and it also may give a clue to the altered character of Kramers' research after he left Copenhagen. The elements mentioned earlier all played a role. Kramers' deference to Bohr's physics was important in guiding Kramers' direction and keeping him from others. Probably equally if not more important was Kramers' deep emotional attachment to the BKS theory. This more than anything else made him willing to constrain theoretical physics to the BKS structure. This infuriated Pauli. All these factors have been stressed throughout the description of this turbulent period. But it is tempting to conjecture that there are in addition other, more individual, personality traits, which also exerted their indirect and intangible influence on Kramers' research. For example, one reason that Kramers did not try the kind of thing that Heisenberg did was that he especially enjoyed special problems and clever tricks. He almost enjoyed a sleight of hand manner to solve problems, more than a systematic, organized procedure.

He liked convoluted arguments and delighted in solutions appearing like magic; unexpected twists and paradoxical situations appealed to him. It almost seems as if part of his whimsical nature was transferred to his mental processes. The correspondence principle had an aura of magic; it could be manipulated by Bohr and Kramers with astounding results. That it was difficult to formalize made it that much more attractive to Kramers. All this would appeal more to Kramers than a heavy-handed, rigorous treatment. This might account for the estrangement of Kramers and Pauli; for Pauli, such vaguely felt ideas were anathema. It might also be the reason why Heisenberg was irritated with Kramers' jokes and supposed lack of seriousness. Kramers, in dealing with a problem, was usually not looking for a structure or for impeccable deductive logic. He was usually looking for a trick, a gimmick, a joke to get at a solution—this in spite of the fact that Kramers had a general philosophy of physics and strong feelings for what physics should be like. But on the actual operating level of research, his formal interest and technical ability always pushed that philosophy to the background. As such, the "translation" of classical formulas into quantum formulas, which required a guessing of the proper form, was very much in Kramers' line. That one could do this guessing once and for all and formalize it at that—as Heisenberg did—would not be a natural direction for Kramers to pursue. In fact, he would be disappointed if such a method would exist, for then the fun of the guessing game would be spoiled! Perhaps Kramers' initial negative reaction to Heisenberg's matrix mechanics was merely the expression of a hope that no general procedure, which would eliminate the need for guessing, would be found. To Kramers, this artful guessing was the true essence of the creative process; it was also what he enjoyed most. So it is possible that the thought that in the future Heisenberg's systematic procedure would make this guessing superfluous, would make him uneasy and disappointed. It is, of course, impossible to know whether such ideas ever entered Kramers' head, but it is not out of the question that such thoughts, however fleeting and ephemeral, affected Kramers' mood, which in turn influenced his receptivity toward Heisenberg's ideas. Of course none of this explains why Kramers did not invent matrix mechanics. After all, Heisenberg's paper was the creative act of a genius, and it is pointless to ask why someone else did not do what an acknowledged genius did. But in Kramers' case, it happened so many times that he failed to carry out his investigation to a decisive conclusion (and this is independent of what others do or do not do), that it is interesting and perhaps informative to comment on some of the possible factors which might well have inhibited his creative surge in this case. The point is not so much that Heisenberg made a profound seminal contribution as that Kramers stopped short of doing so. This discussion raises a much more sensible question: What were the influences *on* and *in* Kramers which kept his most original, most sustained, most committed scientific effort from leading to some major advances? This same pattern in a variety of guises was to be repeated many times. It is not incorrect to say that a good share of the remainder of this book is devoted to an attempt to analyze and understand this recurring pattern, which was probably Kramers' most outstanding characteristic.

CHAPTER 14

The Curious Copenhagen Interlude

I. The Events

The sequence of scientific events described in such detail in the last chapter was bound to have an extraordinary effect on Kramers. The discovery of the Compton effect and its subsequent refinement by the Bothe–Geiger and Compton–Simon experiments, leading to the disproof of Kramers' favorite theoretical scheme—the BKS theory—affected him deeply. The central intellectual reorientation that these results required was an altogether different attitude and approach to the photon concept. Pauli in a letter to Kramers writes succinctly and unequivocally: "It can now be considered as proven—to every objective physicist—that photons are just as physically real (or as unreal!) as electrons."[306] In spite of Pauli's admonition, Kramers had a very difficult time incorporating the photon idea into his physical thinking. Thus, even after he knew that the BKS formulation was untenable, he still writes to Born: "I am convinced that a systematic coupling of individual processes is incompatible with wave optics, but one should try to overcome this contradiction in a less naive manner than Einstein often does."[353] Einstein described these energy- and momentum-conserving processes in terms of photons—and where Kramers grudgingly admitted that energy and momentum were conserved, he did not admit that this implied the reality of photons. Kramers kept on battling against the photon. It will be recalled that several months after matrix mechanics had demonstrated its great power, Kramers, in a postcard to Fowler, reiterated again that in matrix mechanics "not everything is solved, that light quanta are still out."[335]

With such deep-seated antagonism against photons, it can only come as a surprise, not to say a stunning shock, that some 4–5 years earlier (about 1921) Kramers actually sketched a derivation of what later would become known as the Compton effect. Kramers applied the conservation laws of energy and momentum to the scattering of light by electrons; in doing so, he certainly employed photonlike concepts. Kramers' results were never written up and never published.† As might be anticipated, when Kramers informed Bohr of his ideas and results, Bohr was not only unhappy and upset but actually irritated. Bohr immediately set out to convince Kramers of the errors of his ways. Bohr used all his persuasive power to convince Kramers that a consistent photon picture was impossible. Needless to say, Bohr was not about to allow the publication of Kramers' results. Kramers was subjected to an unrelenting series of interminable and acrimonious discussions in which the tenuous and at that time necessarily incomplete character of the photon notion was examined in excruciating detail. The basic incompatibility of the photon with the "wondrous" wave theory of light was mercilessly exploited. To most of the issues raised by Bohr, Kramers simply had no answer, not because he was not smart or had not thought about them, but because at that stage of physics there were no answers to those questions.‡ It is likely that Kramers was excited, because his calculations indicated a frequency shift between the incident and scattered radiation. This shift, a purely quantum effect, if observed would be a striking confirmation of the need for a quantum theory of radiation, a field that always intrigued Kramers. Of course at the time of these discussions, no experimental hints for such a frequency shift were available, so Kramers had to defend a theoretical view (photons) which led to many paradoxes and conflicts, without the benefit of compelling experimental evidence. In the ensuing arguments, Kramers—as Slater, Schrödinger, and Heisenberg would on later occasions—finally capitulated to Bohr's uncompromising onslaught. However, Kramers did not merely acquiesce to Bohr's views; he made Bohr's arguments his own, and in a very short time he joined Bohr in extolling the virtues of the wave theory while emphasizing the difficulties and inconsistencies of the photon. Kramers' complete reversal is a perfect example of the fanatical adherence recent converts often exhibit to their new-found faith. After the Bohr–Kramers discussions, Kramers did not need any further encouragement to criticize and deprecate the Einstein photon idea. He did so on all possible occasions. The discussion of the photon in the popular book by Kramers and Holst is a public testimony of Kramers' official shift to the antiphoton camp. His conversion was complete. After the resolution of this intense series of conflicts, Kramers and Bohr were scientifically closer, more in tune with each other, than before this interlude. It is therefore

† Unfortunately, the detailed documentation for the following claims is extremely sketchy. There are some very rough, incomplete notes in the Kramers files and some diary entries. A good deal of the information comes from interviews and personal recollections. Inconclusive as all this is, a fairly coherent picture still emerges: Kramers did have the Compton effect, but Bohr talked him out of it, kept him from publishing, and totally converted Kramers to his later antiphoton views.

‡ After the Compton experiments and the failure of the BKS theory, Bohr recognized that incompatible features might coexist and be needed for a complete description, but at this time Bohr simply did not wish to entertain the possibility of the existence of physically free photons.

doubtful that Kramers gave much further thought to his original idea during the busy and exciting days of the dispersion relations and the BKS theory. He probably considered it a youthful aberration and was likely grateful that Bohr's intervention saved him from making a fool of himself. So thoroughly was Kramers inbued with the antiphoton philosophy, that when Compton's experiments and the Compton–Debye theoretical explanation were forthcoming (which agreed precisely with his own earlier results), he nevertheless showed no inclination to revert to his original ideas. Quite the contrary, he preferred the somewhat convoluted BKS explanation of the Compton effect over his former straightforward treatment based on the photon idea. When the BKS theory collapsed, Kramers was the last of the three authors to accept the reality of the photons. He did so reluctantly and grudgingly. That is how thoroughly and completely Kramers had turned against his own explanation. There are some indications that after the early upheavals of quantum mechanics were over, and especially after wave mechanics and the principle of complementarity were better understood, Kramers took a less negative view of his own Compton effect efforts. But during the fateful spring and summer of 1925, it might well have seemed to Kramers that his best efforts in physics by some systematic perversity seemed destined to lead him astray. The photon explanation of the Compton effect, which he had abandoned under Bohr's insistence, was correct after all. The reasons for giving up the theory were cogent but not relevant. Yet these same invalid reasons led him to the dispersion theory, which was correct, but it in turn suggested the incorrect BKS approach. Thus, he gave up correct approaches while pursuing misleading clues. All this was done for what seemed to be compelling reasons at the time. What must have been particularly confusing is that the validity and importance of the *concrete* results appeared surprisingly independent of the underlying philosophy. Thus, the curious prologue in Copenhagen, where Kramers came so near to discovering the Compton effect only to abandon it, had no very direct effect on Kramers during the summer of 1925. His scientific views were too rigid at that time. But these same events must have reinforced the inevitable state of depression and confusion from which Kramers suffered that summer. It must have become apparent at that time that he had to rethink his complete approach to physics in a fundamental manner. In that process (which actually took several years), he must surely have pondered the ill-fated personal consequences of his most revolutionary and ultimately correct speculations.

II. Recollections and Some Documentation

It is probably futile to speculate on what prompted Kramers to write down the equations for the conservation of energy and momentum for the collision of an electron and a photon, or for the scattering of x rays from electrons. It is certainly true that just before Kramers made these calculations he was preoccupied with the relativistic Stark effect; thus, he handled relativistic kinematics on a practically daily basis. He was also always interested in new quantum effects and he might well have seen a strikingly new classical feature in the Compton frequency shift. He was and remained interested in x-ray scattering, as his paper on that topic written as early as 1922 indicates.[354] All these observations are correct, but they of course in no way show that Kramers did work on the Compton effect. They merely show that he *might* have done so. After all, Einstein was as familiar with the photon character of radiation and the conservation laws as anyone; he was surely interested in the interaction between radiation and matter, but there is no indication that he ever wrote down the equations for the Compton effect. That Kramers indeed wrote down such equations can be inferred from a few diary entries, some remaining rough notes, a certain amount of general grapevine type information, and a large number of personal recollections.

It seems that some time after Kramers was married to Storm, he came home from the Bohr Institute and, according to his wife, was "insanely excited."* He tried to explain to his wife that he had discovered a "striking, unexpected" new quantum effect. This was unusual for several reasons: Kramers only rarely would tell his wife about the progress of his work and then only in very general terms. It was unusual for Kramers to show great excitement, especially about his own work. It is not surprising that his wife would remember the occasion. At that time, Kramers had apparently no opportunity to discuss this work with Bohr who was preparing for a trip and shortly thereafter left for a hiking trip of several weeks. Kramers did not have a chance to discuss his results with Bohr before he had returned from his vacation trip. By that time Kramers had obtained sufficiently definite results that he considered publication. The indications are that Kramers had obtained the conservation law description of the Compton effect, using the photon description explicitly. If this indeed was the case,† it is obvious that Bohr would object violently to the publication of thse results. According to Mrs. Kramers,‡ Bohr and Kramers immediately started a series of daily, "no holds barred" arguments about the photon which lasted throughout the day and part of the night. In these discussions, Kramers, who had some specific results but no general framework or philosophy, was simply ground down by Bohr who had no specific results, but did have an extraordinary sense of the proper direction for physics to pursue. After or during these discussions which left Kramers exhausted, depressed, and let down, Kramers got sick and spent some time in the hospital. During or after his stay in the hospital, Kramers gave up the

* The information reported comes from several interviews with Maartien Kramers, Agnete Kramers, and Jan Kramers. The information originally came from Mrs. Kramers. Evidently the same story, using even the identical language, was often repeated in the Kramers household.

† See below, for reasons which make this quite believable.

‡ Also according to Klein.[355]

photon notion and the associated calculations altogether. Instead, he soon became violently opposed to the photon notions, and he never let an opportunity pass by to criticize or even ridicule the concept. It is clearly impossible to know whether Kramers' conversion was the result or the cause of his sojourn in the hospital. But after Kramers returned home from the hospital, he disposed of most of the Compton effect papers, although he inadvertently kept a few, early rough notes. A number of years later, well after the Compton effect had been discovered and had gotten its name, he happened to locate these notes. He collected these few remaining notes in a folder and called it the Compton effect folder.† Only then did he tell his wife the whole story. The episode had a long-lasting effect on Storm; she had always had a feeling that Bohr was exploiting Kramers. The Compton effect episode not only reinforced these feelings, but she became convinced that Bohr mistreated Kramers.

It is never very easy to assess the validity and correctness of personal recollections. Whenever strong personal emotional ties are involved, as in the case of Storm Kramers and Hans Kramers, personal memories—however objectively intended—can easily be misleading. If indeed the recollections of Mrs. Kramers relative to the Compton effect were the only information available, it should of course be taken seriously, and the evidence that Kramers did have some version of the Compton effect would be suggestive but not compelling. But there is a considerable amount of additional, albeit presumptive evidence. In an interview, ter Haar,[356] who was one of Kramers' students and closest to the family, reported numerous conversations with Mrs. Kramers in which she told him specifically that Kramers did have the basic formulas of the Compton effect, but that Bohr was adamant in his refusal to let Kramers publish these results. Furthermore, she reported that Bohr persuaded Kramers to give up the ideas altogether. Since this information comes from the identical source as that reported earlier, this does not provide any new evidence; it merely shows a certain consistency of the recollections. However, Korringa,[357] Kramers' last assistant, was positive that he had heard about the Compton effect episode. He was explicit that this information could not be traced back to Mrs. Kramers. It is possible that Korringa learned about Kramers and the Compton effect from Kronig. Kronig was heavily involved in the early development of quantum mechanics and remembered that he heard via Debye that Kramers had obtained the basic expressions for the Compton effect.[358] This is especially pertinent since Debye developed these expressions on his own, quite early and independently of both Compton and Kramers. Several other Dutch physicists—among them Nico Hugenholz, one of Kramers' last students, and H. B. G. Casimir, an early assistant and colleague of Kramers—have recollections of various degrees of precision and concreteness which indicate that they were aware that Kramers had ideas about the Compton effect well before Compton and independent of Debye.

† This folder is in the possession of Maartien Kramers. It is one of the few tangible pieces of evidence of this episode.

They also all agreed that Bohr's opposition kept Kramers from publishing these results. But probably the most compelling evidence comes from Oskar Klein and Pauli. Klein recalls extensive discussions with Kramers in 1923 about the Compton effects.[359] By that time Compton's experiment had actually been carried out. Even so, Bohr was very reluctant to believe the observed frequency shift and was even less inclined to believe the photon interpretation of that effect. During these discussions, Kramers happened to mention that he had obtained similar formulas in Copenhagen during Bohr's absence. At that moment Mrs. Klein, who was present at the interview, interjected: "Oskar also, always did his best work when Bohr was out of town." Ironically, by that time (1923), Kramers no longer believed in the photon interpretation of the Compton effect. Both he and Bohr looked for an interpretation (first suggested by Pauli) in terms of a Doppler shift. But there is no doubt that Kramers told Klein about his early work on the Compton effect, although at that time he no longer believed in it.

There is no evidence to indicate that when the Compton effect was discovered Kramers was bitter or complained about Bohr's refusal to let him publish his results. At that time, Kramers was a staunch believer in the BKS theory and he was still convinced that photons and his own use of photons were unfortunate examples of misdirected physics. As time went on and the evidence for photons became more and more convincing, Kramers like everyone else was compelled to recognize the fundamental significance of the photon notion. In this process in which he scientifically came to terms with the photon idea and its extraordinary significance, he became retroactively angrier and angrier with Bohr for keeping him from publishing his early results on the Compton effect. Only several years later (about 1927), after both the photon character of radiation and the wave character of matter were well established, did Kramers express his great anger at Bohr. The reasons for his anger were quite old: Bohr had forbidden the publication of Kramers' Compton effect work and had in addition talked him out of it. Bohr was in turn upset about Kramers' delayed annoyance and he complained to Pauli about Kramers' unreasonable attitude.† Jost recalls that Pauli reacted in a very characteristic fashion: "I don't quite understand why Kramers became so excited that Bohr kept him from publishing the Compton effect .... After all that was not really such a significant discovery, for otherwise Debye would not have made this same discovery independent of Compton." Where it is certainly possible to doubt Pauli's assessment of the significance of the Compton effect, this letter shows without any doubt that Pauli knew that Kramers had obtained the main results of the Compton effect and that Bohr had kept him from publishing these results. Thus, Pauli's recollections completely confirm Mrs. Kramers' memory of these same events. It is further worth noting

† This information is inferred from Pauli's recollections as remembered by R. Jost.[360] The tentative character of all these recollections will be evident to everyone; even so, the various quotes all have the ring of authenticity.

that Jost, in an official lecture to a Natural Science Society in Zürich, explicitly refers to Kramers' contribution.[361] This lecture was a commemorative lecture reviewing the relation between Zürich and Einstein. In so doing, Jost comments on Einstein's fundamental insight that in the interaction between light and matter a momentum transfer takes place. Then Jost continues: "The quantum of light therefore possesses not only an energy, but also a momentum, its rest mass is zero, it behaves as a particle, it is a photon as we say nowadays. It is well known that these ideas form the basis of the Compton–Debye–Kramers theory of the scattering of x rays from electrons and explain the Compton effect." This is the only time in the open and available literature that Kramers' name is specifically linked with the Compton effect and the photon notion. Combined with all the other incomplete, partial, and personal evidence, this makes it overwhelmingly likely that Kramers did have the basic formulas of the Compton effect in his possession.

III. Conclusions

Even though it must be admitted that actual hard evidence for such a scenario is lacking, the available indirect evidence strongly suggests that such a sequence of events actually took place. It is interesting to inquire what effects such events might have had on Kramers' later development. It appears quite definite that for Kramers the primary concern was the publication of his result.[356] Although there is no direct evidence for it, it is likely that Bohr would under no condition agree to have Kramers publish his results. To Bohr the by-line "Copenhagen" on a publication was extremely precious; he felt *personally* responsible for every work and every idea that was published from the Bohr Institute. Thus, there was no question that Bohr would not agree to the publication of Kramers' results which gave such a prominent role to the photon.

It is noteworthy that several years later, in 1927, Heisenberg, who already had shown his ability as a physicist at that time (it was after his creation of matrix mechanics) was still unwilling (and unable) to publish anything without Bohr's approval. Bohr at that time was unhappy with Heisenberg's celebrated paper on the uncertainty relation,[362] and he exerted enormous pressure on Heisenberg to make substantive alterations in this paper. This was in spite of the fact that Heisenberg had taken the precaution to check practically every step of the argument with Pauli, and Pauli was in sympathy with the general tenor of the paper. But Bohr felt that the wave–particle duality should be made the focal point of all treatments, which Heisenberg did not wish to do. The ensuing discussions were intense, angry, and acrimonious (for a detailed discussion see Mara Beller[156]). The depth of feeling engendered by this controversy may be gauged from comments made by Heisenberg much later. Heisenberg recalls that he and Bohr went through several weeks of strenuous arguing and their relation had become quite tense. Heisenberg remembered that at or near the end of one of these discussions "the meeting ended with my breaking out in tears because I just could not stand the pressure from Bohr."[343] If Heisenberg, who was tough and self-confident, could not stand the pressure that Bohr was exerting on him in 1927, Kramers, who was gentle and accommodating, never had a chance against Bohr in 1921. Furthermore, Heisenberg at that time had a monumental discovery to his credit, while Kramers had nothing comparable at the time of the discussion. Also, Pauli was familiar with and sympathetic to Heisenberg's approach; he mediated the Heisenberg–Bohr conflict. There was no one to take Kramers' side; he had to defend the photon by himself. If nothing else, Heisenberg's recollections show the ferocious intensities of the involvements: Heisenberg burst into tears, Kramers was admitted to the hospital, and Schrödinger became ill under very similar circumstances. In spite of all the pressure, Heisenberg did not give in; he merely agreed to a compromise which consisted of a postscript to Heisenberg's paper. This postscript referred to a forthcoming paper by Bohr which would elucidate and clarify several points in Heisenberg's paper.† Kramers did not withstand Bohr's juggernaut. He was a more malleable person, and his case was not as good as Heisenberg's. As explained earlier, Bohr's objections to Kramers' approach could not be answered satisfactorily at that time. Kramers never submitted his paper, and all the evidence indicates that after some initial struggles he became truly convinced that Bohr was fundamentally right in his unwavering opposition to the photon; therefore, he made that position his own.

It was inevitable that encounters of this intensity should leave their mark on the personal relations between the participants. The Heisenberg–Bohr conflict came very near to an outright break; this was avoided only because of Pauli's continued and skillful mediation. Even so, there was a noticeable estrangement in the Heisenberg–Bohr relations. No such permanent effects remained in the Kramers–Bohr relation: If anything, after the Compton interlude their cooperation increased and their mutual understanding deepened. But unquestionably there must have been tensions between Bohr and Kramers during the summer of 1921. Kramers always had, and maintained, the deepest admiration and respect for Bohr. He wanted Bohr's approval more than anything. He must have been hopeful that Bohr would approve of his new approach to the scattering of x rays from electrons. To have his revered master criticize his work so mercilessly must have been a severe blow to Kramers. For Bohr to have his first and so-far most successful collaborator pursue totally objectionable directions must have been a shock. But after the month-long catharsis of the summer of 1921 and after Kramers' conversion to Bohr's view, their relation reassumed its original character. All later evi-

† Actually, this is a remarkable postscript; it hints very strongly that Heisenberg's arguments leading to the uncertainty relations are very doubtful and that Bohr's later paper will give a better justification.

dence indicates that Kramers felt totally comfortable with the antiphoton view. He espoused that philosophy on every occasion. It was not until much later, really until after a quantum mechanical treatment of the Compton effect had been given, that Kramers blamed Bohr for keeping him from publishing his results on the Compton effect. It is understandable that Bohr in turn would be a bit nonplussed by this delayed reproach. It is not unusual that eminent physicists talk others out of brilliant ideas. Inevitably, if the idea (or something like it) turns out to be correct, there will be tension, irritation, and reproaches, although perhaps less than we imagine. But usually these conflicts surface at the time that the original idea is vindicated. But this was not at all the case with Kramers, Bohr, and the Compton effect. As described so extensively in Chapter 13, Kramers was much more than a technical collaborator in the BKS theory. The obstinate denial of the photon's existence in spite of the evidence of the Compton effect was part and parcel of Kramers' philosophy. He was totally in tune with the nonphoton reinterpretation of the Compton effect. He certainly could not blame Bohr for his own part in the BKS theory. He did this willingly and happily. His irritation with Bohr, expressed so many years later, was probably more an expression of disappointment with himself that he had allowed Bohr not only to talk him out of his results but to imbue in him the antiphoton gospel. Kramers in his discussions with Bohr had not been sufficiently adamant; he had not insisted on the incompleteness of Bohr's viewpoint, but had willingly acknowledged the provisional character of his own constructs. This recognition was probably the main lesson Kramers learned from this interlude: That in presenting his own ideas and investigations to the world of physics, he was not sufficiently single-minded, aggressive, or obstinate. Quite the contrary, in presenting his own ideas he was remarkably self-effacing and tentative. Instead of emphasizing the new results and achievements, while glossing over the deficiencies, he tended to do the opposite, giving the appearance that his accomplishments were tentative and incomplete. Kramers as an introspective person was keenly aware of his own characteristics. He fully realized that in the competitive world of physics these characteristics can become serious handicaps. But they were such essential aspects of his personality that he was unable to shed them. It was totally impossible for him to make a public relations lecture tour, to publicize recent or forthcoming results, or to make exaggerated claims (although it would have been easy to do). He could only operate by letting his work—unembellished and perhaps incomplete—speak for itself. He fully recognized the disadvantages of this approach, but he himself was incapable of doing anything differently. He went through his scientific life cognizant of the limitations that his personality imposed on his scientific reputation. He was not particularly happy with that, and although he could not change his personality, he encouraged his students to be different. Kramers, who very rarely gave anyone advice, in a rare moment of self-revelation told his student Ubbink: "If you are doing something of which you are personally convinced, that it is correct, or promising, then you must push it and pursue it, no matter what any one says, no matter whether it is complete or not .... You must be stubborn[†] and pugnacious .... I myself wasn't always stubborn enough." It is impossible not to recognize in these words a melancholy memory of the Compton interlude.

[†] The interview with Ubbink by M. Dresden was in Dutch. The word translated here as stubborn was *eigenwys*, which has the connotation of being stubborn to the extent of being foolhardy.

Notes for Part 2

[1] A considerable number of these stories are available in the papers of the Kramers family. Later on, H. A. Kramers kept rather irregular diaries. Occasionally, there are poems or short comments in these diaries (referred to as Kramers files).
[2] This particular letter is contained in the Kramers–Romeyn correspondence which is kept in the Institute for Social History, in Amsterdam.
[3] Letter from H. A. Kramers to Jan Romeyn, dated August 26, 1911 (Kramers–Romeyn correspondence).
[4] Entry in Kramers files, undated, 1911.
[5] Interview with G. Kramers, March 1, 1977, by M. Dresden.
[6] Interview with R. ter Haar, January 3, 1977, by M. Dresden.
[7] Interview with H. B. G. Casimir, June 23, 1978, by M. Dresden.
[8] Several of Kramers' notebooks are preserved. See Kramers files.
[9] H. A. Kramers on the occasion of receiving the Lorentz medal, October 30, 1948.
[10] For a most beautiful and incisive analysis, see the biography by Martin Klein entitled *Paul Ehrenfest*. North-Holland, Amsterdam, 1970.
[11] Kramers in a memorial tribute to Paul Ehrenfest. Reprinted in *Physica, Nederlandsch Tydschrift voor Natuurkunde* 1933, 13th year, pp. 273–276.
[12] Interview: Burgers with M. Dresden, October 12, 1975.
[13] Letter from Ehrenfest to Burgers, December 12, 1918.
[14] Interview with Oskar Klein, October 29, 1975, by M. Dresden.
[15] Interview with G. E. Uhlenbeck, September 20, 1975, by M. Dresden.
[16] Interview with Opechovski, September 12, 1978, by M. Dresden.
[17] Interview with Suus Kramers Perk, June 13, 1976, by M. Dresden.
[18] Kramers–Romeyn correspondence, dated April 10, 1916.
[19] Interviews with Opechovski, September 12, 1978, and Mrs. J. Kramers, March 1, 1977, by M. Dresden.
[20] Kramers files, entry dated July 10, 1916.
[21] Interview with Opechovski, September 12, 1978, by M. Dresden.
[22] Kramers–Romeyn correspondence, dated October 3, 1917.
[23] Peter Robertson, *The Early Years*, p. 19, Niels Bohr Institute, Akademisk Forlag, 1979.
[24] Kramers file, diary entry dated August 25, 1916.

[25] Kramers file, diary entry dated September 25, 1916.
[26] Kramers file, diary entry dated September 1, 1916.
[27] Letter from Kramers to Romeyn, dated September 1, 1916.
[28] Postcard from Kramers to Romeyn, dated September 13, 1916.
[28a] Letter from Bohr to Sommerfeld, dated July 17, 1919.
[29] *Niels Bohr Collected Works*, Vol. 3, p. 689 (1920–22) (Kramers file in the Bohr Institute). (Some of these notebooks are in the possession of Maartien Kramers, the Hague, The Netherlands.)
[30] Niels Bohr, *Collected Works*, Vol. 3, pp. 3, 4.
[31] (a) Sommerfeld, *Muench. Ber.* 425–458, 459–500 (1915); (b) Sommerfeld, *Ann. Phys.* **51**, 1–94 (1916).
[32] N. Bohr, Dan. Vid. Selsk. Skr. Naturvid-Mat Afd. 8 Raekke Bd. IV, Nr. 1, 1918.
[33] (a) Oskar Klein in *Niels Bohr*, edited by S. Rozental, John Wiley & Sons, New York, 1967, p. 78; (b) Interview with Oskar Klein, October 29, 1975, with M. Dresden.
[34] Rosenfeld, *Niels Bohr Collected Works*, Vol. 1, p. xxxii.
[35] W. Wilson, *Philos. Mag.* **29**, 795–802 (1915).
[36] J. Ishiwara, *Toyku Sugaku Buturigakkawi* **8**, 106–110 (1915).
[37] Sommerfeld, *Ann. Phys.* **51**, 6 (1916).
[38] A. Einstein, *Z. Phys.* **18**, 121 (1917).
[39] *H. A. Kramers Collected Works*, Vol. I (1956), p. 48.
[40] *Niels Bohr Collected Works*, Vol. 3, p. 656.
[41] *H. A. Kramers Collected Works*, p. 51.
[42] *H. A. Kramers Collected Works*, p. 61.
[43] *H. A. Kramers Collected Works*, p. 72.
[44] *H. A. Kramers Collected Works*, p. 104.
[45] Interview with Suus Kramers Perk, August 3, 1981, by M. Dresden.
[46] Letter from Bohr to Oseen, dated February 28, 1917; *Niels Bohr Collected Works*, Vol. 3, p. 671.
[47] Letter from Bohr to Rutherford, dated December 27, 1917; *Niels Bohr Collected Works*, Vol. 3, p. 682.
[48] Letter from Bohr to Richardson, dated August 15, 1918; *Niels Bohr Collected Works*, Vol. 3, p. 14.
[49] Letter from Epstein to Bohr, dated May 14, 1918; *Niels Bohr Collected Works*, Vol. 3, p. 637.
[50] Letter from Kramers to Bohr, dated March 12, 1917; *Niels Bohr Collected Works*, Vol. 3, p. 654.
[51] As quoted in Peter Robertson, *The Early Years*, p. 51, Akademisk Forlag, Universitets forlageti Kobenhavn, 1979.
[52] Interview with Oskar Klein by M. Dresden.
[53] Postcard from Kramers to Romeyn, dated October 3, 1917.
[54] Interview with J. M. Burgers by M. Dresden; also autobiographical notes of J. M. Burgers.
[55] M. Klein, *Paul Ehrenfest*. North-Holland, Amsterdam, 1970.
[56] *Niels Bohr Collected Works*, Vol. 3, pp. 609–630.
[57] Bohr Institute file, letter from Ehrenfest to Bohr, dated February 24, 1929.
[58] Annie Romeyn, *Omzien in Verwondering*, I, pp. 155–156, Amsterdam, 1970.
[59] Interview with R. de L. Kronig by M. Dresden.
[60] Kramers–Romeyn correspondence, dated November 25, 1917.
[61] Interview with Mrs. Gerda Kramers, March 1, 1977, by M. Dresden.
[62] Kramers–Romeyn correspondence, dated February 4, 1919.
[63] Annie Romeyn interview.
[64] Kramers–Romeyn correspondence, dated November, 1919.
[65] Kramers–Romeyn correspondence, dated February 27, 1920.
[66] Letter from Romeyn to Kramers, spring 1920.

[67] Kramers correspondence, Bohr Institute; letter from Coster to Kramers, dated May 11, 1920.
[68] Interview with Suus Kramers Perk, by M. Dresden.
[69] Postcard from Mrs. Kramers to Romeyn.
[70] Postcard from Kramers to Romeyn, dated January 13, 1921.
[71] *H. A. Kramers Collected Works*, p. 192 (1923).
[72] *H. A. Kramers Collected Works*, pp. 220–221.
[73] Letter from Born to Bohr, dated March 4, 1923 (Bohr files).
[74] Letter from Bohr to Born, dated April 9, 1923 (Bohr files).
[75] Letter from Born to Bohr, dated April 17, 1923 (Bohr files).
[76] Letter from Heisenberg to Pauli, dated December 21, 1921 (Pauli correspondence).
[77] Letter from Heisenberg to Sommerfeld, dated January 15, 1923.
[78] Letter from Heisenberg to Sommerfeld, dated January 4, 1923 (Sommerfeld correspondence).
[79] Letter from Pauli to Bohr, dated February 21, 1924 (Pauli correspondence).
[80] Heisenberg, *Physics and Beyond*, Harper & Row, New York, 1971, p. 37.
[81] *Niels Bohr Collected Works*, Vol. 4, p. 369.
[82] *H. A. Kramers Collected Works*, p. 109. H. A. Kramers, *Z. Phys.* **3**, 199–223 (1920).
[83] Interview with Heisenberg by M. Dresden.
[84] Max Jammer, *The Conceptual Development of Quantum Mechanics*, McGraw-Hill, New York, 1966, p. 197.
[85] van der Waerden, *Sources of Quantum Mechanics*, Dover, New York, 1968, p. 22.
[86] See F. Hund, in *Werner Heisenberg und die Physik unserer Zeit.* (F. Bopp, Ed.), Braunschweig, Fr. Vieweg & Sohn, 1961, pp. 1–7.
[87] Heisenberg, *Physics and Beyond*, Harper & Row, New York, 1971, p. 4.
[88] *Handbook of Physics* (Condon and Odishaw Eds.), McGraw-Hill, New York, 1958, pp. 7–47.
[89] Interview with Oskar Klein by M. Dresden.
[90] H. A. Kramers and Helge Holst, *The Atom and the Bohr Theory of Its Structure*, English translation, Preface, 1923, Knop, New York; German edition, 1925, Springer-Verlag, Berlin.
[91] H. A. Kramers and Helge Holst, *The Atom and the Bohr Theory of Its Structure*, English translation, p. 152.
[92] H. A. Kramers and Helge Holst, *The Atom and the Bohr Theory of Its Structure*, English translation, p. 136.
[93] H. A. Kramers, *Philos. Mag.* **46**, 836–871 (1923).
[94] Letter from Pauli to Sommerfeld, dated June 6, 1923 (Pauli correspondence).
[95] Letter from Eddington to Kramers, dated December 12, 1923 (Kramers' file).
[96] Letter from Kramers to Romeyn, dated January 8, 1924 (Romeyn correspondence).
[97] Interview with Jan Kramers by M. Dresden.
[98] A. Pais, *Rev. Mod. Phys.* **51**, 863 (1979).
[99] H. A. Kramers' diaries. See also Chapter 14.
[100] K. M. Meyer-Abich, Korrespondenz, Individualitat und Komplementarität. No. 5 in the series *Geschichten der Exacten-Wissenschaften.* Edited by Hoffmann, Klemm, and Sticker. Franz Steiner Verlag, Weisbaden, 1965.
[101] N. Bohr, *The Theory of Spectra and Atomic Constitutions*, Cambridge, 1922 p. 22.
[102] See Martin Klein, "The First Phase of the Bohr–Einstein Dialogue," *Hist. Stud. Phys.* **2**, 17 (1970).
[103] N. Bohr, *Z. Phys.* **13**, 117 (1923).
[104] Bohr, Kramers, and Slater, *Philos. Mag.* **47**, 758–802 (1924).
[105] Bohr lecture, February 13, 1920, Copenhagen; *Niels Bohr Collected Works*, Vol. 3, p. 234.
[106] Letter from Darwin to Bohr, dated July 20, 1919 (AHQP).
[107] Draft of letter from Bohr to Darwin, summer 1919 (AHQP).

[108] Darwin, *Nature* **110**, 771 (1923).
[109] N. Bohr, Solvay Report, "Atoms and Electrons," Paris, 1923.
[110] H. A. Kramers and Helge Holst, *The Atom and the Bohr Theory of Its Structure*, English translation, Knop, New York, 1923, pp. 173–175.
[111] Max Born, *Z. Phys.* **38**, 803 (1926), in "*Some strangeness in the proportion*" (Harry Wolf, Ed.), Addison-Wesley, Mass., quoted by Jost in *The Einstein Centennial*, 1980, p. 257.
[112] E. Wigner, in *The Einstein Centennial*, 1980, p. 463.
[113] Letter from Einstein to J. J. Laub, dated November 4, 1910.
[114] Letter from Einstein to J. J. Laub, dated November 11, 1910.
[115] H. A. Kramers and Helge Holst, *The Atom and the Bohr Theory of Its Structure*, English translation, Knop, New York, 1923, p. 175.
[116] *H. A. Kramers Collected Works*, p. 290.
[117] H. A. Lorentz, "Theorie der Electrons," Solvay Conference, 1921.
[118] N. Bohr, "Application of the Quantum Theory to Atomic Structure, Part I," Dan. Vid. Selsk. Skr. Naturvid-mat Afd 8 Raekke Bd. IV, Nr. 1, 1918.
[119] Ladenburg, *Z. Phys.* **4**, 451–468 (1921).
[120] Letter from Ladenburg to Bohr, dated June 14, 1923 (Bohr Institute).
[121] *H. A. Kramers Collected Works*, p. 292.
[122] Letter from Kramers to Ladenburg, dated June 8, 1924 (Kramers files).
[123] H. A. Kramers and W. Heisenberg, *Z. Phys.* **31**, 621 (1925).
[124] As quoted by van der Waerden in *Sources of Quantum Mechanics*, Dover, New York, 1968.
[125] Interview with N. van Kampen by M. Dresden.
[126] Interview with Oskar Klein by M. Dresden.
[127] H. A. Kramers' acceptance of the Lorentz medal, dated October 30, 1948.
[128] *H. A. Kramers Collected Works*, p. 291.
[129] Letter from Slater to his parents, dated November 8, 1923, in Slater's 1968 autobiographical notes (Niels Bohr Library, New York).
[130] Letter from Slater to Kramers, dated December 8, 1923 (AHQP).
[131] Letter from Slater to van Vleck, dated July 27, 1924 (AHQP).
[132] Slater interview, October 3, 1963 (AHQP).
[133] J. C. Slater, *Solid State and Molecular Theory, A Scientific Biography*, Wiley-Interscience, New York, 1975: (a) p. 15; (b) p. 17; (c) p. 18; (d) p. 11; (e) p. 240.
[134] For a beautiful and complete description see R. Stuewer, *Turning Points in Physics*, Science History Publications, New York, 1975.
[135] Letter from Sommerfeld to Bohr, dated January 21, 1923 (AHQP).
[136] Debye, *Z. Phys.* **24**, 165 (April 15, 1923).
[137] Letter from Sommerfeld to Compton, dated October 9, 1923. Quoted in R. Stuewer, *Turning Points in Physics*, Science History Publications, New York, 1975.
[138] K. Stolzenberg, comments on BKS in the introduction to Vol. 4 of *Niels Bohr Collected Works.*
[139] Neil Henry Wasserman, thesis, "The Bohr–Kramers–Slater paper and the development of the quantum theory of radiation in the work of Niels Bohr," Harvard University, June 1981. This contains a very careful and thoughtful discussion of Slater's ideas, likely the best and most complete treatment available.
[140] J. C. Slater, *Nature* **113**, 307 (1924).
[141] J. C. Slater, autobiographical notes, 1968.
[142] Letter from Slater to his mother, dated November 8, 1923.
[143] Letter from Slater to Kramers, dated November 8, 1923 (Kramers file).
[144] Letter from Slater to his parents, dated January 2, 1924, in J. C. Slater, *Excerpts of Personal Letters and ...*, Niels Bohr Library.
[145] Letter from Slater to his parents, dated January 6, 1924, in J. C. Slater, *Excerpts of Personal Letters and ...*, Niels Bohr Library.

[146] Letter from Slater to his parents, dated January 18, 1924.
[147] Letter from Slater to his parents, dated January 22, 1924.
[148] J. C. Slater, *Nature* **116**, 278 (1925).
[149] Letter from Slater to van Vleck, dated July 27, 1924 (SHQP Microfilm No. 49).
[150] Letter from Slater to Bohr, dated July 27, 1924 (Bohr Library).
[151] J. C. Slater, *Phys. Rev.* **25**, 395 (1925).
[152] Letter from Slater to Kramers, dated December 8, 1925 (Kramers file, also SHQP Microfilm).
[153] Letters from Slater to Kramers (1948–1949), Kramers file, Bohr Library.
[154] Neil Wasserman, thesis, "The Bohr–Kramers–Slater paper and the development of the quantum theory of radiation in the work of Niels Bohr," Harvard University, 1981.
[155] K. Stolzenberg, Introduction to Bohr volume, based on his thesis.
[156] Mara Beller, thesis, "Reality and acausality in quantum physics, 1919–1927," University of Maryland, 1980–81.
[157] H. A. Kramers and Helge Holst, *The Atom and the Bohr Theory of Its Structure*, English translation, Knop, New York, 1923, p. 105.
[158] A. Einstein, *Z. Phys.* **18**, 121 (1917).
[159] See Max Born, *My Life*, Charles Scribner & Sons, New York, 1978.
[160] Interview with Jordan, June 19, 1963 (AHQP).
[161] Letter from Schrödinger to Bohr, dated May 24, 1924.
[162] H. A. Kramers and Helge Holst, *The Atom and the Bohr Theory of Its Structure*, English translation, Knop, New York, 1923, p. 164.
[163] Letter from Bohr to Pauli, dated February 16, 1924 (Pauli correspondence).
[164] Letter from Pauli to Bohr, dated February 21, 1924 (Pauli correspondence).
[165] Interview with O. Klein by M. Dresden.
[166] Letter from Klein to Bohr, dated May 6, 1924 (Bohr file).
[167] H. A. Kramers and Helge Holst, *The Atom and the Bohr Theory of Its Structure*, English translation, Knop, New York, 1923, p. 166.
[168] Kramers and Holst, English translation, p. 167.
[169] Kramers and Holst, English translation, p. 168.
[170] R. Stuewer, *Turning Points in Physics*, Science History Publications, New York, 1975.
[171] Kramers and Holst, English translation, pp. 173, 174.
[172] W. Pauli, "Quantentheorie," in *Handbuch der Physik*, Springer-Verlag, Berlin, 1925, p. 83.
[173] Edward N. da Costa Andrade, *The Structure of Atoms*, G. Bell and Sons, London, 1925, p. 697.
[174] R. Feynman, *Physics Today* (August 1966), p. 31.
[175] Kramers and Holst, English translation: (a) p. 133; (b) p. 134.
[176] H. A. Kramers and Helge Holst, *Das Atom und die Bohrsche Theorie seines Baues*, Springer-Verlag, Berlin, 1925; (a) p. 123; (b) p. 139.
[177] Letter from Kramers to Romeyn, dated October 28, 1924 (translation by M. Dresden).
[178] Discussion between Kramers and M. Dresden, Fall 1939.
[179] Handelingen 20, Nederlandsch Natuur en geneeskundig, 1925, pp. 164–167.
[180] Kramers file, Bohr Institute, late 1924.
[181] Schrödinger, "Bohr's new radiation theory," *Naturwissenschaften* **12**, 720–724 (1924).
[182] Letter from Schrödinger to Bohr, dated May 24, 1924 (Bohr file).
[183] Schrödinger, "Bohr's new radiation theory," *Naturwissenschaften* **12**, 720–724 (1924).
[184] Postscript, *Lasse die Sterne in Ruhe*, incomplete letter from Pauli to Kramers, dated December 1924 (Kramers file, Bohr Institute).

[185] H. A. Kramers, unpublished manuscript (Kramers file; also SHQP Microfilm No. 27).
[186] Handwritten notes, Kramers file (in possession of Maartien Kramers, the Hague).
[187] A. Einstein, *Berliner Tageblatt*, April 20, 1924, Beiblatt.
[188] Letter from Born to Bohr, dated April 16, 1924.
[189] Letter from Heisenberg to Pauli, dated March 4, 1924.
[190] Letter from Heisenberg to Sommerfeld, dated November 18, 1924.
[191] Letter from Sommerfeld to Kramers, dated August 3, 1924 (in AHQP), translation by M. Dresden.
[192] Letter from Kramers to Sommerfeld, dated September 6, 1924 (in AHQP), translation by M. Dresden.
[193] Letter from Einstein to Ehrenfest, dated May 31, 1924. See M. J. Klein, *Hist. Stud. Phys.* **2**, 1–39 (1970).
[194] Letter from Einstein to Hedwig Born, dated April 29, 1924.
[195] Letter from Pauli to Bohr, dated October 2, 1924.
[196] Letter from Ladenburg to Kramers, dated June 8, 1924 (AHQP).
[197] Letter from Kramers to Ladenburg, dated July 3, 1924 (AHQP) Kramers file; translation by M. Dresden.
[198] Letter from Pauli to Bohr, dated October 2, 1924; translation by M. Dresden.
[199] Letter from Pauli to Sommerfeld, dated December 6, 1924 (Pauli correspondence).
[200] A. Sommerfeld, *Atombau und Spektrallinien*, Vieweg, Braunschweig, 1924, p. 59.
[201] Letter from K. Joel to A. Einstein, dated October 28, 1924, as quoted by M. Klein, *Hist. Stud. Phys.* **2**, 34 (1970).
[202] Note from A. Einstein to K. Joel, dated November 3, 1924.
[203] Letter from Romeyn to Kramers, dated November 6, 1924; translation by M. Dresden.
[204] Letter from Kramers to *De Telegraaf*, dated October 24, 1924 (Bohr Institute, Kramers file).
[205] W. Bothe and H. Geiger, *Z. Phys.* **26**, 44 (1924).
[206] W. Bothe and H. Geiger, *Naturwissenschaften* **13**, 440 (1925).
[207] W. Bothe and H. Geiger, *Z. Phys.* **32**, 639 (1925).
[208] R. Shankland, *Phys. Rev.* **49**, 8 (1936).
[209] Hofstadter and McIntyre, *Phys. Rev.* **78**, 24 (1950).
[210] Bay, Henri, and McLennon, *Phys. Rev.* **97**, 1710 (1955).
[211] Compton and Simon, *Phys. Rev.* **26**, 189 (1925).
[212] Letter from Born to Bohr, dated January 15, 1925 (Bohr correspondence, AHQP).
[213] Letter from Bohr to Geiger, dated April 21, 1925 (AHQP).
[214] Letter from Bohr to Fowler, dated April 25, 1925, postscript (AHQP).
[215] Letter from Bohr to Franck, dated April 21, 1925 (Bohr file); translation by M. Dresden.
[216] Letter from Franck to Bohr, dated April 24, 1925 (Bohr); translation by M. Dresden.
[217] Letter from Born to Bohr, dated April 24, 1925 (Bohr file); translation by M. Dresden.
[218] N. Bohr, *Nature* (*Suppl.*) **121**, 580–590 (1928).
[219] J. C. Slater, *Nature* **116**, 278 (1925).
[220] Letter from Bohr to Slater, dated January 28, 1926, as quoted by Stolzenberg on the BKS theory.
[221] Letter from Slater to Bohr, dated May 27, 1926, as quoted by Stolzenberg on the BKS theory.
[222] Letter from Kramers to Born, dated May 13, 1925 (Kramers scientific correspondence, SHQP Microfilm No. 8).
[223] Letter from Kramers to Fowler, dated December 9, 1925, in SHQP and also Kramers correspondence.
[224] Letter from Kramers to Urey, dated July 16, 1925 (Kramers correspondence).

225 Postcard from Heisenberg to Pauli, dated June 29, 1925 (Pauli correspondence).
226 Letter from Heisenberg to Bohr, dated August 31, 1925 (Bohr file, Copenhagen); translation by M. Dresden.
227 Letter from Kramers to R. de L. Kronig, dated February 26, 1926 (Kramers correspondence, Bohr Institute).
228 H. A. Kramers and W. Heisenberg, *Z. Phys.* **31**, 601 (1925).
229 W. Heisenberg, *Z. Phys.* **33**, 879 (1925).
230 H. A. Kramers, *Quantentheorie des Elektrons und der Strahlung*, Akademische Verlagsgesellschaft, Leipzig, 1938, footnote to p. 90.
231 W. Heisenberg, *Physics and Beyond*, Harper & Row, New York, 1971, Vol. 42, p. 60.
232 As the final version of this book was prepared, a major study of the history of quantum theory appeared. This important work by Mehra and Rechenberg (*The Historical Development of Quantum Theory*, Springer-Verlag, New York, 1982) necessarily contains material that overlaps material in this book. It is in our opinion helpful and illuminating if these involved and complex times are examined from as many distinct viewpoints as possible. Because of the time this work appeared, references to it are somewhat incomplete and scattered. Only where there are specific arguments or specific differences, or where Mehra and Rechenberg had information (from personal tapes or interviews) not available elsewhere, is that fundamental work referred to in an adequate manner. The quotation attributed to Heisenberg is from Vol. 2, p. 218.
233 M. Born, *Z. Phys.* **26**, 379 (1924).
234 H. A. Kramers, *Verh. Dtsch. Phys. Ges.* **5**(3) 37 (August 15, 1924). Also quoted by Mehra and Rechenberg, Vol. 2, p. 174.
235 N. Bohr, *Naturwissenschaften* **12**, 1116 (1924), footnote 5.
236 Letter from Heisenberg to Bohr, dated May 16, 1925 (Bohr Institute).
237 Letter from Heisenberg to Kronig, as reproduced in Fiertz and Weisskopf, *A Memorial Volume to Wolfgang Pauli*, Interscience, New York, 1960.
238 Max Born, *My Life*, Charles Scribner, New York, 1975.
239 M. Jammer, *The Conceptual Development of Quantum Mechanics*, McGraw-Hill, New York, 1966, p. 193.
240 Mehra and Rechenberg, *The Historical Development of Quantum Theory*, Springer-Verlag, New York, 1982, Vol. II, p. 173.
241 E. Whittaker, *A History of the Theories of Aether and Electricity*, Vol. II, Harper & Row, New York, 1960, p. 204.
242 A. Smekal, *Naturwissenschaften* **11**, 873 (1923).
243 C. V. Raman, *Indian J. Phys.* **2**, 387 (1928).
244 G. Landsberg and L. Mandelstam, *Naturwissenschaften* **16**, 557 (1928).
245 Oral communication from Heisenberg to van der Waerden, as quoted in *Sources of Quantum Mechanics*, Dover, New York, 1968.
246 Interview with Oskar Klein by M. Dresden.
247 In van der Waerden, *Sources of Quantum Mechanics*, Dover, New York, 1968: (a) p. 255; (b) p. 226.
248 Max Born, *Quantum Mechanics*, reprinted by van der Waerden, p. 191.
249 Kramers and Heisenberg, *Z. Phys.* **31**, 698 (1925).
250 Kopferman and Ladenburg, *Z. Phys.* **48**, 26 (1928).
251 Heisenberg, as quoted in Mehra and Rechenberg, *The Historical Development of Quantum Theory*, Vol. II, Springer-Verlag, New York, 1982, p. 189.
252 Kramers and Heisenberg, *Z. Phys.* **31**, 708 (1925).
253 As reprinted in Mehra and Rechenberg, *The Historical Development of Quantum Theory*, Vol. II, Springer-Verlag, New York, 1982, p. 189.
254 Heisenberg interview, AHQP, February 1963.
255 Heisenberg, in *Physics and Beyond*, Harper & Row, New York, 1971, p. 60.
256 Letter from Heisenberg to Pauli, dated June 24, 1925.

[257] Heisenberg, in *Niels Bohr*, edited by S. Rosental, Interscience, New York, 1967, p. 98.
[258] M. Born, *Z. Phys.* **26**, 379 (1924).
[259] M. Born and P. Jordan, *Z. Phys.* **34**, 858 (1925).
[260] Postcard from Heisenberg to Pauli, dated June 29, 1925; translation by M. Dresden.
[261] Postcard from Kramers to Fowler (copy in Bohr file), dated September 1925.
[262] H. A. Kramers, *Physica* **5**, 369 (1925), footnote 1; translation by M. Dresden.
[263] H. A. Kramers, *Collegium Generalum*, private notes, in the possession of Maartien Kramers.
[264] Interview with Klein by M. Dresden, October 29, 1975.
[265] The original German version was: *To Bohr, der für uns alle sucht.*
[266] Pauli, *Science* **103**, 213 (1946).
[267] Heisenberg, in "From a life of physics," *IAEA Bulletin*, 1968.
[268] As reported by Armin Hermann, in *Heisenberg*, Rowohlt Taschenbuch Verlag, Reinbek by Hamburg, 1976, p. 28.
[269] Pauli, *Science* **103**, 213 (1946).
[270] Letter from Pauli to Bohr, dated September 12, 1923 (Pauli correspondence); translation by M. Dresden.
[271] Letter from Hansen to Bohr, dated October 1, 1923 (Bohr Library, Copenhagen).
[272] Letter from Kramers to Bohr, dated October 11, 1923; *Niels Bohr Collected Works*, Vol. 3, p. 661.
[273] Letter from Pauli to Bohr, dated October 2, 1924 (letter 66 in Pauli correspondence). This translation by M. Dresden is almost a paraphrase of the text.
[274] Letter from Pauli to Bohr, dated October 2, 1924. More literal translation by M. Dresden.
[275] Letter from Pauli to Bohr, dated February 21, 1924 (Pauli correspondence).
[276] Letter from Pauli to Bohr, dated December 12, 1924 (Pauli correspondence).
[277] W. Pauli, *Z. Phys.* **31**, 765 (1925).
[278] Letter from Pauli to Bohr, dated December 31, 1924; translation by M. Dresden.
[279] Letter from Pauli to Bohr, dated February 11, 1924; translation by M. Dresden.
[280] Letter from Pauli to Bohr, dated February 21, 1924.
[281] For a personal description of this remarkable incident, see Heisenberg, *Physics and Beyond*, Harper & Row, New York, 1971, p. 234.
[282] Letter from Pauli to Heisenberg, dated April 7, 1958. Translation by M. Dresden, as quoted in Armin Hermann, *Heisenberg*, Rowohlt Taschenbuch Verlag, Reinbek by Hamburg, 1976.
[283] Letter from Heisenberg to Pauli, dated April 13, 1958. Translation by M. Dresden, as quoted in Armin Hermann, *Heisenberg*, Rowohlt Taschenbuch Verlag, Reinbek by Hamburg, 1976.
[284] Heisenberg, *Physics and Beyond*, Harper & Row, New York, 1971, p. 236.
[285] Letter from Heisenberg to Pauli, dated March 6, 1922 (Pauli correspondence).
[286] Bohr and Coster, *Z. Phys.* **13**, 342 (1923).
[287] Kramers and Pauli, *Z. Phys.* **13**, 343 (1923).
[288] Letter from Heisenberg to Sommerfeld, dated January 14, 1923.
[289] Mehra and Rechenberg, *The Historical Development of Quantum Theory*, Springer-Verlag, New York, 1982, Vol. II, p. 139.
[290] Heisenberg interview (AHQP); also Mehra and Rechenberg, *The Historical Development of Quantum Theory*, Springer-Verlag, New York, 1982, Vol. II, p. 150.
[291] Heisenberg interview (AHQP); also Mehra and Rechenberg, *The Historical Development of Quantum Theory*, Springer-Verlag, New York, 1982, Vol. II, p. 151.
[292] Heisenberg, as written in *Niels Bohr*, edited by S. Rosental, Interscience, New York, 1967, p. 96.
[293] Armin Hermann, *Heisenberg*, Rowohlt Taschenbuch Verlag, Reinbek by Hamburg, 1976, p. 8.

[294] Ibid., pp. 8–9.
[295] Interview with V. Weisskopf by M. Dresden.
[296] Letter from Heisenberg to Bohr, dated July 15, 1924 (Bohr correspondence).
[297] Mehra and Rechenberg, *The Historical Development of Quantum Theory*, Springer-Verlag, New York, 1982, Vol. II: (a) pp. 148–149; (b) p. 153; (c) p. 149; (d) p. 272.
[298] Letter from Born to Bohr, dated April 16, 1924 (Bohr file).
[299] Letter from Heisenberg to Sommerfeld, dated November 18, 1924.
[300] Letter from Heisenberg to Pauli, dated June 8, 1924 (Pauli correspondence).
[301] Letter from Pauli to Heisenberg, dated February 28, 1925 (Pauli correspondence).
[302] Letter from Bohr to Pauli, dated January 10, 1925 (Pauli correspondence).
[303] Letter from Pauli to Heisenberg, dated February 28, 1925 (Pauli correspondence).
[304] A. Einstein, *Sitzungsber. Preuss. Akad. Wiss.* (January 8, 1925), p. 3; (January 29, 1925), pp. 18–25.
[305] Letter from Heisenberg to Pauli, dated June 29, 1925 (Pauli correspondence).
[306] Letter from Pauli to Kramers, dated July 27, 1925 (Pauli correspondence); translation by M. Dresden.
[307] W. Pauli, Über die Intensitäten der im elektrischen Feld erscheinenden Kombinationslinien, *Dan. Vid. Selsk. Mat. Fys.* **7**, 3–20 (1925).
[308] Pauli, *Z. Phys.* **31**, 765 (1925).
[309] Letter from Pauli to Bohr, dated February 21, 1924 (Pauli correspondence).
[310] Letter from Pauli to Kramers, dated July 5, 1924 (AHQP); translation by M. Dresden.
[311] This point is correctly emphasized—perhaps first explicitly stated—in the thesis of N. H. Wasserman, Harvard University, June 1981.
[312] Letter from Pauli to Bohr, dated December 12, 1924 (Pauli correspondence).
[313] Letter from Bohr to Pauli, dated December 22, 1924 (Pauli correspondence).
[314] Letter from Pauli to Bohr, dated December 31, 1924.
[315] Letter from Pauli to Kronig, dated May 21, 1924; translation by M. Dresden.
[316] Letter from Pauli to Kronig, dated October 9, 1925 (Pauli correspondence).
[317] Reported in a letter from Heisenberg to Pauli, dated November 3, 1925.
[318] Interview with Oskar Klein by M. Dresden.
[319] Heisenberg conversation as reported in Mehra and Rechenberg, *The Historical Development of Quantum Theory*, Springer-Verlag, New York, 1982, Vol. II, p. 178.
[320] Heisenberg conversations as reported in Mehra and Rechenberg, Vol. II, p. 179.
[321] Bohr, *Naturwissenschaften* **12**, 1115 (1924), footnote 5.
[322] Interview with Oskar Klein by M. Dresden.
[323] Interview with Nico van Kampen by M. Dresden.
[324] Interview with H. C. Kramers by M. Dresden.
[325] Letter from Heisenberg to Kramers, dated January 8, 1925 (Kramers file).
[326] Heisenberg, as reported in Mehra and Rechenberg, *The Historical Development of Quantum Theory*, Springer-Verlag, New York, 1982, Vol. II, p. 180.
[327] J. C. Slater, *Scientific Autobiography*, John Wiley and Sons, New York, 1975, p. 17.
[328] Letter from Romeyn to Kramers, dated July 21, 1922 (Romeyn correspondence).
[329] Letter from Romeyn to Kramers, dated February 20, 1923 (Romeyn correspondence).
[330] Letter from Kramers to Romeyn, dated April 29, 1923 (Romeyn correspondence).
[331] Letter from Romeyn to Kramers, dated November 6, 1924 (Romeyn correspondence).
[332] Letter from Kramers to Romeyn, dated April 11, 1925 (Romeyn correspondence).
[333] Letter from Kramers to Romeyn, dated August 15, 1925.
[334] Letter from Kramers to Urey, dated July 16, 1925 (Bohr file).
[335] Letter from Kramers to Fowler, dated December 9, 1925 (Kramers file, Bohr Institute).
[336] The library in The Institute for Theoretical Physics at the University of Utrecht

has an almost complete collection of lecture notes of most professors. These first notes are in Kramers' own handwriting in a series of notebooks.

337 Peter Robertson, *The Early Years*, Akademisk Forlag, Copenhagen, 1979, pp. 111–112.

338 Interview with Jan Korringa by M. Dresden.

339 Interview with Agnete Kuiper-Kramers by M. Dresden.

340 Interview with Jan Kramers by M. Dresden.

341 Interview with Agnete Kuiper-Kramers by M. Dresden. Agnete seemed to have a very vivid recollection of her mother's reaction.

342 Interview with N. van Kampen by M. Dresden.

343 Interview with Heisenberg, AHQP, February 19, 1963.

344 Van der Waerden, *Source of Q.M.*, Dover, New York, 1968, p. 16.

345 Told by Professor Wassink to S. Dresden (summer of 1975), who told it to M. Dresden.

346 As quoted in the Pauli correspondence XXXII.

347 Letter from Pauli to Heisenberg, dated October 19, 1926 (Bohr file).

348 Private communication from R. Jost (1980).

349 Interview with de Groot and Korringa by M. Dresden.

350 Interview with Casimir by M. Dresden.

351 This description is a composite of the recollections of several persons (Opechovski, de Groot, and others). An event of this type probably took place: Kramers' statement, "This is too simple for you, Pauli," is recalled by all. All other details are uncertain.

352 W. Pauli, The connection between spin and statistics, *Phys. Rev.* **58**, 716 (1940).

353 Letter from Kramers to Born, dated May 13, 1925.

354 H. A. Kramers, Absorption of x-rays, *Fys. Tidss.* **20**, 130–132 (1922).

355 Interview with Oskar Klein by M. Dresden.

356 Interview with D. ter Haar by M. Dresden.

357 Interview with J. J. Korringa by M. Dresden.

358 Interview with R. de L. Kronig by M. Dresden.

359 Interview with Oskar Klein by M. Dresden.

360 Letter from R. Jost to M. Dresden, dated January 12, 1978. It must be understood that this letter contains recollections and memories of discussions between Jost and Pauli. In these discussions the topic of Kramers' unhappiness with Bohr evidently came up. It seems likely that Bohr complained to Pauli, and Pauli's recollections, however incomplete and partial, are recorded in this letter. It should be clear that these comments indicate a general mood rather than pure facts.

361 R. Jost in *Vierteljahrsschrift der naturforschenden Gesellschaft in Zürich*, lecture entitled "Einstein und Zürich," 1979.

362 W. Heisenberg, *Z. Phys.* **43**, 172 (1927).

PART 3
WAITING FOR A REVOLUTION THAT DID NOT HAPPEN

CHAPTER 15

The Search for Identity in Changing Times

I. The Move to Utrecht

Through the death of Professor W. H. Julius in 1925, a faculty position became available at the University of Utrecht in The Netherlands. A committee under the chairmanship of Ehrenfest was appointed to make recommendations for a possible successor to the Utrecht faculty. That body in turn would submit a list of nominees to the Minister of Education, who would make the ultimate decision. The committee recommended Kramers, and the faculty and minister concurred. Kramers' acceptance of the offer was practically guaranteed by the way the system works. He gave his inaugural address on February 15, 1926 and started his lectures in Utrecht on May 5, 1926. Because this is so standard and hackneyed, it seems hardly worth mentioning.

What actually happened, however, was convoluted and unusual and not a bit straightforward. There were a number of twists and turns before the position was offered to Kramers. He himself knew nothing about this. He first learned about the vacancy in April 1925 and mentioned it in a letter to Urey on July 10, 1925, in the summer of his discontent,[1] and again in a letter to Romeyn[2] on August 15. He formally applied for the position sometime in September 1925—which in the Dutch tradition is tantamount to his acceptance if the position were indeed offered.

Ehrenfest, as chairman of the search committee, immediately swung into action. The possibility of bringing a major quantum theorist to Holland was immensely appealing to him. He wrote[3] a lengthy letter to Bohr, hinting (almost insisting) that Bohr personally recommend Kramers for the Utrecht chair. He was both excited and apprehensive at the prospect of Kramers' return to Holland and writes that he is almost "jealous" that Utrecht might obtain a major quantum theorist. He also warns Bohr that if the offer to

Kramers does indeed come through, he (Ehrenfest) should have a private talk with Kramers!" It is too fragile to discuss by letter." [3] Ehrenfest also requested recommendations from Einstein, Planck, and presumably Lorentz. (There are numerous indirect references to this request, but the letter itself appears lost.) This is surely a better than average quartet of referees, for a chair in a small provincial Dutch university.

Einstein wrote an excellent letter of recommendation,[4,5] although usually he had a hard time writing such letters. This time he stated that Kramers always dealt with significant, central questions; next to Bohr, Kramers had made the deepest contributions. He added that any country would be proud to have Kramers as a professor, and surely his native country should be overjoyed to have him. Planck's letter was equally good. Lorentz's response, however, was more guarded, and he did mention that Kramers often came late to his early morning lectures (sometimes he did not come at all).[6] Bohr's letter was vague, guarded, rambling and not a bit helpful. As a consequence, the faculty committee was confused: Should they give the position to an experimentalist or a theorist? If a theorist, who? Kramers certainly did not appear to be the first choice. Furthermore, not everyone in Utrecht was all that thrilled to have a young (probably arrogant) hot shot, in a new fangled, abstract field such as quantum theory. Several physicists in Utrecht and at least one physical chemist (Kruyt) tried to persuade Ehrenfest to accept the chair in Utrecht. How seriously Ehrenfest considered this possibility is hard to say; he mentioned that possibility to two of his Leiden colleagues, de Haas and de Sitter (the famous astronomer), and to his revered master Lorentz.[7] The Utrecht faculty certainly tried to make the offer attractive; they even offered to pay his moving expenses—an unusual concession in those days.

Surprisingly, during this time Ehrenfest's efforts to get Kramers to come to Utrecht continued undiminished. He knew very well that without a really strong letter from Bohr, Kramers would not have a chance. So he wrote a very forceful, pointed letter to Bohr—to explain how Bohr's earlier letter had confused everyone. "Your [Bohr's] previous letters were so vague and indistinct that Kramers and Burgers came out about the same with Burgers slightly higher. Please send me a letter showing unequivocally that Kramers is to be preferred. Vague pronouncements do not help." [8] Bohr answered this letter right away, extolling Kramers' virtues, listing his fundamental contributions, and stressing his great importance—altogther a very strong, positive, unequivocal recommendation and precisely what Ehrenfest had wanted.[9] Even so, Bohr starts this same letter by telling Ehrenfest how excited he (Bohr) is about Heisenberg's recent work in quantum mechanics. Bohr's second letter clearly was decisive. The same day that Ehrenfest received Bohr's letter (October 16), he wrote to Lorentz: "This morning I received an answer from Bohr to a letter I had written Oct. 11. In it he gives an unequivocal and detailed opinion about Kramers and his work .... Now the Utrecht faculty is in the possession of three letters which most definitely and without qualifications name Kramers and only Kramers, the letters of Planck, Bohr, Einstein." [7]

It is hardly surprising that with three such letters from perhaps the outstanding trio of theoretical physicists of the century, the approval of Kramers by all administrâtive layers followed very rapidly. Ehrenfest decided to stay in Leiden and forego any further negotiations about the Utrecht position. In the same letter to Lorentz he writes: "I have today definitely decided not to leave Leiden. I have informed de Haas and de Sitter about this—I also have written to Ornstein and Kruyt."[7] So Ehrenfest, who had severe misgivings about his own ability to understand or teach—let alone do research in quantum theory—stayed in Leiden, while Kramers, the representative of the new wave in physics, came to Utrecht. This defined the new relation between Ehrenfest and Kramers. The former teacher had become the fearful, anxious pupil. Ehrenfest felt (and therefore was) scientifically dependent on Kramers. He and Kramers have had their differences, but he is now "anxious for Kramers' help."[10] And Ehrenfest's request for help and support became more frantic as time went on: "Practically all new theoretical physics stands before me as a completely nonunderstandable wall and I am at a total loss. I no longer know the symbols, the language and I no longer know what the problems are."[11] A little later, Ehrenfest's cry for help is even more pathetic, "Please help me to understand things, the way a schoolmaster should and not a wisdom producer or a copublisher. This I must postpone for the foreseeable future—perhaps for ever."[12] Kramers' answers are generally kind and considerate, but he rarely gave Ehrenfest the detailed information that he literally craved. This was in part because Ehrenfest (as in the case of papers by Dirac and Klein and Jordan) really did not understand what was going on and in part because Kramers did not feel that he could or should engage in a long-term reeducational project.

As exemplified by his relation to Ehrenfest, Kramers was the only modern quantum expert in The Netherlands. To Kramers, this was a matter of sly satisfaction but also considerable concern. In Copenhagen, he could always find out what Bohr thought of it, but in Utrecht he was the final scientific arbiter. In Copenhagen, he was in the nerve center of quantum physics; in Utrecht he was in isolation in a place where quantum theory was barely known and not universally appreciated. In Copenhagen, Kramers was—at least for a while—the heir apparent to Bohr; his move to Utrecht removed him from that highly privileged position. Although Kramers was thrilled by his expected independence in Utrecht, he nevertheless had many concerns and severe misgivings about his move. He once almost plaintively wrote to Kronig about "the many new faces now illuminating the ways of Bohr,"[13] indicating an erosion in his special position. He writes Fowler that "it is hard to leave Bohr and Copenhagen."[14] Evidently, Kramers felt that the time had come to cut away from Bohr, but he was reluctant to do so. A subtle, barely perceptible change in the relation between Kramers and Bohr had started after Heisenberg's discovery of matrix mechanics. It was formalized—in fact it became official—when Heisenberg became Kramers' successor in 1926. It is anyone's guess whether Bohr expressed his limited control of English or some deep seated unconscious reaction when he wrote to Fowler regarding Dirac's visit to Copenhagen: "Dirac's coming will also be a great pleasure to Heisenberg who will arrive here on the first day of May to overtake Kramers' post."[15]

It is hard to escape the impression that, after the BKS debacle and Heisenberg's spectacular successes, Kramers was anxious, maybe even desperate, to find a new basis and start anew on his own, away from Copenhagen and Bohr. The move to Utrecht presented a wonderful opportunity to do just that. Even though this enabled him to be geographically separate from Copenhagen, the scientific and intellectual ties to Copenhagen remained strong. Kramers made many attempts to go his own way in physics; his further intellectual evolution can be viewed as a gradual but never complete emancipation from the Copenhagen influence.

II. The Change in Style and Mood

From the time that Kramers had decided to apply for the position in Utrecht in the fall of 1925 until the Fifth Solvay Conference in Brussels in the fall of 1927, a change became noticeable in the style of Kramers' physics. His general approach was less daring and speculative, but more scholarly and critical. His papers do not contain bold conceptual innovations or revolutionary suggestions; instead, they provide detailed treatments of specific concrete problems, carried out with superlative technical skill. It is not so much that Kramers was not interested in fundamental questions, such as the interpretation of quantum mechanics, he just deliberately stayed away from these topics. He left the difficult and subtle analysis of the quantum mechanical formalism to Heisenberg, Bohr, Pauli, and Schrödinger. Even though he made an early significant technical contribution to Schrödinger's wave mechanics,[16] he never once entered the argument about the proper interpretation of the wavefunction.

This is actually a somewhat puzzling situation. Schrödinger wished to interpret the wavefield as a classical, continuous, causal field; that is, the theory should contain no discontinuous elements. Bohr, Born, and Heisenberg insisted that such an interpretation could never explain all the observations; the field had to have a probability interpretation. Both Born and Heisenberg saw significant elements of the Bohr–Kramers–Slater theory in that interpretation. Born recalls[17] that his probability interpretation was strongly suggested by Einstein's earlier ideas of the "ghost field" or guiding field. In that theory Einstein introduced a field that assigned a probability to finding a number of photons at a given location. The Bohr–Kramers–Slater theory contained a modified version of that same idea. Thus, one might expect that Kramers would pursue and investigate Born's probability interpretation, in depth and detail. He did use it but never participated in the arguments that followed that interpretation. Heisenberg was even more explicit about the direct relation between Born's probability interpretation and the Bohr–Kramers–Slater theory. In a tribute to Pauli he writes: "Born, in the summer of 1926, developed his theory of collision processes, in which he correctly interpreted the Schrödinger wave field in a many dimensional configuration space as a probability wave, basing his considerations on an earlier idea of Bohr, Kramers and Slater."[18] In a later interview, Heisenberg again stresses the important role of the BKS theory for quantum theory: "Still I [Heisenberg] felt that the crucial step was made by Bohr, Kramers, Slater ... The waves ... were a physical reality in the sense that they produced probabilities for decay or emission."[19] It is evident—if only in retrospect—that both Born and Heisenberg felt that the Bohr–Kramers–Slater theory played a significant role in arriving at the probability interpretation of quantum theory. It would appear that with his intense preoccupation with BKS theory Kramers would jump at the chance to save some elements or ideas of that theory by incorporating them in a better theory. But he never commented on it. He never wrote a paper on the interpretation problems of quantum theory. He stayed away from the intense polemics and the sometimes acrimonious discussions surrounding the uncertainty relations and the probability interpretation of quantum mechanics. From his popular lectures and especially from his book, it is evident that he was confortable with the interpretations as developed by Heisenberg and especially Bohr. With the quantum theory and the well-defined unambiguous interpretation supplied by Bohr, Dirac, and Heisenberg, Kramers believed that a formal scheme had been obtained, which allowed the quantitative explanation of many—if not all—phenomena in physics (including chemistry). He chose to devote his efforts to the application, use, and refinement of that scheme, without worrying too much about the philosophical–conceptual basis. Kramers believed at the Solvay meeting (fall of 1927) that only nuclear physics and possibly relativistic phenomena might lie outside that scheme. Everything else should find an explanation in terms of these quantum ideas.

In the period from the fall of 1925 until the Solvay Congress in 1927, Kramers wrote two papers that show he was at "the quantum frontier." The first paper[20]—one of the first "non-Göttingen contributions" to matrix mechanics—was an attempt to give a physical justification for the quantum conditions (or the commutation relations) used by Born and Jordan. Kramers could do this very fast because he realized immediately that the Kramers–Heisenberg dispersion formulas could be written directly in matrix form. By requiring that an atom in a very rapidly oscillating field behave classically (and this is a physical requirement), Kramers obtained the commutation relations of Born and Jordan. This was a significant, nontrivial contribution as may be inferred from the fact that at the 119th meeting of the Kapitza Club in Cambridge, on April 27, 1926, Dirac talked about Kramers' derivation of the quantum conditions[21] and Dirac was not the person to spend time on trivial topics! In this same note, Kramers remarks, almost in passing, that the structure of the Poisson–Brackett algebra is identical with that of the commutator algebra, an observation "which might be useful in other applications." But Kramers, contrary to Dirac, did not recognize that he had discovered the most fundamental connection between classical and quantum theories. It was almost the opposite. To Kramers, his observation only confirmed his already deep conviction that all quantum problems must have a classical counterpart. Each phenomenon should be capable of several levels of description—a purely classical description that has a limited validity, a Bohr quantum theoretical description (where quantum conditions are adjoined to a classical picture) that is almost correct, and a completely correct quantum mechanical description. It was therefore a great joy for Kramers when he could show, in this same period, that an approximate treatment of the Schrödinger equation—in the so-called WKB paper[16]—did lead to quantum conditions of the Bohr type. This paper has become a classic: It required all Kramers' knowledge of differential equations and complex variable theory. The result is mentioned in almost all quantum mechanics textbooks, and the method is used to this day in a large number of research papers. It is still generalized with a certain degree of regularity, to many dimensions, to the Dirac theory, and to field theory. The precise mathematical discussion is rather delicate. The most complete treatment was given by one of Kramers' students in his thesis (Mr. Zwaan[22]). This paper by Kramers was the first of a series of papers devoted to the detailed mathematical elaboration of the quantum mechanical framework.

Kramers' dogged insistence that all phenomena should have a classical counterpart led him in this same time period to a protracted investigation of the electron spin based on the classical picture of an extended rotating charge distribution. The spin, as a particle property, was first proposed by Goudsmit and Uhlenbeck, for purely empirical reasons. The precise fundamental role of this new attribute was not immediately clear:[23] Pauli was convinced that it was an intrinsically nonclassical concept, while Heisenberg hoped for a while that the spin notion could actually be derived from quantum mechanics. Pauli, in a fundamental paper, succeeded in incorporating the spin notion in a quantum formalism.[24] This paper was submitted in May 1927. He did not derive its existence or provide any physical explanation. It seemed very difficult to get a deeper understanding of the spin. At least it appeared so to Heisenberg, who made a bet with Dirac that it would take at least 3 years before one could say anything sensible about the spin. Dirac claimed that it would take only 3 months. The precise conditions of the bet are spelled out in a letter from Heisenberg to Pauli on February 5, 1927.[25] Pauli's paper did not provide a deeper explanation of the spin and its curious couplings. Pauli was actually quite dissatisfied with his paper and especially upset that his treatment was intrinsically nonrelativistic. He had struggled for quite a while with the relativistic formulation of the spin. Pauli fully realized that when dealing with spin questions, relativistic invariance is not merely a "cosmetic" or aesthetic requirement. The spin–orbit interaction and relativistic corrections are of comparable magnitudes, so a nonrelativistic treatment appears almost inconsistent. These problems were very much on his mind when Pauli met Kramers in Copenhagen in June of 1927. They discussed the relativistic problem at length and Pauli now categorically declared that there could not be a relativistic generalization of his two-component spin equation. Furthermore, Pauli claimed that the spin had no classical analogue. Kramers did not believe any of this. He was too strongly influenced by Lorentz's classical theory of electrons as extended charge distributions to accept on faith, without a major effort, that such a relativistic theory was impossible. Furthermore, his personal faith in the universal existence of the classical counterpart of any quantum object "if only viewed correctly"[26] was unshakeable. Pauli then issued a direct challenge to Kramers to construct a relativistic spin theory, presumably starting from classical concepts.[26] In response to this direct challenge, Kramers embarked on one of his most ambitious undertakings. He started from the relativistic generalization of classical rotating charge distributions. He then performed the quantization by first casting this system in an approximate Hamiltonian form. He also developed his own, personal version of spinors, based on the properties of complex null vectors. After a long involved, complex series of arguments, he finally arrived at an (to him) approximate quantum description of a relativistic spinning electron, in terms of a pair of differential equations of the second order in time. These equations can be shown to be equivalent to Dirac's system of four first-order equations. Dirac's fundamental paper appeared January 2, 1928, and Kramers completed his own derivation only a few weeks later. Kramers was deeply disappointed. It was very much against his style to complain, but many years later after he had recovered from his disappointment he mentioned his "almost discovery of the Dirac equations" to several persons including Dirac.[27,28] So disappointed was Kramers with his lack of success, that he withdrew from the field and published practically nothing for several years. It was not until 6 or 7 years later that he began to publish fragments of his considerations in an almost serial form. Since the Dirac equation by that time was well known and often used, these "alternate" and unusual derivations did not generate much interest. The most complete, coherent (and understandable) treatment (including his use of null vectors) was contained in Kramers' great textbook which did not appear until 10–11 years later.[54]

Even though the spin venture, stimulated by Pauli's challenge, was ultimately unsuccessful, it exerted a strong influence on the direction and type of research that Kramers was to undertake. His method of dealing with spinors was eventually developed in an ingenious symbolic algorithm, allowing the calculation of matrix elements in an almost magical manner. Kramers always enjoyed such formal legerdemain and he taught his students, especially Brinkman, how to master this remarkable formalism.[29] Because Kramers' derivation of the Dirac equation, his use of spinors, and his symbolic algorithm were so individual (and so unsystematically and haphazardly published), these papers were not widely read. Most professionals could achieve the same results using more standard methods. Often, as with the Dirac equation, the problems that Kramers treated were no longer considered urgent. This led inevitably to an increasing isolation on the part of Kramers. After leaving Copenhagen, he no longer received the daily stimulation of the activities of that center, nor was he personally aware of the rapidly changing fashions in Copenhagen. He did maintain contacts with Bohr, but as time went on these contacts became less frequent. Kramers' personal contacts remained restricted to those he had made during his Copenhagen sojourn. Although he was well known to almost all physicists, old and young, he had very little serious scientific contact with the newer generation, for example, Bethe, Wigner, and Fermi, and even less with later generations. Kramers tended to select his own problems, and he did not really care whether they were fundamental or important. Because of his vast knowledge of classical physics, Kramers often returned to old, almost totally forgotten, questions. Kramers was intrigued by formal mathematical questions, which were no longer in the center of interest. He tackled problems more because of their individual challenge than for their physical significance. This is very different from the period between the fall of 1925 and the Solvay conference. During that period, Kramers concentrated primarily on very specific and current questions. From that time on, there was a gradual separation of Kramers from the frontier of physics. His work assumes a more diversified character. This in turn implied that Kramers no longer took the lead in the new developments. It was through a combination of these circumstances that Kramers, who dominated Dutch physics since 1927, became somewhat lonely and scientifically isolated. He never wrote a paper in nuclear physics, β decay theory, or cosmic rays. His brilliance enabled him to keep up without any trouble, but he was not guiding or shaping the development.

There is one important exception to Kramers' passive role: He remained passionately interested in the theory of radiation and the electron. That to Kramers was and remained the true frontier. That is where all his research talents should be concentrated. Other problems were interesting, important, and difficult, but scientific greatness could only be obtained by success in those areas. For that reason, Chapter 16 is totally devoted to that work.

The doubts and concerns that had plagued Kramers in the early phases of his career began to recur after he returned from Denmark. He was never in robust health, and he was often bothered by major and minor ailments. Although his scientific contributions were immense and of world class, he was dissatisfied with his own work. He had continual doubts about his scientific accomplishments and future. In a letter to Colby, concerning a visit to Ann Arbor, he writes: "You see Colby this letter is written by a man who is weeping at seeing time pass by without bringing the fulfilment of his scientific ideals." [30] This letter clearly shows Kramers' doubts and misgivings about his scientific future. And this was not an isolated event. His doubts led to extreme swings of mood. Needless to say, Kramers never mentioned much of this to anyone; after all, he was the leading theorist in Holland, but he did admit that he often felt tired, exhausted, and drained. He evidently mentioned his doubts and concerns to his only true confident, the friend of his youth, Jan Romeyn.

Romeyn in a remarkable letter describes the problems in Kramers' life and also gives him advice: "If you do not combat this feeling [of excessive exhaustion] by rest, then I foresee the possibility that while you do not achieve all you should, you will come to the thought (which I also had) 'I am intellectually finished' .... to begin with I would advise you: do not do anything that isn't strictly necessary ... do not drink .. do not smoke more than three cigars a day .. avoid as much as possible any sexual relations."[31]

However, Kramers, a little while after receiving this advice, writes Romeyn that "he is calculating his head off with great pleasure and some success." (He did not say whether he took the advice.) And Kramers vacillated between these extremes for the rest of his life.

III. The Scope of Kramers' Activities

Kramers' scientific activities from the time of the Solvay Conference in 1927 until his death in 1952 covered an unusually broad range of subjects. Although a perusal of his published papers and presented lectures may give the impression that his work was scattered and not especially focused, a more careful examination shows a number of definite trends as well as some unexpected features. Kramers' contributions can naturally be divided into a number of categories. There appear to be five clearly identifiable areas in which Kramers was especially interested. The majority of his total of 50 papers belonged to one of these five central topics. He wrote about eight major papers, investigating the mathematical properties of special examples in nonrelativistic quantum mechanics. In the area of paramagnetism and magnetism at low temperatures, he wrote some 17 papers. Ever since his student days, Kramers had been interested in statistical mechanics and kinetic theory; seven papers are specifically devoted to that topic. Kramers published nine papers which in one way or another were connected with the spin and his special symbolic method to treat these spin problems. The final area, which was of particular importance to Kramers, was radiation theory and field theory. Although he only published five papers which unambiguously can be counted in this area, it was a subject of intense and continual concern. In the second volume of his great textbook, he devotes a great deal of space to this very topic, showing very clearly how fundamental he considered that field. It will be shown in Chapters 16 that most of the material on radiation theory in the book was presented there for the first time, so that it is certainly appropriate to count this as an independent research area. This enumeration shows that of the 48 major papers Kramers wrote, about 36 were in these five areas just listed. Of the remaining papers, three contained the dispersion relations, which are a direct continuation of the work that Kramers did on dispersion theory while still in Copenhagen. He (Kramers) never returned to these Kramers–Kronig relations, so it is better to classify those studies with an earlier period, especially if one looks for changes of style in this later period. The Kramers–Kronig relations belong to the Copenhagen sojourn. There were four papers on diffusing gas mixtures and macromolecular flow, and the remaining papers were scattered.

It is remarkable that the single largest group of papers is in paramagnetism; Kramers' interest in that field was undoubtedly stimulated by his many contacts with the Dutch experimental physicists in that field. He wrote a footnote on superconductivity in 1926, but he did not write a single theoretical paper on superconductivity during the remainder of his career. Instead, in the period 1929–1933, he wrote many papers on the paramagnetic properties of the rare earths. After 1933, Kramers made important theoretical contributions to the study of adiabatic demagnetization of paramagnetic salts. Since adiabatic demagnetization became a very important experimental technique for obtaining ultralow temperature (at that time practically a Dutch monopoly), Kramers was then more directly involved with experiments than at any other time during his life. It is interesting that while Kramers was very active in the field of paramagnetism and low-temperature magnetic phenomena, he did the bulk of that work from 1929 until 1935. He wrote only one more paper on a paramagnetic subject, some 16 years later in 1951. It is characteristic that even though Kramers obtained some of his most lasting results in paramagnetism, he himself considered this work primarily as a set of technical exercises in theoretical physics.[32] It could almost be said that while he took it seriously he did not feel that the road to greatness or to fundamental scientific importance lay in that direction. The rather abrupt cessation of his efforts in paramagnetism in 1936 is also noteworthy. His interest at that time shifted from paramagnetism, where the theory and experiment were in close contact, to ferromagnetism, which at that time was much further removed from experiment. With this shift, Kramers' interest in magnetism merges almost completely with another and older interest—statistical mechanics. As stressed earlier, Kramers had been interested in that field for many years. His mastery of the formal apparatus of physics is nowhere more evident than in statistical physics. In reading these papers, one cannot escape the feeling that Kramers really enjoyed doing problems in statistical mechanics. The papers contain all kinds of clever tricks, unusual twists, and unexpected results. In the papers on statistical mechanics, the responsibility of an acceptable philosophy of physics does not seem to weigh so heavily on Kramers.

Kramers, who was involved with quantum mechanics since its inception, wrote about eight papers which were specifically concerned with quantum mechanics. These papers almost without exception deal with the mathematical elaboration of physical questions. In these, the initial motivation and formulation were always obtained from physics, but the subsequent development, the originality, and many of the results were mathematical in character. Kramers with his mastery of classical analysis could obtain formal results of great elegance; he clearly delighted in the complete mathematical illumination of physical problems. The mathematical problems of quantum mechanics remained of great interest to Kramers all during his career. These papers are distinct from the later papers dealing with the analysis of multiplet situations and spin problems. To be sure, both groups of papers are strongly mathematical, but in the analysis of spin problems, Kramers does not just apply his vast knowledge of classical analysis and assortment of clever personal tricks; instead, he established a systematic symbolic calculus which he hoped could avoid the use of formal group theory. Kramers was especially fond of this symbolic (ξ, η) formalism, which combined formal trickery with great computational power. It was one of his great disappointments that this method was largely ignored; it was too strange for the older generations, and the later generations did not want to avoid group theory. The categories outlined so far—mathematical problems of quantum mechanics, paramagnetism, statistical mechanics, and symbolic properties of spinors—do not lend themselves to the development of a particularly strong, conceptual approach to physics. All emphasize techniques based on established physical principles. The categories mentioned so far comprised a considerable share of Kramers' activities; consequently, a large portion of his efforts were devoted to applications and elaborations. Only one of his major fields of interest in these years was concerned with an extension of the conceptual framework, with a search for new methods, procedures, and possibly new principles. This work includes the studies of the Dirac equation and Kramers' efforts to modify it; it also includes his incisive studies on the interaction of electrons and quantized fields. As might be expected, Kramers was especially anxious to construct a quantum theory of the electromagnetic field, which would indeed possess the Lorentz theory of electrons as a classical correspondence limit.

The five categories mentioned here summarize the bulk of Kramers' scientific activities since the end of the Solvay Conference in 1927. Shortly after the conference and after Kramers' misfortune with his theory of the electron spin, a noticeable withdrawal from the most urgent problems in physics set in. He withdrew to more tangible and concrete questions—but he never could let go of the problems of particles, fields, and their interaction.

IV. Kramers' Contributions to Statistical Mechanics

Kramers' work in statistical mechanics is of extraordinary depth and significance. It is quite possible to argue that his contributions there rank with his extraordinary achievements in dispersion theory and renormalization theory. It is probably useless to conjecture that Kramers would disagree with this assessment; statistical mechanics to him was not a deep or profound subject, just one of forbidding almost irritating technical complexity. It is not surprising that his extraordinary formal power would enable him to make major contributions in that field.

Kramers' deep insights were obtained from the detailed, almost tedious, examination of extremely specific questions. Yet he was often able to extract conclusions of extreme generality and great importance from these detailed computational results. In true Kramers fashion, he neither emphasized these general conclusions nor stressed their broad implications; he let the detailed results speak for themselves.

It is somewhat surprising that Kramers published his first paper on statistical mechanics as late as 1927, after he had returned to Utrecht from Copenhagen, 14 years after he had started as Ehrenfest's student in Leiden. This is surprising because Ehrenfest, who influenced Kramers so strongly during that time, was passionately interested in statistical mechanics. Ehrenfest represented the Boltzmann tradition, which emphasized the approach to equilibrium and the time dependence of physical phenomena. This time dependence was a direct consequence of the underlying dynamics. To understand the irreversible approach to thermal equilibrium, on the basis of a reversible dynamics, was the central question. The basic technical tool was the Boltzmann equation. This approach must be contrasted with Gibbs' more abstract, almost axiomatic, method which seeks a description of the thermal equilibrium state in terms of macroscopic parameters and the dynamics of the system. Kramers' scientific role models, Bohr and Lorentz, really preferred Gibbs' more profound approach to the Boltzmann equation methodology. As a consequence, Kramers became conversant in operational detail with both methods; there were few physicists who had that degree of control. Kramers' personal evolution in statistical mechanics can be inferred from the way he taught the subject at different periods in his life.

During his Copenhagen period, 1921–1924, Kramers' courses were totally devoted to kinetic calculations, to the H theorem of Boltzmann and the Boltzmann equation. Gibbs is never mentioned. However, in a course given in Utrecht—in the year 1929–1930—Kramers adds to this course an appendix on ensemble methods and Gibbs' ideas. He even mentions that this would make Ehrenfest quite unhappy. In later lecture courses, in 1940–1945, Kramers devotes about equal time to the Gibbs and Boltzmann methods. He evidently felt that with the advent of quantum theory, the ensemble approach was superior. Then in 1951—the last time he taught statistical mechanics—he completely reworked the course; he started from the Gibbs ideas and only at the end did he introduce Boltzmann's kinetic procedures as a special example of Gibbs' general approach. This also shows how Kramers throughout his life kept thinking about statistical questions.

Kramers made major contributions to the theory of phase transitions. He was certainly the first person to realize that statistical mechanics can yield discontinuous phase transitions only in the thermodynamic limit (where $N \to \infty$, $V \to \infty$, $V/N = v$ is finite, N = number of particles, V is the volume). Kramers arrived at this fundamental insight in his usual roundabout way. In 1934, he wrote his first paper on ferromagnetism. He had a well-circumscribed purpose—to construct a classical model that would exhibit a ferromagnetic phase transition. He succeeded in developing such a model (very similar to a spherical model) which in three dimensions exhibited such a transition. This made it possible to consider classical models when studying ferromagnetic phase transitions. In the next paper on ferromagnetism, the calculation of the partition function was a veritable nightmare in special function theory. Kramers observed that, in order to get a sensible thermodynamics, he had to go to the limit of an infinite number of spins. More important than that, he showed, by a brutal calculation, that in this thermodynamic limit some of the thermodynamic functions consisted of two separate analytical branches. A little later, at the van der Waals Congress held in 1937 in Amsterdam, the question of phase transitions came up again. The basic issue was whether the partition function could—without further assumptions—explain the shape, including the discontinuities, of a real isotherm. There was considerable confusion about this question; Kramers at that time made his suggestion that discontinuous phase transitions can only occur in the thermodynamic limit. There was no immediate acceptance or rejection of that suggestion; there was doubt and confusion. Kramers was the chairman of that session, and in a whimsical mood he put this question to a vote. The vote was never recorded because it was not an official act of the congress.

It is remarkable that Kramers' offhand comment suggested by his detailed calculations—that phase transitions occur only in the thermodynamic limit—became in later years a well-established dogma. It even became fashionable to prove the existence of the thermodynamic limit, something Kramers had established for his special cases by direct calculation. Interestingly enough, Kramers paid no attention to these developments.

Kramers' next contribution to phase transitions and statistical mechanics is his fundamental paper on the Ising model, written together with Wannier.[33]

This was a seminal paper; it introduced several innovations. It was shown that the partition function could be computed as the largest eigenvalue of a matrix, thereby making it possible to apply powerful matrix techniques to statistical problems. They determined for the first time the numerical value of the transition temperature; they conjectured, against all expectations, that the specific heat would show a logarithmic singularity. Onsager later proved that conjecture. These papers showed many of Kramers' characteristics. There were many explicit calculations with unwieldy matrices, leading to general results and interesting conjectures. It was not too difficult to follow the argument but it was hard to understand their motivation and origin. This paper was soon followed by Onsager's exact and complete solution of the two-dimensional Ising model, which changed the theory of phase transitions. If Onsager caused a revolution in phase transitions, Kramers and Wannier started it. In 1940, Kramers just tossed off a single paper in nonequilibrium statistical mechanics[34] which had a major impact and which, if anything, is more important and useful today than it was 40 years ago. As always, Kramers starts with an extremely specific problem. He wanted to calculate the classical escape probability of a particle bound in a given potential. The particle can escape from the potential through the action of random forces. That is all. It is truly amazing that so many distinct and varied phenomena can be subsumed under this simple picture. The treatment of this classical problem is Kramers at his powerful best. He first sets up an equation for the probability distribution, which in fact is very similar to the Boltzmann equation. (It is easier to solve, but not that easy.) Kramers then shows, in a remarkably clever way, how under suitable circumstances the equation can be approximated by different diffusion-type equations (either in space or energy, but *not* both). It is then not too difficult to use these equations to approximate the escape probabilities in various circumstances.

This paper of Kramers had an extraordinary and long-lasting influence. The succeeding investigations fall into two distinct categories. One seeks to understand, systematize, and refine the approximation procedures that Kramers had formulated in his usual offhand manner. The other effort concentrates on analyzing different physical phenomena in terms of Kramers' model of the escape over a barrier due to random forces. By now this paper is widely used and frequently quoted; it has been unusually fertile in a large variety of quite different physical circumstances. This paper is also extremely useful as a model to elucidate many features of nonequilibrium mechanics. Kramers never did this; in fact, he never returned to this paper at all.

Unfortunately, this is an abbreviated and superficial summary of Kramers' great work in statistical mechanics. It does not and cannot begin to do justice to his extraordinary insight. Even if Kramers was motivated to study statistical problems, by a desire to stay away from deep, controversial, and fundamental questions at the quantum frontier, his own contributions were instrumental in making statistical mechanics a new frontier.

CHAPTER 16

The Recurrent Theme: Electrons and Radiation

I. Kramers' Approach: The Legacy of Lorentz and Bohr

A. Lorentz's Influence

Although Kramers made many distinguished contributions to broad areas in theoretical physics, there was one set of problems that preoccupied him throughout his career. These problems were always related to the proper description, be it classical or quantum mechanical, of the interactions of electrons with an electromagnetic field. In a career as broad and varied as Kramers', it may seem arbitrary to single out one particular set of studies as especially significant, but there is no doubt that Kramers himself considered his investigations in this area to be of unusual importance. This can be inferred more from his persistent activities in this field, extending as they did over many years, than from the number of papers actually published. His early work on dispersion relations (1924, 1925) certainly belongs in this category, as does his work on the relativistic spin theory. Kramers returns to a most serious and meticulous examination of the proper formulation of the interaction between electrons and the electromagnetic field in the second volume of his great textbook published in 1938. No less than 80 pages, out of a total of 280, are devoted to an incredibly detailed treatment of that topic. Even after this intense preoccupation with the subject, he kept on thinking about it, refining his arguments and changing and improving the presentations. He induced several students and collaborators to use his ideas and methods to reexamine and improve the quantum theory of radiation. Not all were equally successful, but from 1940 until 1952, Serpe, Opechovski, and van Kampen all made some contribution to the reformulation of quantum electrodynamics along the lines suggested by Kramers in his book. These studies certainly showed that Kramers' criticism of the conventional treatment of radiation theory was rational and well founded. They demonstrated further that some (but not all) of the difficulties could indeed be removed by Kramers' treatment. After a rather brief summary of Kramers' work in 1938, he published only little in this area. But the basic questions continued to preoccupy him, even haunt him. There are a number of unpublished manuscripts, sketches for talks, and incomplete notes all showing that the problems of quantum electrodynamics and the role of electrons were constantly on his mind. Kramers gave a lecture summarizing his approach at the Shelter Island Conference in 1947. The most systematic and most complete exposition of Kramers' electrodynamics is contained in his presentation to the Solvay Conference in the fall of 1948. Through the customary delays and postponements, the discussions and reports were not published until 1950. By that time, Kramers' methods, procedures, and results were superseded by the explosive development of quantum electrodynamics associated with the names of Schwinger, Feynman, Tomonaga, and Dyson.

Even these few comments should make it evident that Kramers, throughout his life, was fascinated by the problems of the quantum theory of the electromagnetic field. His preoccupation with these questions is evident from his incessant discussions with students and colleagues on these topics (Pais,[35] Belinfante,[36] van Kampen,[37] Uhlenbeck[38]). So intense was Kramers' interest in these problems that his concern almost became an obsession. The correct manner in which to introduce quantum ideas into the description of the electromagnetic field and the role of the electrons were reexamined, reanalyzed, and scrutinized in discussions and protracted arguments—literally for years on end.

It is a great pity that Kramers' important contributions in quantum electrodynamics never received the recognition that they deserved. In fact, his results were not widely known. It was not until after the relativistic formulation of quantum electrodynamics was well on its way toward completion that Kramers' contribution began to be appreciated. Since Kramers published his novel ideas in textbook form, it is not surprising that their impact was rather minor. World War II further retarded and even prohibited any widespread dissemination of Kramers' work. During the war, Kramers worked effectively in total isolation. So it is not surprising that only those in Kramers' immediate environment were aware of his contributions, ideas, and methods. Kramers was so immersed in this set of problems that it would be difficult for those who had daily contact with him not to be continuously reminded of his deep concern. But this, of course, did little for the wider circulation of Kramers' ideas. Perhaps the most important factor in the belated recognition that Kramers' work received was his lackadaisical method of publication. Some ideas were contained in lectures, students' theses, and in his own book. But as usual, he did not present his ideas in an aggressive, forceful manner.

It is most remarkable that although Kramers interest in—or preoccupation with—the correct description of electrons in an electromagnetic field was his personal obsession, he was nevertheless profoundly influenced and forever haunted by the older ideas of Lorentz and Bohr. Since Kramers came of age in quantum theory under the direct tutelage of Bohr, his pervasive influence on Kramers' physics is perhaps not so surprising. But the effect of Lorentz on Kramers, probably equal in importance to Bohr, was more indirect. It would only be a slight exaggeration to say that Kramers' veneration for Lorentz changed but little from his initial reaction. When Kramers saw Lorentz for the first time at the inaugural address of Ehrenfest (1912), he recorded his impression as "I looked at him [Lorentz], searched for him with the sacred devotions, with which as a little child I had stared at the Queen."[41] Some 35 years later in a lecture course on relativity, Kramers taught at Columbia University in the fall of 1946; he commented time and time again on the greatness of Lorentz. His admiration for Lorentz was evident throughout: "He had an immense respect for Lorentz, remarking on one occasion that Lorentz made only one mistake, or was it none? I am not sure."[40] This comment by Frances Low, a student in Kramers' class at the time, certainly catches the mood of Kramers' undiminished admiration for Lorentz. It is especially important to realize that Kramers' admiration was primarily for Lorentz's physics. The personal relation between Kramers and Lorentz was cordial and correct but totally lacked the intense personal involvement that characterized the relation between Bohr and Kramers. When as a student, Kramers, in 1915, had to give a talk in Lorentz's colloquium on the history of electromagnetism, Kramers with his youthful arrogance expressed himself in a somewhat condescending manner about the earlier clumsy approaches of Riemann and Ampere. Lorentz was extremely displeased and reprimanded Kramers severely.[39] Kramers remembered the incident throughout his life and mentioned it again when he accepted the Lorentz medal on October 30, 1948. At that same occasion, Kramers, reflecting on Lorentz, said: "His authority made him an authoritarian. I experienced this and later understood this as natural."[39] Perhaps most surprising in Kramers' tribute to Lorentz is his confession that he did not think that he had learned a great deal in either method or attitude from Lorentz. It is undoubtedly true that Kramers meant this when he said it in 1948. But it is totally certain that, independent of Lorentz's methodology or attitude which might or might not have influenced Kramers, Lorentz's achievements in physics had a lasting, powerful, and even dramatic impact on Kramers' physics. This is evident not only from the deferential manner in which he referred to Lorentz and his work, but even more from the way in which Kramers framed and approached problems in physics. Nowhere is this more evident than in Kramers' efforts to obtain a consistent quantum theory of electrons and radiation. His starting point, method of attack, and definition of success were unquestionably patterned after Lorentz's monumental classical treatment.

To Lorentz, electrons were objects with finite spatial extensions. They possessed a charge and possibly a current distribution. The motion of electrons was to be described as the motion of these extended objects, and the fields they produced were those of extended charges. There is no evidence that Lorentz ever seriously deviated from this basic picture or doubted it in any way. Kramers, in his discussions of the spinning electron, adopts exactly the same picture. Lorentz also pictured the motion of these extended electrons to take place in an all-pervading ether. Kramers never really rejected the idea of the ether but did not use it explicitly either. To Lorentz, the ether functioned as a medium through which the electron moved. The Lorentz picture led to two consequences which were of crucial importance in the subsequent development.

(1) Since the electron is envisaged as a finite charge distribution, each element of an electron exerts a force on each other element. It is not difficult to show that if the net velocity **v** is constant, the net force is zero. Lorentz was pleased with this result: "This shows that if free from external forces an electron, just like a material point, will move with a constant velocity not withstanding the presence of the surrounding ether."[42a] This quote is important; it stresses that although the electron is a finite charge distribution, Lorentz still recognizes that the *mechanical* motion of an electron can be described as that of a point mass. Furthermore, the important distinction between the self-field and the external field is made here for the first time. The difference between these two fields—or better, the precise definition of that difference—plays a key role in Kramers' later investigations.

(2) Another equally important distinction, also crucial for Kramers' later studies and again made by Lorentz, is that between electromagnetic mass and mechanical (or what Lorentz calls "material") mass. In its simplest version, the electromagnetic mass is the inertia that an object possesses by virtue of its charge. A charge moving with a velocity **v** (much smaller than **c**) possesses as a current, a magnetic energy proportional to v^2. Thus, the external force did an amount of work αv^2 in producing the velocity **v**. It is therefore natural to define an electromagnetic mass, so that the kinetic energy $\frac{1}{2}m_{el}v^2 = \alpha v^2$. A simple calculation, made by Lorentz, gives α here: $m_{el} = \frac{2}{3}(e^2/ac^2)$. The coefficient $\frac{2}{3}$ depends on the form of the assumed charge distribution in the electron; a is the radius of that charge distribution. Another way to obtain the electromagnetic mass is by calculating the electromagnetic momentum of a uniformly moving electron:

$$P_x = \frac{1}{4\pi c}\int (\mathbf{E} \times \mathbf{B})_x d^{3x} = \frac{2}{3}\frac{e^2}{ac^2}v_x \tag{1}$$

The identification of $\frac{2}{3}e^2/ac^2$ as m_{el} is again evident from Equation (1). The electromagnetic mass, which depends on the details of the charge distribution, is distinct from the mechanical mass m_0. The total mass or experimental mass is presumably

$$m_{exp} = m_0 + m_{el} \tag{2}$$

$$m_{el} = \frac{2}{3}\frac{e^2}{ac^2} \tag{2a}$$

It was Lorentz's idea that all the mass of the electron was electromagnetic, so that $m_0 = 0$. This was suggested by the experiments which showed that the experimental variations of the electromagnetic mass with velocity closely followed the expression derived by Lorentz for that variation. By putting the mechanical mass equal to zero, the Lorentz electron had become a rather curious object. It had a finite charge extension, with various pieces capable of exerting forces on each other, but no mechanical or inertial mass. Lorentz was well aware of these unusual properties. "After all, by our negation of the existence of material mass, the negative electron has lost much of its substantiality. We must make it preserve just so much of it, that we can speak of forces acting on its parts and that we can consider it as maintaining its form and magnitude. This must be regarded as an inherent property in virtue of which the parts of the electron cannot be torn asunder by the electric forces acting on them."[42b] Thus, the structure of the electron cannot be discussed within the Lorentz theory; it is postulated to possess the properties enumerated. Kramers, in his investigations, follows these Lorentz ideas to the letter, he did not put the mechanical mass equal to zero but he did insist on the importance of a careful distinction between the mechanical (or inertial) mass m_0, the electromagnetic mass m_{el}, and the total mass m.

To study the classical motion of an electron, it is necessary to set up a classical equation of motion. The forces that can act on an electron are, in the Lorentz view, of three types: nonelectrical forces (called $\mathbf{K}$), electrical forces owing to the external field (called $\mathbf{F}_{ext}$), and electrical forces originating from the finite charge distribution and its field (called $\mathbf{F}_{self}$). Newton's equation of motion is

$$m_0 \ddot{\mathbf{x}} = \mathbf{F}_{ext} + \mathbf{F}_{self} + \mathbf{K} \tag{3}$$

It is important that the mass in this equation is the inertial (or mechanical) mass $\mathbf{F}_{ext}$ and $\mathbf{F}_{self}$ are the forces that the various electric and magnetic fields exert on the electron. These forces are given by the usual Lorentz expression, which for a charge distribution ρ in the electron read:

$$\mathbf{F}_{ext} = \int d^3x\rho\left(\mathbf{E}_{ext} + \frac{\mathbf{v} \times \mathbf{B}_{ext}}{c}\right) \tag{4a}$$

$$\mathbf{F}_{self} = \int d^3x\rho\left(\mathbf{E}_{self} + \frac{\mathbf{v} \times \mathbf{B}_{self}}{c}\right) \tag{4b}$$

where $\mathbf{E}_{self}$ and $\mathbf{B}_{self}$ are the electromagnetic fields generated by the electron's own charge and current distribution. Correspondingly, $\mathbf{E}_{ext}$ and $\mathbf{B}_{ext}$ are the electromagnetic fields produced by external agencies. Lorentz sometimes called the electric field $\mathbf{E}_{self}$ the "ether field" to stress that it originated from the interaction of the charge distribution and the surrounding ether. Lorentz, in a justly famous calculation, proceeded to calculate an explicit expression for the self-force.[42c,43] In this calculation, it is necessary to assume that the charge distribution is rigid and spherically symmetric. The result can be expressed as a power series, in the radius of the charge distribution a, as

$$\mathbf{F}_{\text{self}} = -\frac{2}{3c^2}\ddot{\mathbf{x}} \iint \frac{\rho(|\mathbf{x}|)\rho(|\mathbf{x}'|)}{|\mathbf{x}' - \mathbf{x}|} d^3x\, d^3x' + \frac{2}{3}\frac{e^2}{c^3}\dddot{\mathbf{x}} + g\frac{ae^2}{c^4}\ddddot{\mathbf{x}} + \cdots + a^2 + \cdots \tag{5}$$

Here, g is a numerical coefficient that depends on the charge distribution. The first term is just the electrostatic self-energy of the electronic charge distribution. It can be evaluated for any given charge distribution; for a shell distribution it becomes $\frac{4}{3}m_{\text{el}}$, where m_{el} is the electromagnetic mass introduced earlier. (Actually, the numerical coefficients depend on the details of the assumed charge distribution; they do not have a fundamental significance.) It is of great significance that m_{el} diverges as the radius a goes to zero. It is of even greater significance that expression (5) contains only one structure-dependent term—the electromagnetic mass term which can be combined in a natural manner with the inertial mass in the equation of motion. The other terms go to zero as a goes to zero. The radiation reaction term $\frac{2}{3}(e^2/c^2)\dddot{\mathbf{x}}$ is altogether independent of a. The result of the Lorentz analysis shows that in the limit of a point electron ($a \to 0$), the basic equation of motion can be written in the form

$$\left(m_0 + \frac{4}{3}m_{\text{el}}\right)\ddot{\mathbf{x}} = \mathbf{K} + e\left(\mathbf{E}_{\text{ext}} + \frac{\mathbf{v} \times \mathbf{B}_{\text{ext}}}{c}\right) + \frac{2}{3}\frac{e^2}{c^3}\dddot{\mathbf{x}} \tag{6}$$

To obtain this form, it is necessary to assume either that $\mathbf{E}_{\text{ext}}$ and $\mathbf{B}_{\text{ext}}$ change little over distances of order a or that $\mathbf{E}_{\text{ext}}$ and $\mathbf{B}_{\text{ext}}$ are to be considered as averages of the external fields over the charge distribution. In the strict limit $a \to 0$, the term $\frac{4}{3}m_{\text{el}}$, diverges as $1/a$. For a finite charge distribution (a finite), the Lorentz analysis shows that the evaluation of the self-force term leads to an equation of motion which, although of the same form as a Newtonian equation, has an *altered* mass. The change in mass, called the mass renormalization, is [apart from the radiation reaction term $\frac{2}{3}(e^2/c^3)\dddot{\mathbf{x}}$] the only remnant of the self-force. The self-force (4b) leads only to a change in mass and to the radiation reaction. Physically, the only mass that can be observed is the total mass, which includes of course the electromagnetic mass. (One cannot do experiments with electrons whose self-electromagnetic fields have been eliminated.) Thus, the only quantity that should enter the equation of motion is $m_{\text{exp}} = m_0 + \frac{4}{3}m_{\text{el}}$. Actually, the experimental mass is the sum of the mechanical mass and the mass of electromagnetic origin. That is the meaning which must be attributed to $\frac{4}{3}m_{\text{el}}$. The curious factor $\frac{4}{3}$ should not be taken seriously. It appears here because in the definition of m_{el}, given in Equation (2a), a particular model was assumed. Factors of this type, which are model dependent, always appear in a nonrigorous, nonrelativistic treatment. (See Rohrlich[44] for a more precise discussion of this point.) What should be taken seriously is that the evaluation of the self-force inevitably leads to the inclusion of the electromagnetic mass in the force equations. Thus, for the case where $\mathbf{K} = 0$ and where the radiation reaction force can be neglected, one can write Equation (6) as

$$m_{\text{exp}}\ddot{\mathbf{x}} = e\left(\mathbf{E}_{\text{ext}} + \frac{\mathbf{v} \times \mathbf{B}_{\text{ext}}}{c}\right) \tag{7}$$

It is also possible to write the original equation (3) as

$$m_0\ddot{\mathbf{x}} = e\left(\mathbf{E} + \frac{\mathbf{v} \times \mathbf{B}}{c}\right) \tag{8}$$

where now **E** and **B** are the *total* fields, including the self-fields. Equations (7) and (8) obviously possess the identical structure, but it is necessary to specify what fields and what mass enter the equations. Thus, using the experimental mass as in Equation (7) requires that the fields in the Lorentz force equations are the external fields. One can use the inertial mass as in Equation (8), but then the fields must be the total fields. In principle, the Lorentz analysis should be carried out, using Equation (8), to reobtain Equation (7). This same analysis shows that it is not legitimate to use the experimental mass m_{exp} in the equations of motion and use the *total* fields **E** and **B**. In that case, the contribution of the self-field would be counted twice, once because $\mathbf{E}_s$ and $\mathbf{B}_s$ occur in **E** and **B**, and once in m_{el}, which is included in m_{exp}. A number of books (Heitler[45]) are not sufficiently careful in specifying just what fields should be used in the Lorentz force equation (7) or (8). It was a favorite examination question of Kramers to ask his students what masses and what fields were meant in Heitler's book. Many generations of students became confused and flustered by these questions. It eventually became generally known what Kramers might ask, and the answers improved dramatically; everyone knew the answer to "What is wrong in Heitler's book?"[46]

Kramers was profoundly impressed by the Lorentz derivation. In a lecture course on the theory of electrons, delivered in Leiden 1949–1950, he presents a very detailed treatment of these results of Lorentz. Starting from Equation (3), he first rewrites it in the form

$$m_0\ddot{\mathbf{x}} = \mathbf{K} + e\left(\mathbf{E}_{\text{ext}} + \frac{\mathbf{v} \times \mathbf{B}_{\text{ext}}}{c}\right) - f\frac{e^2}{ac}\ddot{\mathbf{x}} + \frac{2e^2}{3c^3}\dddot{\mathbf{x}} + g\frac{ae^2}{c^4}\ddddot{\mathbf{x}} + \cdots \tag{9}$$

where f and g are again numerical coefficients that depend on the details of the charge distribution. By considering the situation when a is very small and incorporating the $f(e^2/ac)$ term as a mass term in the experimental mass, Kramers arrives at the equation which is the basis for the exhaustive treatment of all classical electromagnetic phenomena:

$$m_{\text{exp}}\ddot{\mathbf{x}} = \mathbf{K} + e\left(\mathbf{E}_{\text{ext}} + \frac{\mathbf{v} \times \mathbf{B}_{\text{ext}}}{c}\right) + \frac{2e^2}{3c^3}\dddot{\mathbf{x}} \tag{10}$$

Kramers planned to write a textbook on the classical theory of electrons. The lecture notes of the course that Kramers taught in Leiden are very nearly a final draft of that projected book. It is striking that as late as 1949–1950, Kramers still followed Lorentz's treatment to the letter. There were only occasional elaborations and amplifications, demonstrating once again how persistent and pervasive Lorentz's influence was. Indeed the two salient features in Lorentz's work—the distinction between the experimental and the inertial mass and the distinction between the total fields and the external fields—are of pivotal significance in Kramers' subsequent investigations. In recognizing Lorentz's great influence on Kramers, it is pertinent to recall that Lorentz's considerations were nonrelativistic; that is, he dealt with a finite extended charge distribution. Of necessity, his considerations were classical and not quantum mechanical. By adapting Lorentz's viewpoint so totally, Kramers based his own investigations, which were intended to be quantum mechanical (and perhaps even relativistic), on a completely classical foundation of extended electrons. Kramers was certainly aware of this; this approach is also in line with the philosophy of the Bohr correspondence principle, which demanded that any quantum treatment should possess a presumably smooth classical limit. This was exactly the approach that Kramers pursued over many years. It is clear that adapting this viewpoint might very well have kept him from exploring other, different schemes. But if Kramers was at all concerned about the limitations he himself had imposed on his investigations, such possible misgivings were overbalanced by his dogged insistence that all future developments should be firmly based on the classical Lorentz theory of extended electrons.

B. Kramers' Uneasiness with the Dirac Radiation Theory

It would be hard to overestimate the importance of Dirac in creating the quantum theory of radiation. Other physicists, Pauli, Heisenberg, Jordan, Wigner, Klein, and Fermi, all made notable contributions, but practically all the major advances in quantum field theory and radiation theory between 1927 and 1948 can be traced back to Dirac's seminal contributions. Dirac established the quantum theory of radiation in two fundamental papers, one[47] submitted February 2, 1927 and the other[48] submitted April 4, 1927. By this time the formal structure of quantum theory—in part due to Dirac's own efforts—was beginning to be reasonably well understood. It had become possible within this formalism to understand the induced emission and absorption processes in which atoms under the influence of given external fields make transitions to other states. Thus, the Einstein B coefficients could be understood and indeed calculated in terms of the Schrödinger wavefunctions of the atom. However, before Dirac's papers, the spontaneous emission process so essential for the Planck distribution law was a great mystery. Dirac applied the quantum mechanical formalism to the *combined* system of atom plus radiation field. He quantized the vector potential, considering the vector potential itself as a matrix, or a q number. Although this now seems more or less obvious—or in any case reasonable—it was nonetheless a step of extraordinary scope. Wentzel, a contemporary of Dirac, describes the impact: "Today the novelty and boldness of Dirac's approach to the radiation problem may be hard to appreciate... the disappearance of a photon was not described by this theory [the pre-Dirac theory] nor was there any possibility in this framework of understanding the process of spontaneous emission ... At this point Dirac's explanation in terms of the quantized vector potential came as a revelation."[49] In this paper, Dirac sets up a Hamiltonian for the interaction between the atom and the radiation field and he shows that the Hamiltonian in the quantum theory can be interpreted as an assembly of photons interacting with the atom. From this Hamiltonian, he calculated (to lowest order in e) the probabilities for the emission and absorption of photons. This enabled him to understand the spontaneous emission process and to derive the relation between the Einstein A coefficient for spontaneous emission and the Einstein B coefficients for absorption. This was exactly the relation that had been postulated by Einstein some 10 years earlier. This was certainly a spectacular result. In the second paper, Dirac casts these ideas in a more definite form; he gives a more careful definition of the Hamiltonian. He summarizes his results: "It appears that one can treat a field of radiation as a dynamical system whose interaction with an ordinary atomic system may be described by a Hamiltonian function ... One finds then that the Hamiltonian for the interaction of the field with an atom is of the same form as that for the interaction of an assembly of light quanta with the atom. There is thus a complete formal reconciliation between the wave and the light quantum points of view."[48] The new Hamiltonian constructed in this paper could be used to carry out calculations to second order in e. It is interesting that this Hamiltonian, when applied to the calculation of the dispersion of light, yields the "old"† Kramers–Heisenberg dispersion formula, which had started Heisenberg on his way toward matrix mechanics in the first place. The method originated by Dirac, to quantize the electromagnetic field and the radiation field in particular, has since become the standard procedure of treating the phenomena associated with the interaction of fields and matter. There are a number of rather tricky formal questions that sometimes cause trouble; but the general approach has not changed materially since Dirac's pioneering work.

This state of affairs had considerable influence on physics and on all physicists. The dominance of Dirac's ideas set a style within physics, which, because of the great effectiveness of Dirac's methods, many tried to imitate or emulate (generally with little success). The effect of Dirac on Kramers, even though it was indirect, was particularly profound. Remarkably, Kramers and Dirac frequently were interested in the same or very similar problems. A striking example is their practically simultaneous investigation (unbeknown to each other) of the spin and the relativistic wave equation. Furthermore,

† This "old" result was obtained 2 years earlier, but so much had happened that it must have seemed then (and still does) a long time ago.

Kramers had been concerned for many years with the photon and the quantum theory of radiation, so he was not merely an interested bystander when Dirac formulated his quantum theory of radiation. The unusual turn that Dirac's second quantization method had given to the contradictory wave and particle pictures was of special interest to Kramers. He had struggled with that problem since his Copenhagen years. These are instances where Kramers had thought long and hard about the very problems treated by Dirac. As a profound independent thinker and an accomplished physicist, he of course had his own ideas and approaches to these same subjects. Very often, while recognizing Dirac's penetrating originality and incisive insights, Kramers was not that enamored of Dirac's results. Dirac's style—abstract and general—was quite different from Kramers', which was more computational and specific. Altogether, Dirac's methods and the type of argumentations used were not congenial to Kramers. It is remarkable that, in spite of these different styles, Kramers and Dirac nevertheless often ended up studying the same questions, albeit in characteristically different ways.

To appreciate the somewhat convoluted and contorted trail of Kramers' researches in quantum electrodynamics, it is not only essential to remember his attachment tȯ Lorentz's ideas, but also his strong conviction that both Dirac's radiation theory and the Dirac equation (and their combination) were approximate and had at best a limited validity. Thus, Kramers viewed Dirac's quantum electrodynamics as a possibly useful, heuristic set of prescriptions, often leading to results of practical importance, but he certainly did not consider it as heralding a new theory of fundamental significance. It is surprising that many physicists, even though their criticism was not as detailed as Kramers', still had serious misgivings about quantum electrodynamics. It was criticized and doubted almost since its inception. Even those physicists who participated in constructing quantum field theory frequently expressed their concerns and doubts about the theory. Bohr had often emphasized that quantum mechanics could not be expected to be applicable to phenomena in which distances of the order of the classical electron radius e^2/mc^2 play a role, since quantum theory does not purport to give any account of the structure of the electron itself. That would correspond to energies of about 137 mc^2, and it was widely believed (by Dirac as well) that the theory including the Dirac equation would break down at these energies. In a letter to Bohr, Dirac says exactly that: "A more reasonable assumption to make is that the relativistic wave equation fails for energies of order 137 mc^2."[50] Others were more negative. Thus, Oppenheimer, after calculating the level shift of spectral lines on the basis of Dirac's quantum electrodynamics, concludes: "The theory is however wrong, since it gives a displacement of the spectral lines from the frequency predicted on the basis of the nonrelativistic theory, which is in general infinite. This displacement arises from the infinite interaction of the electron with itself; this interaction depends upon the state of the material system and the difference in the energy for two different states is not in general finite."[51] Oppenheimer expresses here a rather prevalent frustration with a theoretical scheme that, while interesting and at times impressive, often gave wrong or even nonsensical results. An admonition that one must calculate "to the first nonzero, noninfinite approximation" is a prescription alright, but it cannot be considered a satisfactory theory. Pauli had strong misgivings about the Dirac theory and quantum electrodynamics. In his great handbook article—written before the positron was discovered—he writes: "Any attempt to save the Dirac theory in its present form ... appears to be hopeless." [52] Pauli felt that a correct and complete understanding of quantum electrodynamics would necessarily provide a theoretical understanding of the numerical value of the fine structure constant $\alpha = e^2/hc = 1/137$. In particular, he believed that it should be impossible to construct a relativistic quantum theory of electrons and the electromagnetic field for any value of α except the correct empirical value. Consequently, Pauli considered any theory of quantum electrodynamics that did not yield the numerical value of the fine structure constant as preliminary and somewhat superficial.

There was a veritable chorus of objections of all kinds, from detailed to general, to quantum electrodynamics. This of course did not stop those who used the theory for practical computations and generally obtained satisfactory agreement with experiment. Deciding just where quantum electrodynamics would break down was, if not a favorite occupation, still an often discussed topic. Many physicists waited either with anticipation or perhaps anxiety for the radical insights that would transfigure quantum electrodynamics from an incomplete theory with strange rules and irritating ambiguities into a systematic, well-founded, physically transparent scheme.

Kramers was not one of the physicists who withdrew and waited for a new revolution to take place. He had been doubtful about Dirac's radiation theory since its beginning in 1927 and quietly critical of Dirac's electron theory in 1928. It was not in Kramers' character to follow a trend, no matter how fashionable, nor was it his style of physics to expect sudden, grandiose new insights. It was rather by a detailed critical analysis of individual phenomena that he expected progress to be made. This was very much the same situation as that in 1925, when Kramers had given a quantum description of the dispersion relations. Kramers and especially Bohr at that time expected that a similar detailed case-by-case analysis and subsequent quantum transcription would be necessary to arrive at a satisfactory quantum theory. It was precisely Heisenberg's genius which showed that this whole procedure could be done once and for all, leading to the new autonomous matrix mechanical scheme. Kramers, in attempting to construct quantum electrodynamics, again pursued a step-by-step, detailed procedure. He at no time wished to make too sudden or too radical alterations from the classical Lorentz theory of electrons unless compelled to do so by computational evidence. The imposition of general, a priori physical or philosophical principles was alien to his style of physics. It is thus not surprising that Kramers was continually surprised and at times exasperated with the success of Dirac's physics. But the successes of Dirac's physics, which Kramers always acknowledged, did not keep him from expressing his doubts about the basic ideas of Dirac. In lectures, in private conversations, in papers, and in his great textbook, Kramers let no opportunity pass to express gently but unmistakably his general disapproval of the Dirac theory. This continued criticism of Dirac's radiation theory lasted until the end of Kramers' life, a period of over 20 years. During this time, the nature of the criticism changed but little. Even the successes of the theory did not remove Kramer's misgivings about Dirac's ideas. At best, they made him tolerate the Dirac theory as a primitive initial stage of a later, more fundamental theory. For example, in his inaugural address as Extraordinary Professor at the Technical University in Delft, Kramers comments in a most guarded manner on the Dirac theory of radiation: "The concepts created by Dirac are sufficient for everyday uses, for most purposes the photon idea of Einstein is incorporated in an acceptable manner."[53] This is certainly not an enthusiastic endorsement, nor does Kramers appear elated at the remarkable integration of the photon concept in quantum mechanics, a problem that preoccupied him so totally only a few years earlier. Quite the contrary, in this same address Kramers stresses what is left unsettled by the Dirac theory: "The problem of principle—which is the complete synthesis of quantum theory and relativity—remains unsolved and is left untouched."[53] Furthermore, Kramers again refers to the problems of the electromagnetic mass of the electrons and the question of the finite size of a Lorentzlike charge distribution. These same themes recur time and time again in Kramers' comments on the Dirac theory.

Nor was this explicit and pointed criticism of the Dirac theory a one-time occurrence in a nontechnical, popular lecture. In the introduction to his textbook, Kramers writes: "We emphasize especially that Dirac's radiation theory cannot be considered by itself to be a quantisation of the classical theory of electrons, as it is able to describe—in contra-distinction to this theory—only the secular interactions between radiations and particles."[54] This was written in August 1937. In the textbook itself, the limited character of quantum electrodynamics is again emphasized: "The difficulties of the classical theory of electrons connected with the particle field dualism ... also crop up in quantum electrodynamics. Quantum electrodynamics is, as the classical theory of the electrons was, an approximate theory and the expectations that the known methods of quantum theory would, because of this very nature, be able to give more information about the nature of elementary particles has by and large not been fulfilled."[55] Even as late as 1950, when Kramers knew about the Lamb shift and about Bethe's calculation of that experimental result, Kramers was not convinced. This in spite of the fact that he was well aware that Bethe's successful calculation was little more than an adaptation of his own method. He writes in 1950: "The results obtained in this way from Dirac's theory of 1928 to the beautiful results of the very last year† are certainly impressive .... Still the fundaments on which they rest ... are derived from correspondence ... first of all the device of introducing

† Kramers refers here to the various calculations of the Lamb shift.

the interactions of radiation with a particle by adding to the total momentum the terms $-(e/c)\vec{A}$.... We have seen that in the nonrelativistic approximation, the elimination of structure does not simply consist of changing this $\vec{A}$ into $\vec{A}_1$.... Thus one might say that the famous $e\vec{\alpha}\vec{A}$ interaction in the Dirac theory doesn't even ensure that simple unrelativistic effects are well rendered by the theory ... although it is known to work for the simplest effects."[56] These remarks, written in 1950, were the last Kramers ever made on his favorite topic. It is evident that his negative attitude survived even the spectacular successes of the quantum electrodynamics of 1946–1950 of Schwinger, Dyson, and Feynman.

It is interesting and more than a little ironic that Kramers' continual criticism of the Dirac theory was in fact correct and pertinent. His insistence on the proper separation of bare and electromagnetic mass and his continual preoccupation with the elimination of structure-dependent features were indeed key ideas in the development of renormalization theory. Unfortunately, instead of concentrating his criticism on subtle details and incomplete features (see Sections III and IV) of the Dirac theory, Kramers rejected Dirac's relativistic framework altogether. In fact, he went so far as to deny the possibility of a consistent, Hamiltonian, relativistic classical theory of electrons. Thus, Kramers in his own investigations worked throughout with an explicit nonrelativistic formalism, with specific models of the electron. Consequently, his results were, of necessity, nonrelativistic and incomplete. Where logically, Kramers' concern and criticism were perfectly legitimate, he concentrated so single-mindedly on the shortcomings of the theory (which were of a subtle and detailed character) that he failed to recognize the major advances and new insights that the relativistic treatment provided.

It appears that Kramers' forceful rejection of Dirac's relativistic theory was based not so much on a deep analysis of physics but on the historical example of the Lorentz model of an extended electron. It is, of course, hard to know and probably not even sensible to inquire just how this overall skeptical attitude influenced his own investigation. But there can be little doubt that Kramers' strongly held opinions, perhaps unduly influenced by Lorentz and Bohr, eliminated certain approaches, thereby forcing his research in other directions. His personal opinions certainly limited his scientific options. Kramers' insistence on a nonrelativistic treatment caused him to miss the general nature of the renormalization idea, which his incisive treatment of the role of the electromagnetic mass had uncovered. There is no evidence that Kramers ever realized the importance or existence of charge and wavefunction renormalization: This in spite of the fact that, in his paper on charge conjugation in 1937,[57] Kramers suggests a Hamiltonian which in a heuristic, nonrelativistic manner describes electromagnetic interactions in a hole theory. In so doing, Kramers comes very close to the idea of charge renormalization and the polarization of the vacuum, but he does not phrase it in this manner, nor did he see any parallel between this redefinition of the vacuum and the redefinition of mass. His conclusion once again was a criticism of Dirac: "As a result we expect that a correction must be applied to the energy values of the stationary states of the hydrogen atom as given by the Dirac theory of 1928."[58] It is evident that Kramers was for a long time groping for a general understanding of the renormalization process. It is also clear that he did not achieve it. His rejection of a fully relativistic treatment was a crippling handicap in the systematic development of renormalization ideas. Kramers never developed his ideas beyond specific nonrelativistic models. Perhaps because Kramers was so aware of the imperfections of his theoretical ideas, he never attempted to compare his theory with experiment. He never even put in numbers to estimate what size effects could be expected. He considered all such efforts premature and even presumptuous. So no confrontation of his theory and experiment ever took place. In this case, as in many others, Kramers by a careful and meticulous examination of specific problems had arrived at new fundamental insights. But once again the universal importance of his results was recognized and formalized by others and they, not Kramers carried his ideas to their ultimate successful conclusions.

C. The Chronology of an Obsession

It would not be wrong to say that all through his scientific career Kramers struggled with the problems of radiation theory. His early work on the Compton effect (see Chapter 14), his intense preoccupation with the BKS theory and the photon, and his initial doubt that quantum mechanics could provide a satisfactory theory of the emission and absorption of light, all these attest to his deep concern with the quantum theory of radiation. After Dirac had given an altogether new orientation to these problems, his interest did not diminish, nor did he find, as explained in detail in the previous section, that an especially profound advance had been achieved. Thus, Kramers' interest in the problems of radiation theory remained intense for a period of at least 20 years. Although he did other work and wrote significant papers in many other areas, his interest in the problems of radiation theory never waned. So intense and persistent was this interest and so strong was Kramers' preconceptions that it is not wrong to characterize Kramers' attitude toward the quantum theory of radiation as an obsession. The very intensity of his feelings may indeed be one of the reasons why Kramers' own work in this area was not widely recognized. His arguments always started with a criticism of the Dirac theory; this by itself, even if correct, did not lead to any striking new insights. Consequently, many physicists did not pay attention to his generally involved criticism. The actual procedures that Kramers himself suggested were based on models and ideas of Bohr and Lorentz, which, without detailed analysis, were often considered old fashioned and inappropriate. Kramers' method of publication did little to make his version of radiation theory accessible and understandable to the physics community. He did not publish his theory of the electron until 6 or 7 years after the Dirac theory. He expressed his criticism of the Dirac radiation theory in discussion in lectures but only occasionally in papers. He wrote only one paper on hole theory, published in 1937.[59] The most complete and systematic description of his personal version of electrodynamics is contained in the second volume of his great textbook. The discussion starts at about page 400 and covers some 60 pages. It is generally not a good idea to publish new research in a textbook. Few active researchers take the trouble to plow through a textbook through material they know (or believe they know in one form or another) just to read an alternate theory. In addition, very few people will fight their way through a 60 page paper. This combination almost guaranteed that no one would read Kramers' exposition. The same year (1938) that Kramers' second volume appeared, he published a short note in *Il Nuovo Cimento*, summarizing the new theory that had just appeared in his book.[60] This note was so brief that it is hard to appreciate the subtlety and importance of its comments. The further elaborations, indeed the first new applications of Kramers' approach, were contained in two papers in *Physica* in 1940—not by Kramers, but by some of his collaborators and students. The important later developments during World War II in The Netherlands and in the United States in 1946–1947 were not published until years later in the not too accessible Solvay reports. Thus, there is not a single paper under Kramers' name on this topic in any of the standard journals, *Physica, The Physical Review, Zeitschrift für Physik*, the Danish journals, or the British journals. It is interesting and quite characteristic that although Kramers considered this work especially important, he did absolutely nothing to promote its dissemination. Quite the contrary, if he had wished to hide his work, he could not have done it much more effectively. Of course, the outbreak of World War II so soon after the publication of his book, which contained the only complete version of his theory, made it almost impossible for Kramers' novel ideas to reach a broad audience.

The starting point of Kramers' considerations was always a critique of the Dirac theory. Although the detailed objections varied somewhat, there were three (possibly four) basic objections that recurred again and again.

(1) The occurrence of divergencies in the Dirac theory was objectionable to Kramers. He was unhappy and concerned about the divergence of the zero-point energy, but he was especially critical of the result (first obtained by Oppenheimer[51]) that the Dirac Hamiltonian and the Dirac theory lead to an infinite shift of the spectral lines of an atom in a radiation field.

(2) To Kramers, who felt with Bohr that the roots of quantum mechanics *must* lie in classical physics, it was particularly upsetting that the relation between the Dirac theory of the electron and the classical Lorentz electron was very tenuous. A naive application of the Bohr correspondence principle to the Dirac theory does not yield the correct correspondence limit. This point bothered Kramers a lot; the correspondence principle had been a reliable and trusted guide in early quantum theory. From the almost miraculous applications that Kramers and Bohr had been able to make,

it had evolved from a useful heuristic guide to a sacrosanct principle, and Kramers was constitutionally unable to give it up or contemplate a limited role for that principle.

(3) Kramers was enormously impressed by Lorentz's discussion of the electromagnetic mass of an electron. He felt that Dirac had not made a sufficiently precise distinction between the electromagnetic mass and the experimental mass. A good share of Kramers' criticism of the Dirac theory was focused on this very point. To Kramers, all these difficulties were connected; as time went on, other specific, unsatisfactory appearing features were added, such as the persistence of divergencies in the hole theory and meson theory. But the lack of correspondencelike connection between the Lorentz theory of an extended electron and the Dirac theory was undoubtedly the core of Kramers' severe misgivings about the Dirac theory of radiation. He simply could not accept a theory in which the famous Lorentz radiation term $\frac{2}{3}(e^2/c^3)\dddot{\mathbf{x}}$, which classically is responsible for electromagnetic radiation, would not have a simple straightforward quantum mechanical interpretation.[61]

It is not a coincidence, and far from irrelevant, that both Bohr and Pauli were for a long time violently opposed to many aspects of the Dirac theory. Even after Anderson's experimental discovery of the positron, Bohr remained doubtful. He at first was critical of the experiments themselves, and he did not really change his mind even after the experiments had become quite definite. Bohr is reported to have said: "Even if all this [the experiments] turns out to be true, of one thing I am certain, that it had nothing to do with Dirac's theory of holes."[62] Pauli was equally unconvinced. In a letter to Dirac he writes: "I do not believe in your perception of 'holes' even if the existence of the 'antielectron' is proved."[63] Apart from a number of specific points, the objections of Bohr and Pauli were based on a philosophical bias about the ultimate nature of a theory of electrons. Bohr in numerous lectures stressed that the quantum theory (of that time) could only be expected to be consistent for distances larger than the classical electron radius $r_0 = e^2/mc^2$. Since the classical theory treats electrons as points, yet assigns them a radius r_0, one can only trust (according to Bohr) a quantum description involving distances larger than r_0. "The present quantum mechanics cannot be expected to apply to phenomena in which distances of the order of the classical radius of the electron e^2/mc^2 are important, since the present theory cannot give any account of the structure of the electron."[64] Bohr and Kramers both thought that in a truly fundamental theory elementary particles and the quantum of action would appear in an intrinsically inseparable manner. Pauli strongly felt that such a theory should at the same time explain the numerical value of the fine structure constant. No wonder they considered Dirac's theory a crude preliminary effort; nor is it surprising that all three tended to emphasize the difficulties of the theory, such as the negative mass states and the divergencies, rather than its successes. Clearly, Kramers was not alone in his criticism of the Dirac theory. He must have been greatly encouraged in his efforts to construct an alternate theory of radiation by the knowledge that he had the sympathetic approval of his venerated teacher Bohr. Although Pauli and Kramers agreed in their critique of the Dirac theory, Pauli was not enthralled with the radiation theory that Kramers eventually produced. But for Kramers, the knowledge that Pauli—generally considered the "conscience of theoretical physics"—was adamantly opposed to the Dirac theory must have been at once reassuring and stimulating.

Kramers' critique of the Dirac theory combined with his own preconceptions defined and shaped the program that Kramers was to follow in search of a consistent quantum theory of radiation. The program was never formalized as such, but it probably had assumed a fairly definite form as early as 1934. It involved these individual steps:

(1) Start from a classical model of an extended electron.
(2) Construct an exact or approximate Hamiltonian for the system consisting of electrons and radiation.

Steps (1) and (2) are of course traditional. Other persons, not just Kramers, had started in the same manner. The next step, however, is peculiar to Kramers' procedure. Starting as he did from an extended electron with a charge distribution $\rho(x)$, the Hamiltonian depends on the assumed electronic structure. There are several ways in which the structure of the electron manifests itself: The charge distribution will produce an electromagnetic field of its own. There are forces within that charge distribution itself which Kramers called the "self-field" or "proper" field. The total field would be split into a self-field and an external field $\mathbf{E} = \mathbf{E}_{\text{ext}} + \mathbf{E}_{\text{self}}$. The response of this extended charge distribution to external forces will depend on the charge distribution. Consequently, the mechanical inertia, the mass, will depend on the charge distribution. According to Bohr's injunction, a quantum theory can be expected to be consistent only for distances $r > r_0$; in that domain, there is presumably no, or only a very slight, dependence on the internal properties of the electron. These ideas were incorporated in the next steps of Kramers program:

(3) The separation of the description of the system into structure-dependent and structure-independent parts. Directly connected with this separation was the next point.
(4) The elimination of the structure-dependent features by a canonical transformation.

The hope was that, if these steps could be carried out either exactly or approximately, an autonomous scheme involving just the structure-independent features would emerge. The structure-independent features would be just the experimental charge and the experimental mass of the electron; in particular, there should be *no* dependence on the radius a of the charge distribution. Consequently, the limit $a \to 0$ could then be taken in the structure-independent scheme; the divergencies might or might not be present in the structure-dependent terms. But by exclusively concentrating on the structure-independent terms—which contain all the physics—the divergencies, which can only occur in the structure-dependent terms, have been eliminated. It must be emphasized that the elimination of the structure-dependent terms also requires a splitting of the field into two parts corresponding to the external field and the field and effect of the charge distribution itself. In spite of its intuitive character, the splitting of $\mathbf{E} = \mathbf{E}_{\text{self}} + \mathbf{E}_{\text{ext}}$ is actually not such a simple process; the precise definitions require care but it is necessary to do this to implement the elimination of the structure. Thus, the next point was

(5) The definition and elimination of the self-field.

Points (1)–(5) give a broad outline of Kramers' program. He carried it out over a period of some 20 years. At different stages, he made different simplifying assumptions but he worked throughout in a nonrelativistic framework. Furthermore, he always introduced a cutoff, a, so that only waves of wavelength $\lambda > a$ were taken into account. This restriction was actually necessary. To have a truly structure-independent theory means that it must be impossible to probe the interior of an electron of radius a. Consequently, only wavelengths $\lambda > a$ are allowed to enter the theory. Kramers felt that it was this limitation that gave his scheme consistency and coherence. The logical structure of Kramers' procedure can be summarized in this scheme:

Classical Lorentz model for an extended charge (a)
↓
Hamiltonian
↓
Eliminate the structure-
dependent terms ($\lambda > a$)
↓
Structure-independent
Hamiltonian ($a \to 0$?)
↓
Canonical quantization
↓
Applications

Although the actual implementation of this sequence of steps is technically very difficult, Kramers was very pleased with the logical coherence of the scheme and he was totally convinced that this was the correct way to proceed. A glance at this scheme shows that it incorporated exactly all those features (enumerated earlier) which Kramers felt were fundamental. The idea of the approximate elimination of the structure allowed Kramers to start with a classical model of extended charges, as Lorentz would have liked, and still end up with a quantum theory in which the internal structure of the electron did not play a role and could not be probed, in exact harmony with Bohr's admonitions. To Kramers, "the elimination of the structure" almost became a slogan; it was the key concept that allowed a systematic transition from classical electrodynamics to a true quantum theory of radiation. Kramers was encouraged that a systematic elimination of the structural features would indeed be possible by an old result of Lorentz. This result, already mentioned in Equations (9) and (10), shows that the effect of the internal structure is to change the Lorentz equations by adding a radiation term and changing the mechanical mass m_0 to

$$m_{\text{exp}} = m_0 + f\frac{e^2}{ac} \tag{11}$$

Kramers hoped that the general effect would be a replacement from m_0 to m_{exp} by an additive term $m_{\varepsilon,M}$, so that

$$m_{\text{exp}} = m_0 + m_{\varepsilon,M} \tag{12}$$

Furthermore, by insisting that the elimination of the structure could be accomplished by a canonical transformation, Kramers could guarantee that the structure-independent system would again be a Hamiltonian system. It would then be possible to apply the canonical quantization rules to this system, which would undoubtedly give a quantum theory satisfying the correspondence principle. Thus, it is not surprising that Kramers thought that a canonical structure elimination was an indispensable preliminary for the construction of a quantum theory of interacting charged particles. Difficult and complicated as this program was, Kramers pursued it with great skill and unusual tenacity. Kramers' program proceeded along very difficult lines from the commonly used (if not accepted) Dirac theory. Consequently, in the pursuit of his program, Kramers rapidly became isolated; he became a somewhat lonely figure, pursuing unusual and out of the way lines of research, without the prospect of immediate success. But Kramers was sure that his method was sound. Instead of starting with the wrong Hamiltonian as he insisted Dirac did, quantizing it and making amends later on,[65] he felt confident that he at least had started from a classical Hamiltonian of impeccable pedigree. That the structure elimination was tricky was perhaps unfortunate, but not really a serious objection. In fact, Kramers was rather pleased when he showed that the elimination of the structure for a free electron merely resulted in a redefinition (a renormalization) of the mass. He was especially pleased when he showed that the structure elimination for a particle in an external field does *not* allow the interaction to be described merely by the gauge replacement of $\mathbf{p}$ by $\mathbf{p} - (e/c)\mathbf{A}_{\text{free}}$. Since this, is of course, precisely what Dirac did, Kramers was overjoyed that his procedure gave results different from Dirac's.

An important question that Kramers toyed with for a long time, but did not really formalize until much later, is whether the elimination of the structure gives rise to new effects not described by the Dirac theory. Kramers knew that for a free electron the only effect of the elimination of the structure and the self-field would be a change in mass from m_0 to m_{exp}. Since this is an additive effect, one can write it as

$$m_{\text{exp}} = m_0 + W_s(m_0, 0) \tag{13}$$

where W_s is the self-energy of a free electron.†

If one considers now an electron in a potential field V instead of a free electron, there will again be a self-energy $W_s(m, V)$ and Kramers argued that the difference should lead to observable effects. He further believed that this difference $W_s(m, V) - W_s(m, 0)$ would be finite. He could have calculated this as early as 1940. But he did not calculate it until after the Shelter Island meeting in 1947. This is the final step in which he calculated the corrections to the hydrogen spectrum, which were found by Lamb.[66] The experimental discovery of the Lamb shift, a deviation from the Dirac equations, was to Kramers a vindication of his ideas. It was one of the ironies so frequent in physics and especially in Kramers' life that before Kramers' contribution was recognized, it was already superseded (see especially Section IV). Kramers had worked in solitude on radiation theory during the war years and he could present a fairly complete, but of course nonrelativistic, discussion at the Shelter Island Conference in June 1947. By that time, several physicists—Weisskopf, Bethe, and Schwinger—were interested in these problems; the experiments by Lamb had just been performed. The reaction to Kramers' discussion was quite mixed: Some paid no attention to it at all,[67] but a number of others immediately seized upon Kramers' ideas and initiated what was to become the explosive development of the renormalization program in quantum field theory. In this further development, Kramers did not participate, and the importance of his ideas was not publicly recognized until much later (see Section IV). Kramers never again dealt with problems of comparable depth and profundity.

II. The Struggle Toward a Consistent Theory‡

A. The Construction of the "First" Kramers' Hamiltonian

Kramers was probably as aware as anyone that the final chapter of a lengthy textbook is not an ideal place to publish new research. However, Kramers also felt that nothing short of a systematic, complete rethinking and reanalysis of the foundations of the quantum theory of radiation were necessary before a theory meeting all his demands could be constructed (see Section I.C). It therefore made sense to present his own considerations in the context of an organized, systematic framework: This his textbook provided. Indeed, the

† For a more precise definition see Section II and especially Section III.

‡ This section is pretty formal.

chapter on electromagnetic radiation starts out with a clear, perhaps somewhat belabored, but quite traditional exposition of the quantum theory of the free radiation field. It is only after some 50 pages that Kramers presents his own material starting with paragraph 89 "The Equations of the Classical Electron Theory as Canonical System." As might be anticipated, Kramers emphasizes the construction of a classical Hamiltonian; he devotes 12 densely written pages to this construction, but only four pages to a much more casual discussion of the quantization.

Kramers describes the electromagnetic field in terms of the complex vector $\mathbf{F} = \mathbf{E} + i\mathbf{B}$. The system (electrons and radiation) is enclosed in a box of size L with the periodic boundary conditions

$$\mathbf{F}(\tfrac{1}{2}L, y, z) = \mathbf{F}(-\tfrac{1}{2}L, y, z), \quad \text{same for } y \text{ and } z \tag{14}$$

This periodicity requirement gives for the wavevector $\mathbf{K}$ the values

$$\mathbf{K} = \mathbf{n}\left(\frac{2\pi}{L}\right) \tag{15}$$

The vector $\mathbf{n}$ has integer components n_x, n_y, n_z. (Kramers uses $\boldsymbol{\sigma}$ instead of $\mathbf{K}$; $\mathbf{K} = 2\pi\boldsymbol{\sigma}$. He uses λ for n_x, n_y, n_z and occasionally uses the frequency $\nu_n = c\sigma_n$, where the angular frequency $\omega_n = 2\pi\nu_n$ will be used more often.) Any vector field $\mathbf{F}$ satisfying the periodicity conditions can always be split into longitudinal and transversal parts. Kramers uses a somewhat unusual decomposition. He first introduces for each n a complete null† vector $\mathbf{U}_n$ specified by

$$(\mathbf{U}_n \cdot \mathbf{U}_n) = 0 \qquad (\mathbf{U}_n \cdot \mathbf{K}_n) = 0 \qquad (\mathbf{U}_n^* \cdot \mathbf{U}_n) = \frac{K^2}{4\pi^2} \tag{16}$$

Kramers now splits $\mathbf{F}$ into three pieces, two of which are transversal and one of which is longitudinal:

$$\mathbf{F} = \mathbf{F}_{\mathrm{I}} + \mathbf{F}_{\mathrm{II}} + \mathbf{F}_{\mathrm{III}} \tag{17}$$

$$\operatorname{div} \mathbf{F}_{\mathrm{I}} = \operatorname{div} \mathbf{F}_{\mathrm{II}} = 0 \tag{17a}$$

$$\operatorname{curl} \mathbf{F}_{\mathrm{III}} = 0 \tag{17b}$$

The quantities of special physical importance are the Fourier coefficients of $\mathbf{F}_{\mathrm{I}}$, $\mathbf{F}_{\mathrm{II}}$, and $\mathbf{F}_{\mathrm{III}}$:

$$\mathbf{F}_{\mathrm{I}} = \frac{1}{L^{3/2}} \sum_n a_n \mathbf{U}_n e^{i(\mathbf{K}_n \cdot \mathbf{x})} \tag{18a}$$

$$\mathbf{F}_{\mathrm{II}} = \frac{1}{L^{3/2}} \sum_n b_n \mathbf{U}_n e^{-i(\mathbf{K}_n \cdot \mathbf{x})} \tag{18b}$$

$$\mathbf{F}_{\mathrm{III}} = \frac{1}{L^{3/2}} \sum_n c_n \mathbf{K}_n e^{i\mathbf{K}_n \cdot \mathbf{x}} \tag{18c}$$

† Kramers was fond of the complex combination $\mathbf{F} = \mathbf{E} + i\mathbf{B}$, but he especially enjoyed the use of complex null vectors. He used them extensively in his derivation of the Dirac equation. In classes and seminars, he often showed how derivations were simplified by their use. But only he could use them effectively![68]

The coefficients a, b, and c are, in general, functions of the time. The properties of the field $\mathbf{F}$ are determined by the Fourier coefficients a, b, and c. (There is a factor 2π difference between the c_n used here and the c_n used by Kramers. It is not important for the following and should not distress anyone. c_n used here is Kramers' c_n divided by 2π. But it seems silly to write it as $¢_n$!) Making a specific rule about the orientations (if $\mathbf{U}_n = \mathbf{e}_{1,n} + i\mathbf{e}_{2,n}$, with $\mathbf{e}_1$ and $\mathbf{e}_2$ real, then $\mathbf{e}_1$, $\mathbf{e}_2$, and $\mathbf{K}$ are oriented as the coordinate axis x, y, z) allows the identification of $\mathbf{F}_{\text{I}}$ as a superposition of *right* circularly polarized waves. Similarly, $\mathbf{F}_{\text{II}}$ consists of *left* circularly polarized waves. This decomposition of $\mathbf{F}$ as given by (17) and (18) is a general decomposition, valid for all vector fields. Kramers always used this particular form; others are of course possible and were known, but Kramers used this decomposition exclusively.

If the field $\mathbf{F}$ satisfies particular field equations, this, via the decompositions (17) and (18), leads to a set of coupled equations for the coefficients a, b, and c. In the simple case of the pure radiation field, $\mathbf{F}$ satisfies

$$\operatorname{curl} \mathbf{F} = \frac{i}{c}\frac{\partial \mathbf{F}}{\partial t} \tag{19a}$$

$$\operatorname{div} \mathbf{F} = 0 \tag{19b}$$

Since the total field is transversal, it is only necessary to use $\mathbf{F}_{\text{I}}$ and $\mathbf{F}_{\text{II}}$ of (18a) and (18b) in the expansions. The energy density of the field is, as always, given by

$$\frac{1}{8\pi}(\mathbf{E}^2 + \mathbf{B}^2) = \frac{1}{8\pi}\mathbf{F}^* \cdot \mathbf{F} \tag{20}$$

The total energy can now be expressed very simply in terms of the Fourier coefficients a and b as†

$$H = \int \mathbf{F}^* \cdot \mathbf{F} = \sum_n \mathbf{K}_n^2 (a_n^* a_n + b_n^* b_n) \tag{21}$$

The equations of motion of the a and b coefficients are just harmonic oscillator equations, which can be immediately solved:

$$a_n(t) = a_n(0)e^{-i\omega_n t} \qquad b_n(t) = b_n(0)e^{+i\omega_n t} \tag{22a}$$

$$\omega_n \equiv cK_n \tag{22b}$$

This shows the familiar equivalence between a pure radiation field and an infinite set of noninteracting harmonic oscillators. All this was well known; Dirac had emphasized this equivalence many times, but Kramers rederived it, using his own method and notation. He especially stressed that Equation (21) showed that the pure radiation field could be considered a classical canonical system. The next step (natural but nontrivial) was to obtain a canonical description of the coupled system of radiation and charges. It was

† From now on, factors 2π, $\frac{1}{2}$, and so on are omitted.

there that Kramers began to deviate from the established pattern. For that system one can still employ the same development—Equations (17) and (18). However, the field now satisfies the Maxwell equations with matter terms instead in Equation (19):

$$\operatorname{curl} \mathbf{F} - \frac{i}{c}\frac{\partial \mathbf{F}}{\partial t} = 4\pi i \frac{\mathbf{j}}{c} \tag{23a}$$

$$\operatorname{div} \mathbf{F} = 4\pi\rho \tag{23b}$$

where ρ is the charge density and $\mathbf{j}$ is the current density vector.

As in the case of the pure radiation field, the field $\mathbf{F}$ now satisfying Equations (23a) and (23b) is decomposed into two transversal and one longitudinal fields. But the decomposition is slightly different from before:

$$\mathbf{F} = \mathbf{F}_{\mathrm{I}} + \mathbf{F}_{\mathrm{II}} + \mathbf{F}_{\mathrm{III}} \tag{24}$$

$$\operatorname{div} \mathbf{F}_{\mathrm{I}} = \operatorname{div} \mathbf{F}_{\mathrm{II}} = 0 \tag{24a}$$

$$\operatorname{curl} \mathbf{F}_{\mathrm{III}} = 0 \qquad \operatorname{div} \mathbf{F}_{\mathrm{III}} = 4\pi\rho \tag{24b}$$

An exactly analogous decomposition is made of the current $\mathbf{j}$.

$$\mathbf{j} = \mathbf{j}_{\mathrm{I}} + \mathbf{j}_{\mathrm{II}} + \mathbf{j}_{\mathrm{III}} \tag{25}$$

$$\operatorname{div} \mathbf{j}_{\mathrm{I}} = \operatorname{div} \mathbf{j}_{\mathrm{II}} = 0 \tag{25a}$$

$$\operatorname{curl} \mathbf{j}_{\mathrm{III}} = 0 \tag{25b}$$

The current, like the field, is split into transversal and longitudinal parts. It must be stressed, and Kramers does this, that the splittings (24) and (25), although mathematically well defined and computationally useful, are *not* Lorentz invariant (at all). Two such splittings in distinct Lorentz frames bear no simple relation to each other. By using this procedure throughout, Kramers abandons the possibility of employing these same methods in a relativistic situation. Following the pattern outlined before, $\mathbf{F}_{\mathrm{I}}$, $\mathbf{F}_{\mathrm{II}}$, and $\mathbf{F}_{\mathrm{III}}$ are developed in Fourier series as Equation (18). This, via the Maxwell equations (23a) and (23b), leads to a set of equations of the type:

$$\dot{a}_n = -i\omega_n a_n - \frac{C}{K_n^2}\int (\mathbf{j}\cdot\mathbf{U}_n^*)e^{-i\mathbf{K}_n\cdot\mathbf{x}} \tag{26}$$

C is a constant; it contains factors π and $L^{3/2}$. There are similar equations for b. The longitudinal field, as Equations (24b) clearly show, is completely determined by the charge distribution ρ. The crucial decision to be made at this point is whether the charge density ρ and the current density $\mathbf{j}$ are prescribed given functions of t, or whether they in turn depend and respond to the radiation field. Consider first the case where $\mathbf{j}$ and ρ are given fixed functions. Physically, this corresponds to the calculation of the electromagnetic fields $\mathbf{E}$ and $\mathbf{B}$ from a given charge and current distribution. Stated differently, in the situation now considered, the current $\mathbf{j}$ and the charge density ρ do not participate in the dynamics. They are given and fixed once and for all. (They could depend on the time, but again in a preassigned manner.) In that case, the equations for a and b can be solved explicitly. The solution, which can be obtained directly from Equation (26), is

$$a_n(t) = e^{-i\omega_0 t}\left(a_n(0) - \frac{C}{K_n^2}\int_0^t [\mathbf{j}(t')\mathbf{U}_n^*]e^{-i(\mathbf{K}_n\cdot\mathbf{x}-\omega_n t')}\,dt'\right) \tag{27}$$

If a_n is known at time $t = 0$, and if $\mathbf{j}$ is known as a *function* of t in the interval 0–t, $a_n(t)$ follows from Equation (27). If $|a_n(t)|$ increases with t, the source emits right polarized light of wavenumber $\mathbf{K}_n$. Kramers notes that the field $(\mathbf{F}_{\mathrm{II}} + \mathbf{F}_{\mathrm{I}})$ obtained by using the $a_n(t)$ in Equation (27) [and $b_n(t)$ from a similar equation] is exactly the radiation field obtained using the usual retarded or advanced potentials for the given charge and current distributions. Thus, the equations for the Fourier coefficients a (and also b) have the same physical content as the Maxwell equations, as indeed they must. The field $\mathbf{F}_{\mathrm{III}}$, and consequently its Fourier coefficient c_n, plays a different role. It follows immediately from Equations (24b) that $\mathbf{F}_{\mathrm{III}}$ is determined uniquely by the given charge distribution ρ; $\mathbf{F}_{\mathrm{III}}$ is the electrostatic field produced by that charge distribution. As such, there is at this stage no equation of motion for the coefficients c_n. Kramers next writes the equations of motion for the a_n and b_n in Hamiltonian form. More accurately, by requiring that the canonical variable pairs in the theory with given $\mathbf{j}$ and ρ be the same as those in the previous theory where $\mathbf{j}$ and ρ were zero, he could by a rather lengthy calculation derive an explicit expression for the Hamiltonian of that system:

$$H = \sum_n K_n^2(a_n^* a_n + b_n^* b_n) + \int F_{\mathrm{III}}^* F_{\mathrm{III}} + H' \tag{28a}$$

$$H' = -\frac{1}{c}\int d^3x(\mathbf{j}\cdot\mathbf{A}') \tag{28b}$$

In Equation (28a),† the first term is the energy of the transversal waves and the second term is the Coulomb energy of the charge distribution ρ, while H' is the interaction energy between the current and the radiation field. $\mathbf{A}'$ is known and depends exclusively on the transversal field. The derivation shows that it is given by

$$\mathbf{A}' = +\mathrm{Im}\left[\sum_n \frac{1}{K_n}(a_n\mathbf{U}_n e^{i\mathbf{K}_n\cdot\mathbf{x}} - b_n\mathbf{U}_n e^{-i\mathbf{K}_n\cdot\mathbf{x}})\right] \tag{29}$$

It is clear that $\mathbf{A}'$ as the imaginary part of a complex expression is real. It is also straightforward to check that

$$\operatorname{div}\mathbf{A}' = 0 \qquad \operatorname{curl}\mathbf{A}' = \mathbf{B} \tag{30}$$

Consequently, $\mathbf{A}'$ is just the divergence-free vector potential. Kramers was especially pleased with this derivation of the Hamiltonian (28) which he

† As usual, numerical coefficients have been omitted in Equations (28a), (28b), and (29).

obtained without an appeal to the usual replacement of $\mathbf{p}$ by $\mathbf{p} - (e/c)\mathbf{A}$. He writes: "With this we have *effortlessly* written the equation of the classical electron theory as a canonical system of equations in infinitely many variables."[69] It is perhaps questionable whether Kramers' derivations were totally effortless, but it is clear that Equations (28a) and (28b) are a Hamiltonian system for the transversal field variables with $\mathbf{j}$ and ρ as given functions.

The discussion given so far is incomplete because the current density $\mathbf{j}$ and the charge density ρ were considered as given prescribed functions. Since the essence of the Lorentz electron theory is contained in the recognition that currents are moving charged particles, which are influenced by fields, it is artificial and unrealistic to consider the particle motion as given and unaffected by the fields. Instead, it is necessary to have a canonical description of the total system consisting of both the radiation field and electrically charged particles. Only in that context can one discuss the radiation reaction and the influence of the fields on the motion of the charged particles, which in turn generate the field. Formally, for point particles, one has

$$\rho(\mathbf{x}, t) = \sum_i e_i \delta(\mathbf{x} - \mathbf{x}_i(t)) \tag{31a}$$

$$\mathbf{j}(\mathbf{x}, t) = \sum_i e_i \mathbf{v}_i \delta(\mathbf{x} - \mathbf{x}_i(t)) \tag{31b}$$

where e_i is the charge of particle i and $\mathbf{x}_i(t)$ and $\mathbf{v}_i(t)$ are the positions and velocities, respectively, of the charged particles. The derivation of the equations for the Fourier coefficients a and b from the Maxwell equations (23a) and (23b) proceeds as before [but now ρ and $\mathbf{j}$ are given by Equations (31a) and (31b)]. For the coefficients a_n, for example, one obtains instead of (26), the equation

$$\dot{a}_n = -i\omega_n a_n - \frac{C}{K_n^3} \sum_i e_i (\mathbf{v}_i \cdot \mathbf{U}_n^*) e^{-i\mathbf{K}_n \cdot \mathbf{x}_i} \tag{32}$$

But where previously in Equation (26) $\mathbf{j}$ and ρ were given functions of time, now neither $v_i(t)$ nor $\mathbf{x}_i(t)$ are known functions of time. Their time dependence must be determined from their equations of motion, which of course involve the Lorentz force on particle i. This force in turn depends on the fields $\mathbf{E}$ and $\mathbf{B}$ or $\mathbf{F}$ at x_i; these fields, via the Fourier expansion, must be expressed in terms of a and b. The complete system is therefore described by a coupled set of equations, relating a_n, b_n, and $\mathbf{x}_i(t)$. This is an extremely complicated system. In line with Kramers' program, the next task would be to write this coupled system in Hamiltonian form. Kramers evidently believed that it was impossible to obtain an exact Hamiltonian description of the combined classical system of charges and electromagnetic radiation. He states flatly: "An exact solution of this problem is impossible. This is connected with the well known difficulties which the particle–field duality introduces in the classical theory of electrons."[69] Kramers does not elaborate further on precisely what the "well known" difficulties of the "particle–field duality" actually are. Perhaps he referred to the intrinsic nonlinear coupling of the system; perhaps he wanted to stress that the existence of point particles in a classical field necessarily introduced singularities and divergencies. Perhaps he referred to the necessity of introducing nonelectrical forces for finite objects. Whatever these difficulties were, Kramers' immediate goals were more modest. He wanted to set up a nonrelativistic Hamiltonian which in an approximate manner would describe the response of the sources of the field (i.e., the moving charges) to the field itself. This means that in the total Hamiltonian there would not only be field variables, such as a_n and b_n, but also particle or matter variables, such as $\mathbf{x}_i$ and $\mathbf{v}_i$. Inspection shows that H' of Equation (28b) is a term containing both the field variables and the particle variables. The Hamiltonian of the complete system should contain, apart from the terms in H [Equation (28a)], at least another term that describes the matter: Call it H^{matter}. Then by the properties of classical canonical point mechanics, the particle velocities are given by

$$v_{xi} = \frac{\partial H^{\text{matter}}}{\partial p_{xi}} \tag{33}$$

where v_{xi} is the x component of the velocity of particle i; it is exactly the quantity that occurs in Equation (32). Because of gauge invariance, the matter Hamiltonian H^{matter} contains the momentum $\mathbf{p}_i$ always in the combination $\mathbf{p}_i - (e/c)\mathbf{A}^{\text{ext}}(\mathbf{x}_i)$. Here $\mathbf{A}^{\text{ext}}(\mathbf{x}_i)$ is the *external* vector potential evaluated at the position of particle i. In general, div $\mathbf{A}^{\text{ext}}$ need not be zero; $\mathbf{A}^{\text{ext}}$ is, or could be, different from $\mathbf{A}'$ in Equation (29b). The particular manner in which $\mathbf{p}_{xi}$ enters H^{matter} allows Equation (33) to be rewritten:

$$v_{xi} = -\frac{c}{e}\frac{\partial H^{\text{matter}}}{\partial A_{xi}^{\text{ext}}} \tag{34}$$

The vector potential in the expression $\mathbf{p}_i - (e/c)\mathbf{A}^{\text{ext}}(\mathbf{x}_i)$ must be that of the *external* field, because the gauge substitution of $\mathbf{p}_i$ by $\mathbf{p}_i - (e/c)\mathbf{A}^{\text{ext}}$ involves just the external field. This is so because $\mathbf{p}_i$ in the field-free Hamiltonian refers to the momentum of an object—a particle, for example. It is this identical object which in an *additional* field, by the principle of gauge invariance, will be described by the identical Hamiltonian, with, however, $\mathbf{p}$ replaced by $\mathbf{p} - (e/c)\mathbf{A}$. Thus, the $\mathbf{A}$ term must refer to a physical agent that is not present when the particle is by itself but that is present in the new situation. Since an electron is necessarily surrounded by its own field, the $\mathbf{A}$ term in the gauge substitution must be in addition to the field that is already present for a free electron. Consequently, the $\mathbf{A}$ term must be $\mathbf{A}^{\text{ext}}$, which excludes the vector potential of the self-field. Kramers was very insistent on this distinction; it marks his point of departure from the usual radiation theory. This raises the obvious question of just how the self-field and the external field are to be defined. Specifically, if one deals with a total field $\mathbf{F}$, or the vector potential of the total field $\mathbf{A}'$, as, for example, given by Equation (29), what part of $\mathbf{F}$ or $\mathbf{A}'$ should be attributed to the self-field and what part to the external field? Kramers resolves this problem by splitting $\mathbf{F}$ and $\mathbf{A}$ additively:

$$\mathbf{F} = \mathbf{F}^{\text{self}} + \mathbf{F}^{\text{ext}} \tag{35a}$$

$$\mathbf{A}' = \mathbf{A}'^{\text{self}} + \mathbf{A}'^{\text{ext}} \tag{35b}$$

In Equations (35a) and (35b) $\mathbf{F}$ and $\mathbf{A}'$ refer to the total field and are presumed known. Kramers chose to make the splitting meaningful by giving an independent definition of the self-fields. The self-field and the charged particle form an indissoluble unit; when the particle moves, the self-field (possibly distorted) moves along. Indeed, Kramers refers to the self-field as that part of the field "dragged" along by the particle. The precise definition of the self-field of particle i is that part of the field which in a reference system in which particle i is at rest is a pure Coulomb (electrostatic) field. This is certainly a definition that agrees with the intuitive idea of a self-field. An equivalent, and perhaps physically more immediate, definition is that the self-fields $\mathbf{E}$ and $\mathbf{B}$ at a general point $\mathbf{x}$ at time t due to a particle i at $\mathbf{x}_i$, moving with a velocity $\mathbf{v}_i$, are just those fields produced by particle i, if it *were* at $\mathbf{x}_i$ and moving uniformly with constant velocity $\mathbf{v}_i$. The Kramers description of electromagnetic fields now involves four transversal fields—$\mathbf{F}_{\text{I}}^{\text{ext}}$, $\mathbf{F}_{\text{I}}^{\text{self}}$, $\mathbf{F}_{\text{II}}^{\text{self}}$, and $\mathbf{F}_{\text{II}}^{\text{ext}}$—and one longitudinal field—$\mathbf{F}_{\text{III}}$. Each one of these fields can be developed in a Fourier series. The development coefficients of the self-fields are called a_n'' and b_n''; those of the external fields are called a_n' and b_n'. One has for the total transversal self-field and the total transversal external field, respectively,

$$\mathbf{F}_{\text{I}}^{\text{self}} + \mathbf{F}_{\text{II}}^{\text{self}} = \text{constant} \times \left(\sum_n a_n''\mathbf{U}_n e^{i\mathbf{K}_n\cdot\mathbf{x}} + \sum_n b_n''\mathbf{U}_n e^{-i\mathbf{K}_n\cdot\mathbf{x}}\right) \tag{36a}$$

$$\mathbf{F}_{\text{I}}^{\text{ext}} + \mathbf{F}_{\text{II}}^{\text{ext}} = \text{constant} \times \left(\sum_n a_n'\mathbf{U}_n e^{i\mathbf{K}_n\cdot\mathbf{x}} + \sum_n b_n'\mathbf{U}_n e^{-i\mathbf{K}_n\cdot\mathbf{x}}\right) \tag{36b}$$

A similar decomposition in terms of a', b' and a'', b'' can be made for the vector potential $\mathbf{A}'$. Since the matter Hamiltonian depends only on $\mathbf{A}^{\text{ext}}$, it can depend only on the external coefficients a_n' and b_n'. The distinction between the total field, the self-field, and the external field, in the language of the Fourier coefficients, is that between (a_n, b_n), (a_n', b_n'), and (a_n'', b_n''). It was pointed out in one of Kramers' earliest papers[70] that in the usual radiation theory this distinction is not maintained consistently. Equation (32) is an equation of motion for the Fourier components of the total field. One can use Equation (34) for the particle velocity in that equation. Equation (34) involves a derivative with respect to $\mathbf{A}^{\text{ext}}$. By using the expansion of $\mathbf{A}^{\text{ext}}$ in a_n' and b_n', one can rewrite this derivative with respect to a_n'. Using this relation, one can write Equation (32) for the case of one electron as

$$\dot{a}_n + i\omega_n a_n = -i\frac{C'}{K_n}\frac{\partial H^{\text{matter}}}{\partial a_n'} \tag{37}$$

where C' like C in Equation (32) is a numerical constant. Clearly, Equation (37) involves both a_n (the total field) and a_n' (the external field). In the usual treatment, a_n and a_n' are just identified. This is precisely what Kramers objected to. *If* one makes this identification, it is not difficult to show that the resulting a and b equations can be derived from a Hamiltonian:

$$H_1 = \sum K_n^2(a_n^* a_n + b_n^* b_n) + H^{\text{matter}} \tag{38}$$

The quantization of this Hamiltonian yields the usual (Dirac) theory of radiation. Kramers objected to the uncritical identification of $\mathbf{A}$ and $\mathbf{A}^{\text{ext}}$, or a_n and a_n'; consequently, he did not accept the Hamiltonian (38) as a legitimate starting point for a quantum theory of radiation. [Kramers had additional objections to the Hamiltonian (38), which will be discussed in the following paragraphs.]

Instead of just identifying $\mathbf{A}$ and $\mathbf{A}^{\text{ext}}$, Kramers argues that one must systematically analyze the coupled equations for the Fourier coefficients a_n', b_n' and a_n'', b_n''. These equations can be obtained by using expansions such as (36) in the Maxwell equations. It must be remembered that as Kramers defines it, the self-field due to particle i is completely specified. The total self-field—the sum of the individual self-fields over all particles—is therefore explicitly known. It could be calculated directly, since the electromagnetic fields of a uniformly moving charge are well known. Consequently, the Fourier coefficients a_n'' and b_n'' are also known. They could also be obtained directly from the equations which a_n'' and b_n'' satisfy. Both derivations give of course the same result; most important is that a_n'' and b_n'' are known (and this is a direct consequence of their definition).

$$a_n'' = i\frac{C''}{K_n^2}\sum_i \frac{e_i(\mathbf{v}_i\mathbf{U}_n^*)e^{-i(\mathbf{K}_n\cdot\mathbf{x}_i)}}{K_n - (1/c)(\mathbf{v}_i\cdot\mathbf{K}_n)} \tag{39a}$$

$$b_n'' = -i\frac{C''}{K_n^2}\sum_i \frac{e_i(\mathbf{v}_i\mathbf{U}_n^*)e^{+i(\mathbf{K}_n\cdot\mathbf{x}_i)}}{K_n - (1/c)(\mathbf{v}_i\cdot\mathbf{K}_n)} \tag{39b}$$

With these expressions for the Fourier coefficients a_n'' and b_n'',† the equations for the Fourier coefficients of the external fields (a_n', b_n') can now be obtained from the known equations (32) for the total field. This is a somewhat involved and lengthy calculation, which Kramers carries out with his usual virtuosity, but it does not involve any new physical ideas. The resulting equation for a_n', which is still exact, is

$$\dot{a}_n' + i\omega_n a_n' = -i\frac{C''}{K_n^2}\sum_i e_i\left[\frac{\dot{\mathbf{v}}_i\mathbf{U}_n^*}{K_n - (1/c)(\mathbf{v}_i\mathbf{K}_n)} + \frac{(\mathbf{v}_i\mathbf{U}_n^*)(\dot{\mathbf{v}}_i\mathbf{U}_n)}{K_n - (1/c)(\mathbf{K}_n\cdot\mathbf{v}_i)^2}\right]e^{-i\mathbf{K}_n\cdot\mathbf{x}_i} \tag{40}$$

There is a similar set of equations for the b_n' coefficients. Equation (40) connects the Fourier components of the external radiation field (the Fourier components of the transversal field) directly with the particle motions. The complete dynamical system contains, in addition to Equation (40) for a_n' and

† The coefficient C'' again contains numerical factors, the volume, and factors c.

the equation for b_n', the Lorentz force equations for the particle positions $\mathbf{x}_i(t)$. Of course, in those equations the fields $\mathbf{E}$ and $\mathbf{B}$ must be expressed in terms of their Fourier coefficients. Since the $\mathbf{E}$ and $\mathbf{B}$ occurring in these equations are the external fields (if the mass used is the experimental mass), only the a' and b' coefficients enter. Therefore, in this system of equations, the self-field no longer occurs explicitly, but it is not neglected either. Its disappearance is due to an "honest" elimination of the self-fields via Equation (38). Thus, Kramers considers that system (40), supplemented by the Lorentz force equations, is a sound, suitable starting point for further discussion. (In any case, better than the usual theory.) It is further satisfactory that it can be shown that the system (40) is precisely equivalent to the equations which $\mathbf{F}_{\mathrm{I}}$ and $\mathbf{F}_{\mathrm{II}}$ (the transversal part of the field) satisfy according to Equations (23) and (24). It is also interesting to observe that Equation (40) shows that the radiation field (a', b') is generated by the particle accelerations $\dot{\mathbf{v}}_i$. All these are satisfactory features, but this system of equations which Kramers needs to analyze is of gruesome complexity. In particular, it appears impossible to write the complete system in a Hamiltonian form. The occurrence of $\dot{\mathbf{v}}_i$ in the equation of motion already points in that direction. This is the same $\dot{\mathbf{v}}_i$ term that is essential in producing the radiation field. Consequently, one should not expect that these terms can be removed by some formal trick. But they appear to prohibit a straightforward canonical formulation of the coupled system.

It is possible, however, to obtain a sensible approximation, starting from the coupled equation of motion. In one of the intermediate stages of the derivation of Equation (40), an equation connecting a_n' and a_n'' occurs (obtained as always by substituting the series developments in the Maxwell equations).

$$\dot{a}_n' = -i\omega_n a_n' - i\frac{C'}{K_n}\frac{\partial H^{\text{matter}}}{\partial a_n'} - (\dot{a}_n'' + i\omega_n a_n'') \tag{41}$$

In fact, subsitution of the expression for the self-field a'' in Equation (41) leads directly (or better eventually) to Equation (40). Equation (41), which is still exact, can be written in the form

$$\frac{d}{dt}(a_n' e^{i\omega_n t}) = -i\frac{C'}{K_n}\frac{\partial H^{\text{matter}}}{\partial a_n'} - \frac{d}{dt}(a_n'' e^{i\omega_n t}) \tag{42}$$

Both Equations (41) and (42) show that the terms depending on the self-field act as forcing terms for the external radiation field a_n'. Without the a_n'' terms, the system could easily be written in Hamiltonian form. The a_n'' terms act as rapidly fluctuating forcing terms on a Hamiltonian system of a_n'. In many applications of radiation theory, one is primarily interested in phenomena that occur in times long compared to the characteristic periods of the self-field. An electron bound in an atom emitting radiation should be studied for times large compared to the classical period of revolution. Similarly, for an electron emitting radiation when passing by a nucleus, one is interested in times large compared to the time it spends near the nucleus. Consequently, one is often interested in the long-term (secular) behavior of the external field variables $a_n' e^{i\omega_n t}$. Since this long-term behavior is of primary importance, it is natural to average Equation (42) over time. The time interval over which the equation is to be averaged should be long compared to the periods during which a_n'' varies and short compared to the times during which a_n' varies. The particular splitting of terms that Kramers employed had as a consequence that a time-independent or slowly varying magnetic field was counted as part of the self-field. For that reason it is not proper to leave the average of the a_n'' term out altogether, as it would have been if a_n'' consisted exclusively of rapid fluctuations. But it is reasonable to replace the a_a'' by a time average. Once this is done, Equation (41) becomes

$$\dot{a}_n' = -i\omega_n a_n' - i\frac{C'}{K_n}\frac{\partial H^{\text{matter}}}{\partial a_n'} - i\omega_n \bar{a}_n'' \tag{43}$$

In Equation (43) $\bar{a}_n''$ is given; it is no longer part of the dynamical system. This equation can be expected to give a reasonably good description of the radiation field in interaction with charges, for long measurement times. For short times, the description will break down, because the rapid changes in a_n'' are not included in Equation (43); that is, they are averaged out. Equation (43) describes just the secular change in a_n'. Thus, for short times, the fast periodic variations of a_n' will not be correctly given by Equation (43). This in turn will cause errors in the description of the particle motion, because the particle motion is determined by $\mathbf{A}^{\text{ext}}$, which depends on a' and b'. Thus, Equation (43) will give only an approximate description of the effect of the radiation field on the particle motion. However, this effect in general will be small. In the first discussion [see Equations (28) and (32)], the influence of the fields on the particle motion was neglected altogether; this amounted to neglecting the terms $\bar{a}_n''$ in Equation (43) [see also Equation (41)]. In Kramers' approximation (43), the influence of the interactions between the radiation field and the changes is taken into account by replacing the actual self-field by its time average. This can be viewed as the first step in a systematic self-consistent approximation scheme. In his book, Kramers does not go beyond this first step. Indeed, the replacement of the self-field by its time average [the replacement of Equation (42) by (43)] or the restriction to the slow, secular changes of a_n' is the basic approximation in Kramers' theory. Since the self-field no longer plays a dynamical role in the system, it is not surprising that this system can now be written in a Hamiltonian form. It is in fact very easy to check that the Hamiltonian can be written (in terms of Fourier coefficients) as

$$\begin{aligned} H = \sum_n K_n^2(a_n' a_n'^* + b_n' b_n'^*) + \sum K_n^2(a_n' \bar{a}_n''^* + a_n'^* \bar{a}_n'') \\ + b_n'^* \bar{b}_n'' + \bar{b}_n''^* b_n') + H^{\text{matter}} \end{aligned} \tag{44}$$

The Hamiltonian in terms of the fields has a suggestively simple form:

$$H = \int d^3x(F^{*\text{ext}}F^{\text{ext}} + F^{*\text{ext}}\bar{F}^{\text{self}} + \bar{F}^{*\text{self}}F^{\text{ext}}) + H^{\text{matter}} \tag{45}$$

Both forms show that only the time averages of the self-field appear. Comparison with the Hamiltonian H_1 in Equation (38), which Kramers rejected because in its derivation no distinction was made between the external field and the total field, shows that indeed the radiation field occurring in Equation (44) (the first term) is that of the external field and not the total field. There is an additional term—an interference between the external field and the self-field. If that term can be neglected (i.e., if $\bar{F}^{\text{self}}$ can be approximated by zero), Kramers' Hamiltonian differs from the conventional Hamiltonian, merely by the replacement of the *total* field by the *external* field. The last term in Equation (45) is the matter Hamiltonian. Through the vector potential, it depends on the Fourier coefficients a_n' and b_n' of the external vector potential. Since the matter Hamiltonian depends on the external vector potential $\mathbf{A}^{\text{ext}}$, it is necessary to specify the gauge. Kramers throughout assumes a transversal vector potential $\mathbf{A}^1$, so that

$$\operatorname{div} \mathbf{A}^1 = 0 \tag{46}$$

His choice of potentials Φ' and $\mathbf{A}'$ is not totally conventional. The equations for the potentials in his gauge read:

$$\Delta\Phi' = 4\pi\rho \tag{47a}$$

$$\Delta\mathbf{A}' - \frac{1}{c^2}\frac{\partial^2 \mathbf{A}'}{\partial t^2} = \frac{4\pi\rho\mathbf{v}}{c} + \frac{1}{c}\frac{\partial}{\partial t}\nabla\Phi' \tag{47b}$$

The fields in terms of these potentials are given by

$$\mathbf{F}_{\text{I,II}} = -\frac{1}{c}\frac{\partial \mathbf{A}'}{\partial t} + i \operatorname{curl} \mathbf{A}' \tag{48a}$$

$$\mathbf{F}_{\text{III}} = -\nabla\Phi' \tag{48b}$$

This particular choice of potentials is very suitable for the splitting that Kramers made in the self-field and external field. It is, of course, not Lorentz invariant, the potentials are a little awkward to work with, and the various split fields do not satisfy very transparent equations, but for Kramers' purposes they are especially useful. This gauge choice leads to a matter Hamiltonian (or a particle Hamiltonian) which for the case of one particle assumes the form

$$H^{\text{matter}} = m_1 c^2 + e_1 \Phi'^{\text{ext}} + \frac{1}{2m_1}\left(\mathbf{p}_1 - \frac{e_1}{c}\mathbf{A}'^{\text{ext}}\right)^2 \tag{49}$$

H^{matter} as an expression for the particle energy must implicitly contain the energy of the self-field. In a purely electromagnetic theory of the mass (where the inertial mass would be zero and the experimental mass equals the electromagnetic mass), the mass itself must be expressible in terms of the self-field. In that case, H^{matter} would be precisely equal to the energy of the self-field.

The longitudinal part of $\mathbf{F}_{\mathrm{III}}^{\mathrm{self}}$ contributes primarily to the mass energy m_1c^2. (If the particle were at rest, the energy of the longitudinal self-field would be exactly m_1c^2.) The kinetic energy terms [the second term in Equation (49)] contain the transversal self-field energy which is given by

$$\sum_n K_n^2(a_n''^*a_n + b_n''^*b_n'') = \int \mathbf{F}_{\mathrm{I,II}}^{*\mathrm{self}}\mathbf{F}_{\mathrm{I,II}}^{\mathrm{self}} \tag{50}$$

Consequently, the total Hamiltonian (45), through H^{matter} as given by Equation (49), contains the total energy in Kramers' approach. It is interesting that in spite of the rather involved subdivisions that Kramers makes, he ends up with a final Hamiltonian (45) that contains all the expected sources of energy. The energy of the self-field—which upon a casual inspection of Equation (45) appears to be absent but is nevertheless present—is a part of the particle energy (just as argued). It should still be stressed that if the mass were not purely electromagnetic, so that there is a difference between the experimental mass and the electromagnetic mass, the mass occurring in the Hamiltonian must be the experimental mass, because it is the experimental mass that occurs in the equation of motion of a charged particle when the involved electromagnetic fields are the external fields. [See especially Section I.A, Equations (7)–(10)]. The Hamiltonian H^{matter} must be so constructed that its canonical equations are the particle equations of motion. This requires the mass to be the experimental mass. This was yet another objection that Kramers had against the Hamiltonian (38), which employed the total field. It is clear from equation (9) that the classical equation of motion is

$$\left(m_0 + f\frac{e^2}{ac}\right)\ddot{\mathbf{x}} = e\left(\mathbf{E}_{\mathrm{ext}} + \frac{\mathbf{v} \times \mathbf{B}_{\mathrm{ext}}}{c}\right) \tag{51}$$

The $f(e^2/ac)$ term comes directly from the self-field. The experimental mass m_{exp} is given by

$$m_{\mathrm{exp}} = m_0 + f\frac{e^2}{ac} \tag{52}$$

so that the correct equation of motion reads

$$m_{\mathrm{exp}}\ddot{\mathbf{x}} = e\left(\mathbf{E}_{\mathrm{ext}} + \frac{\mathbf{v} \times \mathbf{B}_{\mathrm{ext}}}{c}\right) \tag{53}$$

The equation of motion that would follow from the Hamiltonian (38), which contained the total fields $\mathbf{E}_{\mathrm{total}}$ and $\mathbf{B}_{\mathrm{total}}$ instead of the external fields, would be

$$m_{\mathrm{exp}}\ddot{\mathbf{x}} = e\left(\mathbf{E}_{\mathrm{total}} + \frac{\mathbf{v} \times \mathbf{B}_{\mathrm{total}}}{c}\right) \tag{53a}$$

Kramers (and evidently also Pauli[70]) objected to Equation (53a). If in the true Lorentz tradition the total fields were split again in the self-field and the external field, the contribution of the self-field would once again be a factor

$f(e^2/ac)\ddot{\mathbf{x}}$, which then should be added to the mass term in Equation (53a). This amounts to a double counting of the electromagnetic mass. Kramers' Hamiltonians (44) and (45), with the mass terms in H^{matter} defined as the experimental mass, do not suffer from this difficulty. It is still worth noting that the energy of the self-field diverges for point masses. This corresponds precisely to the divergence of the electromagnetic mass of point electrons. Thus, Kramers warns that classically the particles should be considered as possessing a finite extent. In spite of these difficulties and in spite of the somewhat tortuous and approximate method of derivation, Kramers happily adopted his Hamiltonian as the basis for a nonrelativistic treatment of the quantum theory of radiation. Whatever its shortcomings, his theory did distinguish between the self-field and the external field, it did use the experimental mass, and it was a legitimate, classical Hamiltonian theory of electrons and radiation which contained the reaction of the external field on the particle motion. This Hamiltonian was to Kramers a promising start on the way toward a consistent theory, free from the objectionable features of the conventional theory.

B. Further Elaborations of Kramers' Ideas

It was totally characteristic for Kramers to feel that with the construction of the classical Hamiltonians (44) and (45) the basic problem had been solved. He observes that if the Fourier components of the radiation field, a_n', $a_n'^*$, b_n', $b_n'^*$ are quantized, as if they were free fields, the Dirac radiation theory is obtained. Since Kramers went through some pains to stress that his Hamiltonian was approximate, the inference that Dirac's radiation theory is also approximate is clear. The approximate character of his scheme is further emphasized by the manner in which the self-field is treated. The longitudinal part of the self-field $\mathbf{F}_{\text{III}}^{\text{self}}$ is just the Coulomb field of the charges, and it is not quantized.† In the quantum transcription, Kramers replaces the external fields by their time averages, because the classical canonical equations for the combined system are valid only in that approximation. This leads to commutation rules for the total fields, which are the same as those for free fields. The resulting quantum system, which Kramers chose to describe in terms of a combined Schrödinger equation, is treated in perturbation theory. This Kramers again felt was appropriate and necessary since the construction of the Hamiltonian itself is correct only if the interaction between matter and radiation could be considered small. There was certainly no point in treating the quantum problem with more accuracy than the underlying Hamiltonian had in the first place. The upshot of this analysis was that, for the one-electron

† It is a little bit surprising that Kramers does not refer to Fermi in this connection. As early as 1932, Fermi wrote a well-known and celebrated paper[71] on quantum electrodynamics in which he showed that the longitudinal field just gives the Coulomb field.

problem, the quantum operator giving the interaction between a charged particle and a field can be written

$$H^{\text{int}} = e(\boldsymbol{\alpha} \cdot \mathbf{A}'^{\text{ext}}) \tag{54}$$

Here $\boldsymbol{\alpha}$ is the Dirac $\boldsymbol{\alpha}$ matrix; $c\boldsymbol{\alpha}$ is the Dirac velocity operator, $\mathbf{A}'$ is the transversal total vector potential operator, and $\mathbf{A}'^{\text{ext}}$ is the external part of that quantity as defined earlier. It is particularly interesting to compare and contrast this interaction term with the term obtained previously for the system of the electrons and radiation, with *given* current and charge distributions. According to Equation (28b) this term was

$$H' = -\frac{1}{c}\int d^3x(\rho\mathbf{v} \cdot \mathbf{A}') \tag{55}$$

Apart from a difference in sign (owing to the choice of sign of the electronic charge) and the fact that H' in (55) is an energy, while H^{int} in Equation (54) is an energy density, the Kramers expression (54) is the obvious quantum transcription of the classical expression (55) with the fundamental difference, so often stressed by Kramers, that the *total* transversal field $\mathbf{A}'$ in Equation (55) is replaced by the external transversal field $\mathbf{A}'^{\text{ext}}$ in Equation (54). To Kramers, this was not a minor matter; it showed a wholly different approach to radiation theory. In typical understated fashion, Kramers summarizes his result as: "The difference between two expressions is so to say 'only' that in the one equation there occurred the vector potential of the total field."[72] While Kramers stressed the difference, he was pleased that he could pinpoint the difference between his own and Dirac's formalism so precisely. There is little doubt that he was glad that in so doing he had demonstrated to his satisfaction that both the Dirac theory and his own were intrinsically approximate. He considered his own method as one of successive approximations and the Hamiltonian he had found should be considered as just the first step in that scheme. Consistent with that view was his insistence that the quantum version of his theory could only be trusted in perturbation approximation.

Although Kramers' main physical ideas—the need to introduce the experimental mass in the Hamiltonian and the necessity to distinguish between the total field and the external field—were eventually recognized as fundamental (see Section IV), Kramers' formalism was not widely appreciated and actually almost totally ignored. One important reason was undoubtedly the awkward and cumbersome splitting of the fields into external and self-fields. Not only are these splittings not Lorentz invariant, but the individual fields satisfy unfamiliar equations. Even the nomenclature is liable to cause confusion. The definition of external field was

$$\mathbf{E}_{\text{ext}} = \mathbf{E}_{\text{total}} - \mathbf{E}_{\text{self}} \tag{56}$$

where $\mathbf{E}_{\text{self}}$ at P is the field that would be generated at P by a charge at $\mathbf{x}$ moving uniformly with the instantaneous velocity at $\mathbf{x}$. This means that for a single electron under the influence of *elastic* (nonelectromagnetic) forces, there would still be a splitting of the field into an external field, defined by Equation (56), and a self-field, defined by the prescription for $\mathbf{E}_{\text{self}}$. This even though there is no *outside* electromagnetic field at all! There is clearly nothing wrong with that; it is a legitimate splitting of the field. But the individual terms do not have a simple pictorial interpretation. Furthermore, if for a system of electrons and radiation, one has a genuine external agency, say the field of an outside electromagnet, this field in Kramers' classification is counted as belonging to the self-field.[73] With so many different fields and with an unusual terminology, it is difficult to get an immediate intuitive understanding of these fields and their precise meaning. The most reliable guide is the formalism itself, but since this is approximate and will, for the case of point electrons, contain divergencies, great care is needed to interpret all the terms correctly. So it is not surprising that occasionally confusion arose. Thus, in his book, Kramers notes that the series of a'' and b'' [Equation (39), the Fourier components] diverge at the particle locations.[74] He further observes that both the total field and the self-field diverge at these same locations. However, he also comments in an almost offhand manner that their difference (by definition the external field) remains finite.[74] No use is made of this observation. A further and more detailed investigation by Kramers and Opechowski[75] showed that the actual situation is considerably more delicate. Contrary to Kramers' initial claim, the external field *does* diverge at the position of the electron. (Kramers' original remark is indeed omitted from the English edition.[76]) It is true that both $\mathbf{B}_{\text{ext}}$ and $\partial\mathbf{A}'^{\text{ext}}/\partial t$ are finite at the particle position; however, $\mathbf{E}_{\text{ext}}$, the external electric field (as defined earlier), actually diverges. It is not just proportional to $\partial\mathbf{A}'^{\text{ext}}/\partial t$, it contains an additional term proportional to the acceleration of the particle position. At the particle position, this term diverges as $1/r$; consequently, $\mathbf{E}_{\text{ext}}$ diverges at the particle position.[76] Proofs of these statements are not presented formally but are scattered through several papers. These difficulties show that Kramers' ideas at that time were more in the nature of a *program* than a complete deductive scheme. Kramers' program required analysis, clarification, and considerable formal extension before it could be applied effectively. Because of the highly personal style and somewhat unwieldy formalism, only Kramers' very close associates and immediate students took an active part in this development.

Kramers continued to work on these problems, although he published little. He continually searched for a more powerful formulation of his physical ideas, for example, he used variational methods to carry out the structure elimination by means of canonical transformation. In some unpublished notes, he explicitly introduces the electron as a finite charge distribution. Thus, his Hamiltonian contains terms such as $(e/c)(\dot{\mathbf{R}}\cdot\mathbf{A}_s)$ with

$$\mathbf{A}_s = \frac{1}{e}\int \mathbf{A}\rho\, d^3x \tag{57}$$

In Equation (57), $\mathbf{A}$ is again the transversal vector potential and ρ is the charge distribution of the electron. On many different occasions, such as invited lectures and unofficial talks, Kramers returned to the problems of radiation theory. He evidently was well aware that his own considerations needed further elaboration and, above all, clarification and simplification. But although Kramers was heavily preoccupied with these matters, he actually published nothing, at least not personally. It would not be an exaggeration to say that in the publication of these ideas, as in the work by Serpe[77] and Opechowski,[75] Kramers published his ideas by proxy. The first paper by Serpe[77a] treats the interaction of a one-dimensional harmonic oscillator with a radiation field, both according to the traditional Dirac theory and according to Kramers' version of radiation theory. If Kramers' book were used as a textbook in a graduate course, this problem would be a typical exercise left to the students. The exercise illuminates and clarifies many of Kramers' innovations in an especially simple context. Although these papers[75,77] are not part of Kramers' bibliography, the problems they touch on could not have been conceived, nor the papers written, without Kramers' pioneering work. It is especially unfortunate that these papers were published in the middle of a disastrous war, which severely hampered their dissemination.

Many of the themes that Kramers had stressed for so long in his discussions of radiation theory already show up in the simple example of an oscillator interacting with a radiation field. This problem had been treated earlier by Oppenheimer[51] and by Weisskopf[78] and Wigner using the usual radiation theory. For a zero cutoff—for point particles—there is an infinite shift of the spectral lines. The analysis of this same problem following the Kramers philosophy shows that this divergence is intimately related to the divergence of the classical electromagnetic mass. Since, according to Kramers, the correct starting Hamiltonian should contain the *experimental* mass, this difficulty and its resolution support Kramers' contention that the Dirac theory starts from the wrong classical Hamiltonian. The detailed calculation of Serpe starts from a harmonic oscillator of experimental mass m and natural frequency ω_0, oscillating in the z direction, with amplitude a_0, and interacting with its own electromagnetic field. The *total* electromagnetic is described by $\mathbf{F} = \mathbf{E}_{tr} + i\mathbf{B}$. ($\mathbf{E}_{tr}$ is the *transversal* part of the field.†) $\mathbf{F}$, as well as the transversal vector potential $\mathbf{A}$, is developed in Fourier series.

The divergence-free vector potential is called $\mathbf{A}$ here. (Since it is the only vector potential occurring in this discussion, it is called $\mathbf{A}$, whereas earlier it was called $\mathbf{A}'$.) The Hamiltonian of the system in the traditional (non-Kramers) version (as used by Weisskopf and Wigner) is

$$H = \frac{1}{2}m\omega_0^2 z^2 + \frac{p^2}{2m} - \frac{e}{mc}(p\mathbf{A}) + \sum \frac{1}{2}K_n^2(a_n^* a_n + b_n^* b_n) \qquad (58)$$

In Equation (58), m is the experimental mass, $\mathbf{A}$ is the total field, and the last term is the total field energy. Strictly speaking, Equation (58) yields an infinite

† The notation used here is *exactly* that used earlier in this chapter. Occasionally, formulas are repeated to make the treatment readable.

set of equations, corresponding to the infinite set of frequencies $\omega_n = cK_n$. A cutoff frequency is introduced, $\omega_{\max} = CK_{\max}$, where the maximum frequency is chosen as

$$\frac{c}{a_0} = \omega_{\max} \tag{59}$$

In other words, electromagnetic waves with wavelengths less than the oscillator amplitude a_0 are excluded. Bohr's dictum—that quantum field theory cannot be expected to be valid in domains where the electromagnetic field could probe the internal structure of the electron—is transcribed here as limiting the electromagnetic waves to those of wavelengths longer than a_0. In doing so, one expects the theory to yield finite results throughout. The equations of motion, which couple matter (in this case just the oscillator) and the radiation field, can easily be written down. They are special cases of Kramers' Equation (32):

$$\dot{a}_n + i\omega_n a_n = -\frac{C}{K_n^2}\dot{z}u_{n,z}^* \tag{60a}$$

$$-m\ddot{z} = m\omega_0^2 z + \frac{d}{dt}\left[C'\left(\sum_n \frac{a_n u_{n,z}}{\omega_n} - \sum_n \frac{b_n u_{n,z}}{\omega_n}\right)\right] + \text{c.c.} \tag{60b}$$

$u_{n,z}$ is the z component of the unit vector $\mathbf{u}_n$. C and C' are constants. To analyze this problem further, it is necessary to find an approximate solution for the oscillator coordinate $z(t)$. Physically, one expects that the field absorbs energy from the oscillator, so that the motion of the oscillator under the influence of the field should be damped. In addition, in the Weisskopf–Wigner treatment, the oscillator experiences a frequency shift. Thus, a reasonable ansatz for $z(t)$ in Equation (60) is

$$z \simeq a_0(\cos\omega_1 t)e^{-\gamma t} \tag{61}$$

In Equation (61), γ is real; it is necessary that $\gamma \ll \omega_1$ to maintain the oscillatory character of the motion. The frequency shift $\Delta\omega_0$ is defined by

$$\Delta\omega_0 = \omega_1 - \omega_0 \tag{62}$$

Substituting Equation (61) in (60) shows that expression (61) for z is indeed an approximate solution if

$$\gamma = \frac{4\pi}{3}\frac{e^2}{mc^3}\omega_1^2 \quad \text{or} \quad \gamma \simeq \omega_1^2 \tag{63a}$$

$$\Delta\omega_0 = -\frac{4}{3}\frac{e^2\omega_1}{mc^2 a_0} \quad \text{or} \quad \Delta\omega \simeq \frac{\omega_1}{a_0} \tag{63b}$$

Consequently,

$$\frac{\Delta\omega_0}{\gamma} = \frac{C}{a_0\omega_1} \tag{63c}$$

These equations fix the parameters γ and ω_1 in the solution for z. It is clear that the result obtained here is completely classical; it follows straightforwardly from the (traditional) classical Hamiltonian (58). This Hamiltonian itself is again approximate; the $\mathbf{A}^2$ term is omitted from Equation (58). It is therefore interesting that the results for the frequency shift are of the same order as those obtained by Weisskopf and Wigner in their quantum treatment.[78] It is particularly striking that for *point oscillators*, $a_0 = 0$, the frequency shift diverges. These same questions are now reconsidered from the Kramers standpoint. Before doing this, it is pertinent to observe that the basic equations (60a) and (60b) can be solved for the Fourier components a_n, as well as for z. The solution shows two quite distinct types of term which can be identified directly as the Fourier components of Kramers' *external field* and *self-field*, respectively. Apart from numerical constants, the splitting obtained is

$$a_n = \left(\frac{e^{i\omega_1 t - \gamma t/2} - e^{-i\omega_n t}}{\omega_n + \omega_1 + i\gamma/2} + \frac{e^{-i\omega_1 t - \gamma t/2} - e^{-i\omega_n t}}{\omega_n + \omega_1 + i\gamma/2} \right) - e^{-\gamma t/2}(e^{i\omega_1 t} - e^{-i\omega_1 t}) \quad (64)$$

The second term in Equation (64) can be shown to correspond exactly to Kramers' self-field (in the notations used previously it is a_n''). The first term, the resonance term, is therefore a_n'; it is the Fourier coefficient of the external field. From the now known Fourier coefficients a_n' and a_n'' the fields $\mathbf{A}'$ and $\mathbf{A}''$, here called $\mathbf{A}^{\text{ext}}$ and $\mathbf{A}^{\text{self}}$, can now be reconstructed. This will involve a sum over all the modes of the field. It is worth noting that Equation (64) is the simplest explicit example of the Kramers splitting into what he called the *self*-field and *external*-field.

The results so far obtained for the frequency shift are obtained from the usual (and to Kramers the *unacceptable*) Hamiltonian (58). Two separate comments will clarify Kramers' objections to the usual procedure in this example. The first observation is that Equations (60a) and (60b), which in the usual theory determine the frequency shift, allow a *formal* solution in which there is no frequency shift at all, but instead the mass of the oscillator is changed from m to $m + \Delta m$. Instead of Equation (61), if one writes

$$z \simeq a_0(\cos \omega_0 t)e^{-\gamma t} \quad (65)$$

and uses this in system (60), it is very easy to verify that Equation (65) is indeed a solution provided that the oscillator experiences a change of mass Δm, given by

$$\Delta m = -\frac{8}{3}\frac{e^2}{c^2 a_0} = -m_{\text{el}} \quad (66)$$

It is also simple to check that Δm (apart from a sign) is exactly the contribution to the electromagnetic mass of electromagnetic waves of frequencies less than ω_{max}. Simple manipulation of the formulas yields

$$\omega_1 \simeq \omega_0\left(1 + \frac{1}{2}\frac{\Delta m}{m}\right) \quad (67)$$

Consequently, the divergent frequency shift is related directly to the divergent electromagnetic mass. If there were no electromagnetic contributions to the mass, or if this were somehow already incorporated in the mass used, Δm would be zero and there would be no frequency shift. This is all very much in the spirit of Kramers' approach, where the correct identification of the mass and fields was of special importance. If, as was done in Equation (58), it were automatically assumed that m was the experimental mass and that $\mathbf{A}$ was the total field, it would be very contrived to even contemplate a solution of the type (65), which describes an object with a changed mass. In the Kramers approach, where it is recognized that the self-field contributes to the mass, such a solution is more natural, showing that even at this early stage (remember this paper was written in 1939) the idea of our present mass renormalization was very much on Kramers' mind. The second criticism refers again to an old point. With the Hamiltonian used,

$$H = \frac{1}{2m}\left(\mathbf{p} - \frac{e}{c}\mathbf{A}\right)^2 + \frac{1}{2}\sum K_n^2(a_n^* a_n + b_n^* b_n) \tag{68}$$

where m is the experimental mass and $\mathbf{A}$ is the total field, the electromagnetic mass is counted twice. Kramers had made this point numerous times before, but the simplicity of the example discussed by Serpe allows an explicit demonstration of this fact. The equations of motion of the particle described by the Hamiltonian (68) is

$$m\ddot{\mathbf{x}}_1 = -\frac{e}{c}\dot{\mathbf{A}} = -\frac{e}{c}(\dot{\mathbf{A}}^{\text{ext}} + \dot{\mathbf{A}}^{\text{self}}) \tag{69}$$

($\mathbf{x}_1$ is the particle position.) In Equation (69) the Kramers splitting of the field into external and self parts is made. In the case under consideration, $\mathbf{A}^{\text{self}}$ is known from formula (64). A nonrelativistic ($v/c \ll 1$) approximation of the equations that couple the Fourier coefficients a_n' to $\ddot{\mathbf{x}}_1$ [see Equations (43) and (41) for the general case, Equation (60) for this case] shows that *near* the particle position

$$\mathbf{A}^{\text{ext}} \simeq -\frac{2}{3}\frac{e}{c^2}\ddot{\mathbf{x}}_1 \tag{70a}$$

while the expression for $\mathbf{A}^{\text{self}}$ becomes

$$\frac{e}{c}\mathbf{A}^{\text{self}} = \dot{\mathbf{x}}\frac{8}{3}\frac{e^2}{c^3 a_0} = m_{\text{el}}\dot{\mathbf{x}}_1 \tag{70b}$$

Combining Equations (70a), (70b), and (69) shows that the equation of motion becomes

$$m\ddot{\mathbf{x}}_1 + m_{\text{el}}\ddot{\mathbf{x}}_1 = \frac{2}{3}\frac{e^2}{c^3}\dddot{\mathbf{x}}_1 \tag{71}$$

The term $\frac{2}{3}(e^2/c^3)\dddot{\mathbf{x}}_1$ is the familiar radiation reaction term; however, if m in the Hamiltonian were the total mass, which already includes the electromagnetic mass, Equation (71) shows clearly that the Hamiltonian would count the electromagnetic mass twice. Kramers' contention, which of course goes back to Lorentz, is now very obvious: The use of the total field **A** in the Hamiltonian is compatible only with the use of the mechanical mass m_0 in that Hamiltonian. If that is done, Equation (71) shows that the equation for the oscillator contains $m_0 + m_{el} = m_{exp}$, the experimental mass. Since the frequency shift is tied directly to the mass shift (67), it can be anticipated that the consistent use of the experimental mass and the external fields would not yield a divergent frequency shift. Indeed, the correct treatment of a particle in a potential U and its own field is, according to Kramers, described by†

$$H = \frac{\mathbf{p}^2}{2m} - \frac{e}{mc}(\mathbf{p}\cdot\mathbf{A}^{\text{ext}}) + U + \frac{1}{2}\sum K_n^2(a_n^* a_n + b_n^* b_n) \tag{72}$$

Here m is the experimental mass, as before, but $\mathbf{A}^{\text{ext}}$ is just the external field. $\mathbf{A}^{\text{ext}}$ is expanded as usual into Fourier coefficients a_n' and b_n'. U is the potential energy of the oscillator. Equation (72) yields again a coupled system of equations for the Fourier coefficients a_n', b_n' and the oscillator variables. The general equations (actually contained in Kramers' book) become for this special case

$$\dot{a}_n' + i\omega_n a_n' = -\frac{C}{K_n^3}(\ddot{\mathbf{x}}_1\cdot\mathbf{U}_n^*)e^{-i\mathbf{K}_n\cdot\mathbf{x}_1} \tag{73a}$$

$$m\ddot{\mathbf{x}}_1 = -\nabla U - \frac{e}{c}\dot{\mathbf{A}}^{\text{ext}} \tag{73b}$$

These are the basic equations from which the further analysis proceeds. It is from the approximate treatment of these equations that the results (70) for $\mathbf{A}^{\text{ext}}$ and $\mathbf{A}^{\text{self}}$ were actually obtained. The equation of motion for the particle (the oscillator) emerging from this analysis is very familiar:

$$m\ddot{\mathbf{x}}_1 = -\nabla U + \frac{2}{8}\frac{e^2}{c^3}\dddot{\mathbf{x}}_1 \tag{74}$$

The equation contains the experimental mass, the nonelectrical forces (∇U), and the radiation reaction. Applied to the special case of the oscillator vibrating in the z direction (74), the solution is of the form conjected earlier (65):

$$z = z_0(\cos\omega_0 t)e^{-\gamma t} \tag{75}$$

with

$$\gamma = \frac{4\pi}{3}\frac{e^2}{mc^3}\omega_0^2 \tag{75a}$$

† It will be clear that Equation (72) is actually an approximation to—as well as reinterpretation of—Equation (68): The $\mathbf{A}^2$ term in Equation (68) should give rise to a $(\mathbf{A}^{\text{ext}})^2$ term in Equation (72). Serpe omitted this term.

Thus, in the Kramers theory there is *no* frequency shift there are no divergencies. The spectral distribution $J(\omega)$ of the light emitted by this damped oscillation (the lineshape) has the usual Weisskopf–Wigner form:

$$J(\omega)\,d\omega = C'' - \frac{d\omega}{(\omega - \omega_0)^2 + \gamma^2/4} \tag{76}$$

Thus, the example of the oscillator interacting with its own electromagnetic field, worked out in detail by Serpe, illustrated several of Kramers' points. The double counting of the electromagnetic mass was exhibited explicitly; the same modes of the electromagnetic field which gave rise to the infinite level shift in the usual Dirac theory were shown to be the cause for the divergence of the electromagnetic mass. In a theory that operates with the experimental mass, this divergence should and did (or appeared to) disappear. The equations of motion, derived from the Kramers coupled equations for the Fourier coefficients and the particle coordinates, contain an explicit mass renormalization as Equation (71) shows. This same set of coupled equations, such as (73a) and (73b), shows that there are solutions where the frequency shift is zero and no divergencies remain. It therefore appears as if Serpe's explicit treatment of this simple case supports all Kramers' contentions. This must have been satisfactory to Kramers. Even so, the principal results of Serpe's paper contained in Equation (75) is perhaps a bit of an anticlimax. That Kramers' incisive and deep analysis of the description of electrons in electromagnetic fields would yield no more than a classical damped oscillator—even if not so surprising—is still faintly disappointing. The main change that Kramers' formalism yielded was of course the change from the mechanical mass to the experimental mass in the Hamiltonian. As such, it was the first instance of a *mass renormalization*, but there is no indication that at this stage Kramers recognized this as a *general* principle of great importance rather than as the result of a particular technical procedure (which was approximate at that). Serpe's paper contains just the computations without further comment; there is no hint of any kind that this procedure might possess a degree of generality. Even though this paper illustrated many of Kramers' ideas, it was not extensively read and rarely quoted. The final disappointment was that Serpe's paper was incomplete in an important respect. It is quite correct that for a harmonic oscillator the frequency shift calculated from the Kramers Hamiltonian (72) is zero and that no divergencies remain. However, Serpe never calculated the effect of the $(\mathbf{A}^{\text{ext}})^2$ term, and this term, which should be considered [see Equation (68)], leads to a logarithmic divergence (for point particles). This is an important point because it demonstrates that merely incorporating the (possibly divergent) electromagnetic mass in the Hamiltonian does not of and by itself remove all divergencies, not even for harmonic oscillators. Thus, Serpe's results were superseded by the later developments before they ever became generally known.

The discussion given so far—based on Serpe's paper[77a]—is primarily an application of Kramers' formalism as it was presented in his book. But apart from some applications, this paper did not carry the investigation much beyond the point reached there. In two later papers (Serpe[77b] and Opechowski[75]) and also in Kramers' private notes,[79] serious efforts were made to improve the canonical formulation of Kramers' ideas. That Kramers was highly interested in that feature can be inferred from a letter by Rosenfeld to Kramers,[80] in which the (approximate) canonical character of the structure elimination is discussed extensively. But again, Kramers himself published little or nothing on these topics although his subsequent publications, lectures, and private conversations show very clearly that these questions continued to prey on his mind. There can be no doubt that he directly suggested the research direction of his associates. The central question was and remained the Hamiltonian formulation of the system of equations that Kramers had derived earlier. These equations, which couple the Fourier coefficients a_n' of the external field and the particle coordinates, could not be put into Hamiltonian form and a variety of approximate schemes were suggested. [See the discussion in Section II, after Equation (40).] Both Serpe and Opechowski (and also Kramers) addressed this same question. Serpe considered the simplest case possible: a charged harmonic oscillator interacting with its own field. In the case that the velocity **v** is much less than **c**, so that only dipole radiation need be considered, it is indeed possible to obtain a sensible yet approximate Hamiltonian form for this system. The transversal vector potential **A** in a Fourier series is

$$\mathbf{A}(\mathbf{x}, t) = \frac{1}{L^{3/2}} \sum_n \mathbf{u}_n (q_n e^{i\mathbf{K}_n \cdot \mathbf{x}} + q_n^* e^{-i\mathbf{K}_n \cdot \mathbf{x}}) \tag{77}$$

[Note that this development is distinct from that made by Kramers, who developed the *fields* **F** in coefficients a; see Equation (18a).] In Equation (77) $\mathbf{u}_n$ is a unit polarization vector; the other symbols have their usual meaning. The Maxwell equations transcribe to an infinite set:

$$\ddot{q}_n + \omega_n^2 q_n = C(\dot{\mathbf{x}}_1 \cdot \mathbf{u}_n) e^{-i\mathbf{K}_n \cdot \mathbf{x}} \simeq C(\dot{\mathbf{x}}_1 \cdot \mathbf{u}_n) \tag{78}$$

$\mathbf{x}_1$ is the oscillator's position and C a constant. The last approximate equality expresses that $\mathbf{x}_1$, the oscillators amplitude, is small compared to the wavelengths considered. This is the *dipole* approximation referred to earlier. Following Kramers' procedure, the total field **A** must be decomposed into $\mathbf{A}^{\text{ext}}$ and $\mathbf{A}^{\text{self}}$ (also called $\mathbf{A}'$ and $\mathbf{A}''$). The respective Fourier coefficients of the self-field and external field are q_n'' and q_n'. The self-field is defined as before [see the discussion after Equation (35a)]. Since these fields are known (by their definitions), their Fourier components q_n'' are also known:

$$q_n'' = \frac{2\pi e c}{\omega_n^2} \frac{\dot{\mathbf{x}}_1 \cdot \mathbf{u}_n}{L^{3/2}} \tag{79}$$

The combination of Equations (79) and (78) gives an equation coupling the Fourier components of the external field and the particle coordinate $\dot{\mathbf{x}}$:

$$\ddot{q}_n' + \omega_n^2 q_n' = -\frac{2\pi ec}{\omega_n^2}\frac{\dddot{\mathbf{x}}_1 \mathbf{u}_n}{\sqrt{L^{3/2}}} \tag{80}$$

The further approximation made at this juncture is that for an oscillator, even a damped oscillator, it is legitimate to replace $\dddot{\mathbf{x}}_1$ by $\omega_0^2 \dot{\mathbf{x}}_1$. This means that the damping is considered small. This approximation only works for an oscillator. It is parenthetically the occurrence of the triple derivative in the equations of motion, which makes a rigorous canonical formulation awkward. With the assumption that $\dddot{\mathbf{x}}_1 \simeq \dot{\mathbf{x}}_1$, the resulting system can then be brought into a Hamiltonian form without trouble. The equations of motion for an oscillator, moving only in the z direction, are [compare Equation (80)]

$$\ddot{q}_n' + \omega_n^2 q_n' = \frac{2\pi ec\omega_0^2}{\omega_n^2\sqrt{L^{3/2}}}\dot{z}u_{nz} \tag{81a}$$

$$\ddot{z} + \omega_0^2 z = -\frac{e}{mc}\dot{A}_z^{\text{ext}} \tag{81b}$$

If one introduces

$$Q_n \equiv \frac{1}{2c\sqrt{\pi}}\frac{\omega_n}{\omega_0}(q_n' + q_n'^*) \tag{82a}$$

$$P_n \equiv \dot{Q}_n \tag{82b}$$

Equation (81a) is just the canonical equation of the Hamiltonian

$$H = \frac{1}{2m}\left(p - \frac{e}{c}A_z^{\text{ext}}\right)^2 + \frac{1}{2}m\omega_0^2 z^2 + \sum_n \frac{1}{2}(P_n^2 + \omega_n^2 Q_n^2) \tag{82c}$$

The interpretation of this Hamiltonian is reasonably clear: The first two terms are the kinetic and potential energies of the material oscillator, while the third term is the field energy. This field energy in turn can be shown to be the sum of the external field energy in the sense of Kramers, and a term that depends on the square of the acceleration which is just the interference energy between the external field and the self-field. This same term is already contained in formula (45) of Kramers. Thus, in the special case of the harmonic oscillator, the complicated system of equations (40) coupling the Fourier coefficients of the field to the particle coordinates can indeed be cast—at least approximately—in a Hamiltonian form. With this accomplished, the quantum theory can be obtained in the usual manner. The net transition probability obtained via the Kramers theory differs by a factor of ω_0^2/ω^2 from that calculated by Weisskopf and Wigner. It is this difference that causes the frequency shift to be zero in the Kramers approach, in contrast to that of Weisskopf and Wigner. But both theories give the same shape of the emitted spectral line [as given by Equation (76)]. Thus, for a harmonic oscillator, Kramers' program could be carried out completely. Starting from the correct classical Hamiltonian with the electromagnetic mass (i.e., the electronic structure) eliminated, a theory was obtained in which the divergence in the frequency shift had disappeared. It was unfortunately very difficult, in fact impossible, to extend this same program to more general systems. Kramers evidently spent an enormous effort on this very problem. Repeating a now familiar pattern, he actually published very little on this topic until some 6–8 years later. In a footnote in a paper published in 1941, Opechowski[75] refers to a paper by Kramers that is "to appear shortly" which contains a Hamiltonian identical with that obtained by Opechowski. The same footnote ascribes other related results to Kramers. But this paper never appeared, nor are there references to these results in Kramers' notebooks or papers. Presumably, the investigations of that period are incorporated in the more detailed treatment which did not appear until some 10 years later.[65] It is evident that Kramers was deeply preoccupied with these questions around 1940 and 1941. Since he was in daily contact with Opechowski at that time, it is reasonable to assume that Opechowski's investigations were very much in line with Kramers' own views. This is also strongly suggested by Opechowski's remark that Kramers "starting from the Lagrangian of an electron in an electromagnetic field obtained a Hamiltonian identical to Hamiltonian (29) [in the Opechowski paper] as far as first order processes are concerned."[81] It is therefore a safe inference that a good share of Kramers' thinking at that time may be gleaned from an examination of Opechowski's published work.[75]

This work is based explicitly on Kramers' formulation of radiation theory as outlined in Section II.A. In these later studies, the finite size of the electron was introduced in a more explicit manner. Let the charge distribution of the rigid electron be $\rho(\mathbf{x})$ with

$$\int \rho\, d^3x = 1 \tag{83}$$

Then the average of any quantity Q over the electron is defined as

$$\int \rho Q\, d^3x = \langle Q \rangle \tag{84}$$

The mechanical equation of the electron is written

$$m\ddot{\mathbf{x}}_0 = -\nabla_{\mathbf{x}_0} V - \frac{e}{c}\frac{\partial \langle \mathbf{A}^{\text{ext}} \rangle}{\partial t} \tag{85}$$

In Equation (85), $\mathbf{x}_0$ is the center of the charge distribution, and V is the potential energy.

The total field $\mathbf{F}$ is split again in the usual manner as a sum of $\mathbf{F}^{\text{ext}}$ and $\mathbf{F}^{\text{self}}$; each one is developed in a Fourier series. All this follows the previous pattern exactly. The equations that couple the particle motion and the Fourier coefficients become [compare Equation (32)]

$$\dot{a}_n' + i\omega_n a_n' = -\frac{C}{K_n^3}(\ddot{\mathbf{x}}_0 \mathbf{u}_n^*) g_n^* e^{-i(\mathbf{K}_n \cdot \mathbf{x}_0)} \tag{86a}$$

$$\dot{b}_n' - i\omega_n b_n' = +\frac{C}{K_n^3}(\ddot{\mathbf{x}}_0 \mathbf{u}_n^*) g_n e^{+i(\mathbf{K}_n \cdot \mathbf{x}_0)} \tag{86b}$$

C is again one of the catch-all constants. The g_n are the structure factors of the charge distribution:

$$g_n = \int d^3x\, e^{i\mathbf{K}_n\cdot\mathbf{x}}\rho(\mathbf{x}) \tag{87}$$

For a point electron, the structure factors g_n are all equal to 1. *Structure independence* therefore means that the theory remains well defined (finite) in the limit that $g_n \to 1$. The total energy according to Kramers is given by [see also Equation (44)]

$$\begin{aligned} W = {} & \tfrac{1}{2}\sum_n K_n^2(a_n'a_n'^* + b_n'b_n'^*) \\ & + \tfrac{1}{2}\sum K_n^2(a_n'a_n''^* + b_n'b_n''^*) + \text{c.c.} \\ & + \tfrac{1}{2}m\dot{\mathbf{x}}_0^2 + V(\mathbf{x}_0) \end{aligned} \tag{88}$$

The first line represents the energy of the transversal external field, the last line is the energy of the material particle (now no longer an oscillator), while the middle line is the interference energy of the external field and self-field. Equation (88) is the same as the formulae obtained earlier by Kramers, that is, Equation (45).

$$W = \int \mathbf{F}^{*\text{ext}}\mathbf{F}^{\text{ext}} + \int \mathbf{F}^{*\text{ext}}\mathbf{F}^{\text{self}} + \text{c.c.} + H^{\text{matter}} \tag{89}$$

The problem is now to write the system of equations (85) and (86) in Hamiltonian form, so that the numerical value of the Hamiltonian is given by W in Equation (88). This is the problem Opechowski attacked in his paper; it was also the problem that Kramers struggled with for many years. Eventually, he developed a procedure that led to a reasonably satisfactory, albeit still approximate, solution. Opechowski's work reported here is the first in a long series of investigations all devoted to the construction of a suitable Hamiltonian for the system (85) and (86). These investigations culminated in van Kampen's work in 1952.[82] The main new feature that allows the approximate Hamiltonian formulation of the system is the introduction of a new set of canonical variables which—and this is the novel aspect—are combinations of the field variables a', b', *and* the particle variables $\mathbf{x}_0$, $\mathbf{p}_0$. This same idea was employed earlier in a fundamental paper by Pauli and Fierz,[83] although the treatment of the Kramers school is distinctly different.

The canonical variables of the pure radiation fields (which *can* be put in Hamiltonian form) are the a_n', b_n' variables together with their complex conjugates. It is assumed throughout that in the general case these variables will remain the canonical variables. The particle canonical variables are $\mathbf{x}_0$ and the momentum $\mathbf{p}_0$:

$$\mathbf{p}_0 = m\dot{\mathbf{x}}_0 + \frac{e}{c}\langle\mathbf{A}\rangle \tag{90}$$

The basic idea is to introduce a new positional coordinate $\mathbf{x}_0'$, defined by

$$\mathbf{x}_0 = \mathbf{x}_0' + \langle\mathbf{Q}\rangle \tag{91}$$

$\langle \mathbf{Q} \rangle$ is a function of $\mathbf{x}_0$ and of all the field variables a'_n, a'^*_n, b'_n, and b'^*_n but not of $\dot{\mathbf{x}}_0$. $\mathbf{Q}$ is then to be chosen so that the Hamiltonian H defined by

$$H = \frac{1}{2} \sum K_n^2 (a'_n a'^*_n + b'_n b'^*_n) + \frac{1}{2m} \mathbf{p}_0^2 + V(\mathbf{x}'_0 - \langle \mathbf{Q} \rangle) \tag{92}$$

reproduces the canonical equations (85) and (86) with a'_n, a'^*_n, b'_n, b'^*_n, $\mathbf{x}'_0$, and $\mathbf{p}_0$ as canonical variables. The whole point, of course, is to demonstrate that such a function Q can indeed be found. It takes some calculation to show this. Some approximations are necessary to obtain an explicit form for Q. It is interesting and perhaps surprising that Q so obtained can be related directly to the Hertz vector $\mathbf{Z}$ of the external field. The transversal Hertz vector is defined by the relations

$$\mathbf{E}_{\text{ext}} = \operatorname{curl} \operatorname{curl} \mathbf{Z} \tag{93a}$$

$$\operatorname{div} \mathbf{Z} = 0 \tag{93b}$$

$$\langle \mathbf{Z} \rangle = \int \rho \mathbf{Z} \, d^3x \tag{93c}$$

With this definition, Q can be shown to be given by

$$\langle Q \rangle = \frac{e}{mc^2} \langle \mathbf{Z} \rangle \tag{94}$$

The result of this somewhat intricate analysis is that a system consisting of an electron in a *static* field V, interacting with its radiation field, can be written in a Hamiltonian form:

$$H = \frac{1}{2} \sum_n K_n^2 (a'_n a'^*_n + b'_n b'^*_n) + \frac{1}{2m} \mathbf{p}_0^2 + V\left(\mathbf{x}'_0 - \frac{e}{mc^2} \langle \mathbf{Z} \rangle\right) \tag{95}$$

Thus, the net result of the interactions of the electron with the radiation field can be described succinctly by a change in the apparent position of the electron in the field. The shift itself, given by the $(e/mc^2)\langle \mathbf{Z} \rangle$ term, depends on the field variables. It should be stressed that in spite of its intuitive appeal, the simple form of this result depends on a number of approximations.

(1) All $\mathbf{A}_{\text{ext}}^2$ terms are neglected.
(2) Only first-order terms in v/c are kept.
(3) The radiation reaction term is small.
(4) $\langle \mathbf{Z} \rangle$ changes little over distances of order e^2/mc^2.

The validity—and indeed the form—of the result obtained in this discussion is tied to the presence of a static field V in the Hamiltonian. As Equation (95) shows, in the absence of a field V, the Hamiltonian is just the sum of a radiation field and a particle kinetic energy. The analogue of Equation (86) in that case is just $\dot{a}'_n + i\omega_n a'_n = 0$. Thus, the method used is too crude to describe the interaction between a *free* electron and its own radiation field. The main interest of this investigation lies in the method suggested rather than the results obtained. It is worth noting that Pauli and Fierz[83] obtained a Hamiltonian very similar to Equation (95), but they had to assume that the electromagnetic mass m_{el} was much smaller than the mechanical mass m_0. The derivation sketched here shows that it is possible and indeed more systematic to eliminate the structure and deal with the external field throughout.

As always, once the Hamiltonian is obtained, the quantization can be carried out in the canonical time-honored fashion. It is interesting that the canonical formalism presented here, although different in spirit and method from that carried out by Serpe [see formulas (80)–(82)], gives effectively the same results. For a harmonic oscillator $V(\mathbf{x}) = \frac{1}{2}m\omega_0 x^2$; consequently, Equation (95) can be written immediately as

$$H = H_{\text{rad}} + H_{\text{mat}} + H_{\text{int}} \tag{96}$$

with the obvious identifications

$$H_{\text{rad}} = \frac{1}{2}\sum_n K_n^2(a_n' a_n'^* + b_n' b_n'^*) \tag{96a}$$

$$H_{\text{mat}} = \frac{1}{2m}\mathbf{p}_0^2 + \frac{1}{2}m\omega_0 \mathbf{x}_0'^2 \tag{96b}$$

$$H_{\text{int}}(K) = -\frac{e}{c^2}\omega_0(\mathbf{x}_0'\langle\mathbf{Z}\rangle) \tag{96c}$$

The quadratic terms $\langle\mathbf{Z}\rangle^2$, which should be included in H_{int} in Equation (96c) have been neglected. Thus, Equation (96c) is restricted to first-order (one-photon) processes only. This interaction term $H_{\text{int}}(K)$ involving the Hertz potential should, according to Kramers, replace the usual Dirac term for the interactions between an electron and a radiation field $H_{\text{int}}(\text{Dirac})$:

$$H_{\text{int}}(\text{Dirac}) = -\frac{e}{mc}(\mathbf{p}_{\text{matter}} \cdot \mathbf{A}_{\text{total}}) \tag{97}$$

In Equation (97), $\mathbf{p}$ is the usual canonical momentum and $\mathbf{A}_{\text{total}}$ is the vector potential of the total field. $\langle\mathbf{Z}\rangle$ in Equation (96c) is the Hertz potential of the external field. It is now a simple matter to compare the matrix elements of a transition $|n, 0\rangle$ to $|n-1, l_k\rangle$, for the two interaction Hamiltonians. This is a transition where, in the initial state, the oscillator is in state n and there are no photons; in the final state, the oscillator is in state $(n-1)$, and one photon of wavevector $\mathbf{K}$ is present. The calculation using Equations (97) and (96c) gives for the ratio of these matrix elements (in dipole approximation)

$$\frac{\langle n-1, l_k| H_{\text{int}}(K)|n, 0\rangle}{\langle n-1, l_k| H_{\text{int}}(D)|n, 0\rangle} = \frac{\omega_0}{\omega_k} \tag{98}$$

This ratio of the matrix elements leads to the factor ω_0^2/ω_k^2 found by Serpe for the transition probabilities. Thus, for the case of the harmonic oscillator, the two procedures outlined here—Serpe's method [Equations (80) and (82)] and Opechowski's method [Equations (91) and (95)]—give effectively the same results. These two investigations, completed by the end of 1941, constitute the totality of studies devoted to Kramers' formulation of radiation theory. Kramers himself actively pursued the further development of his ideas—but it is certainly true that his work did not create much of a stir in the scientific world. The subsequent development rather followed the direction of the Opechowski study than that of Serpe. Kramers was especially fond of the idea that the interaction of an electron with its own radiation field could be described in terms of an alteration in the position of the electron. Serpe's method appeared incapable of further generalization, while Kramers in his later investigations employed the idea of the potential at a shifted position to great advantage. This same idea was to reappear in the work of van Kampen,[82] perhaps the most definitive and certainly the most coherent formulation of Kramers' ideas. But all this was still to come in 1941. It is certain that Kramers worked at these problems during the war, but he did not publish any part of that work. In a lecture given on April 14, 1944,[84] he refers in a somewhat guarded manner to the work he and Opechowski had been doing, but he does not give any details. Such further details were not forthcoming until the Shelter Island Conference in 1947 and the Solvay Conference in 1948. The material described in this section gives a pretty complete picture of the early elaborations of Kramers' work on radiation theory.

III. The Story of Two Conferences: Shelter Island and Solvay

A. The Pre-Shelter Island Status: The April 1944 Lecture

The lecture that Kramers presented at the Symposium on Elementary Particles in Utrecht on April 14, 1944 provides an illuminating insight into Kramers' thinking about the problems of radiation theory. It is particularly fortunate that just 2 days before that lecture, on April 12, Kramers wrote a long letter to Heisenberg. In this letter,[85a] which for Kramers is surprisingly candid, he comments briefly on his own activities and state of mind; the bulk of the letter is devoted to Heisenberg's new approach to fundamental particle theory, the S matrix theory. In spite of the brevity of Kramers' personal comments, they are extremely revealing; combined with the lecture on "Fundamental Difficulties of a Theory of Particles" delivered 2 days later at the symposium, they give a vivid picture of Kramers' scientific and personal state of mind the time.[†] There is a slightly brooding, somewhat melancholy, mood to the

[†] Kramers' letter[85a] is of great significance in a number of different contexts. The scientific aspects dealing with the S matrix will be discussed on page 378. The rare comments Kramers makes about himself will be analyzed in Chapter 19 on "Kramers' Self-Image." Here, just the aspects relevant to the "recurrent theme" are mentioned.

personal comments in this letter. The letter itself was in answer to a note from Heisenberg which contained a manuscript of Heisenberg's latest paper on the *S* matrix theory. Kramers had received this manuscript sometime in January, but he did not start to look at it until about a week before he was obliged to give the talk for the symposium. Heisenberg had visited Leiden in the fall of 1943 to give a talk (the first) on *S* matrix theory. Kramers' letter was a reaction to Heisenberg's talk; from that letter it can be inferred that he had done very little physics in the preceding year (from 1942 to 1943) because he writes to Heisenberg: "Last Fall you have given my scientific life a tremendous 'shot in the arm'. Since that time I have worked intensively and hard but unfortunately not on fundamental physical problems.... everything I did was applied ... quantum chemistry, chemical physics, astrophysics."[85b] Soon afterward, still in 1944, Kramers published several notes on the quantum theory of molecular structure.[86] Since Kramers certainly considered the problems of the quantum theory of radiation as fundamental, the quote just given indicates that he did not work on those problems even after he returned to active research. This agrees with the total lack of published materials on the problems of the electron and radiation in that period. It appears that from the time of the publication of Opechowski's paper in late 1941 until the April 14, 1944 lecture, little progress was made in developing Kramers' ideas in the quantum theory of radiation. It is certain that Kramers remained vitally concerned with these questions. He considered his own approach as correct, sound, and fundamental. But progress along the lines he wished and insisted on was painfully slow and not always positive. Possibly the slightly resigned tone in the letter to Heisenberg was related to this lack of progress. In spite of these worries, Kramers in the lecture of April 14, on the difficulties of a fundamental theory of particles, mentions his approach as a viable research direction for the elimination of the current difficulties.

Kramers' lecture was beautiful, well organized, up to date, and informative. Whatever misgivings Kramers might have about physics, about fundamental particles, or perhaps about his own efforts, none of these concerns were apparent in his lecture. It was a strong, confident lecture, given by a person who understood and controlled his subject totally. From the lecture it was evident that Kramers had thought long and hard about these fundamental questions. He reported with great candor on his personal views; these were the views of a seasoned professional, who recognized both the strengths of the current theory and the seriousness of the current problems. In addition, Kramers had a remarkable sense of the context of basic questions and he understood the conceptual development of physics as few others did. In this lecture, which might almost be considered a model for all assessment or summarizing lectures, Kramers starts out by stating a general principle: the distinction between *real* (class A) and *formal* (class B) difficulties of a theory. He then proceeds to analyze the theories on the basis of this distinction.

The real difficulties (class A) are those facts or circumstances which, although related to the explained phenomena, are themselves *not* explained by the theory and might even be in conflict with the theory.

The formal difficulties (class B) are a lack of coherence of the theory, which manifests itself either as an intrinsic incompleteness of the theory or as a logical inconsistency within the theory. Kramers then classifies the difficulties of both the old classical electron theory and the—in 1944—"modern" particle quantum theory in terms of the A, B distinction. The subsequent discussion shows once again just how great an influence the Lorentz electron theory exerted on Kramers, even at this stage of his life. There are certainly electromagnetic phenomena that cannot be explained with the classical Lorentz electron theory. Examples of these real (class A) difficulties are the Balmer series and the anomalous Zeeman effect. There are also formal difficulties having to do with the nature and structure of the electron. Such questions cannot really be answered within the confines of the Lorentz electron theory itself, and Kramers classifies the resulting incompleteness as difficulties of class B. He also emphasizes an old familiar feature (old and familiar to Kramers, but *not* to his audience and certainly not to the physics community[87]) that the Lorentz formulas when used in practice require an identification

$$m_{\mathrm{exp}} = m_0 + f\frac{e^2}{ac^2} \qquad (99)$$

In Equation (99), m_{exp} is the experimental mass, m_0 is the mechanical mass, a is the radius of the electron, and f is a numerical factor that depends on the details of the charge distribution. Within the Lorentz theory, m_0 is just an undetermined parameter; with the advent of relativity and the acceptance of the Lorentz contractile electron, all possibilities to find out anything about m_0 either via theory or experiment disappeared. Thus, m_0 remains an undetermined element in the theory. It is logical, and of course possible, to just leave it at that: The Lorentz theory does not determine m_0. This is no way impairs its practical utility. But other approaches are also possible. Kramers comments on a common one: "This [the unobservability of m_0] suggests that the impossibility of finding out anything about the structure of the electron should be elevated to a principle."[88] One way to implement this principle is to modify the original Lorentz theory, so that the electrons become genuine mathematical points. That "trivially" eliminates the internal structure. Kramers mentions several such attempts, including the most recent and most elegant treatment due to Dirac,[89] who studies the limiting process $m_0 \to -\infty, a \to 0$. But Kramers clearly did not care for any of these attempts. He did not believe that such theories could lead to any tangible physical results. But most important, he felt that such procedures violated "the spirit of the original Lorentz theory." Although the "spirit" of a theory is not so easy to define, it is clear that to Kramers an electron theory, patterned after Lorentz's idea, should start by considering the electron as an extended charge distribution and the *point limit* should not even be contemplated. The problem, as Kramers saw it, was to show that in spite of this extended structure, the physically significant aspects of the theory could either approximately or asymptotically be described by a few parameters independent of the detailed structure.

Kramers next subjects the difficulties of the particle physics of 1944 to the same analysis. In so doing, many of Kramers' ideas on electrons and radiation, chronicled in this chapter, recur again. Thus, among the class A difficulties, Kramers mentions that the value of the fine structure constant e^2/hc has not been (and possibly cannot be) determined within the quantum electrodynamics. Perhaps more surprising and certainly more revealing is Kramers' assertion that the Hamiltonian of the relativistic *many-body* problem can only be determined up to order v^2/c^2. Only for one-body problems, as Kramers claims, can one trust the Dirac theory. Specifically, this means that Kramers anticipated (or expected, or hoped) that v^4/c^4 effects in the wavelengths of spectral lines of heavy atoms, or v^2/c^2 effects in their intensities, would actually be given incorrectly by the Dirac theory. The doubts and misgivings that Kramers expressed earlier about the Dirac theory (see Section I.B) evidently had not diminished over the years; because of the unquestioned successes of the one-body Dirac theory, they were now transferred to the many-body theory.

In this lecture, Kramers takes it for granted that everyone realizes that the quantum theory of particles has many class B problems, difficulties of coherence and completeness. He actually makes a much stronger statement: "Theoretically, the present day quantum theory of particles is not only incomplete and unfinished, No, the theory is even logically inconsistent."[89] The theory is inconsistent because of the occurrence of the divergencies, and Kramers makes it very clear that he considers the removal or elimination of these divergencies as the central problem, a problem of the greatest fundamental significance. Kramers actually seemed somewhat perplexed that, in spite of these divergencies, it was still possible, by invoking appropriate subtraction rules, to use the theoretical scheme for computational purposes. For example, the fluctuations of the electric field at a point, if just one photon is present, diverge, that is, $\langle 1|\mathbf{E}^2(x)|1\rangle$ diverges [$\langle 1|$ is a one-photon state and $\mathbf{E}(x)$ is the electric field at x]; the fluctuations in the vacuum state, $\langle 0|\mathbf{E}^2(x)|0\rangle$, also diverge. But their difference, $\langle 1|\mathbf{E}^2(x)|1\rangle - \langle 0|\mathbf{E}^2(x)|0\rangle$, is a finite quantity. Thus, the field fluctuations of the one-photon state compared to those of the vacuum state are finite. Kramers states that such a "quantity can be used with confidence, not in the least because it shows automatically the desired correspondence with the analogous classical situation."[90] It is remarkable that, to Kramers, a quantum quantity processing the correct classical limit was almost legitimate even if it was defined as the difference of two divergent quantum expressions. It demonstrates once again the enormous importance that Kramers attributed to the Bohr correspondence principle. Another aspect, again related to the classical limit, is that Kramers stressed that many—but not all—of the quantum divergencies were related directly to the classical divergence of the electromagnetic mass. In this connection, Kramers refers to the explicit result of Serpe, who established this for a harmonic oscillator (see Section II.B). It will be clear that, in this lecture, Kramers stated publicly many of the ideas that had been guiding his thinking for many years. Another favorite item, already hinted at in his book but emphasized here, is that the hope that a systematic application of the mathematical procedures of quantum theory would eliminate the difficulties of a classical relativistic electron theory was in vain. The classical problems do *not* disappear automatically when the quantum notions are applied. *Quantum mechanics does not cure the problems of classical electron theory.* This was a very important point to Kramers; there can be no doubt that this insight strongly influenced the direction in which he personally sought the solutions of the divergence problem.

This lecture established without doubt that to Kramers, the most basic problem in physics was the elimination of the divergencies. A particularly interesting feature of this address is that Kramers not only summarizes the various attempts made to resolve the divergence difficulties, but he also gives his personal (and quite frank) assessment of the importance of these attempts. Since judgments of incomplete theoretical proposals inevitably have a strong personal bias, the discussion provides a rare insight into Kramers' private scientific expectations of that time.

The attempts to resolve the divergence difficulties by the introduction of a space–time lattice—or a granular geometry—are dismissed by Kramers. His reasons for this cavalier disregard of this alternative are not made very clear; he merely writes: "I have the feeling that it does not exhibit the necessary promise as a starting point for a methodological investigation; it is for me so to say too mathematical, it has too few contacts with experiment."[91] The strong personal bias is evident.

Kramers feels that there were three distinct legitimate systematic attempts to eliminate the divergencies.

(1) The introduction of additional fields.
(2) A renewed analysis of classical particle theory.
(3) The Heisenberg *S* matrix theory.

Although Kramers was willing to consider all three as legitimate efforts, it is certainly not true that he considered all three as equally promising.

Introduction of Additional Fields. Several investigators (e.g., Bopp[92] and Stueckelberg[93]) introduced new fields—and corresponding new particles. The hope was that the divergencies generated by the new fields would exactly cancel those of the electrons without the need for any subtraction. It has been possible to achieve this compensation for some of the divergencies in this manner. To Kramers, the introduction of auxiliary fields—for no other reason than that they could cancel divergencies—was a little bit artificial and ad hoc. It was also not clear to him that the introduction of auxiliary fields would necessarily cancel *all* the divergencies. He was not enthusiastic about this approach, but he was willing to admit that these investigations might suggest useful ideas. The only possible connection with physics that Kramers saw in these compensation schemes was that they might give some insight into the problem of the proton–neutron mass difference. Kramers did not think that this was a promising approach.

Renewed Analysis of Classical Particle Theory. The reanalysis of classical particle theory seemed to Kramers, better motivated in physics, more systematic and altogether more desirable than any other approach. However, Kramers was most doubtful about investigations, such as those by Born and Infeld[94] and Dirac,[95] which considered the particles (the electrons) as mathematical points. He was especially impressed with the difficulties of quantizing these theories, so much so that he was most doubtful that such theories would lead to any advance in the divergence problem. Kramers wrote: "These theories [point limit] violate the spirit of the original classical theory." As already mentioned, it is always ambiguous to decide just what the "spirit" of a theory is, but evidently Kramers felt that the limit $a \to 0$, for a classical electron, is not a natural process to contemplate and no physically or formally simple structure can be expected to emerge from such a limit. Whatever the logical justification of these views are, with these comments Kramers effectively discarded a whole class of theories. Since Kramers insisted on a precise parallelism of the quantum and classical treatment, that left as the only alternative his own theory expounded earlier in Section II. Many of the points which were so important in that discussion are repeated in this lecture: the difference between the total and external fields and the lack of a full correspondence between the Dirac radiation theory and the classical theory of electrons. As a special example, which turned out to be important for the future,† Kramers stresses that the prevalent (Dirac) quantum theory was unable to formulate a stationary state in which an electron (bound or free) scatters strictly monochromatic light. This was a very simple problem in the classical theory of electrons. To Kramers, such a striking difference was objectionable, paradoxical, and had to be wrong; a correctly formulated quantum theory of radiation could not possibly be so different from the corresponding classical theory. While Kramers' criticism of the Dirac radiation theory is forceful, he is rather guarded about the claims he makes for his own results. It should be noted that in this public talk, Kramers attributes this work to Opechowski and himself. It was stressed earlier (in Section II.B) that Opechowski's (and also Serpe's) work would have been altogether impossible without Kramers' continual help and support. In commenting on these investigations, Kramers describes them as a "modest attempt." It is clear that he considers the basis as totally sound, but the implementation of his ideas was still incomplete, not to say inadequate. In this lecture, Kramers stresses the crucial importance of relativistic invariance for the first time. He emphasized strongly that further investigations were necessary to see whether it would be possible to formulate a satisfactory relativistically invariant theory incorporating Kramers' physical ideas. This insistence on a relativistic theory is a little strange, because all Kramers' detailed studies are explicitly based on noninvariant models. The extensions of Kramers' work by students and

† This problem was brilliantly solved by van Kampen[82] in his thesis (in 1952) (see Section III.D), on the basis of Kramers' formulation of radiation theory.

collaborators always remained nonrelativistic. In fact, in all the studies of Kramers' radiation theory, the relativistic requirements are either ignored or considered of secondary importance. Because Kramers noted that the subtraction procedures can destroy relativistic invariance, it made him more conscious of the need for relativistic considerations in his own theory. In spite of his own admonition that it was important to carry out his program relativistically, he never did it.

The Heisenberg S Matrix. As a third possible viable method to eliminate the divergencies, Kramers mentions Heisenberg's S matrix (then called the η matrix).[97] This matrix was to replace the Hamiltonian in describing the dynamics of a system, although just how this was to be done was not clear at the time. Relativistic and correspondence considerations surely would play a role in this determination. Kramers was pleased that correspondence arguments would play such an important role and he was positively enthralled when he discovered that the stationary states of a system were related to the poles and zeros of the S matrix, considered as a function of the complex energy. Kramers was also impressed that Heisenberg could construct a relativistic two-particle model, whose stationary state was determined and with no divergencies occurring. But it is hard to escape the impression that Kramers took this approach especially seriously, because it was Heisenberg who suggested it. Kramers always maintained an inordinate respect for Heisenberg's originality and insight. (Actually, Wheeler had suggested this idea of the scattering matrix some 5 years earlier,[98] but few people, including Kramers, paid much attention to it.) In any case, Kramers was sufficiently taken with Heisenberg's idea to list it as one of the promising new directions. He was, of course, aware of its extremely preliminary character and he expressed doubts about the S matrix description of photons and the electromagnetic field, but his overall assessment of the S matrix program was positive. It is not atypical that Kramers, in spite of his early knowledge of the S matrix ideas, his initial enthusiasm, and his brilliant suggestion, considered the S matrix as a function of complex energy and never studied the topic in a systematic detailed manner. Instead, in his fundamental studies he concentrated exclusively on refinements and improvements of his own radiation theory.

Two conclusions emerge with great clarity from Kramers' lecture. He considered the divergence problem of quantum electrodynamics as *the* truly fundamental problem of physics. With the possible exception of nuclear physics and mesons, everything else, he felt, was computational, numerical, and certainly important, but *not* fundamental. He considered his personal approach as the only systematic, viable method to arrive at an acceptable quantum theory of radiation. Elaborations of the Dirac theory were clearly—so he felt—on the wrong track. Even though Kramers was well aware that progress along the lines he proposed had been painfully slow, and even though he might occasionally be discouraged and depressed, he was totally convinced that the path he had taken was the right one and probably the only correct one. These were the convictions that shaped his subsequent activities in dealing with his scientific obsession.

B. The Shelter Island Conference: Kramers' Presentation

Kramers presented his personal approach to radiation theory on April 14, 1944, in the middle of World War II, to a group of Dutch physicists in Utrecht. The next public presentation of his favorite topic was several years later at the Shelter Island Conference, held June 2–4, 1947, at the Ram's Head Inn on Shelter Island—more than 100 miles from New York City—literally in the middle of nowhere. This time, Kramers' audience consisted of some 20–30 internationally known theoretical physicists, especially selected for this conference. Although World War II had been over for some time, travel, especially from Europe, was not all that easy and it is not immediately obvious why these eminent theorists were all assembled on the eastern tip of Long Island, nor how Kramers happened to be there. Since this conference was of great importance for the development of postwar physics and of special significance for Kramers personally, it is of interest to inquire just how all this came about.

The war years in The Netherlands after 1944 were especially difficult. The last winter of the war—the so-called "hunger winter"—caused incredible hardships. Kramers and his family suffered badly and Kramers' health, never too robust, deteriorated sharply. In addition, Kramers tried his best against extraordinary odds to help others (students who had gone underground) with whatever food he could scrounge. That under these impossible circumstances, Kramers maintained his energy and interest to do physics research, is nothing short of miraculous. In this period, Kramers concentrated on the exact solutions of the two- (and three-) dimensional Ising model;[99] the problems of quantum electrodynamics were left alone. After the war was finally over on May 5, 1945, Kramers, as the leading physicist in The Netherlands, became immediately involved in the rebuilding and reorganization of university life in The Netherlands. Because of his international reputation and personal acquaintance with most of the European physicists, he also assumed a major role in the redevelopment of European physics. He did this out of a sense of obligation and duty, certainly not because he enjoyed these activities. It was therefore not particularly surprising that in the spring of 1946 Kramers was asked to be the delegate of The Netherlands to the Scientific and Technological Committee of the United States Atomic Energy Committee. After some vacillation and considerable misgivings, Kramers accepted this assignment. The government was "properly" appreciative, but when Kramers requested a leave of absence with pay—the Ministry of Education expressed its willingness to pay Kramers with the stipulation that Kramers should pay his replacement at the University of Leiden:[100] a remarkable example of a truly generous government! Kramers reluctantly accepted these conditions, but as a consequence, he was in continual financial trouble during his sojourn in the

United States. To supplement his income, Kramers taught at Columbia University in the fall of 1946 and again in the spring of 1947.[101,102] Kramers was elected to the chairmanship of the committee, and this imposed additional demands on him. After a truly backbreaking effort, the committee finished its work by January 1947. Their report was submitted on February 8, 1947. The committee's activities, the interminable debates, and the meetings were grueling and Kramers was exhausted physically and mentally. Apart from his teaching obligations at Columbia, Kramers did not do anything else, certainly no physics. He therefore was thrilled when he got an opportunity to spend the second semester (spring 1947) at the Institute for Advanced Study in Princeton. In so doing, Kramers could return to physics, and it would still be easy for him to meet his teaching obligations in New York. On January 14, 1947, Kramers asked for an extension of his leave from Leiden, which was eventually granted on the same magnanimous terms as the original leave.[103] On June 31, 1947, Kramers received a curt, perfunctory note[104] from the government thanking him for his services: so ended Kramers' sojourn in the service of international political understanding. To Kramers—if not to the world—the best result of these exacting activities was that he could spend some time in the stimulating scientific atmosphere of the Institute for Advanced Study in Princeton. He had many friends and acquaintances there, the visit to the institute enabled him to renew contacts with physicists that had been interrupted during the war. Equally important, the visit gave him time to reflect on the problems of physics in an atmosphere of peace and quiet. During his stay at the institute, Kramers returned to an intensive study of the problems of the quantum theory of radiation, problems he had left untouched since 1944. This is how Kramers happened to be in Princeton in the spring of 1947; he extended his "tour of duty" with the United Nations, to recover from its exhaustive demands in the peaceful atmosphere of the Institute for Advanced Study. While there, he returned to the quantum theory of radiation. It was through Kramers' presence in Princeton that he was eventually invited to the Shelter Island Conference; it was through his investigations on the problem of electrons and radiation that he was asked to be one of the speakers. This series of coincidences explains Kramers' presence and role at the Shelter Island Conference.

As early as January 1946, K. K. Darrow, the perennial secretary of The American Physical Society, wrote to W. Pauli (who was in Princeton at the time) to ask his help in the organization of a conference patterned after the earlier Solvay Conferences. Within the physics community right after World War II, there was a strong desire to get away from applications and technology, and instead return to pure physics. But to do this, it would be necessary to make an evaluation—an assessment of the current status of physics. For this purpose, a conference (or series of conferences) on basic, fundamental topics would be most useful. This was of course precisely the role of the Solvay Conferences, and that was why Darrow in his letter referred to the Solvay Conferences as examples to emulate. Actually, the idea to organize a small conference came not from Darrow but from Duncan McGinnes, the president of the New York Academy of Science.† McGinnes especially stressed the importance of small conferences, with only *invited* participants (at most 30), devoted to very specific current problems. All those invited should be actively engaged in research in the area of the conference. Only then, McGinnes felt, could serious critical discussions take place; large conferences were too unwieldy and generally too diffuse. The idea to have a conference of active researchers, where basic questions of physics could be discussed, was sympathetic to many physicists. The American physics community certainly had not suffered during the war; it was not fragmented and decimated as was its European counterpart. Quite the contrary, the large war projects in places such as Los Alamos and The Massachusetts Institute of Technology had given physics a visibility that it never had before. Nevertheless, many physicists, returning from the military projects to academic or purely scientific pursuits, welcomed an assessment of the status of physics, an opportunity to identify new and promising research areas. A small conference where a few topics could be discussed in a leisurely manner, with colleagues struggling with these same problems, appeared to be an ideal vehicle to achieve new insights. Even though there was general support for the idea of such a conference, it took some time to decide who should be invited, where the conference should be, and what its main focus should be. As usual, this evolved slowly. The organizers, Darrow and McGinnes, eventually contacted John A. Wheeler from Princeton to begin to make detailed suggestions. Wheeler had returned some time earlier from the Los Alamos project, in which he played a significant role. At this time (1946–1947), he was anxious to return to pure physics. He had made an auspicious start by a brilliant study of positronium. For this study, Wheeler was awarded a prize by the New York Academy of Sciences. Wheeler was always (and remained) interested in fundamental questions of quantum mechanics; to have a conference on this topic certainly appealed to him. Because of his close association with the leading physicists of Los Alamos, it was natural that he would immediately think of them in connection with this conference. Of the 20 persons that Wheeler first suggested for the conference, no less than 12 had been in Los Alamos. Foremost among them, of course, was J. R. Oppenheimer. Oppenheimer's influence on physics was so great at that time that his presence and participation were absolutely essential for a successful conference. Oppenheimer's attendance was considered so vital that the dates for the conference were specifically picked to accommodate his schedule. Although the invitations to the conference were issued by Duncan McGinnes, Wheeler and Oppenheimer functioned as an unofficial nominating committee; their suggestions carried a great deal of weight. There were additions to—and deletions from—Wheeler's initial list; some persons asked could not come, while others

† A very interesting and extensive discussion of this conference is given by S. Schweber.[105] Only those aspects directly related to Kramers' role are recalled here. For details, the reader should consult Schweber.

were most unhappy because they were not invited. The final list was not all that different from Wheeler's original list. It included about twelve persons who had been associated in some capacity with Los Almos, about five who had been connected with the Massachusetts Institute of Technology Radiation Laboratory, three or four scientists from Princeton, a few others from scattered places, and then there was Kramers. It is interesting that Kramers was the single foreign visitor. It is not so surprising that Kramers was invited. He was in the New York–Princeton area in 1946–1947. He was a man of great eminence and his reputation and visibility were further enhanced by his activities for the AEC. More important, Wheeler, who played such a key role in selecting the members of the conference, was a long-time friend and admirer of Kramers. Finally, Oppenheimer always had the greatest respect for Kramers, ever since they met in Europe. All these circumstances, but especially his physical presence in the New York area, make it understandable that Kramers would be invited to this prestigious conference. With the choice of the major persons for the conference, the general character of the subject matter had effectively been settled. The official title—"Problems in Quantum Mechanics"—meant that the organizers were totally free in their choice of subjects. Thus, the topics actually selected represented the concerns of the theoretical physics community at that time. It is certainly true that all physicists were profoundly impressed with the accuracy and predictive power of quantum mechanics. This was even true in areas where it was not so obvious that quantum mechanics would be effective, such as in the relativistic domain. Everyone knew that there were fundamental questions dealing with infinite self-energies—the very problems that Kramers had been considering for so long. The vast majority of physicists were so enthralled with the computational successes of quantum electrodynamics that they did not worry too much about the remaining questions. They had learned by ingenious means to extract meaningful results from an imperfect theory, so many persons tended to ignore the imperfections in the expectation that a "future, more complete theory" would remedy these defects. Still, there remained nagging doubts about the sense and justification of the procedures employed. The imperfections in the theory became more pronounced and more serious as physics moved into the new areas of meson physics, nuclear physics, cosmic ray physics, or generally high-energy physics. There was a general suspicion that the somewhat makeshift procedures customarily employed might be legitimate or justifiable at low energies, but that they would lose their validity at higher energies. The conspicuous divergencies appeared as a nuisance at lower energies; they could not be circumvented by special clever tricks at higher energies. In meson theory they became more serious, and for a systematic development of the theory they were a crippling handicap. This led to the rather widespread conviction that a critical reexamination of the foundations of quantum theory, and especially of quantum field theory, was very much in order. Such a reexamination or reconsideration was the motivating theme of the conference. Weisskopf surely expressed the prevailing feeling when he wrote to Darrow: "It is a good sign that somebody is again interested in discussing the foundations of quantum mechanics, instead of only thinking of high voltage machines or how to make mesons."[106] The difficulties and imperfections of quantum electrodynamics, which began to be appreciated at the time of the Shelter Island Conference, were nothing new to Kramers. He had been concerned with these questions for some 15 years. As was extensively recorded in this chapter, he had gone his own way in attempting to resolve these difficulties in a systematic, organized way, not in a makeshift opportunistic manner. It therefore became clear that the questions that Kramers had been grappling with for so many years were directly relevant for the *critical reexamination of quantum field theory*, which by unspoken agreement was the central scientific theme of the Shelter Island Conference.

The organizers of the conference hoped, or more precisely expected, that a monograph would result from the deliberations and discussions of the conference (see Schweber[105]). When this turned out to be impractical, they suggested instead that a few, three or four, persons would be asked to write "position papers." These papers would sharply define the problems to be considered and, if possible, suggest new directions for solutions. These outlines would be distributed to the participants before the conference. In this way, the discussion would be focused and tangible results were more likely to emerge. The first person to be contacted to speak was Oppenheimer. He readily agreed (provided the date of the conference could be arranged); after some further negotiations and discussions between Oppenheimer, Darrow, and Wheeler, the other speakers selected were Kramers and Weisskopf. Kramers' work on the quantum theory of radiation was not particularly well known; it is doubtful that many physicists had studied his work in any degree of technical detail. During Kramers' stay in Princeton in the spring of 1947, he discussed his personal approach to the divergence problem with a number of his colleagues, including John Wheeler. Although it is unclear just how detailed these discussions were, it became clear that Kramers had new and important things to say. As always, such new information, or rumors about new information, percolated very rapidly through the physics community. Since the physics world was really starved for new approaches, it was very natural that Kramers would be selected as one of the "rapporteurs" for the Shelter Island Conference. Rumors about Kramers' work were evidently in the air, although few people seemed to know the details. Weisskopf, one of the rapporteurs to the Shelter Island Conference, wrote to Bethe: "I would like to hear from Kramers in great detail about his new theory."[107] In a letter written slightly later, Weisskopf again refers to Kramers' theory. In that letter, Weisskopf comments that his rapporteur's outline "was written in this manner only because I had no specific idea ready to discuss.... Since Kramers and to some extent Oppenheimer have something to say, let us give them all the time they need to do so."[108] This clearly expresses the general expectation that Kramers had concrete novel and worthwhile ideas to present.

The three rapporteurs, Oppenheimer, Weisskopf, and Kramers, each approached their charge—to identify the basic difficulties in quantum theory and to suggest new directions—in characteristically different ways. Weisskopf's outline expresses his own unhappiness and frustration with the status of "particle physics" in a remarkably candid manner. He starts his outline with these words: "The theory of elementary particles has reached an impasse. Certain well known attempts have been made in the last fifteen years to overcome a series of fundamental problems. All these attempts seem to have failed at an early stage ... the list [of failures] which follows will be well known to everyone and will probably invoke a feeling of knocking a sore head against the same old wall."[109] Weisskopf's low regard for particle theory is further evident from his final sentence: "In view of the failure of the present theories to represent the facts and the small probability that this conference may produce a new theoretical idea, part C of this agenda [proposed experiments] could become the most useful part of this Conference."[109] Clearly, Weisskopf is not very optimistic about the immediate future of particle physics. He does not expect any major new insights from the conference, he does not anticipate any major reorientation, and he does not make any specific new proposals. To him, promising new developments most likely would come from experiment or from theoretical studies, intimately related to experiments.

Oppenheimer's assessment was different. He was not all that sanguine about field theory either, but he did seem to feel that some field theoretic methods might be helpful in formulating an acceptable meson theory. But the meson theory itself and the high-energy cosmic ray phenomena, when treated according to the then current field theory, led to a variety of puzzles and paradoxes. Oppenheimer very tentatively suggests that the difficulties in the theory of multiple scattering of mesons are just an amplified version of similar difficulties encountered in radiation theory. Bloch and Nordsieck ran into such difficulties in their treatment of the emission of radiation by electrons during scattering. He consequently hints that an examination of the theory of radiation reaction (the Lorentz theory) in electromagnetic radiation might be helpful in understanding multiple meson production. But Oppenheimer, like Weisskopf, puts most stock in new experimental results.

It is interesting that both Weisskopf and Oppenheimer expect further progress to come from experiments, from areas *outside* quantum electrodynamics. It is also interesting that while both recognized the difficulties of the quantum theory of radiation, neither expressed serious doubts about the basic validity of the Dirac approach to that theory. The vast majority of physicists accepted the Dirac methodology, the Dirac Hamiltonian, as a sound and secure starting point for whatever developments were to come.

For Kramers, the situation was altogether different. He had never believed that the quantum theory of radiation, accepted since 1928, was anything but a first and crude approximation. He also did not believe that the difficulties would be clarified or eventually removed by experiments or theories *outside* quantum electrodynamics. Instead, he was firmly convinced that it was first of all necessary to understand the quantum theory of electromagnetic phenomena. Meson phenomena were of a later concern. Furthermore, Kramers' whole philosophy was based on the belief that a necessary prerequisite for the understanding of the quantum theory of electromagnetic phenomena was a removal of the difficulties and ambiguities in the classical theory of electrons. It was for that reason that he had introduced the separation of external and proper (self) fields. It was in this process that he stressed the important distinction between experimental mass and electromagnetic mass. The analysis of the classical problems led Kramers to the important notion of a structure-independent theory. Although far from completely successful, Kramers, in contradiction to both Weisskopf and Oppenheimer, had a program, a method for implementation and specific interesting results, in the nonrelativistic domain. Consequently, at the time of the conference, Kramers was not quite as pessimistic about the future of quantum electrodynamics as either Weisskopf or Oppenheimer. Remarkably enough, in a year or so, in part due to Kramers' own work, that situation would be reversed.

It is unfortunate that the outline which Kramers presented to the conference primarily emphasized his criticism of the then current Dirac quantum electrodynamics.[110] He only hints at the positive, constructive modifications that his research had produced in quantum electrodynamics. Presumably, these were discussed more extensively in Kramers' oral presentation at the conference, but no written record is available, not even private notes or personal reactions. It is only from occasional letters that some of the contents of Kramers' presentations may be inferred. Presumably, the actual technical content of Kramers' Shelter Island Conference report is not all that different from the report he presented at the Solvay Conference, September 27, 1948. It is extremely unlikely that during this intervening year Kramers spent a great deal of time on the problem, because he was very ill from August to about December 1947, and even after that time he recovered only slowly. Since the proceedings of the Solvay Conference were published (very much delayed), it is best to discuss the more technical aspects of the Shelter Island report in conjunction with the published information of the Solvay report (see Section III.C). For the same reason, the discussion of the reception and eventual appreciation of Kramers' efforts will be postponed until Section IV. The somewhat tortuous and intricate interrelations can be appreciated best when they are presented against the background of published materials. Kramers' lecture at the Shelter Island Conference had an almost didactic quality. He wanted to explain to his audience, or convince his audience, that there were compelling reasons to question the Dirac theory of radiation. This theme has been stressed throughout this chapter and the enumeration of his misgivings to be given here is no more than a systematic collection of the worries that Kramers had expressed over many years. But it is still useful to summarize them here, if for no other reason than to stress how pervasive and long lasting his objections were. Kramers starts his talk with these words: "At almost every important stage of the development of quantum mechanics, not only were new positive results added to what had already been achieved, but also certain 'defects' revealed themselves." In some cases, such defects were remedied by the next step; in other cases, they just stayed or even were emphasized more strongly.[110] It is clear from this phraseology that Kramers is considering the basic problems of quantum theory on a rather grand scale. This is something he had not done since the early days of the BKS theory. At that time too, he ventured into broad generalities regarding the character of quantum theory. But since that time he had purposely or by chance (who can say) avoided considerations of such sweeping scope. In the Shelter Island lecture, Kramers starts out by listing what he felt were the major successes and attendant difficulties of various stages in quantum theory. Actually, all the specific items that Kramers mentions are from the quantum theory of radiation. Since this is largely a repetition of points mentioned numerous times before in this chapter, it is sufficient to enumerate the individual points. The list provides an illuminating and concise survey of Kramers' thinking at that time. In addition, Kramers comments very briefly that "the meson theory showed encouraging features, but also brought new divergence sorrows."[110] Even a superficial glance at this list shows that it is just a compilation of the many misgivings with quantum electrodynamics that Kramers had expressed so often. Here his disenchantment is merely put in the context of a general dictum that "with each success comes an accompanying difficulty." Whether the pertinence of this maxim was due to the unsatisfactory basis of the Dirac theory, the necessarily limited scope of *any* theory, or the intrinsic and tragic limitations of the human mind, Kramers does not say. But with his penchant for oraclelike pronouncements, he likely meant them all. Kramers' remarks about his own modifications of radiation theory are very brief indeed, and they too are repetitions of comments made earlier during 1937–1944; they are recorded in Section II of this chapter. He reiterates that the Dirac Lagrangian does not show the proper correspondence with the classical Lorentz theory. He asserts that a variation principle, which correctly describes the interactions of an electron (characterized by its experimental mass) with an external electromagnetic field, does *not* lead to the Dirac Lagrangian. Furthermore, the Dirac Lagrangian, even classically, often leads to divergent results, while the Lagrangian advocated by Kramers removes these divergencies. In this connection Kramers refers to Serpe's result (described in Section II.B. In so doing, he touches on one of the points that would have a major impact on the conference (and the future development): this divergence is nothing but the divergence of the electromagnetic mass. Finally, Kramers insists that in his theory, which he calls "a correspondence theory," the correct interaction Hamiltonian is given by $H_K = -(e/c^2)(\dot{\mathbf{v}} \cdot \mathbf{Z})$ instead of the customary Dirac Hamiltonian $H_D = -(e/c)(\mathbf{v} \cdot \mathbf{A})$. Again in simple examples, the use of H_K eliminates some of the divergencies. This is effectively the context of the Opechowski paper also discussed in Section II.B. It will be recognized that this series of remarks is indeed a repetition of comments made over the years. But if it is recalled that Kramers' work was not widely read, was expressed in an unfamiliar notation, and started from an unusual and unfashionable viewpoint, it is easy to see that Kramers' presentations would evoke mixed reactions and considerable confusion. It was only much later that Kramers' contribution was fully understood and generally appreciated (see especially Section IV). But even at this "first hearing" it was clear that something important was happening and although not all details were crystal clear, certain features stood out as being especially significant.

Successes†	Difficulties
1926 Quantization of the Maxwell field	Zero-point energy (Kramers considered this an unresolved problem)
1927 The Dirac theory of emission and absorption of light	(i) Divergencies in second-order calculations (ii) Infinite shift of spectral lines (iii) Impossibility of steady state of an atom in a radiation field (iv) Impossibility to get a dispersion formula with the correct phase (v) Radiation reaction
1928 (i) Dirac theory of the electron spin (ii) Klein–Nishina formula (iii) Sommerfeld formula (Kramers is not convinced these require the quantization of the radiation field)	(i) Second-order divergencies remain (ii) Negative energy states
1931–1934 (i) Positron theory (ii) Positrons (iii) Pair production (Kramers does not think these necessarily require the infinite sea)	(i) The "sea" requires the scattering of light by light that has not been calculated and has not been seen (ii) The self-energy diverges logarithmically

† Kramers' private views or personal comments are written in parentheses.

Kramers' studies (especially as applied by Serpe) showed convincingly that at least some of the divergencies of field theory had their origins in the divergence of the electromagnetic mass of the electron. The electromagnetic mass, or the corresponding energy, resulted from the interaction energy of the charge distribution with its own field. This diverges for a point charge. That the divergencies of field theory were a manifestation of this divergence of the electromagnetic mass was suspected by many physicists, but Kramers made it quite explicit. One of the central points of Kramers' analysis was his insistence that only the experimental mass, that is, the sum of the mechanical mass m_0 and the electromagnetic mass m_{el}, had any physical significance. Only $m_{exp} = m_0 + m_{el}$ has a meaning; m_0 and m_{el} are not separately observable, and it does not make much difference whether they are separately infinite or finite. The systematic reformulation of all equations in terms of the experimental mass m_{exp} and the elimination of the self-field and m_{el} from physical equations was the core of Kramers' program. It was the earliest example of what now is called a mass renormalization procedure. It is likely that the audience at the Shelter Island Conference got the general idea of what Kramers was driving at, although the details were probably difficult to understand. In his presentations, Kramers represented an electron as a finite charge distribution. It is then, of course, possible to calculate the interaction energy of this charge distribution with the field. As anticipated, this gives merely an addition to the mechanical mass (which actually depends on the model assumed). The elimination of the structure, demanded by Kramers, yields a modified mass, which according to Kramers must be identified with the experimental mass. The main moral to be drawn from this analysis is that if one describes an electron by its empirical mass, one has included a good share of the interaction energy of the charge distribution with its own field, because m_{exp} contains the electromagnetic mass, which is a measure of that interaction energy. It is unclear just how impressed the Shelter Island Conference was with this particular observation. But this simple, almost self-evident observation has far reaching consequences. It follows almost immediately that the self-energy of a bound and free electron are different. Since the electromagnetic field of a bound electron is different from that of a free electron, it is not so surprising that their respective interaction energies with a charge distribution would be different. Formally, both are infinite for point electrons and model dependent for finite charge distribution. But Kramers argued that only their difference would have physical significance and he suspected that this difference might well be finite. Thus, Kramers reasoned that the energy of an atomic system would be modified by an amount equal to the difference of the self-energy in that state and the self-energy of the free electron. Consequently, Kramers' analysis of the self-energy problem led to the prediction that there would be energy shifts in the energy spectrum of atomic systems, such as the hydrogen atom, from the values predicted by the Dirac theory. This was a genuine novelty. Up to Kramers' detailed analysis, it had tacitly been assumed that divergencies, when properly treated or otherwise circumvented, would disappear and leave no trace. Their elimination would presumably remove a nuisance, giving a more sensible tractable theory, but few persons suspected that new observable effects would result from this elimination. Kramers' analysis showed, on the other hand, that new phenomena should be expected from his more systematic theory. Stated succinctly, the Dirac theory had to be modified by the subtraction of self-energy terms. Since the self-energy terms in different atomic energy states would have different values, different levels would be affected differently. In particular, distinct levels, degenerate in energy before the subtraction, might no longer coincide after the subtraction. The technical implementation of a subtraction type of procedure can be very tricky, especially since one has to manipulate divergent quantities.

Different, presumably equivalent (but not obviously equivalent) procedures exist, and Kramers' presentation of his own procedure was convoluted and very difficult to follow. Still, in one form or another, Kramers' subtraction procedure, or the elimination of the structure of the electron, was central to the reorientation of the thinking about the divergence problem. It is hard to know precisely how the audience at the conference reacted to these ideas. [It is relatively easy to know at this moment (in 1987) how they think they reacted at that time! The knowledge of present successes always introduces distortions in the perception of the past. These distortions are especially severe in this case!] It appears that many people got a general impression of Kramers' program, without fully appreciating its full scope.[111] (In retrospect, this was even true of Kramers himself.) But it was clear that Kramers' theory could accommodate deviations from the Dirac theory in a natural manner. To Kramers, this was of course no surprise; he had mentioned the possibility of corrections to the Dirac theory as early as 1937; "As a result we expect that a correction must be applied to the energy values of the stationary states of the hydrogen atom as given by the Dirac theory of 1928."[112] But to most participants, Kramers' suggestion that there should be deviations from the Dirac theory was quite surprising. A few of the participants recognized in Kramers' procedure the beginnings of a significant new development in field theory, of which the corrections to the Dirac theory were only the first manifestation. It is one of those remarkable coincidences in the history of physics that at the same time—in the spring of 1947—that Kramers' theories suggested deviations from the Dirac theory, a brilliant experiment by Lamb precisely demonstrated such an effect. According to the Dirac theory, two levels, $2P_{1/2}$ and $2S_{1/2}$, must have exactly the same energy. The experiment by Lamb seemed to indicate instead that these levels were shifted relative to each other—thus providing clear evidence that the uncorrected Dirac theory was *not* exact. The experimental results were not yet published at the time of the Shelter Island Conference, but rumors of these results had been circulating well before the meeting. Lamb himself was present at the conference and he presented his still somewhat preliminary but yet quite definite findings. The results were astounding; the level splitting (the *Lamb·shift*) was real and the Dirac equation did need correction. It is not clear just how familiar Kramers was with the details of the Lamb shift, nor when he first heard of it. Such knowledge would certainly not have changed his approach to radiation theory, nor any of the details of his development. After all, these were based on general principles that Kramers trusted implicitly; he never tried to imitate or emulate the Dirac theory. Kramers had been convinced for so long of the approximate, incomplete character of the Dirac theory that an experimental verification to that effect almost came as an anticlimax. Conceivably, had Kramers known about the Lamb shift, he might have computed it. He certainly did not do that; but he did produce such a calculation about a week after the Shelter Island Conference (June 10, 1947).[113] Even if the experimental deviations from the exact Dirac equations were not much of a shock to

Kramers, it is certain that they created enormous excitement among the physicists present. Lamb's result showed unequivocally that experiments demanded modifications, however subtle, of the Dirac equation. This equation was based so immediately on relativity and quantum mechanics that the need for alterations came as quite a shock. Such modifications could only be arrived at after the most careful scrutiny of the basic concepts and methods of the quantum theory of electrons and radiation. It was precisely this set of problems that Kramers had started to ponder some 10–15 years earlier. He had set himself the task to construct a systematic quantum theory of the electron and radiation, starting from a sound classical basis—a task to which he had devoted almost all his creative efforts. His presentation at the Shelter Island Conference was in the nature of a meditation on this theme. He presented both his concerns and the results he had obtained in his lengthy study of these fundamental questions. It might appear as a happy coincidence that the very time that Kramers reported on his ideas, experimental results actually supported his contentions. Up to that time, Kramers had worked in isolation; his ideas had not attracted much attention. Even though his book, which contained an extensive discussion of his method, was widely praised, the parts containing Kramers' formulation of quantum electrodynamics were largely ignored as being "too difficult," "too original," and too "far out."[114] Because of this, Kramers often felt somewhat unappreciated, lonely, and neglected.[115] To have experimental evidence strongly implying that the modifications required by Kramers' approach were really there, might appear to be the best thing in the world for Kramers. This is true in part, but only in part. With the new experiments reported at the Shelter Island Conference, Kramers' ideas and methods—in short, Kramers' philosophy of quantum field theory—became of central significance almost overnight. However, the transition from neglect to center stage almost came too fast. Immediately after the Shelter Island Conference, Bethe[116] made a most ingenious application of Kramers' subtraction procedure to obtain the first theoretical value for the Lamb shift. Very soon afterward, Schwinger[117] developed a powerful, elegant formalism that replaced the rather cumbersome mathematical procedures employed by Kramers. Furthermore, the Schwinger formalism was relativistically invariant and this invariance was maintained throughout all the calculations—something Kramers could not do. The Feynman formalism,[118] developed about the same time, was even more elegant and had the additional advantage of computational accessibility. It would be an exaggeration to say that Kramers was all but forgotten in the explosive developments that followed the Shelter Island Conference. But it is certainly true that Kramers' original physical ideas receded almost immediately into the background and were instantly superseded by the brilliant technical exploitations of these same ideas. Physicists were understandably dazzled by the extraordinary power and elegance of the new theoretical developments which were capable of explaining delicate and subtle experimental effects. Not surprisingly, the physics community was impressed by—and paid more attention to—the methods of implementation, than the physical motivation and methodology leading to these developments. Kramers' procedure and indeed his philosophy went just the other way around. In his own work and in the Shelter Island Conference, he emphasized general principles and a general philosophy. His technical implementation was neither elegant, nor polished, nor even complete. He never completed a calculation, so that his result could never be compared with experiment. No wonder that to many physicists the results and means of implementation of Kramers' program were better known, more important, and more relevant than the physical or philosophical foundations of that program. And so it happened that under the compelling stimulus of Lamb's experiments, Kramers' general programs were catapulted from relative obscurity into the limelight, while Kramers' detailed personal contributions remained about as unknown and nonunderstood as they had ever been. During this exciting and productive period of the development of the renormalization scheme in the Schwinger–Feynman–Dyson electrodynamics, it was far from universally recognized that the motivation for a good share of the procedures employed was due to Kramers. His earlier studies certainly stimulated the subsequent developments, but this often was not stated explicitly. Thus, it was not an unmixed blessing for Kramers that his ideas had become so fashionable because of the experimental discovery of the Lamb shift. In the excitement of the succeeding development, Kramers' somewhat tentative treatment was literally swept away by the avalanche of new results. In keeping with his general style, Kramers had considered only a very specific, nonrelativistic problem for a detailed treatment, and this he had solved in an approximate manner. In so doing, Kramers had actually discovered principles of great generality. The scope of his results was much larger than might appear from his approximate treatment. In the period immediately following the Shelter Island Conference, the generality of Kramers' results began to be recognized, and the central and simplifying role of relativistic invariance became apparent very soon. But the methods and formulation of the resulting theory were so different from Kramers' original version that its connection with Kramers' work was no longer clear and Kramers' initial efforts were no longer mentioned. It was not until several years later (actually after Kramers' death in 1952) that Kramers' contributions began to be recognized generally. The story of this belated, retroactive recognition will be told in Section IV.

C. The Solvay Report: Culmination and Termination of Kramers' Program

The Solvay report[119] was published in the *Proceedings of the Solvay Conference*. The conference was held in September 1948 but publication of the proceedings was excessively delayed; they did not appear until 1950, a good 2 years after the lecture was actually delivered. This report contains Kramers' last and most complete exposition of his ideas on quantum electrodynamics. It is striking how similar the basic physical ideas expressed in this lecture are to those contained in his earliest publications on this topic. These were written some 10 years earlier; the first public presentation of these ideas was at the Galvani Congress in Bologna in 1937.[120] Kramers, in his Solvay lecture, used the old Lorentz model, representing the electron as a finite, rigid charge distribution. His scientific goals were also unchanged: first seek a mathematical formulation of the structure-independent features of classical electron theory. Once that is accomplished cast the resulting equations in a Hamiltonian form. Kramers never wavered in his conviction that this was the only trustworthy basis for a quantum theory of radiation; he considered other methods and procedures at best as opportunistic and very likely as superficial and misleading. It was never in Kramers' style to circumvent a physical difficulty by artfully dodging it, something he suspected other methods did with style and skill. So convinced was Kramers of the soundness of his approach that he was quite willing to accept the difficulties, limitations, and restrictions which a systematic implementation of his program required. He was more than willing to restrict his efforts to a strictly nonrelativistic description if this meant that he could satisfy the correspondence principle. These and other limitations are spelled out in the Solvay Conference report; they were only hinted at in the Shelter Island presentation. That report concentrated mainly on the critique of the conventional radiation theory. The physical ideas and general principles in the Solvay presentations are not changed much from the Shelter Island talk, but the technical treatment is much more complete. It represents a systematic summary of the various partial results that Kramers had obtained over some 10 years. The resulting formalism is not particularly simple or elegant, but it does represent the furthest advance that Kramers made in radiation theory. Kramers is quite candid about the limitations he himself had imposed on his treatment, and he is explicit about the approximations and simplifying assumptions.

The first topic treated in great detail is the elimination of the electronic structure from the equations of motion. The same two ingredients used before reappear again: the introduction of the experimental mass m_{exp} instead of the mechanical mass and the splitting of the fields into an *external* field and *self* field. This discussion contains a number of technical differences and formal improvements. Kramers no longer uses his earlier favorite combination $\mathbf{F} = \mathbf{E} + i\mathbf{B}$; instead he works with $\mathbf{E}$ and $\mathbf{B}$ directly. The Fourier expansions, so prominent in earlier discussions, are absent; Kramers now works directly with the (approximate) solutions of the wave equation for the potentials. Finally, the splitting of the fields into their self and external parts is carried out in terms of the vector potential $\mathbf{A}$, not in terms of $\mathbf{F}$, or $\mathbf{E}$ and $\mathbf{B}$, directly. It must be stressed that since such splittings are not Lorentz, nongauge invariant, they tend to be a bit contrived. There is not really a "unique" or "correct" splitting; one can only hope that the splitting eventually adopted captures some of the intuitive physical ideas expected of the notions of self-field and external field. Because of the approximate character of the theory, one can reasonably expect only a simple interpretation of these terms to lowest order in v/c. Kramers is more careful and circumspect about these issues than he was in previous investigations. (It is likely that Opechowski's earlier, somewhat cryptic, reference to ongoing work of Kramers as early as 1941 dealt with exactly these issues.)

The starting equations are the Maxwell equations for the electric fields **E** and **B**. The electrons are rigid bodies of finite extent, of mechanical mass m_0 and charge e. The mass and charge distribution are spherically symmetric, with radius a. In addition to the customary restrictions that $v/c \ll 1$, it will further be required that only electromagnetic modes of wavelength $\lambda \gg a$ play a role in the phenomena to be described. Under these conditions, Kramers anticipates that a nonrelativistic, structure-independent description will be possible. The equation of motion for an electron (position **R**, velocity $\dot{\mathbf{R}}$), is†

$$m_0 \ddot{\mathbf{R}} = e\langle \mathbf{E} \rangle + e[\dot{\mathbf{R}} \times \langle \mathbf{B} \rangle] - \nabla_R U \tag{100}$$

Here $\langle \mathbf{E} \rangle$ and $\langle \mathbf{B} \rangle$ are the average values of the total fields over the finite charge distribution. Generally, $\langle Q \rangle$ is defined by

$$e\langle Q \rangle = \int d^3x \, \rho Q \tag{101a}$$

The current **j**, occurring in the Maxwell equation, is given by

$$\mathbf{j} = \rho \dot{\mathbf{R}} \tag{101b}$$

In Equation (100), U is the potential energy resulting from nonelectrical forces. U could also be the potential energy of an electron in a fixed system of charges, where these charges are not participating in the dynamics. In that case, **E** in Equation (100) does *not* include the field strength resulting from these charges. The potentials **A** and ϕ are introduced in the usual way:

$$\mathbf{B} = \operatorname{curl} \mathbf{A} \quad \text{and} \quad \mathbf{E} = -\nabla\phi - \dot{\mathbf{A}} \tag{102a}$$

The gauge is chosen such that the vector potential is transversal:

$$\operatorname{div} \mathbf{A} = 0 \tag{102b}$$

This leads in the standard manner to the wave equations for the potentials:

$$\Delta \mathbf{A} - \ddot{\mathbf{A}} = \Box \mathbf{A} = 4\pi \operatorname{Tr} \mathbf{j} = -4\pi\rho\dot{\mathbf{R}} + \nabla\dot{\phi} \tag{103a}$$

$$\Delta\phi = -4\pi\rho \tag{103b}$$

The symbol Tr **Q** stands for the transversal part of the vector field **Q**.‡

All this is totally standard. Equation (100), according to Kramers, is not a physically meaningful equation, since it contains the mechanical mass instead

† We follow Kramers' notations to the extent that c is put equal to 1.

‡ Tr **Q** is clearly a vector; it is specifically *not* the trace!

of the experimental mass. The implementation of Kramers' program now requires both the separation of the field into self-field and external field and the introduction of the experimental mass m_{exp}. According to Kramers, m_{exp} is the sum of the mechanical and electromagnetic mass:

$$m_{\text{exp}} = m_0 + m_{\text{el}} \tag{104a}$$

The electromagnetic mass is unambiguously fixed for this model:

$$m_{\text{el}} = \theta \frac{e^2}{a} \tag{104b}$$

θ is a numerical factor that depends on the particular nature of the charge distribution ρ. For a surface charged sphere, $\theta = \frac{2}{3}$.

The definition of the self-fields is, as was stressed earlier, tricky and certainly not unique. A "good," useful splitting is one which either agrees with the physical intuition or can be given a simple pictorial interpretation. It is simplest to first carry out such a splitting for the vector potential $\mathbf{A}$:

$$\mathbf{A} = \mathbf{A}_{\text{ext}} + \mathbf{A}_{\text{self}} \equiv (\mathbf{A}' + \mathbf{A}'') \tag{105}$$

Occasionally, $\mathbf{A}_{\text{ext}}$ will be written as $\mathbf{A}'$ and $\mathbf{A}_{\text{self}}$ as $\mathbf{A}''$. $\mathbf{A}_{\text{self}}$ is defined by the stipulations

$$\operatorname{curl}\operatorname{curl}\mathbf{A}_{\text{self}} = 4\pi \operatorname{Tr} \rho \dot{\mathbf{R}} \tag{106a}$$

$$\operatorname{div}\mathbf{A}_{\text{self}} = 0 \tag{106b}$$

Equations (106a) and (106b) can be solved directly, so $\mathbf{A}_{\text{self}}$ is known. Inspection of these equations shows that $\mathbf{A}_{\text{self}}$ up to order $\dot{R}$ (i.e., order v/c) is the same as the divergence-free vector potential of an electron which at time t has a position $\mathbf{R}$ and moves uniformly with velocity $\dot{\mathbf{R}}$. It is this fact which suggests that $\mathbf{A}_{\text{self}}$ is a suitable choice for the vector potential of the proper field (or self-field) of the electron.

The definition just given makes it explicit that $\mathbf{A}_{\text{self}}$ agrees with the vector potential generated by a uniformly moving charge, only up to order v/c. It is $\mathbf{A}_{\text{self}}$ as determined by (106a) and (106b) that "correctly" defines the self-field: The self-field is not just the field produced by the uniformly moving charge; these two fields are only approximately equal. It is not surprising, but worth mentioning anyhow, that $\mathbf{A}_{\text{self}}$ for a point electron diverges at the electron. From now on, Equations (106a) and (106b) are the adopted definitions of the self-field.

The splitting of the vector potential into an external field and a self-field, specified by Equations (105) and (106a) can now be used to define a corresponding splitting in the fields. The self-magnetic-field is simply defined by

$$\mathbf{B}_{\text{self}} = \operatorname{curl}\mathbf{A}_{\text{self}} \tag{107}$$

Since $\mathbf{A}_{\text{self}}$ is determined, $\mathbf{B}_{\text{self}}$ is fixed by Equation (107). The external field $\mathbf{B}_{\text{ext}}$ is defined in the obvious manner:

$$\mathbf{B}_{\text{ext}} \equiv \mathbf{B} - \mathbf{B}_{\text{self}} = \operatorname{curl}\mathbf{A}_{\text{ext}} \tag{108}$$

It is evident, but still important, that the splitting of the fields actually amounts to a particular subtraction procedure. The self-fields, say $\mathbf{B}_{\text{self}}$, are defined autonomously by Equations (106a), (106b), and (109); the external field as in (108) is then obtained by subtraction. The definition of $\mathbf{B}_{\text{self}}$ via (107) was almost obvious and quite straightforward. The definition of $\mathbf{E}_{\text{self}}$ is considerably more subtle. It is clearly true that the total electric field can be written

$$\mathbf{E} = -\nabla\varphi - \dot{\mathbf{A}}_{\text{self}} - \dot{\mathbf{A}}_{\text{ext}} \tag{109}$$

By its defining equations (106a) and (106b), $\mathbf{A}_{\text{self}}$ depends on the time through its dependence on $\mathbf{R}$ and $\dot{\mathbf{R}}$. Consequently, (109) can be written

$$\mathbf{E} = -\nabla\varphi - (\dot{\mathbf{R}}\cdot\nabla_R)\mathbf{A}_{\text{self}} - (\ddot{\mathbf{R}}\cdot\nabla_{\dot{R}})\mathbf{A}_{\text{self}} - \dot{\mathbf{A}}_{\text{ext}} \tag{109a}$$

Kramers now includes the first $\mathbf{A}_{\text{self}}$ term in (109a) in the self-electric-field and the second term in the external electric field. Thus,

$$\mathbf{E}_{\text{self}} = -\nabla\varphi - (\dot{\mathbf{R}}\cdot\nabla_R)\mathbf{A}_{\text{self}} \equiv -\nabla\varphi + \mathbf{E}_0 \tag{110}$$

The external electric field is defined by

$$\mathbf{E}_{\text{ext}} \equiv -(\ddot{\mathbf{R}}\cdot\nabla_{\dot{R}})\mathbf{A}_{\text{self}} - \dot{\mathbf{A}}_{\text{ext}} = -\dot{\mathbf{A}}_{\text{ext}} + \mathbf{F}_0 \tag{111}$$

$\mathbf{E}_0$ and $\mathbf{F}_0$ are defined by these equations; both are clearly known in terms of $\mathbf{A}_{\text{self}}$.

The subdivision adopted by Kramers is $\mathbf{E} = \mathbf{E}_{\text{self}} + \mathbf{E}_{\text{ext}}$. Using this splitting it can be shown that

$$\langle\mathbf{E}_{\text{ext}}\rangle = -\langle\dot{\mathbf{A}}_{\text{ext}}\rangle - \frac{m_{\text{el}}}{e}\ddot{\mathbf{R}} \tag{112}$$

The average value of the external field diverges for a point electron, since $\langle\mathbf{A}_{\text{ext}}\rangle$ diverges. This divergence of the external field is of course a consequence of the particular splitting made. [In his earlier studies Kramers had forgotten the second term in Equation (112), and that led to all kinds of contradictions and paradoxes.]

With this particular splitting, the field equations and the equations of motion can be written in the form:

$$m\ddot{\mathbf{R}} = -e\langle\dot{\mathbf{A}}_{\text{ext}}\rangle + e[\dot{\mathbf{R}} \times \langle\text{curl}\,\mathbf{A}_{\text{ext}}\rangle] - \nabla_R U \tag{113}$$

$$\Box\mathbf{A}_{\text{ext}} = \ddot{\mathbf{A}}_{\text{self}} = [\dddot{R}\nabla_{\dot{R}} + 2(\ddot{R}\nabla_{\dot{R}})(\dot{R}\nabla_R) + (\dot{R}\nabla_R) + (\dot{R}\nabla_R)^2]\mathbf{A}_{\text{self}} \tag{114}$$

It is important to stress that m in (113) is indeed the experimental mass, $m = m_0 + \theta e^2/a$. It arises "naturally" from the splitting that Kramers had made. With these equations,† Kramers had completed the first part of his program. The equations of motion (113) contained just the experimental mass and the external field. The field equation (114) is an inhomogeneous wave equation for $\mathbf{A}_{\text{ext}}$, where the inhomogeneous terms depend on the known self-field only. [The self-field itself is specified by Equations (106a) and (106b).]

† In the only published version of these equations, there are numerous misprints.

Thus, one might hope that the system (113) and (114) would provide a structure-independent description of the classical system of electrons and radiation. It can readily be guessed that such a formalism would have enormous appeal for Kramers. The precise issue is the following: Suppose $\mathbf{A}_{ext}$ and $\dot{\mathbf{A}}$ are given initially in a way compatible with the inevitable singularities in $\mathbf{A}_{self}$ but without further divergencies at the positions of the electrons. Then the basic question is whether or not the equations for $\mathbf{A}_{ext}$ and $\mathbf{R}$ [Equations (113) and (114)] will lead to a time dependence in which the motion of the charges and the evolution of the fields do not depend on the peculiarities of the charge distribution ρ, but just on the experimental mass m_{exp} and the charge e. The theory is structure independent if the answer to this basic question is affirmative. Kramers' discussion of this fundamental question is surprisingly brief. His answer is that the theory is not and cannot rigorously be structure independent, but he never states this as such. He does state categorically that "the vector potential $\vec{A}$ in the course of time will develop a structure dependent singularity near the electron."[121] This might well appear as a devastating blow to Kramers' hopes and expectations but he appeared unconcerned and totally unperturbed. While the theory as formulated might not be structure independent, Kramers nevertheless argues that the only physically meaningful statements that the theory can make must be structure independent. It is possible and even likely that only certain approximate results can be structure independent. These results, according to Kramers, are then singled out as having physical significance. Kramers almost turns the argument around. The approximate structure-independent solutions of his basic equations are of physical importance. Structure independence becomes the criteria for the selection of physically meaningful solutions. The "basic equations" are basic only in so far as they provide a systematic starting point for the construction of approximate structure-independent solutions. The approximations in Kramers' view are more important than (or at least as important as) the general principles. This attitude was, as stressed so often, highly characteristic of Kramers.

The next major point, also extensively discussed for the first time at the Solvay Conference, was the reformulation of the basic equations into Hamiltonian form. This again could not be accomplished rigorously. Kramers knew this very well from many years experience. Since, to Kramers, the whole formalism was approximate anyhow, additional approximations allowing a Hamiltonian formulation were very much in the spirit of his approach. Thus, the next effort was directed toward the construction of an approximately structure-independent, approximately Hamiltonian formalism. In doing this, it is useful to recall the definitions of the Hertz potentials:

$$\dot{\mathbf{Z}}_{ext} = \mathbf{A}_{ext} \tag{115a}$$

$$\operatorname{div} \mathbf{Z}_{ext} = 0 \tag{115b}$$

$$\operatorname{curl}\operatorname{curl} \mathbf{Z}_{ext} = -\Delta \mathbf{Z}_{ext} = \mathbf{E}_{ext} + \mathbf{E}_0 \tag{115c}$$

Inspection of (115c) shows the $\mathbf{Z}_{\text{ext}}$ diverges at the location of a point electron, since $\mathbf{E}_0$ diverges as $1/r^2$ and $\mathbf{E}_{\text{ext}}$ as $1/r$.

It is not difficult to verify that the starting equations of Kramers' theory—(100), (103a), and (103b)—can be derived from a variation principle. Thus, the equations

$$m_0\ddot{\mathbf{R}} = e\langle \mathbf{E}\rangle + [\dot{\mathbf{R}} \times \langle \mathbf{B}\rangle] - \nabla_R U \tag{116}$$

$$\Delta\mathbf{A} - \ddot{\mathbf{A}} = \Box\mathbf{A} = -4\pi\rho\dot{\mathbf{R}} + \nabla\dot{\phi} \tag{117}$$

$$\Delta\phi = -4\pi\rho \tag{118}$$

can be derived by varying a Lagrangian L_0 independently with respect to $\mathbf{A}$ and $\mathbf{R}$, subject to the constraint

$$\delta(\operatorname{div}\mathbf{A}) = 0 \tag{119}$$

Here L_0 is given by

$$\begin{aligned} L_0 &= \tfrac{1}{2}m_0\dot{\mathbf{R}}^2 + e[\dot{\mathbf{R}}\langle \mathbf{A}\rangle - U(R)] \\ &\quad - \frac{1}{\delta\pi}\int [\mathbf{B}^2 - (\operatorname{Tr}\mathbf{E})^2]d^3x \end{aligned} \tag{120}$$

L_0 is a function of $\mathbf{R}$, $\dot{\mathbf{R}}$, $\mathbf{A}$, and $\dot{\mathbf{A}}$.

It might appear now that to obtain a Hamiltonian form of Kramers' structure-independent equations it would only be necessary to rewrite $\mathbf{A} = \mathbf{A}_{\text{ext}} + \mathbf{A}_{\text{self}}$ and introduce m_{ext} instead of m_0 in the Lagrangian. This is actually a rather subtle and involved process, because the field $\mathbf{A}_{\text{self}}$, by its definition, depends on the particle position and velocity ($\mathbf{R}$ and $\dot{\mathbf{R}}$), so that in this substitution, the field and particle coordinates become mixed up. Consequently, the proper definition of canonical variables must be carried out with considerable care. Kramers carries out this lengthy analysis in excruciating detail; here just the main steps are summarized. First, define $\mathbf{P}$ by the prescription

$$\mathbf{P} = \nabla_{\dot{R}} L_0 \tag{121}$$

In this relation, $\dot{\mathbf{R}}$ can be solved in terms of $\mathbf{R}$, $\mathbf{P}$, $\dot{\mathbf{P}}$, $\mathbf{A}$, and $\dot{\mathbf{A}}$. Second, call the $\dot{\mathbf{R}}$ so expressed $\mathbf{V}$. It is easy to check that

$$m_0\mathbf{V} = \mathbf{P} - e\langle \mathbf{A}\rangle \tag{122}$$

In terms of these variables, the field equations can be derived from a Lagrangian L_1, given by

$$L_1 = -\tfrac{1}{2}m_0\mathbf{V}^2 - U - (\dot{\mathbf{P}}\cdot\mathbf{R}) - \frac{1}{8\pi}\int [\mathbf{B}^2 - (\operatorname{Tr}\mathbf{E})^2]d^3x \tag{123}$$

Using these variables, one can substitute the decomposition of $\mathbf{B}$ and $\mathbf{E}$ in terms of the self-fields and the external fields. A very suggestive and interesting consequence is that

$$m\mathbf{V} = \mathbf{P} - e\langle \mathbf{A}_{\text{ext}}\rangle \tag{124}$$

where now m is the experimental mass. Clearly, to Kramers, (124) would be a "perfect" formula because it connects two quantities that should enter his description: the experimental mass and the external field. The result of these manipulations is

$$\begin{aligned} L_2 = -\frac{P^2}{2m} + \frac{e^2}{2m}(\langle \mathbf{A}_{ext} \rangle)^2 - U - \dot{\mathbf{P}}\cdot\mathbf{R} \\ - \frac{1}{8\pi}\int (\mathbf{B}_{ext}^2 - \mathbf{E}_{ext}^2 - 2(\mathbf{E}_{ext}\cdot\mathbf{E}_0)d^3x \end{aligned} \tag{125}$$

As written, the Lagrangian L_2 follows almost directly from the substitutions and the form of L_1. Just the term $\int \mathbf{E}_0^2 d^3x$ has been omitted. This term would diverge for a point electron, and it would be proportional to $\mathbf{V}^4$ if a were kept finite. Kramers insists that it must be neglected, since it refers to an effect that only a truly relativistic theory could describe correctly. The Lagrangian L_2 contains only structure-independent features,† but it still cannot be cast into Hamiltonian form, since it depends on $\mathbf{P}$, $\dot{\mathbf{P}}$, $\mathbf{A}_{ext}$, $\dot{\mathbf{A}}_{ext}$, $\dot{\mathbf{R}}$, and $\mathbf{R}$ and not all of these variables are independent. Because of the various constraints between the variables, Equation (125) cannot immediately be used to write the equations of motion in canonical form. For example, $\mathbf{V}$ through Equation (124) depends on $\mathbf{P}$, $\mathbf{A}_{ext}$, and $\mathbf{R}$; the quantity $\mathbf{E}_0$ which occurs in $\mathbf{E}_{self}$ depends on $\mathbf{P}$, $\mathbf{A}_{ext}$, $\mathbf{R}$, and $\dot{\mathbf{R}}$. All these dependencies give rise to constraints. In the process of eliminating the redundant variables and identifying a proper set of independent canonical variables, it is necessary to make several approximations. The basic approximation made is the restriction to dipole radiation; the wavelengths λ of the relevant modes of the electromagnetic field is large compared to the dimensions of the charge distribution a. As a consequence, $\mathbf{A}_{ext}$ may be considered independent of $\mathbf{R}$, while the $\mathbf{E}_0$ term can be omitted altogether. To construct the canonical form, it is necessary to find the *vector* conjugate to $\mathbf{P}$ and the *vector field* conjugate to $\mathbf{A}_{ext}$.

Denote the vector conjugate to $\mathbf{P}$ by $-\mathbf{X}$ and the vector field conjugate to $\mathbf{A}_{ext}$ by $-\mathbf{E}_2/4\pi$.‡ Then a rather lengthy but pretty straightforward calculation shows that

$$\mathbf{X} = \mathbf{R} + \frac{e}{m}\langle \mathbf{Z}_{ext} \rangle \tag{126a}$$

$$\mathbf{E}_2 = \mathbf{E}_{ext} - \frac{4\pi\Delta e^2}{m}\langle \mathbf{Z}_{ext} \rangle \tag{126b}$$

It is at this point that the Hertz vector, defined earlier [cf. Equation 293], reappears. (It is interesting that the variable canonical to $\mathbf{P}$, variable $\mathbf{X}$,

† Recall that $\mathbf{E}_0$ is defined in terms of $\mathbf{A}_{self}$, which depends on $\mathbf{R}$ and $\dot{\mathbf{R}}$.

‡ This rather curious notation is chosen so that the final formulas acquire a suggestive appearance.

contains the field variables explicitly through $\mathbf{Z}_{\text{ext}}$, while the field conjugate to $\mathbf{A}_2$ contains the particle coordinates through $\langle \mathbf{Z}_{\text{ext}} \rangle$.) Equation (126b) is written in terms of Δ, the scaled charge distribution defined by $\rho = e\Delta$. One has $\int d^3x\, \Delta = 1$; taking Δ as a Dirac δ function amounts to the transition to a point, that is, a structure less electron. In that case, $\mathbf{F}_2$ has a δ function as well as $1/r$ singularity at the origin. Once the canonical variables are known, the Hamiltonian can easily be obtained:

$$H = \frac{\mathbf{P}^2 - e^2(\langle \mathbf{A}_{\text{ext}} \rangle)^2}{2m} + U(\mathbf{R}) + \frac{1}{8\pi}\int d^3x(\mathbf{E}_{\text{ext}}^2 + \mathbf{B}_{\text{ext}}^2) \tag{127}$$

The canonical variables are $\mathbf{P}$ and $-\mathbf{X}$, while the canonical fields are $\mathbf{A}_{\text{ext}}$ and $-\mathbf{E}_2/4\pi$. It can be checked directly that the canonical equations of the Hamiltonian (127), together with the definitions (126a) and (126b), yield exactly the original particle and field equation in the required approximation.† To have a consistent canonical formulation, H in Equation (127) should be expressed in terms of the canonical variables $\mathbf{P}$ and $\mathbf{X}$, $\mathbf{A}_{\text{ext}}$ and $\mathbf{E}_2$. This can easily be accomplished using Equations (126a) and (126b). Straight substitutions give

$$\begin{aligned} H = {} & \frac{\mathbf{P} - e^2\langle \mathbf{A}_{\text{ext}} \rangle^2}{2m} + U\left(\mathbf{X} - \frac{e}{m}\langle \mathbf{Z}_{\text{ext}} \rangle\right) \\ & + \frac{1}{8\pi}\int d^3x\left(\mathbf{B}_{\text{ext}}^2 + \mathbf{E}_2^2 + 8\pi\frac{e^2}{m}\Delta\mathbf{E}_2\cdot\langle \mathbf{Z}_{\text{ext}} \rangle + 16\pi^2\Delta^2\frac{e^4}{m^2}\langle \mathbf{Z}_{\text{ext}} \rangle^2\right) \end{aligned} \tag{128}$$

The Hamiltonian that Kramers finally adopts is an approximation to (128). One of the most striking features of Kramers' analysis is that the positional coordinate $\mathbf{X}$ is shifted by an amount (a small amount) that depends on the field. Presumably, $\langle \mathbf{Z}_{\text{ext}} \rangle$ is different from zero, only in domain of order a, where the shape factor Δ is nonvanishing. This implies [from Equation (126a)] that $\mathbf{X}$ is not too different from $\mathbf{R}$, the original particle position, and that [from Equation (126b)] $\mathbf{E}_2$ is not too different from $\mathbf{E}_{\text{ext}}$ (although $\mathbf{E}_2$ would possess a δ function singularity, which $\mathbf{E}_{\text{ext}}$ does not have). These observations suggest that it is reasonable to expand U in Equation (128) in a Taylor series around $\mathbf{X}$. Kramers goes as far as e^2 in that expansion. He consequently also neglects the e^4 terms in the integral in (128). Finally, the charge distribution factor Δ could either be taken as a δ function, in which case the $\mathbf{E}_2$ term can be taken out of the integral, or be expressed directly in terms of the average of $\mathbf{E}_2$ over the electron. The resulting approximate Hamiltonian is then

† It should be stressed that in the published version[122] there are numerous misprints, average value symbols come and go, subscripts are changed, as are signs. Miraculously, to the approximation to which Kramers works even the misprints cancel out.

$$H = \frac{\mathbf{P}^2}{2m} - \frac{e}{m}(\langle \mathbf{Z}_{\text{ext}} \rangle \cdot \nabla_X)U + \frac{1}{2}\frac{e^2}{m}(\langle \mathbf{Z}_{\text{ext}} \rangle \cdot \nabla_X)^2 U$$
$$+ \frac{e^2}{2m}(-\langle \mathbf{A}_{\text{ext}} \rangle^2 + 2\langle \mathbf{E}_2 \rangle \cdot \langle \mathbf{Z}_{\text{ext}} \rangle) \tag{129}$$
$$+ \frac{1}{8\pi}\int [(\text{curl } \mathbf{A}_{\text{ext}})^2 + \mathbf{E}_2^2]$$

This, as far as Kramers is concerned, is the end of the road. He does something rather atypical at this point; in the published version (also in the lecture delivered) he rewrites Equation (129), leaving out all subscripts, so that H assumes the seductive form:

$$H = \frac{\mathbf{P}^2}{2m} - \frac{e}{m}(\langle \mathbf{Z} \rangle \cdot \nabla_X)U + \frac{1}{2}\frac{e^2}{2m}(\langle \mathbf{Z} \rangle \cdot \nabla_X)^2 U$$
$$+ \frac{e^2}{2m}(-\langle \mathbf{A} \rangle^2 + 2\mathbf{E} \cdot \langle \mathbf{Z} \rangle) \tag{130}$$
$$+ \frac{1}{8\pi}\int [(\text{curl } \mathbf{A})^2 + \mathbf{E}^2] d^3x$$

It is this Hamiltonian or the Hamiltonian written in this language which Kramers compares and contrasts with the conventional Dirac Hamiltonian

$$H = \frac{\mathbf{P}^2}{2m} - \frac{e}{m}(\langle \mathbf{A} \rangle \cdot \mathbf{P}) + \frac{e^2}{2m}\langle \mathbf{A} \rangle^2 + \frac{1}{8\pi}\int [(\text{curl } \mathbf{A})^2 + \mathbf{E}^2] d^3x \tag{131}$$

The transition from (129) to (130) might appear as just a matter of notational convenience,† which by itself has no physical or mathematical significance. That is certainly correct, but it is nevertheless curious that Kramers, who went through such pains to stress the subtle distinction between the various fields $\mathbf{E}_1, \mathbf{E}_2$, and $\mathbf{E}_{\text{ext}}$, would choose a notation as in (130) which tended to obliterate the very distinctions he had struggled so long to uncover. Thus, Kramers compares Equations (130) and (131); however, the same letters in the two equations *do not* have the same meaning. For example, $\mathbf{E}$ in (131) is presumably the external field $\mathbf{E}_{\text{ext}}$, while the same symbol $\mathbf{E}$ in (130) is in fact $\mathbf{E}_2$, which although related to $\mathbf{E}_{\text{ext}}$ is not identical with $\mathbf{E}_{\text{ext}}$. [It is defined by Equation (126b).] There is no doubt that Kramers with his many years experience (and his superb technical skill) kept these distinctions clearly in mind, but there is considerable doubt about just how an audience generally unfamiliar with Kramers' philosophy and totally unfamiliar with his formalism would react to Kramers' legerdemain. Even the streamlined and abbreviated

† Actually, the situation is even more confusing. There clearly is an important difference between $\mathbf{R}$, the electron's position, and $\mathbf{X}$ [Equation (126a)], the altered position. Kramers calls $\mathbf{X}$, $\mathbf{R}_1$. In the simplified notations (130), he writes $\mathbf{R}_1$ as $\mathbf{R}$ again. But it really means $\mathbf{X}$!

derivation presented here is indirect, tortuous, and tricky. To have such a long analysis capped by a sudden change of notation might well have caused considerable confusion in the audience.

It appears that many of those attending the Solvay Conference were only interested in the general thrust of Kramers' ideas and cared but little about the complicated details.[123] But to Kramers, the Hamiltonians (129) and (130) were the long-sought structure-independent starting point for the study of the interactions between charges and radiation. Of course, in the nonrelativistic dipole approximation, the first term $-(e/m)(\langle \mathbf{Z} \rangle \cdot \nabla_X)U$ in (130) was already contained in Opechowski's 1941 paper; the next term is new with Kramers. The physical consequences of both Equations (130) and (131) are obtained from a perturbation analysis. The lowest-order perturbation in e in Kramers' case is $H_{\mathrm{K}}^{(1)} = -(e/mc)(\mathbf{Z} \cdot \nabla_X)U$; for the Dirac Hamiltonian it is $H_{\mathrm{D}}^{(1)} = (e/mc)(\mathbf{A} \cdot \mathbf{P})$. It is not difficult to show (but it takes some calculation) that the matrix elements of $H_{\mathrm{K}}^{(1)}$ and $H_{\mathrm{D}}^{(1)}$ are the same for states of the same energy. This in particular implies that Kramers' Hamiltonian would give the same values for the Einstein A and B coefficients as the calculation based on the Dirac Hamiltonian. This is certainly satisfactory. The study of the scattering of light by *free* electrons requires a second-order calculation. For free electrons in the Dirac case, only $H_{\mathrm{D}}^{(2)} = (e^2/2m)\mathbf{A}^2$ contributes; the only contribution to that scattering using Kramers' Hamiltonian comes from the terms $H_{\mathrm{K}}^{(2)} = (e^2/2m)(-\mathbf{A}^2 + 2\mathbf{E}_2 \cdot \mathbf{Z})$. These contributions are again the same so that the scattering of light of free electrons is described in the same way by both the traditional Dirac and the decidedly nontraditional Kramers Hamiltonian. The scattering of light by bound electrons requires again a second-order calculation; this time the terms $H_{\mathrm{K}}^{(1)}$ and $H_{\mathrm{D}}^{(1)}$ do contribute differently in second order. However, that difference is just compensated by the term $\frac{1}{2}(e^2/2m)(\mathbf{Z} \cdot \nabla_X)^2 U$, which has no counterpart in the Dirac Hamiltonian (131). Thus, for the A and B coefficients for the scattering of light, the Kramers and Dirac Hamiltonians, in spite of their different formal appearances, give the same physical results. It should be stressed that these results actually established the validity of Kramers' methods. However, this identity is not true in general; it is not even true for all second-order calculations. This is certainly what Kramers expected and hoped. He expressed these expectations as early as 1937.[124] He certainly intimated as much in his Shelter Island discussions and more than likely said so publicly.[125] In spite of this, in the surviving written documentations of the Shelter Island Conference, there are no examples of calculations in which the Kramers and Dirac Hamiltonians [(130) and (131)] actually give different results. It is merely suggested that some examples existed. In the Solvay Conference Lecture, delivered about a year after the Shelter Island Conference (also about a year after Bethe had published his derivation of the Lamb shift), Kramers presents an example of a second-order calculation, in which the two Hamiltonians do give different answers. The use of Kramers' Hamiltonian automatically yields a correction to the expression obtained by the usual Dirac Hamiltonian. This correction is precisely that introduced by Bethe in a more or less ad hoc manner to remove the divergence encountered in the calculation of the Lamb shift. A more detailed analysis of these results is given in Section III.D). Here it is pertinent to stress that a careful reading of the Solvay lectures shows that Kramers was pleased that he had been able to develop his cherished ideas on the quantum theory of radiation into a systematic, coherent scheme. It was no doubt a source of deep satisfaction to him that his consideration showed that Dirac's theory often—but not always—would give correct answers. The implementation of Kramers' ideas required a number of approximations; for this reason, Kramers did not consider specific and necessarily approximate calculations of detailed effects as especially significant. But he did feel strongly that his method of analysis and his procedure for setting up calculations were of fundamental importance.

D. Culmination or Oblivion: Van Kampen's Thesis

The development just described almost seems like an "epic" of many Hamiltonians. It is striking how close—in spite of many twists and turns and refinements in his arguments—Kramers stuck to his original central ideas. As Kramers had argued since 1936, it was necessary to distinguish carefully between the external field and the self-field. Furthermore, it was essential to distinguish between the total mass, which included the electromagnetic mass, and the mechanical mass. In the equations of motion, the experimental mass and the external field should occur. Once this was accomplished, a Hamiltonian formulation, either approximate or exact, had to be constructed. These two ideas remained unchanged throughout Kramers' work toward a final Hamiltonian. In the course of his investigations, Kramers tried several different approaches but these two general principles were maintained throughout. The systematic implementation of these principles took a long time. It took Kramers 5 years to generalize his results to all orders in e. This derivation was nonrelativistic and in the dipole approximation, led to the Hamiltonian (128). As a glance at Equation (128) shows, the result is neither simple nor physically transparent. Only if one is satisfied with a theory to lowest order in e is it possible to give a simple rule for the interaction between an electron and an electromagnetic field. Instead of the usual Dirac interaction, given by

$$H_{\mathrm{D}} = -\frac{e}{mc}(\mathbf{p}\cdot\mathbf{A}_{\text{total}}) \tag{132a}$$

Opechowski and Kramers obtain (to order e)

$$H_{\mathrm{K}} = -\frac{e}{c^2}(\mathbf{v}\cdot\mathbf{Z}_{\text{ext}}) = -\frac{e}{mc^2}(\mathbf{p}\cdot\mathbf{Z}_{\text{ext}}) \tag{132b}$$

Here Z_{ext} is the Hertz potential for the external field. Although more complicated than H_{D}, and without the attractive gauge-invariant feature of H_{D},

(132b) is still a rule, which can easily be implemented. The Hamiltonian (128), which is in principle valid to all orders in e, is much more complicated and no simple rule exists for its construction. As was explained in the last section, Kramers did not use that Hamiltonian for any applications. He did develop the Hamiltonian up to order e^2 as in (130) and all the applications mentioned before, as well as those to be discussed later, are based on that approximate Hamiltonian. It is operationally correct to say that the Hamiltonian (130) represents the final result of Kramers' efforts at constructing a structure-independent formulation of quantum field based on his fundamental and "nonnegotiable" assumptions. It will no doubt be evident that the derivation is lengthy and not very straightforward, nor is the physical interpretation particularly direct. Furthermore, Kramers' results and his methods appear peculiarly individualistic; they seem only remotely related to other efforts to construct suitable Hamiltonians describing the interactions between electrons and radiation. (See, for example, Bloch and Nordsieck[126] and Pauli and Fierz.[127]) Dirac expressed his surprise and almost bewilderment at Kramers' results after Kramers had delivered his talk at the Solvay Conference with the following comment: "I think that it should be possible to set up an exact contact transformation which will make Kramers' Hamiltonian more comparable with the usual one."[122]

It is illuminating to compare and contrast Kramers' Hamiltonian with that obtained later by van Kampen in his thesis.[82] Van Kampen was one of Kramers' last and most gifted students. That Kramers entrusted van Kampen with a significant study in his personally favorite topic shows beyond question that Kramers had a great deal of confidence in van Kampen. Thus, van Kampen's thesis dealt with extensions and applications of Kramers' approach to the quantum theory of radiation. Although there is no doubt that the initial impetus for these investigations came from Kramers, the execution is certainly due to van Kampen. In his studies, van Kampen reformulated Kramers' ideas but added a number of significant original touches. Compared to Kramers' investigations, the derivation of the Hamiltonian and the splitting of the fields in van Kampen's thesis is a model of simplicity and elegance. Kramers' ideas certainly pervade this work, but the technical implementation by van Kampen is both simple and direct. Consequently, the physical interpretation of the various manipulations is much easier to survey than was possible in Kramers' involved formalism. Van Kampen's procedure also allows a direct comparison of his derivation and assumptions with other studies, so that through van Kampen's work the physical importance of Kramers' approach became much more evident. The results in van Kampen's thesis can legitimately be considered as the culmination of Kramers' program.

First, van Kampen begins with the Hamiltonian†

† The notation used here is identical with that used in Section III.C. This should facilitate the comparison with Kramers' derivation given there. However, this notation is different from that used by Kramers, or van Kampen originally, so that a comparison with the original papers takes some care.

$$H_1 = \frac{1}{2m_0}(\mathbf{P} - e\langle\mathbf{A}\rangle)^2 + U(\mathbf{R}) + \frac{1}{8\pi}\int[\mathbf{E}^2 + (\operatorname{curl}\mathbf{A})^2]d^3x \tag{133}$$

H_1 is the Hamiltonian of the system, consisting of the field plus the electron. The electron's position is $\mathbf{R}$, its momentum is $\mathbf{P}_1$, and m_0 in (133) is the mechanical mass. It is assumed that in (133) the longitudinal field has been eliminated (it could be contained in U), so that both $\mathbf{E}$ and $\mathbf{A}$ are transversal. $\mathbf{A}$ and $\mathbf{E}$ both refer to *total* fields; no splitting of any kind is made in (133). The equations of motion obtained from (133) correspond precisely to Kramers' equation (100). These equations could be derived from the Lagrangian (120). The corresponding Hamiltonian would again be (133). Thus, van Kampen's starting point is exactly the same as Kramers.

Second, van Kampen introduces the restriction to dipole radiation at an early stage. The quantity $e\langle\mathbf{A}\rangle$ occurring in (133) is defined in the usual way:

$$e\langle\mathbf{A}\rangle = \int A(\mathbf{x})\rho(\mathbf{x} - \mathbf{R})d^3x = \int A(\mathbf{R} + \mathbf{x})\rho(\mathbf{x})d^3x \tag{134}$$

$\rho(x)$ is as usual the charge distribution in the finite electron; it depends on (or in fact defines) the structure of the electron. Equation (134) shows that generally $\langle\mathbf{A}\rangle$ depends on $\mathbf{R}$. If the electron considered is bound in an atom, and if the size of the atom is small compared to the relevant wavelengths in the external field, $e\langle\mathbf{A}\rangle$ is practically independent of $\mathbf{R}$. Formally, we can expand $e\langle\mathbf{A}\rangle$ in a Fourier series:

$$e\langle\mathbf{A}\rangle = \sum_k \int d^3x\, A_k e^{i\mathbf{K}\cdot\mathbf{R} + i\mathbf{K}\cdot\mathbf{x}}\rho(\mathbf{x}) \tag{135a}$$

If the wavelength $\lambda > R$, the usual dipole approximation yields

$$e\langle\mathbf{A}\rangle \cong \int d^3x\, \rho(\mathbf{x})A(\mathbf{x}) \tag{135b}$$

The dipole approximation (135b) shows via (133) that

$$\dot{\mathbf{P}} = -\nabla_R U \tag{136}$$

Hence, in the dipole approximation, the transport of momentum from the transverse field to the electron is neglected.

Third, since both van Kampen and Kramers restrict their considerations to dipole radiation, it makes sense to exploit the simplification that this approximation produces. This can be done most effectively by making a multipole expansion of $\mathbf{A}$. Since all multipole waves, other than dipole waves, vanish at the origin, their contribution to (135b), where $\rho(\mathbf{x})$ is effectively $\delta(\mathbf{x})$, is very small. In a sphere of radius L, the multipole expansion of $\mathbf{A}$ in the dipole approximation can be written

$$\mathbf{A}(\mathbf{x}) = \operatorname{Tr}\sum_{n=1}^{\infty}\sqrt{\frac{3}{L}}\,\mathbf{q}_n\frac{\sin\nu_n r}{r} \tag{137a}$$

$$\nu_n = \frac{n\pi}{L} \tag{137b}$$

Tr as usual stands for the transversal part; the $\mathbf{q}_n$ are oscillator-type coordinates for the transversal field. Since $-(1/4\pi)\mathbf{E}$ is canonically conjugate to $\mathbf{A}$ via the Hamiltonian (133), $\mathbf{E}$ can be developed in terms of coefficients $\mathbf{p}_n$, which are canonically conjugate to $\mathbf{q}_n$:

$$-\mathbf{E}(\mathbf{x}) = \mathrm{Tr} \sum \sqrt{\frac{3}{L}} \mathbf{p}_n \frac{\sin \nu_n r}{r} \tag{138}$$

One can express $e\langle\mathbf{A}\rangle$ in terms of these coordinates $\mathbf{p}_n$ and $\mathbf{q}_n$ as

$$e\langle\mathbf{A}\rangle = \sum_n \varepsilon_n \mathbf{q}_n \tag{139}$$

The coefficients ε_n depend explicitly on the structure of the electron:

$$\varepsilon_n = \sqrt{\frac{4}{3L}} \int_0^\infty 4\pi r \, dr \, \rho(r) \sin \nu_n r \tag{140}$$

It is very convenient to define a new set of coefficients δ_n, which also depend on the structure of the electron, by

$$\varepsilon_n = \delta_n \nu_n \sqrt{\frac{4e^2}{3L}} \tag{141}$$

The δ_n structure factors have the nice property that they approach unity in the limit of a point electron. The Hamiltonian in terms of the canonical variable pairs $(\mathbf{P}, \mathbf{R})$ and $(\mathbf{p}_n, \mathbf{q}_n)$ becomes

$$\begin{aligned} H_2 = {} & \frac{1}{2m_0}\mathbf{p}^2 + U(\mathbf{R}) - \frac{1}{m_0}\mathbf{P}\left(\sum \varepsilon_n \mathbf{q}_n\right) + \frac{1}{2m_0}\left(\sum \varepsilon_n \mathbf{q}_n\right)^2 \\ & + \frac{1}{2}\sum (\mathbf{p}_n^2 + \nu_n^2 \mathbf{q}_n^2) \end{aligned} \tag{142}$$

It is clear that H_2 is a strongly structure-dependent expression. Both m_0, the mechanical mass, and ε_n through Equation (140) depend explicitly on the structure.

Fourth, inspection of Equation (142) shows that the particle coordinates $\mathbf{P}$ and $\mathbf{R}$ are coupled bilinearly to the field coordinates $\mathbf{p}_n$ and $\mathbf{q}_n$. In fact, the only coupling terms is $\mathbf{P} \cdot (\sum \varepsilon_n \mathbf{q}_n)$. (This, of course, is due to the dipole approximation.) This suggests immediately that it should be possible to eliminate this term by a suitable linear transformation. This can indeed be accomplished by introducing $(\mathbf{P}', \mathbf{X})$ and $(\mathbf{p}_n', \mathbf{q}_n')$ as canonical variable pairs, where

$$\mathbf{p}_n' = \mathbf{p}_n \tag{143a}$$

$$\mathbf{q}_n' = \mathbf{q}_n - \frac{\varepsilon_n}{m\nu_n^2}\mathbf{P}' \tag{143b}$$

$$\mathbf{P}' = \mathbf{P} \tag{143c}$$

$$\mathbf{X} = \mathbf{R} - \sum_n \frac{\varepsilon_n}{m\nu_n^2}\mathbf{p}_n' \tag{143d}$$

The field oscillator coordinates $\mathbf{q}_n$ are shifted by an amount depending on the particle momentum $\mathbf{P}$; the particle position $\mathbf{R}$ is shifted by an amount depending on the field momentum. m in Equations (143b) and (143d) is to be identified with the experimental mass and is given by

$$m = m_0 + \sum_n \frac{\varepsilon_n^2}{\nu_n^2} = m_0 + \frac{4e^2}{3L} \sum_n \delta_n^2 \tag{144}$$

It is not too difficult to show that $\sum (\varepsilon_n^2/\nu_n^2)$ in (144) is just the electromagnetic mass; consequently, it is perfectly in line with Lorentz's and Kramers' ideas to identify m in (144) with the electromagnetic mass. It is not surprising, but yet instructive to observe, that for a point electron, where $\delta_n \to 1$, the electromagnetic mass term diverges, as it should for a point electron. Substitution of (143) in (142) gives the penultimate Hamiltonian

$$H_3 = \frac{1}{2m}(\mathbf{P}')^2 + U\left(\mathbf{X} + \sum \frac{\varepsilon_n}{m\nu_n^2}\mathbf{p}_n'\right) + \frac{1}{2}m_0\left(\sum \varepsilon_n \mathbf{q}_n'\right)^2 + \frac{1}{2}\sum (\mathbf{p}_n' + \nu_n^2 \mathbf{q}_n'^2) \tag{145}$$

This Hamiltonian is similar, but not identical, to Kramers' Hamiltonian (128). It too contains the potential energy at a shifted position, but Kramers' expression is considerably more involved; it also contains a separation in the external field and self-field, which (145) does not have.

Fifth, the final transformation is just a principal axis transformation of the two quadratic terms in (145). At the same time, m defined by (144) is introduced, resulting in

$$H_4 = \frac{\mathbf{P}'^2}{2m} + U\left(\mathbf{X} + \frac{e}{m}\sqrt{\frac{4}{3L_n}}\frac{\cos\eta_n}{k_n}\mathbf{p}_n''\right) + \frac{1}{2}\sum (\mathbf{p}_n''^2 + k_n^2 \mathbf{q}_n''^2) \tag{146}$$

In (146), $\mathbf{p}_n''$ and $\mathbf{q}_n''$ are linear combinations of $\mathbf{p}_n'$ and $\mathbf{q}_n'$, respectively; these are the new canonical field variables. The k_n are the roots of a characteristic equation; they do contain the structure of the electron:

$$m = k^2 \sum \frac{\varepsilon_n^2}{\nu_n^2(k^2 - \nu_n^2)} \tag{147}$$

All other terms in (146) are defined in terms of the solution k_n of the characteristic equation (147). The phase shifts η_n are given by

$$Lk_n = \eta_n + n\pi \qquad 0 < \eta_n < \pi/2 \tag{148}$$

The lengths L_n are

$$L_n = L - (\cos\eta_n)^2\left(\frac{2e^2}{3m}\right) = L - r_0(\cos\eta_n)^2 \tag{149}$$

In (149) $r_0 = 2e^2/3mc^2$ is of the order of the classical electron radius. If L is a macroscopic length, (149) shows that L_n must be very nearly L.

It is an interesting observation that the characteristic equation (147) retains a well-defined meaning in the limit of a point electron; that is, when $\delta_n = 1$,

$\varepsilon_n = v_n\sqrt{4e^2/3L}$. Not only does the series (147) converge in that case, but it can be summed explicitly, giving

$$m = \frac{2e^2}{3}k\left(\cot kL - \frac{1}{kL}\right) \tag{150}$$

For L macroscopic, (150) leads to a very simple equation for k:

$$\tan Lk = kr_0 \tag{151}$$

A glance at (151), (149), and (148) shows that for wavelengths much larger than e^2/m, $kr_0 = (1/\lambda)(e^2/m) \simeq 0$, so that $k_n \simeq n\eta$. The structure-dependent effects can expect to be small. Thus, (146) is indeed the approximately structure-independent Hamiltonian that Kramers had sought for such a long time. The distinct elements of Kramers' philosophy are also clearly discernible in this analysis. As (144) shows, the limit to a point particle $\delta_n \to 1$ causes the electromagnetic mass to diverge, but Kramers argues that this divergence is of no consequence, since one should use the experimental mass anyhow. The remaining interactions between the electron and the field are described (in the dipole approximation) by the shifted position of the electron in (146). It is possible in principle to make ordinary Schrödinger-type wave mechanical calculations starting from H_4. This is precisely what van Kampen does in his thesis; thus, the derivation leading to (146) replaces the earlier, more involved Kramers derivation.

It is remarkable that the splitting in external field and self-field, which caused Kramers so much grief, does not seem to occur in van Kampen's derivation at all. This splitting is actually present in this formalism as well, but it enters in such an unobtrusive manner that it takes some analysis to recognize it as the separation into self- and external fields. If equations (143a–d) and the final (5th) transformation are combined with the expansion of the vector potential, we obtain

$$\mathbf{A}(\mathbf{x}) = \mathrm{Tr}\sum_{n=1}\sqrt{\frac{3}{L}}\mathbf{q}'_n\frac{\sin v_n r}{r} + \mathrm{Tr}\sum_{n=1}^{\infty}\sqrt{\frac{3}{L}}\frac{\varepsilon_n}{mv_n^2}\mathbf{P}'\frac{\sin v_n r}{r} \tag{152}$$

This is very similar to (or at least approximately so) Kramers' splitting of the external field and self-field. Define $\mathbf{A}_{\text{self}}$ by

$$\mathbf{A}_{\text{self}} \equiv \mathrm{Tr}\sum_{n=1}^{\infty}\sqrt{\frac{3}{L}}\frac{\varepsilon_n}{mv_n^2}\mathbf{P}'\frac{\sin v_n r}{r} \tag{153}$$

This automatically gives the external field as

$$\mathbf{A}_{\text{ext}} = \mathrm{Tr}\sum_{n=1}^{\infty}\sqrt{\frac{3}{L}}\mathbf{q}'_n\frac{\sin v_n r}{r} \tag{154}$$

The surprising part about these trivial rearrangements is that $\mathbf{A}_{\text{self}}$, as defined, not only can be evaluated explicitly but indeed gives a physically reasonable expression for the self-field. The calculation of (153), using (140) and (37b) gives

$$\mathbf{A}_{\text{self}} = \operatorname{Tr}\frac{\mathbf{P}'}{m}\int\frac{\rho(\mathbf{r}')d^3r'}{|\mathbf{r}-\mathbf{r}'|} \tag{155a}$$

The integral is immediately recognized as a solution of the Poisson equation $\Delta\phi = 4\pi\rho$, which for a point charge becomes e/r. Thus, the self-field (155a) is

$$\mathbf{A}_{\text{self}} = \operatorname{Tr}\frac{\mathbf{P}'}{m}\left(\frac{e}{r}\right) \tag{155b}$$

This expression for the vector potential of the self-field can be shown to be

$$\mathbf{A}_{\text{self}} = \operatorname{Tr}\left(\dot{\mathbf{R}}\frac{e}{r} - \frac{e}{r}\sum_n \frac{\varepsilon_n}{mv_n^2}\dot{\mathbf{p}}_n'\right) \tag{156}$$

Thus, the self-field as defined by (153) and by Kramers in (106) agree up to order e but are different in higher orders. It was emphasized in the previous discussion that there is no unique definition of the self-field anyhow. Physically, one wants to characterize that portion of the field of a moving (accelerating) electron which can legitimately be considered as being swept along by the electron. As such, it should reduce to the Coulumb field when the electron is at rest, but these stipulations in the absence of invariance principles do not suffice to define the self-field uniquely. One could demand, as Kramers did in his earliest studies, that this field to first order in e gives the electric and magnetic fields of an electron at $\mathbf{R}$, moving with constant velocity $\dot{\mathbf{R}}$, but this still allows any number of expressions, differing in order e^2. The choice made by van Kampen leads to the Hamiltonian H_4 [Equation (146)], which is considerably simpler than Kramers' Hamiltonian (128), and the corresponding choice of the self-field is every bit as legitimate as Kramers' choice. Kramers' definition of the self-field was based on his vast experience with computations of the fields of classical electrons. As such, it is a bit more pictorial and appeals more immediately to the physical intuition than van Kampen's choice, which is more formal. But as mentioned numerous times, the actual definition of the self-field adopted is somewhat arbitrary anyhow. What is important in Kramers' program is the elimination of the self-field and the structural properties, and this can be accomplished by either the Kramers or van Kampen choice—which agree up to order e. But the van Kampen choice is formally much simpler.

The choice of $\mathbf{A}_{\text{ext}}$ in (154) also allows a transparent interpretation of the phase shifts occurring in the Hamiltonian H_4. The expression of $\mathbf{A}_{\text{ext}}$ in terms of the $\mathbf{q}_n''$ has the form

$$\mathbf{A}_{\text{ext}} = \operatorname{Tr}\sum_{n=1}^{\infty}\sqrt{\frac{3}{L}}\mathbf{q}_n'\frac{\sin\nu_n r}{r} = \operatorname{Tr}\sum\sqrt{\frac{3}{L}}\mathbf{q}_n''\frac{\sin(k_n r - \eta_n)}{r} \tag{157}$$

The orthogonal transformation Q, which implements the change of variables from q', p' to q'', p'',

$$q_n' = \sum Q_{nn'}q_{n'}'' \qquad p_n' = \sum Q_{nn'}p_{n'}'' \tag{158}$$

is just a change of basic functions for the expansion $\mathbf{A}''_{\text{ext}'}$ as (157) shows explicitly. Consequently, the final Hamiltonian H_4 [Equation (146)] depends on the external field [by (157) expressed in terms of q''_n and η_n] and on the experimental mass m, just as Kramers wished. However, $\mathbf{P}'$ and $\mathbf{X}$ are no longer the velocity and position of the electron $\dot{\mathbf{R}}$ and $\mathbf{R}$. But none of this really matters, for the Hamiltonian H_4 can be used as it stands to compute all physical effects.

The transformations employed by van Kampen in obtaining his final Hamiltonian H_4 can be compared directly with those used earlier by Bloch and Nordsieck[126] and Pauli and Fierz[127] in their investigations of the proper mathematical formulation of an electron interacting with a radiation field. All these transformations possess the common feature that the particle coordinates are modified by variables which describe the modes of the field [see Equations (143a–d), for example]. Physically, the new variables are intended to describe a particle as an object surrounded by a cloud of photons. The motivation for all these transformations is the same and always goes back to Kramers' idea of the elimination of the self-field. It is first necessary to identify what properly should be called the self-field; then this can be eliminated at a later stage. This was accomplished in formally different ways by different authors, but all had to construct canonical variables, which were combinations of particle coordinates and field variables.

The most definitive and complete implementation of this program in a nonrelativistic theory, in dipole approximation, was given by van Kampen in his Hamiltonian (146). This result genuinely represents the culmination of Kramers' program and van Kampen indeed used this Hamiltonian to describe and clarify a number of questions posed for so long and so insistently by Kramers. It was one of these not so rare ironies of the development of science that at the same time (or even before) that Kramers' program achieved its conclusions, it was superseded by the monumental results of Schwinger.[128] As Schwinger later freely admitted,[129] these seminal papers are based on ideas very similar to those of Kramers. But Schwinger's powerful formalism enabled him to construct a fully relativistic theory that was not restricted to the dipole approximation. In this respect, Schwinger went far beyond the work of Kramers and van Kampen. Schwinger, as van Kampen before him, had to construct a series of canonical transformations to isolate the structure-independent terms in the Hamiltonian. This was done in successive approximations by making an expansion in e. In his earlier papers, Schwinger only computed to the first power in e; later, higher approximations (in e) were calculated. The Schwinger quantum electrodynamics was certainly a vast improvement over previous formulations and as recast by Feynman, it rapidly became the standard version of quantum electrodynamics; Kramers' heroic attempt was all but forgotten. The culmination of Kramers' program in van Kampen's thesis was certainly an anticlimax and a disappointment, eclipsed as it was by the technical brilliance of the Schwinger–Feynman–Dyson formalism. With the availability of this formalism, most physicists concentrated on detailed, specific calculations rather than delving into the physical and conceptual foundations of the formalism. It is therefore not surprising that the investigations of Kramers and van Kampen, which dealt with a nonrelativistic formalism that leaned heavily on classical ideas, were largely ignored. (The relative inaccessibility of their papers did not help either.) This is actually most unfortunate; there are few places where the physical ideas underlying the renormalization theory are explained more clearly, and there are no places where the relation between the classical and quantum theory of scattering is treated as well. To the world at large, the work of Kramers and van Kampen did not have a tremendous impact—at that time in any case—but an examination of van Kampen's thesis shows that it was possible to construct a consistent quantum theory of radiation, based on the ideas that Kramers had propounded for so long. Within its domain of applicability, the Kramers–van Kampen treatment allowed an analysis of emission, absorption, and scattering of light by atoms which supplemented and clarified the traditional treatment. Thus, Kramers' Solvay paper and van Kampen's thesis *did* present substantial progress; questions not settled earlier with Kramers' program were resolved in these papers. The results obtained represented the culmination of Kramers' program. Within the self-imposed limit of that program, van Kampen's thesis was as far as the analysis could be carried out. Independent of the degree of recognition or the type of appreciation this work engendered by the physics community.

Kramers' scheme of starting with a finite electron, eliminating the self-field and the structure, and going to a canonical form, worked in the nonrelativistic dipole approximation. There was, as expected and demanded by Kramers, a close parallelism between the classical and quantum treatment. A start had been made with the understanding and elimination of divergencies. New phenomena, not describable by the Dirac theory—remnants of the elimination process of the structure-dependent terms—could be anticipated. The succeeding development, based in general terms on Kramers' ideas, proceeded in quite different directions. Kramers, with his customary incisive insight, had indeed recognized and identified some fundamental problems. In a characteristic manner, he set out to investigate this problem in a detailed, specific, circumscribed context. In so doing, he prepared the road for the new explosive developments. But he did not travel that road. The limitations which he had imposed on himself were his undoing. The culmination of his program led directly to its oblivion.

IV. Retrospect: The Postponed Recognition

A. The Lamb Shift: How High the Stakes

By the time of the Shelter Island Conference, June 1947, Kramers had been studying the quantum theory of radiation for some 10–12 years. He had devoted a major part of his textbook to that topic, and he had guided a number of his students (especially the promising ones) in that direction. Although many physicists were generally aware of the circle of problems that Kramers attacked, especially the self-energy problem, most of them paid little attention to those questions and even less to Kramers' efforts to deal with them. Even in The Netherlands, where Kramers' influence was surely the greatest, only Kramers' students and immediate associates were familiar with the details of his investigations. Elsewhere, Kramers' work was hardly known at all. Yet in the summer of 1947, Kramers' work was suddenly in the center of interest. However implicit his ideas on mass renormalization might have been in his Shelter Island presentation, many of those present—such as Weisskopf, Oppenheimer, Bethe, and Schwinger—were totally clear that Kramers' dictum to "eliminate the electronic structure" would lead to experimentally observable deviations from the Dirac equation. Even though this is interesting enough in its own right, this by itself would hardly have generated the excitement that pervaded physics that summer. It is, after all, hardly surprising that different theories give rise to different experimental results. What caused the commotion was the experimental discovery of the Lamb shift, the first definitive indication of the inadequacy of the Dirac equation. Kramers' scheme did suggest such deviations, and many persons seized upon Kramers' ideas to explain the new observations. The major stimulation for the frantic developments following the Shelter Island Conference was the need to produce a number for the Lamb shift that could be compared with experiments. This indeed was accomplished by Bethe, not more than 2 weeks after the conference. The succeeding investigations, especially by Schwinger and Feynman (but also by French and Weisskopf[130]), were applications, systematizations, subtle refinements, and vast extensions of these basic ideas. About a year later, Oppenheimer, referring to these developments, stated: "These developments [mass and charge renormalization] which could have been carried out at any time during the last fifteen years, required the impetus of experiments to stimulate and verify."[131] Oppenheimer was partially correct in his assessment; the experimental results of Lamb (and later of Kusch and Foley) unquestionably provided a major impetus for these studies. But Oppenheimer was also wrong; Kramers perhaps more than most physicists did concern himself with the conceptual problems of radiation theory, for about exactly the same 15 years mentioned by Oppenheimer. And he did not just worry about the problem; he made substantial contributions to its resolution. He did stress the need to introduce the experimental mass of the electron in the basic equations and in so doing he introduced the idea of mass renormalization. Oppenheimer was not correct in claiming that theorists had ignored the renormalization problems; it is more correct to say that the physics community ignored those who worked at it. But even though Kramers thought long and hard about these questions, it is doubtful that he ever anticipated that these questions and his approach would be thrust into the center of physics. This had another consequence: Kramers usually worked by himself in his own way at his own pace; he had not encountered direct scientific competition since his days in Copenhagen and his bet with Pauli on what was to become the Dirac theory. But with the sudden interest in self-energy questions, the field, far from being Kramers' semiprivate concern, became of vital importance to a large number of physicists. Indeed, many physicists literally descended on quantum electrodynamics and Kramers had to contend with a large number of brilliant, eager competitors. This was not much in his style, and he responded in characteristic fashion by withdrawing into his own set of ideas, ignoring the competition. It is hard to know how far a stroke in the late summer of 1947 contributed to this reaction. It is even harder to know whether the inevitable pressure brought on by the ferocious competition following his Shelter Island presentation had anything to do with his subsequent illness and depression. All that is clear is that he wrote during 1947–1948 an elaborate version of his Shelter Island talk—which summarized and completed his earlier studies.

But it remains remarkable how little attention Kramers paid to the Lamb shift. He realized early that his procedure would lead to results different from the Dirac theory. Within his formalism, Kramers could easily have computed the level shift, but he did so only in a rather haphazard manner, for rather special models. It never occurred to him[132] to put in numbers to estimate the order of magnitude of the deviations expected. To Kramers, his theory was too incomplete, too fragmentary, to warrant a serious numerical treatment. A serious comparison with experiment seemed presumptuous to Kramers; he lacked the *gotspe* for such comparisons. Whatever the precise motivations for Kramers' lack of attention might be, he did not make a serious calculation of the Lamb shift. It is interesting and typical that even after Bethe had carried out his calculation, Kramers in his comments on Bethe's work emphasized its shortcomings rather than its successes. With all this clearly understood, it is still pertinent to show how Kramers' formalism leads directly to the Lamb shift. Especially in the version given by van Kampen, the calculation of the level shift is quite smooth and proceeds without arbitrary cut-off and subtraction rules. The special feature of Kramers' procedure, lucidly presented in van Kampen's treatment, is that in the final Hamiltonian H_4 it is possible to go to the limit of a point electron and obtain a scheme with which one can calculate without ambiguities. This of course was one of the key elements in Kramers' philosophy: Once the self-field is defined and properly eliminated, the remainder of the theory should be finite and unambiguous. This is exactly the way things worked out. However, the transition to the point electron, where $m_0 \to \infty$, could not be carried out on the level of H_3 [as given by Equation (145)]; it contained an explicit dependence on m_0. The transformation to the final Hamiltonian, which included the phases η_n, was essential to eliminate the last vestige of the m_0 dependence. The resulting Hamiltonian depends on the parameters k_n, L_n, and η_n. A knowledge of k_n implies η_n by (148), while η_n in turn fixes L_n by (149). Thus, only the k_n needs to be calculated. By either (150) or (151), they have a well-defined meaning for a point electron; the only remnant of the electron structure remaining is the radius $r_0 = 2e^2/3mc^2$. Thus, the Kramers procedure works like a charm; the essential novelty is the occurrence of the $\cos \eta_n$ terms in (151). These are precisely the terms that eventually will produce convergence.

The advantage of the van Kampen formalism is just that within a well-defined context one can carry out a systematic mathematical analysis. The phases $\cos^2 \eta(k)$ always cause the convergence of integrals. These same phases, in exactly the same sense, "cause" the convergence of the positional fluctuation. A free electron in a photon vacuum will execute a fluctuating motion (perhaps superimposed on a uniform motion), because of the zero-point fluctuation of the quantum field. In the usual formulation of quantum field theory, these positional fluctuations given by $\langle (\mathbf{R} - \mathbf{X})^2 \rangle$ (where $\langle \ \rangle$ is the ground-state quantum average) are known to diverge.† However, the van Kampen Hamiltonian (146) gives for these fluctuations the finite expression

$$\langle (\mathbf{R} - \mathbf{X})^2 \rangle = \frac{2e^2}{3m^2 L} \sum_n \left(\frac{\cos^2 \eta_n}{k_n} \right) \tag{159}$$

It is clear that without the $\cos^2 \eta_n$ term the sum would be divergent; however, the $\cos^2 \eta_n$ term again causes convergence. This is a general feature; the Kramers methodology of amalgamating m_0 in m, as applied by van Kampen, leads to a finite formalism, where no further subtraction or cut-offs are needed—or allowed. This is particularly true for the level shift. It is especially interesting that this expression is very similar to the divergent expression obtained for the level shift by Bethe (116) which contains a K integral of the form

$$\int dK \frac{1}{K + h\nu_{nm}} \tag{160}$$

Bethe presented arguments suggesting that the divergent integral should be cut off at the inverse Compton wavelength $K_C = mc/\hbar$. The level shift computed with that cut-off is quite near the experimental value. It should be emphasized that in the Kramers–van Kampen formalism, there is no question of a cut-off; the integral for the level shift is finite and there is no need or opportunity for a further cut-off. It is worth pointing out that the value of the convergent integral for the level shift in Kramers theory is practically the same as that obtained if the divergent integral (160) were cut off at the wavenumber corresponding to the classical electron radius $K_0 = mc^2/e^2$:

$$\int^{K_0} dK \frac{1}{K + h\nu_{mn}} \cong \int_0^\infty dK \frac{\cos^2 \eta(k)}{K + h\nu_{mn}} \tag{161}$$

It is particularly interesting that the value of the level shift as given by the systematic and organized development of van Kampen and Kramers (without cut-off or subtractions) is not nearly as close to the experimental value as

† Welton[133] has used these positional fluctuations to give a charming and intuitive derivation of the Lamb shift.

Bethe's answer obtained using the Compton wavelength as a cut-off. This illustrates that Kramers' method, even though it can be carried out consistently and completely within this framework, had made too many restrictive assumptions which vitiated its applicability to realistic physical situations. In this discussion, all relativistic and recoil effects have been neglected, so it is not surprising that the numerical agreement is not very good. Although some of the aspects of the Lamb shift are capable of a "pseudo" nonrelativistic quantum mechanical description (as was shown, for example, by the Welton treatment[133]), other features require a precise relativistic treatment. This is in fact one of the instances (for others see Section IV.D) where a naive correspondence argument can give misleading results. Even a qualitative discussion requires some elements of the full relativistic theory. In retrospect, Bethe was quite correct in cutting off his divergent integral at the Compton wavelength. This incorporates just enough of the relativistic theory to lead to an excellent first approximation. For finer details, a fully relativistic theory must be employed.

It was Bethe's[134] great insight to recognize immediately that Kramers' ideas were the key to the understanding of the level shift observed by Lamb. Bethe, who was always especially interested in—and brilliantly adept at—obtaining experimental numbers from mathematical considerations, summarized Kramers' cogitations in a few simple pragmatic principles which he then applied in a most imaginative way to obtain a numerical value for the Lamb shift. He attributes the observed level shift directly to the interaction between the electron and its own (quantized) electromagnetic field. This effect was, of course, old and well known, but in all previous treatments it came out to be infinite and it was usually discarded or otherwise ignored (not by Kramers). Bethe instead recognizes this interaction as the physical mechanism for the level shifts. To obtain or extract finite results, he accepts Kramers' ideas that a good share (if not all) of the divergence is due to the divergence of the electromagnetic mass of the electron. Bethe starts by writing down the self-energy of an electron in a quantum state s. This is given by a standard formula as

$$W_s = -\frac{2e^2}{3\pi\hbar c^3}\int_0^{\varepsilon_0} \varepsilon\, d\varepsilon \sum_n \frac{|\mathbf{v}_{sn}|^2}{E_n - E_s + \varepsilon} \tag{162}$$

In this formula, $\mathbf{v}_{sn}$ is the s, n matrix element of the velocity operator $\mathbf{v} = \mathbf{p}/m$, E_n is the energy of state n, $\varepsilon = \hbar\omega$ is the energy of an emitted quantum, and ε_0 is a cut-off in the energy integral. Formula (162) necessarily involves the mass of the electron. If that mass is taken as the experimental mass of the electron, as was customary, expression (162) cannot be the physical value of the level shift resulting from the interaction of the electron with its own field. Recall that the experimental mass contains the electromagnetic mass, and, as emphasized for so long by Kramers, the electromagnetic mass is just the manifestation of the interaction of the *free* electron with its own electromagnetic field. As was often mentioned by Kramers and as used explicitly by Bethe here, the only effect that the interaction of a self-field with an electron could produce is this change of mass. For a *free* electron the interaction energy corresponding to W_s would be W_0, given by

$$W_0 = -\frac{2e^2}{3\pi\hbar c^3} \int \varepsilon \, d\varepsilon \frac{|\mathbf{v}|^2}{\varepsilon} \tag{163}$$

This is the change in kinetic energy of a free electron (for fixed momentum) (i.e., the change of mass) resulting from its interaction with its own electromagnetic field. But by using the experimental mass in (162), the electromagnetic mass is already included. Thus, to obtain the physical level shift, the contribution of the free electron self-energy W_0 must be subtracted from the expression W_s. The physical value of the level shift is just the difference between (162) and (163);

$$\Delta W_s = \frac{2e^2}{3\pi\hbar c^3} \int_0^{\varepsilon_0} \sum_n \frac{|\mathbf{v}_{sn}|^2 (E_n - E_s)}{E_n - E_s + \varepsilon} = W_s - W_0 \tag{164}$$

Some further manipulations (Bethe was a true virtuoso in these calculations), together with the assumption that ε_0, the cut-off, is much larger than the energy differences $E_s - E_n$, leads to the result

$$\Delta E_s = -\frac{2e^2}{3\pi\hbar c^3} \sum_n |\mathbf{v}_{sn}|^2 (E_n - E_s) \log \frac{\varepsilon_0}{|E_n - E_s|} \tag{165}$$

It is evident that this result still diverges for $\varepsilon_0 \to \infty$. Bethe argues that relativity provides a natural cut-off, so he chooses $\varepsilon_0 = mc^2$ as a cut-off. Since ε_0 is large compared to $|E_n - E_s|$, Bethe treats the log in (165) as constant (independent of n). This enabled him to evaluate the sum and to obtain an explicit numerical value for ΔE_s. The resulting numerical value of 1040 megacycles for the Lamb shift in hydrogen is in impressive agreement with the observed value of 1000 megacycles.

Even a casual examination of Bethe's[134] procedure shows clearly that it contains many of the elements of the quantum theory of radiation that Kramers had been belaboring for so many years. It is therefore more than a little surprising that when Kramers did refer to Bethe's contribution, in his Solvay report, he did so in an off-hand manner.[119] In this report, which was described in detail in Section III.C, Kramers finally arrived at a structure-independent Hamiltonian that satisfied all his conditions and constraints [formulas (128)–(130)]. A perturbation analysis of this Hamiltonian (which was markedly different from the conventional Dirac Hamiltonian) had shown that for many processes it gave the same answers as did the customary Dirac treatment. However, if Kramers' Hamiltonian was used to calculate the self-energy of an electron, different results were obtained. A straightforward second-order perturbation calculation using Kramers' Hamiltonian gives a result equivalent to (165), but *no* subtraction of the type that Bethe employed in (164) is invoked at all. Kramers does get two terms: Since he calculates to order e^2, one term arises from the e^2 term in (130) and the other from a second-order calculation of the e terms. The combination is equivalent to Bethe's subtracted result but Kramers emphasizes that his calculation involved no subtraction. It appears that Kramers was not all that enthralled with Bethe's procedure. He comments: "It is difficult to make this [Bethe's] argument quite rigorous, but it has a certain physical plausibility."[135] This is hardly an enthusiastic endorsement. In both Bethe's and Kramers' treatments, the resulting integral is logarithmically divergent. Kramers was not particularly happy with that either; he ends his discussion of the Lamb shift with these words: "On the question of the upper limit of the logarithmically divergent integral this [Kramers'] treatment throws of course as little light as Bethe's [first] treatment referred to, [this] theory being restricted to electric dipole radiation."[136] He was evidently not impressed by Bethe's argument[134] that relativity would provide a natural cut-off, leading to a cut-off of mc^2 for ε_0. [This is the same as a cut-off at the inverse Compton wavelength in Equation (160).] Recall from the discussion given earlier [(159)] that a systematic application of Kramers' ideas, as implemented by van Kampen, gives a finite value for the Lamb shift, without any subtraction or cut-off. Unfortunately, the numerical value obtained this way is not nearly as close to the experimental value as that of Bethe. An understanding of Bethe's success, as contrasted with the failure of the Kramers–van Kampen formula, required a physical explanation of why the inverse Compton wavelength in the K integral was a more appropriate cut-off than a cut-off at the (inverse) classical electron radius which was more natural in the Kramers–van Kampen formalism [see Equation (161)]. In the absence of such an explanation, Kramers considered Bethe's procedure as highly arbitrary. It was clear that since Kramers' treatment was nonrelativistic and neglected recoil effects, the numerical results could not be trusted. But it was not clear that Bethe's treatment—which was also nonrelativistic—could be made qualitatively correct merely by introducing a cut-off. In later studies (Kroll and Lamb[137]), it was indeed shown that the inclusion of relativistic and recoil effects led just to the cut-off that Bethe had assumed. However, this was not known at the time Kramers wrote his comments, and he clearly remained skeptical.

It is not at all easy to assess, let alone understand, Kramers' attitude toward the new experimental and theoretical developments which followed the Shelter Island Conference. It was certainly clear to Kramers—and to a number of others—that he had made a seminal contribution. The later developments, while more sophisticated and of extraordinary technical brilliance, were based squarely on the physical ideas that Kramers had been espousing for so long. It might even be expected that Kramers would feel vindicated by the experimental successes of Lamb and Kusch and Foley. His ideas, ignored for so long, now began to play an increasingly important role. But in his lectures, classes (as late as 1951), and publications, Kramers was most guarded in his assessment of the new developments. Only once did he express admiration for the advances made, and at that time he took pains to repeat once again his unchanged criticism of the Dirac radiation theory. In summarizing his talk for the Solvay Conference (delivered in the fall of 1948) Kramers writes: "The results obtained in this way from Dirac's theory of 1928 to the beautiful results of the last year [1947] are certainly impressive.... Still the fundaments on which they rest are naturally for a large deal derived from correspondence so first of all the device of introducing the interaction of radiation with a particle adding to the momentum **p**, the term ($-e\mathbf{A}$). Now we have seen ... that the introduction of a proper field ... involves a much profound change in the structure of the Hamiltonian."[138] This quotation appears to indicate that Kramers considered the successful calculation of the Lamb shift as a triumph for the Dirac theory, and that in view of such additional successes, it took a special justification to consider other approaches such as his own. It almost appears from this quotation that Kramers is amazed and even a little peeved that a theory based on such tentative foundation (the replacement of **p** by $\mathbf{p} - e\mathbf{A}$) could describe such subtle phenomena. In Kramers' later public lectures and publications, there is no hint or indication of any kind that the Bethe–Lamb shift calculation and the later relativistic refinements are based in any way on, or have anything to do with, Kramers' earlier investigations. After the Solvay Conference report in 1948, Kramers simply dropped the topic: He did not write a paper on it, he did not discuss it in his popular lectures, and he did not teach a course in it. He did discuss it with two of his favorite students, van Kampen and Hugenholtz, but his was a purely vicarious involvement. This indeed is a strange ending of an obsession, which preoccupied, indeed possessed, him for so long, at the very time that his guiding ideas seemed to be leading physics into new and exciting directions.

It is more than likely that besides the purely scientific concerns and inevitable technical difficulties, there were other profoundly personal factors that caused the abrupt cessation of Kramers' activities in quantum electrodynamics after the Solvay Conference. After his presentation at the Shelter Island conference, Kramers was well aware that his long-time favorite personal theme was now at the heart of the emerging developments. He had originally planned to spend most of the summer in Princeton (after he had completed his lectures in Ann Arbor), to concentrate—in peace and quiet—on the further elaboration of his theory. He returned to The Netherlands rather earlier than he had planned to be present at the church-wedding ceremony of his daughter Agnete. There had been a civil wedding about a month earlier, but Kramers was not there. Although he was anxious to stay in Princeton and work, he changed his plans so that he could join his family in the church celebration. By a sadly ironic twist, he suffered a stroke right after his return, just before the celebration, and this prevented him from attending the ceremony. So his reluctant early return did not achieve its main purpose. To the consternation and dismay of his family,[139] Kramers had not used the ocean voyage on his return trip to relax and rest (as he had promised); instead, he worked all the time on board ship. This succession of events shows how preoccupied Kramers was with the renormalization theory. He clearly did not intend to just stand by and watch its further development.

Kramers was certainly not the only person who recognized the basic role of the renormalization development. Oppenheimer, in particular, saw in renormalization theory the beginnings of a major new synthesis. So rapidly did the theory develop, that Oppenheimer felt that, with the work of Schwinger in the spring of 1948, this synthesis was nearing completion. How strongly Oppenheimer felt can be inferred from an unimportant but interesting event at the Washington meeting of the American Physical Society in the spring of 1948.

When Schwinger was about to start his presentation of a 10 minute paper at this meeting (April 29–May 1), Oppenheimer, who was in the audience, requested that "in view of the importance of the subject, Dr. Schwinger be allowed 20 minutes, instead of the customary ten." The chairman of the session, Dr. David Inglis, ruled that "in principle, a request for more time could only be made *after* a paper had been delivered. However in view of the position of the person who requested the extra time, I nevertheless grant the request."[140] Oppenheimer at that time was president of the American Physical Society! This tiny incident clearly shows the significance that Oppenheimer attributed to the relativistic, gauge-invariant method of calculation developed by Schwinger. These same ideas are expressed, except more completely and more forcefully, in Oppenheimer's report to the Solvay conference.[141] By that time, Oppenheimer was of the opinion that a fully consistent relativistic and gauge-invariant formulation of quantum electrodynamics had been achieved—a formulation that was also computationally powerful. It allowed the surprisingly accurate numerical calculation of many new delicate effects. It thus appears, at least in the mind of one thoughtful, leading physicist, that Kramers' obsessive theme (a consistent formulation of a nonrelativistic quantum electrodynamics) was, by the fall of 1948, no longer a problem and no longer a question of central interest in physics.

Enormous progress had been achieved in the *relativistic* formulation of quantum electrodynamics, while Kramers had deliberately restricted himself to nonrelativistic considerations. Even though by 1948 not all questions of relativistic electrodynamics (especially mathematical questions) were totally settled, it was clearly Oppenheimer's view that a consistent relativistic formulation of quantum electrodynamics was nearing completion. How general this view was does not really matter much; it does matter that there was a marked shift of interest in the physics community from electrodynamics to nuclear, meson, and high-energy phenomena. In this shift, Kramers' nonrelativistic version of quantum electrodynamics did not play a major role. It stood as a remnant of an incomplete earlier theory and it was soon forgotten.

The rapid development of quantum electrodynamics following the Shelter Island conference had a direct, major, and lasting effect on Kramers. There are several indications which suggest that Kramers' subsequent scientific behavior (which at times was quite erratic) must be viewed as an intensely personal reaction to the events outlined. Several features stand out with great clarity. First, Kramers was deeply unhappy with the lack of recognition of his significant role in mass renormalization. Second, Kramers was impressed with the formal brilliance of the work by Schwinger and Dyson and correspondingly concerned about his own ability to keep up with and contribute to this field. Third, although somewhat shaken in his convictions, Kramers still maintained a firm confidence that the path he had chosen—the construction of a consistent, nonrelativistic classical theory with subsequent quantum and relativistic generalization—was fundamentally sound. Evidently, these developments affected Kramers, because his activities in field theory diminished sharply. After the Solvay lecture in the fall of 1948, he did not publish anything in quantum electrodynamics. He had a few students—van Kampen being the most important—who continued studies in this field, but Kramers himself appeared to have dropped the subject altogether. This might be due in part to Kramers' deteriorating health. Still, Kramers in this period was a "promoter" (thesis director) of nine to ten students; he published some 10 papers, taught classes, and gave many lectures. He clearly could and did do research, in spite of his weakened health. But he did not touch quantum electrodynamics again and this appears as a conscious and deliberate choice. It is of course hard, if not impossible, to know why he chose to withdraw from active participation in the renormalization theory. As mentioned, Kramers was extremely disappointed that his contributions to renormalization theory were largely ignored. It was not in Kramers' style to complain publicly about a lack of appreciation. It had happened before and he was usually extremely circumspect about showing his true feelings in such matters. He assumed a lofty posture: "As long as the work is done it doesn't matter who does it, it matters even less who gets the credit." It is therefore striking that Kramers expressed his disappointment on many different occasions. Shortly after Bethe's paper appeared, Kramers told his favorite student van Kampen that he had been thinking for some time about performing a similar calculation, but the ambiguities and arbitrariness bothered him. He then added: "Bethe always calculates faster, works harder and worries less than I do."[142] Echo's of resentment are evident in this quote. And this was only one of many comments of this type. Another occasion was a Santa Claus party on December 5, 1950. It is a hallowed Dutch tradition that at such a party each gift must be accompanied by a poem that is amusing and that usually gently chides the idiosyncrasies of the recipient. Occasionally, a laudatory poem is written, but then it should have a tongue-in-cheek quality. In any case, it should not be taken too seriously. At this party Jan Korringa, Kramers' assistant at the time, wrote a traditional poem in which he mentioned that Kramers had discovered the need for mass renormalization. From Kramers' reaction, Korringa[143] inferred that Kramers himself thought that this was a very important piece of work, which furthermore the physics community up to that time had almost totally ignored. It was more than a little unusual for Kramers to be so touchy; he usually took such things in stride. If nothing else, it demonstrates that some 3 years after the Shelter Island Conference and after the appearance of Bethe's paper, he still resented the neglect of his work in renormalization theory.

His general mood of doubt and depression (in part due to his weakened physical condition) was sustained, if not aggravated, by his realization that technically the Schwinger–Dyson–Feynman methodology was far superior to his own. It took Kramers (as it did Pauli) a considerable amount of time to come to terms with the fundamental Feynman approach; even at that, there is no indication that he ever studied it or used it. What familiarity he had with the new developments came from the seminars of Dyson. (It was at the instigation of Kramers that the physicists in Eindhoven invited Dyson to give one of the first expository talks on quantum electrodynamics.[144] Kramers was generally aware of what was going on, but he certainly did not follow the ongoing development in any detail. (He very rarely read papers anyhow.) But he appeared puzzled and nonplussed that the major contributions to these new developments came from a new generation of physicists in the United States rather than from the established persons such as Pauli, Heisenberg, and Bohr in Europe. He expressed his general surprise rather frequently in a talk with Wouthuysen:[145] "How can these Jewish Boys develop such physics in that bleak [intellectual] desert."[†] Kramers was so impressed with the superiority of the Dyson–Schwinger formalism that he withheld his own rather more pedestrian calculation of the Lamb shift for a very considerable time. It was, of course, nothing new for Kramers to procrastinate in publishing his results; that had been a pattern all his life. In principle, nothing had prevented him from publishing his preliminary results on the Lamb shift, as early as June 10, 1947. He certainly had a completed version as early as July 5, the last day of his stay in Ann Arbor. While there, he had shown his derivation to Weisskopf in sufficient detail that Weisskopf in turn could explain it to Oppenheimer.[146] So Kramers' calculation must have been completed well before that time. Bethe's paper was not submitted to *The Physical Review* until June 27, 1947. There was nothing to keep Kramers from publishing his results, but he certainly did not do it. It is hard to know why he did not submit it. He certainly must have recognized the importance of his result. Perhaps he felt that his treatment was still too sketchy and incomplete with too many loose ends. Perhaps he believed that the theoretical framework was too lacking in rigor and detail for publication at that time. Perhaps he was put off by the announcement of Bethe's paper (as a participant in the Shelter Island Conference he must have received it about June 11). Perhaps after working on these problems for 10 years he did not wish to be coerced into hasty premature publication. Whatever the reasons for Kramers' inaction

[†] It would be quite wrong to see any antisemitic tendencies in this comment. Wouthuysen, to whom Kramers was speaking, was himself Jewish. Kramers was especially fond of Wouthuysen and helped and supported him, before, during, and after World War II. The comment must be seen as an expression of great surprise that these young physicists, without the benefit of the European cultural and scientific tradition, could produce such impressive physics.

that summer, in this instance his habit of publishing major results in his own good time, combined with the breakneck advances of the field itself, caused his results to be superseded before they were published in full. Even so, in earlier papers and certainly at Shelter Island, Kramers had expounded his views in sufficient detail so that the importance of mass renormalization was clear to all active workers in the field. Kramers resented it deeply that his early efforts were ignored, while his results were incorporated in a more powerful scheme, often without referring to his pioneering contribution. It is clear that this would be especially upsetting to a sensitive person, who strongly believed that recognition should be forthcoming for visible and concrete accomplishment, without the necessity of mounting a public relation compaign. It is not surprising that Kramers would be deeply hurt to receive so little appreciation for what he considered to be a major contribution. That the new calculational techniques developed by Schwinger and Dyson were more powerful, more useful, and more elegant than his own limited, and somewhat belabored, methods must have come as an additional blow. Under these circumstances, it becomes understandable that there would be a strong tendency to withdraw altogether from active participation in the problems of quantum electrodynamics. And this is exactly what Kramers did.

B. The Convoluted Recognition

On June 9, 1947, less than a week after the Shelter Island Conference, Bethe completed his fundamental paper on the calculation of the level shift. In the accompanying letter to Oppenheimer,[147] he refers in passing to a Kramers procedure for a similar computation. Although Kramers in his only recorded reaction to Bethe's work was somewhat skeptical, even a casual look at Bethe's methods will show the strong influence of Kramers' work. In fact, both Kramers and Bethe effectively employed a subtraction method. Kramers' method consisted of a systematic elimination of the self-fields from the Hamiltonian. Although there is no overt subtraction involved, the Hamiltonian is written in terms of fields from which the self-fields are subtracted out [Equations (110) and (111)].

Bethe, on the other hand, started from the traditional (nonrenormalized) Hamiltonian and carried out the subtraction in the perturbation calculation [see Equation (164)]. The physical reasons for both subtractions are identical: By using the experimental mass, which of course includes the electromagnetic mass, a large share of the interaction of the electron with its own field is already included. This was a point of long standing for Kramers. Bethe, in summarizing his paper, wrote:

(1) The level shift is due to the interactions with radiation; it is a real effect and is of finite magnitude.
(2) The effect of the infinite electromagnetic mass of a point electron can be eliminated by the proper identification of terms in the Dirac radiation theory.

Both these points are explicitly contained in Kramers' earlier work. Kramers did not really know for sure that the calculated shift would be finite, nor had he considered relativistic refinements, but he certainly had written about the existence of the shift. He had been nagging the world of physics about the electromagnetic mass for 10 years. It is therefore puzzling that Bethe in this paper in no way refers to Kramers. He credits Schwinger and Weisskopf for the suggestion that a relativistic calculation might give finite results, and he acknowledges extensive and stimulating discussions at the Shelter Island Conference—but no mention is made of Kramers, his presence, his presentations, his ideas, or his earlier work. It appears that Kramers, in his usual controlled way, was hurt and upset by this oversight. It is worth nothing (and Kramers surely noted) that of the three groups attempting a more detailed relativistic calculation of the Lamb shift in the summer of 1947, only Schwinger referred to Kramers. Weisskopf and French, in an extensive and important paper submitted on December 10, 1948, refer in connection with the mass renormalization idea to Bethe, Schwinger, Weisskopf, Koba, Tati, Tomonaga, and Feynman, but never to Kramers.[148] In another important paper, Kroll and Lamb[137] compute the self-energy of a bound electron. That calculation is based less directly on the mass renormalization idea. But again Kramers' original work is not mentioned. Schwinger, on the other hand, in describing the general philosophy of the renormalization program in his usual incisive manner, gives full credit to Kramers' contribution. Schwinger, by a series of transformations, recast the Hamiltonian of the Dirac hole theory in a physically more transparent form. He comments: The new Hamiltonian is superior to the original one in essentially three ways: it involves the experimental electron mass, rather than the unobservable mechanical mass, an electron now interacts with the radiation field only in the presence of an external field, that is, only an accelerated electron can emit or absorb a quantum, the interaction energy of an electron with an external field is now subject only to a finite radiative correction."[117] This is a beautiful and succinct statement of the renormalization idea and Kramers' influence is clearly noticeable. Schwinger appends a footnote to this comment (after the word quantum in this quote): "A classical nonrelativistic theory of this type was discussed by H. A. Kramers at the Shelter Island Conference held in June 1947, under the auspices of the National Academy of Science." This is as complete and as fair an acknowledgment of Kramers' contributions as one could reasonably expect. But it is one of the very few instances where such explicit credit is given.† In his report to the Solvay Conference, delivered in October 1948, Oppenheimer summarizes the developments in quantum electrodynamics in the "wonderful year" since the Shelter Island Conference in the summer of 1947. Oppenheimer[141] very briefly refers to Kramers' work—

† The book by Pais[149] is an exception to this general observation. Here Kramers' role is carefully examined and his contributions are recognized. But given the special personal relation between Kramers and Pais, this example cannot be used to conclude anything about the general appreciation of Kramers' work.

although he actually refers mainly to the material in Kramers' talk at this same conference. But apart from these remarks, very few of the papers of that time make any mention of Kramers' work at all.

Thus, although there was certainly no conspiracy of silence, Kramers' work was not recognized (at least at that time) as the starting point of an important new development. In the excitement of new understanding and new insight, the future goals—believed to be within reach—are more important and indeed more relevant than the precise historical pedigree. Reasonable as this may be, it is equally reasonable that Kramers would have perceived this as an inadvertent but unmistakable slight of his work. Whether Kramers was overly sensitive or not, it is certainly true that during the first 3 or 4 years after the Shelter Island Conference, Kramers' contributions, which played such an important part in initiating the whole development, were rarely if ever mentioned publicly.

What makes this whole story so curious and remarkable is that soon after the renormalization method had become an established part of field theory, a rather sharp reversion occurred. Perhaps the most dramatic example is in the textbook by Bethe, Schweber, and de Hoffman.[150] This book contains the first organized discussion of renormalization theory. Section 21a deals specifically with mass renormalization and the nonrelativistic Lamb shift. Where up to that time, Kramers' name and ideas were hardly ever mentioned, this section reads like an ode of praise to Kramers: "The fact that the self energy of a free particle can only lead to a change of its mass is the modern solution of the divergence difficulties. The original suggestion was made by Kramers in 1947."[150] This is how the authors introduce the idea of mass renormalization: "Kramers therefore suggested, returning to Heisenberg's original view of quantum mechanics, that only observable (experimental) mass should play a role and not the separate quantities m_0 (the mechanical mass) or δm (the electromagnetic mass). This principle is known as the principle of mass renormalization."[150] Not only do Bethe, Schweber, and de Hoffman refer often to Kramers, they give a very cogent and clean description of the ideas that Kramers had been expounding for 10 years: "We now accept Kramers' viewpoint: we recognize δm as the electromagnetic mass which the particle has acquired by virtue of its interaction with the radiation field.... We identify $m_0 + \delta m$ as the observed mass .... All through this section, Kramers' name comes up again and again. (In this section of about 9 pages, there are at least 9 direct references to Kramers' ideas and suggestions.) The importance of Kramers' ideas for the subsequent relativistic development (which of course Kramers himself never attempted) is also mentioned: "Subsequent relativistic calculations confirmed that the level shift calculated by Kramers' renormalization principle indeed converges."[150] The profusion of references to Kramers is certainly noteworthy and perhaps a little surprising in view of the almost complete neglect of such references in earlier papers. The comments made about Kramers' role in the Lamb shift are especially interesting: "When Kramers made his suggestion [of mass renormalization] in 1947, Lamb and

Retherford had just measured the ^2S level displacement from the ^2P level in hydrogen. Stimulated by Kramers' ideas, Bethe interpreted Lamb's effect as an effect of the interactions of the electron with the radiation field. In fact, Bethe simply calculated the self energy of an electron bound in the atom. But according to Kramers' idea a large part of this self energy has been taken into account if we use the observed mass m in the calculation rather than the bare mass m."[150] This is certainly a lucid explanation of the idea of mass renormalization. It is also the first acknowledgment in print that Bethe's calculation was based directly on Kramers' ideas. Several years later (in 1968), Bethe, in his recollections of the Shelter Island Conference and his calculation of the Lamb shift, mentions the great significance of Kramers' ideas again: "Of course people had struggled with self energy infinities for a long time, in particular Oppenheimer and Kramers. Kramers suggested that what one really ought to do was to renormalize the mass of the electron, taking into account its interaction with its own electromagnetic field. Then only those parts of the self energy which are not contained in the mass of the particle would be observable and amenable to experimentation." "I found this suggestion very interesting and I thought that it ought to be possible to get Lamb's result by applying the idea of Kramers."[151] A more explicit statement of the direct influence of Kramers' ideas is hard to imagine. In this same Trieste lecture, Bethe recalls that the result he obtained for the level shift was logarithmically divergent (this was explained in Section IV.A). Bethe hoped and expected that a relativistic theory would remove this divergence, so that a finite result would emerge. Bethe recalls: "Stupidly or boldly, I just assumed that the higher energy [upper limit of the logarithmic energy integral] was mc^2 and with this assumption I got about the right answer."[151] It is interesting that, as the conversation between Kramers with van Kampen shows,[144] Kramers thought about making similar arguments—but he never did it, because the ambiguities bothered him and he did not have the brashness to make an ad hoc guess. But Bethe's recollections make unmistakingly clear that he fully recognized the seminal importance of Kramers' contributions. This same issue came up again at the second anniversary of the Shelter Island Conference held (at the same place as the first) in June, 1984. Many of the original participants were there, including Lamb and Bethe. Lamb, reviewing the first conference, mentioned that Kramers gave a long involved talk, which was especially difficult to understand—and Lamb added that only a few people paid much attention. At this point Bethe interrupted Lamb and said "But I did."[152] This shows once again how aware Bethe was of the importance of Kramers' work for his own investigations. Schwinger referred to Kramers' work earlier than almost anyone else.[117] That reference was purely informational: He merely reported that Kramers had carried out these studies, but he made no value judgment of that work. Considerably later, when Schwinger was reviewing the development in quantum electrodynamics, he once again referred to Kramers' efforts, but this time he did comment on the significance of Kramers' work.[153]

> Of course the concept of electromagnetic self action of electromagnetic mass, had not entirely died out in that age of subtraction physics; it had gone underground to surface occasionally. Kramers must be mentioned in this connection. In a book published in 1938 he suggested that the correspondence principle foundation of quantum electrodynamics was unsatisfactory because it was not related to a classical theory that already included the electromagnetic mass and referred to the physical electron. He proposed to produce such a classical theory by eliminating the proper field of the electron, the field associated with uniform motion. Very good—if we lived in a nonrelativistic would ... but the relativistic electromagnetic mass problem is beyond the reach of the Correspondence principle.... Nevertheless I must give Kramers very high marks for his recognition that the theory should have a structure independent character. The relativistic counterpart of that was to be my guiding principle.

The candid comments by Schwinger show several interesting features. It is clear that Schwinger fully recognizes the fundamental importance of Kramers' work on mass renormalization. But more than that, Schwinger is one of the very few physicists who was aware that Kramers' ideas on mass renormalization were of long standing. He clearly read and understood Kramers' attempt to eliminate the self-field, as it was contained in his book (see Sections II.A and II.B), and just what this attempt had to do with the Lamb shift. Thus, it appears that Schwinger realized that Kramers' presentation at Shelter Island was indeed the logical conclusion of a systematic approach toward quantum electrodynamics. At the same time Schwinger correctly stresses the inherent weakness in Kramers' approach: his deliberate but artificial restriction to a nonrelativistic theory. Schwinger's positive assessment of Kramers' contributions must be viewed in the context of Schwinger's own formidable contribution to, and profound understanding of, field theory.

Recently, Schweber wrote a book entitled *Some Chapters for a History of Quantum Field Theory 1938–1952.*[154] In it, Schweber analyzes the significance of Kramers' contribution in terms of the general development of quantum field theory and fully recognizes Kramers' fundamental role. He gives the most complete description of Kramers' participation at the Shelter Island Conference yet available. After discussing Kramers' observation that the divergencies were directly related to the divergencies of the electromagnetic mass, he comments: "Kramers' remarks proved to be central to the Conference."[155]

In discussing the general importance of the three conferences on quantum electrodynamics (Shelter Island, Pocono, Oldstone[154]), of which Shelter Island was the first, Schweber writes: "Shelter Island indicated the problems and possible solutions, primarily by virtue of the focus which Kramers' insight had given."[156] This certainly attributes as central a role to Kramers as could ever be expected. Unfortunately and regrettably, most of these words of appreciation and tribute came long after Kramers' death. He died in the spring of 1952; the book by Bethe, Schweber, and de Hoffman did not appear until 1955. All other laudatory comments came even later. Kramers never knew the central importance that the physics community would eventually ascribe to his idea of structure independence as a key element in the development of renormalization theory. This was probably the most public recognition of Kramers' fundamental contributions to renormalization theory. Unfortunately, this recognition came too late for Kramers. He died fully convinced that the world of physics had neither understood nor appreciated the profound significance of his heroic struggles with his great obsession.

C. What Went Wrong? Can One Be Too Principled?

The story of Kramers' involvement in the quantum theory of electrons and radiation certainly does not have a happy ending. In fact, it really does not have an ending at all. After the initial flurry of excitement generated by Kramers' ideas at the Shelter Island Conference, the full relativistic formalism pervaded the subsequent development to such an extent that Kramers' role was only occasionally recalled. Kramers did not participate in any way in this further development. Consequently, Kramers had no part in the discovery of some of the most striking results of renormalization theory. It is generally futile to speculate on why a particular scientist did not solve a problem or did not push a promising development to a successful conclusion. In this case, however, Kramers had not only recognized the difficulties and ambiguities in radiation theory earlier than most physicists, but his investigations and studies had brought him very close to a definitive solution of these questions. His failure to achieve the full relativistic renormalization theory was not due to a lack of understanding of the basic problems or to technical limitations; rather, it was the result of a priori limitations that Kramers himself had imposed on his investigations. Even though an examination of these self-imposed restrictions will not really explain why Kramers did not pursue his own ideas to the relativistic domain (which was physically the most relevant and exciting), such an examination will nevertheless throw some light on the person, the importance of his scientific associates, and his style of doing physics.

Throughout Kramers' protracted investigations of the electron and the quantum theory of radiation, he adhered almost religiously to two basic principles. One was the universal validity and unrestricted applicability of Bohr's correspondence principle. The other was that the only reliable starting point for a quantum theory of electrons and radiation was the *classical* Lorentz theory of extended charges. This automatically implied (as detailed in Section I) that the notion of electromagnetic mass acquired a fundamental significance in his considerations. Although perhaps never stated as an independent principle, Kramers had quite early come to the conviction that the difficulties and ambiguities of the quantum theory of electrons and radiation were of a classical origin. He hinted frequently, and occasionally stated explicitly, that the early expectations that quantum theory would give new insights into the nature of elementary particles, electrons in particular, had not been fulfilled.[157] It is therefore understandable that Kramers concentrated most of his efforts on constructing a consistent and divergence-free classical theory, before ever attempting a quantum version. It is not obvious (as often stressed by Pais[149]) that in the subsequent quantization process no new ambiguities or new divergencies would occur. Kramers was well aware of this possibility, with the zero-point energy and the field fluctuations being obvious examples, but he appeared to feel that these "new" quantum divergencies could be handled once a consistent classical theory was obtained. Thus, the construction of a prior divergence-free classical theory became a sine qua non, and Kramers' whole effort over a period of some 10 years was based on the following maxim: *First* construct a classical theory and *then* quantize. Kramers held unwaveringly to these ideas. Only once, in the fall of 1948 in his Solvay Conference lecture—and this was well after the relativistic version of the renormalization scheme was formulated—did he express any doubts that the path he had chosen was the only correct one: "It may of course be that point model and fluctuation divergencies can and must ultimately be traced to the same source from the point of view of a future complete theory and that one therefore should not think too hard of the device: First quantizing a wrong Hamiltonian and trying to make amends later on. Still something might be learned from a theory of the sort to which the present paper aims and in which it is tried to analyse the structure independent features of classical theory first and to see whether and to what extent quantization can learn us† something new."[158]

It is clear from this quote, that Kramers is both modest about his own accomplishments ("something might be learned") and critical of the customary Dirac theory ("one should not think too hard of the device ... trying to make amends later on"). But apart from this almost casual comment, Kramers remained totally firm in his conviction that the classical theory should be repaired before any quantum ideas were introduced. This strongly held opinion also forced Kramers into a treatment that was necessarily nonrelativistic. He frequently expressed the opinion that it was impossible to construct an exact canonical theory of classical fields and charges. As late as 1948, in the discussion following his Solvay lecture, Kramers stated explicitly: "I think there does not exist a really consistent classical relativity theory."[159] So that there was no doubt about Kramers' views, he ends his talk at the Solvay Conference with the following admonition: "The establishment of a relativistic theory of charged particles has hitherto been and perhaps will remain so for some time, a story of artful and ingenious guessing."[159] Kramers made these comments in the fall of 1948. Schwinger's fundamental work on relativistic renormalization theory had appeared more than 6 months earlier, and Dyson's incisive analysis of the relation between the relativistic quantum field theories of Schwinger, Feynman, and Tomonaga was in press.[160] The latter was

† There is something most puzzling and disquieting about this gross grammatical error. Kramers was a superb linguist and it is hard to believe that he would make such a crude mistake. But this is the way it is written in the paper.

submitted October 6, 1948, just 3 weeks before the Solvay Conference, and its general contents, if not all the details, were known to the experts. It is not clear whether Kramers was familiar with these developments, but since Oppenheimer reported on these matters at the same Solvay meeting, he surely was generally aware of these momentous results. It does not appear that any of these developments changed Kramers' opinions materially. For example, a month after the Shelter Island Conference, on July 13, 1947, Kramers writes a letter to Jost,[161] in which he asks for some information about the *classical* correspondence aspects of the Bloch–Nordsieck procedure. Kramers refers briefly to his own calculations, which dealt with a description of the interaction of the electron and the radiation field. Much along the lines of the calculations sketched in Section II, Kramers again introduces a cut-off: He only considers wavelengths larger than a Compton wavelength—a deliberately nonrelativistic consideration. The two major themes of Kramers' work, the correspondence principle and the Lorentz theory of extended electrons, emerge unchanged even after the intense Shelter Island experience, where relativistic suggestions were very much in the air.

How strongly Kramers held on to his principles is perhaps most vividly illustrated by his treatment of the classical theory of electrons and radiation theory, in a lecture course he gave in the academic year 1949–1950 in Leiden. It must be recalled that when this course started, in the fall of 1949, relativistic quantum electrodynamics through the published work of Schwinger, Feynman, Tomonaga, and Dyson had achieved a well-defined structure. Yet Kramers' discussion follows the original Lorentz discussion to the letter. He devotes a considerable amount of time to the definition and calculation of the electromagnetic mass. He derives the Lagrangian of the particle in a field, closely following the derivation given first in his book and extensively described in Section II.A.

In his lectures, he again considers an extended electron and repeats his favorite refrain that it is necessary to distinguish carefully between the self-field and the external field. He lists several ways in which the separation between the self-field and the total field can be carried out: by introducing the structure of the electron and the Poincaré stresses which hold it together or by a mass renormalization. But he is extremely sketchy about these points. They appear more as incidental remarks than as profound problems with which he had struggled for many years. Even stranger is the almost cryptic comment: "Merely subtracting the field of the electron is not permissible, it is not a Lorentz invariant procedure."[162a,163] Kramers' remark is, of course, perfectly correct, although it is not something he had stressed before. But perhaps more surprising is Kramers' complete silence about the great progress made in these very topics in the last 2 years. This total disregard of the ongoing development is certainly most remarkable. Even stranger is that in these same lectures he refers to a recent paper by Schwinger on synchrotron radiation[162b] published in 1949. Thus, the recent journals were available and Kramers was surely familiar (or could be familiar) with Schwinger's relativistic quantum electrodynamics, but he did not mention it.

It is unclear whether Kramers was unaware of the relativistic renormalization procedures, or whether he considered these new methods wrong or incomplete, or whether he felt they had no place in a lecture course for students. However, it is totally clear from these lectures and actions that his basic views about the proper foundations of a theory of electrons and radiation were unaffected by the newer developments. It is possible that Kramers in the privacy of his study pondered about whether, and in what way, these new ideas would affect his own program for the construction of a quantum theory of electrons and radiation. But he never expressed whatever concerns he might have had publicly. He certainly never gave any inkling that the development of relativistic quantum electrodynamics was a reason to reconsider or modify his own approach or to change his basic position. His position was (and remained) that the only legitimate starting point for a theory of electrons and radiation was a classical description of an extended electron in a classical electromagnetic field. Both relativity and quantum theory were modifications and refinements grafted on to an essentially classical pictorial world view. Crucial to Kramers was the existence of correspondence principles, which guaranteed the possibility of smooth, continuous transitions from the relativistic as well as the quantum domains to the purely classical regimes. Kramers certainly expected and probably believed that transitions were "seamless", that each and every quantum or relativistic feature should have its classical counterpart. He most likely would feel somewhat ill at ease with a *freestanding* or *autonomous* quantum or relativistic notion. He would almost automatically look for classical aspects or classical analogues in all such situations. It was this unwillingness, or inability, to detach himself from classical visualizable physics or to make a decisive, irreversible break with classical concepts that ultimately defined the limitations which he imposed on himself and his work.

In the development of quantum mechanics and relativity, two distinct and somewhat contradictory trends must be distinguished (see especially Pais[149] for this distinction). One is the correspondence idea: Both these theories contain their respective classical predecessor as a limiting case. The Bohr correspondence principle is an elaborate formal extension of this same general idea. However, of equal significance is the opposite limiting case. There are circumstances where systems can exhibit a type of behavior which can only be described in quantum or relativistic terms; no satisfactory purely classical theory is possible.

These particular circumstances demand quantum or relativistic concepts which do not possess a classical counterpart. The spin is an obvious example. Pais calls this domain,[149] in which novel features require a reexamination of the classical notions, the *domain of disparity*. This term will not be used here, but it is nevertheless important to stress that both quantum theory and relativity possess (relative to classical physics) their domains of correspondence—where the classical laws and concepts retain an approximate or asymptotic validity—and their domains of *anticorrespondence*—where both the classical laws and the classical concepts lose their unambiguous meaning. Kramers, throughout his life, had considerable trouble in coming to terms with the *anticorrespondence domains*. His first impulse was always to look for classical analogues. Because of his great facility and vast knowledge of classical physics, he was almost always able to accomplish this even if the resulting constructions were at times quite contrived: it is only necessary to recall to what extremes Kramers was willing to go to eliminate the photon notion in the Bohr–Kramers–Slater theory (see Chapter 13, Sections I and IV). This is precisely because the photon lies outside the correspondence-type description of the electromagnetic field. There is no classical limit of a single photon. Both the Compton effect and the photoelectric effect are perfect examples of anticorrespondence phenomena, requiring descriptions that cannot be based on classical electrodynamical concepts.

It should not come as a surprise that novel features appear in the relativistic quantum formalism. The physical framework of relativistic quantum field theory contains many nonclassical notions. The impossibility of defining a single particle problem, the vacuum effects (the boiling, turbulent vacuum), and the related idea of vacuum polarization are all intrinsic to the relativistic quantum field theoretic domain and have no classical analogues. Since all these features play an important role in the calculation of the electromagnetic mass, a limitation of the discussion of electromagnetic mass to a framework based exclusively on classical concepts (as Kramers' program was) is an artificial and unwarranted restriction of the relevant theoretical structure. Such a limitation necessarily omits important features.

This was really the heart of the matter. Kramers apparently had not realized that with his insistence on a classical, nonrelativistic starting point, and the unlimited validity of the correspondence principle, he had defined a theoretical structure that lacked some of the essential ingredients of a relativistic quantum field theory. It was not until much later (about 1951) that Kramers began to recognize that the quantum fluctuations effectively destroy the unlimited validity of the correspondence principle and that consequently his own theory lacked some crucial features. But that realization came rather late, and Kramers was most reluctant to give up his cherished principles, and he never really did.

It is probably futile to inquire further into the origins of Kramers' preconceptions. Perhaps Kramers was so insistent on the proper correspondence limit of the radiation theory because many years earlier he had obtained his most important scientific discovery, the dispersion relation, by insisting that Ladenburg's dispersion relation be modified so as to satisfy the correspondence principle. It is certainly possible that hopes or thoughts of a similar success using a similar method motivated his insistence on the validity of the correspondence principle in this instance. But he never wrote or said anything about it. Perhaps the validity of the correspondence idea was so ingrained that it was unnecessary to reiterate it.

It is impossible to separate Kramers' personal endeavors from a dislike of the directions that Dirac appeared to give physics. Pauli, who had an enormous influence on Kramers' taste, abhorred Dirac's filled up vacuum. Pauli called his famous paper with Weisskopf[164] on the quantization of the scalar field the "anti-Dirac paper." This dislike of Dirac's work pervades all Kramers' efforts. The last word Kramers said in public about radiation theory (at the end of his Solvay lecture) is a final criticism of the Dirac theory. "Thus one might say that the famous [$e\boldsymbol{\alpha} \cdot \mathbf{A}$] interaction in Dirac's theory does not even ensure that certain simple unrelativistic effects will be well rendered by the theory, although it is known to work for the very simplest effects like low energy scattering."[165a] This is not the language of a physicist who repents of his sin of ever doubting the Dirac radiation theory.

The language is surprisingly similar to that used in his earliest critical discussions of the Dirac radiation theory. Neither his long struggles, with their incomplete successes, nor the brilliant and spectacular accomplishments of the relativistic theory caused Kramers to change his mind. When, after his presentation at the Solvay Conference, where he came tantalizingly close to breaking the subject of renormalization theory wide open, somebody asked him why he did not treat the theory relativistically, he revealed in his reply perhaps the real reasons for his self-imposed limitations: "The first reason is that in a relativistic theory there are so many things which are not correspondence—like in the ordinary way. Secondly I think there does not exist a really consistent classical relativity theory.... It would be a disillusionment for Lorentz, who liked to work with precise models, that the work he did with rigid electrons could not be done in a simple way with a contractible election."[165b] This clearly is no explanation of a scientific choice, nor a justification of the adopted scientific strategy. By giving this explanation, he makes it appear that to go beyond the correspondence ideas is practically a criminal act. The implication, however veiled, still made loud and clear that going beyond the correspondence principle limit would constitute explicit disloyalty to Bohr. This to Kramers was impossible, even inconceivable. In a completely analogous manner, going beyond the Lorentz rigid electron would constitute a lack of respect for Lorentz. This was unthinkable! Kramers' debt, obligation, and almost subservience to Lorentz and Lorentz's physics are made explicit in this last quote. It is hard to believe and almost impossible to understand why a person of Kramers' power, vast knowledge, and deep understanding would give up—or refuse to explore—a promising avenue of research merely because the anticipated results might be a disappointment to a predecessor, however eminent. There can be little doubt that Kramers in answering the question "Why did he not work relativistically" was candid. He had no reason to give the particular explanation he did, if it were not true. He could have ignored the question if he so wished. (He was exceptionally skilled in deflecting questions that he did not want to answer.) Since he did not do so, he must have felt that an explanation of some kind, to the world of physics and perhaps to himself, was in order.

His explanation was partly in terms of physics but mainly in terms of the expectations of particular physicists. A crucial element in Kramers' scientific choices depended on his personal relations with Bohr and Lorentz. This involves a complex and tangled interplay of emotions, scientific dependence, scientific vision, and individual hopes. It is probably impossible to understand the resulting scientific attitude exclusively in personal terms, but there is no doubt that the feelings so generated were largely responsible for the direction of Kramers' work and for the choices that he did not make. In the many years that Kramers struggled with the problems first raised by Lorentz and Bohr, he not only had to contend with difficult and involved physics and to overcome indifference and lack of understanding by his contemporaries, but he also had to struggle with demons of his own making. The ghosts of his scientific past kept on haunting him. He was unable to disengage himself from the expectations that he sensed Bohr and Lorentz had of him. It almost seemed that Kramers felt that Bohr and Lorentz were continually looking over his shoulder, that he owed them an explanation and justification for what he did or tried to do.

The impression is overwhelming that Kramers felt obliged to show Bohr and Lorentz their greatness through his own accomplishments, building on but never challenging the foundations that they had established. Kramers knew perfectly well that neither Bohr nor Lorentz needed his achievements to demonstrate their genius. But he felt in true Calvinist style that the privilege of his personal association with two such towering figures conferred on him the eternal obligation of continuing physics in their vein. That was his duty, his responsibility, his true obsession.

Notes for Part 3

[1] Letter from Kramers to Urey, dated July 15, 1925 (Bohr Institute, Kramers file).
[2] Letter from Kramers to Romeyn, dated August 15, 1925 (Romeyn–Kramers correspondence, Amsterdam).
[3] Letter from Ehrenfest to Bohr, dated September 11, 1925 (Bohr Institute, Copenhagen).
[4] Letter from Einstein to Ehrenfest, dated September 18, 1925 (Einstein archives, Princeton, N.J.).
[5] Letter from Einstein to Ehrenfest, dated December 26, 1925 (Einstein archives, Princeton, N.J.).
[6] Private communications from S. A. Goudsmit to M. Dresden. Goudsmit recalled this from discussions with Ehrenfest. There presumably was a letter to this effect, but it appears lost.
[7] Letter from Ehrenfest to Lorentz, dated October 6, 1925, translation by M. Dresden (Lorentz archives).
[8] Letter from Ehrenfest to Bohr, dated October 11, 1925 (Bohr file, Copenhagen).
[9] Letter from Bohr to Ehrenfest, dated October 14, 1925 (Bohr file).
[10] Letter from Ehrenfest to Kramers, dated spring 1926 (Kramers file, Copenhagen).
[11] Letter from Ehrenfest to Kramers, dated August 24, 1928 (Kramers file, Boerhaave Museum Collection, Leiden).
[12] Letter from Ehrenfest to Kramers, dated November 4, 1928 (Boerhaave Museum Collection, Leiden).
[13] Letter from Kramers to Kronig, dated February 26, 1926 (Kramers file, Copenhagen).
[14] Letter from Kramers to Fowler, dated December 9, 1925 (Kramers file, Copenhagen).
[15] Letter from Bohr to Fowler, dated April 14, 1926 (Bohr file, Copenhagen); also quoted in Mehra and Rechenberg, *The Historical Development of Quantum Mechanics*, Springer-Verlag, New York, 1982, Vol. 4, p. 222.
[16] H. A. Kramers, *Z. Phys.* **39**, 828 (1926).
[17] Max Born, *My Life*, Charles Scribner and Sons, New York, 1978, p. 232.
[18] Heisenberg, in *Theoretical Physics in the 20th Century*, edited by M. Fierz and V. Weisskopf, Interscience, New York, 1960.

[19] SHQT interview with Heisenberg, February 19, 1963, p. 22.
[20] H. A. Kramers, "Eenige opmerkingen over de quantum mechanica van Heisenberg," *Physica* **5**, 369–376 (November 25, 1925). For some reason, this paper is not included in Kramers' collected works. It was written in Dutch.
[21] See, for example, Mehra and Rechenberg, *The Historical Development of Quantum Mechanics*, Springer-Verlag, New York, 1982, Vol. 4, p. 248.
[22] A. Zwaan, Dissertation, 1926, University of Utrecht.
[23] S. A. Goudsmit and G. E. Uhlenbeck, *Physics Today* (June 1976), p. 43.
[24] W. Pauli, *Z. Phys.* **43**, 601 (1927).
[25] Letter from Heisenberg to Pauli, dated February 5, 1927 (*Pauli-Heisenberg Correspondence*, Vol. I, Springer-Verlag, New York, 1979).
[26] Interviews with O. Klein and G. E. Uhlenbeck by M. Dresden.
[27] P. A. M. Dirac, in "Recollections of an exciting era," Course LVII, *Proceedings of the International School of Physics Enrico Fermi*, Vol. 57 (1977), p. 109.
[28] Interviews with Uhlenbeck, Hugenholtz, and ter Haar by M. Dresden.
[29] H. C. Brinkman, *Application of Spinor Invariants in Atomic Physics*, North-Holland, Amsterdam, 1956.
[30] Letter from Kramers to Colby, dated February 17, 1927 (Kramers file, Bohr Institute, Copenhagen).
[31] Letter from Romeyn to Kramers, dated May 11, 1934 (Romeyn correspondence, Amsterdam).
[32] Private discussions between M. Dresden and H. A. Kramers; interview with Hugenholtz by M. Dresden.
[33] Kramers and Wannier, *Phys. Rev.* **92**, 252–276 (1941).
[34] H. A. Kramers, *Physica* **1**, 284–304 (1940).
[35] Interview with A. Pais by M. Dresden.
[36] Interview with F. J. Belinfante by M. Dresden.
[37] Interview with N. van Kampen by M. Dresden.
[38] Interview with G. E. Uhlenbeck by M. Dresden.
[39] H. A. Kramers, acceptance lecture Lorentz medal, October 30, 1948, Royal Academy of Sciences, Amsterdam.
[40] Frances Low, letter to M. Dresden, November 3, 1981.
[41] Kramers in the eulogy for Ehrenfest.
[42] H. A. Lorentz, *Theory of Electrons*, Dover, New York, 1952 (a) p. 37; (b) p. 43; (c) p. 252 and the famous note 18.
[43] A very nice presentation of this derivation is contained in J. D. Jackson, *Classical Electrodynamics*, Wiley, New York, 1975, p. 787.
[44] F. Rohrlich, *Classical Charged Particles*, Addison-Wesley, Reading, MA, 1965.
[45] W. Heitler, *Quantum Theory of Radiation*, Oxford University Press, 1938.
[46] Interviews with J. Korringa and F. J. Belinfante by M. Dresden.
[47] P. A. M. Dirac, *Proc. R. Soc. London Ser. A* **114**, 243 (1927).
[48] P. A. M. Dirac, *Proc. R. Soc. London Ser. A* **114**, 710 (1927).
[49] G. Wentzel, "Quantum Theory of Fields," in *Theoretical Physics in the 20th Century*, edited by M. Fierz and V. Weisskopf, Interscience, New York, 1960.
[50] Letter from Dirac to Bohr, dated March 1934 (Bohr scientific correspondence).
[51] J. R. Oppenheimer, *Phys. Rev.* **35**, 461 (1930).
[52] W. Pauli, *Handbuch der Physik*, Vol. 23, Springer-Verlag, Berlin, 1933.
[53] H. A. Kramers, "Werkelyk heid en begrippen vorming," Inaugural, Delft, October 30, 1931; translation by M. Dresden.
[54] H. A. Kramers, *Theorien des Aufbaues der Materie II*, Akademische Verlagsgesellschaft, Leipzig, 1938, p. VI.
[55] H. A. Kramers, *Quantum Mechanics*, Dover, New York, 1964, pp. 394–395.
[56] *H. A. Kramers Collected Works*, p. 898.
[57] *H. A. Kramers Collected Works*, p. 707.
[58] *H. A. Kramers Collected Works*, p. 706.

[59] *H. A. Kramers Collected Works*, p. 697.
[60] *H. A. Kramers Collected Works*, p. 831.
[61] H. A. Kramers, notes for the Shelter Island Conference, June 1947.
[62] *AHQP*, March 13, 1962, Max Delbruck to Thomas S. Kuhn. Also quoted by S. Schweber, "Some chapters in a history of quantum field theory," Les Houches lectures, 1983.
[63] Letter from Pauli to Dirac, dated May 1, 1933, Churchill College Archives. Also quoted by Schweber (note 62).
[64] Niels Bohr, "Lectures in Rome," October 1931; lectures on atomic stability and conservation laws, March 12, 1931.
[65] H. A. Kramers, "Discussion Remark," in *Solvay Conference Report*, Vol. 8, Gauthier-Villars, Paris, 1950.
[66] W. E. Lamb, et al. Phys. Rev. 72, 241 (1947).
[67] W. E. Lamb, discussion remark, Second Shelter Island Conference, June 1983, reported by A. S. Goldhaber.
[68] H. A. Kramers, in seminar discussions, 1939.
[69] H. A. Kramers, *Theorien des Aufbaues der Materie II*, Akademische Verlagsgesellschaft, Leipzig, 1938, p. 452.
[70] H. A. Kramers, *Nuovo Cimento* **15**, 108–114 (1938). In this paper, Kramers attributes this "double counting" argument to a discussion remark of Pauli.
[71] E. Fermi, *Rev. Mod. Phys.* **4**, 87 (1932).
[72] H. A. Kramers, *Theorien des Aufbaues der Materie II*, Akademische Verlagsgesellschaft, Leipzig, 1938, p. 464; translation by M. Dresden. The translation in the English version does not quite catch the ironic innuendo.
[73] H. A. Kramers, *Theorien des Aufbaues der Materie II*, Akademische Verlagsgesellschaft, Leipzig, 1938, p. 461, footnote 1.
[74] Ibid., p. 454.
[75] W. Opechowski, *Physica* **VIII**, 164 (February 1941). In a footnote on p. 165, Opechowski mentions that the statement in Kramers' book—that the external field is finite at the position of the electron—is incorrect. He mentions that it was Kramers himself who recognized that.
[76] H. A. Kramers, *Quantum Mechanics*, Dover, New York, 1964, p. 446.
[77] (a) J. Serpe, *Physica* **VII**, 133 (February 1940); (b) **VIII**, 226 (February 1941).
[78] V. Weisskopf and E. Wigner, *Z. Phys.* **63**, 54 (1930); **65**, 18 (1930).
[79] Manuscript (November 1940), "Celebration of a Sexagenarium" (Kramers file).
[80] Letter from Rosenfeld to Kramers; xeroxed copy from Belinfante from mimeographed copies of the letter by E. Guth, 1947, Physics Department, Purdue University.
[81] See W. Opechowski, *Physica* **VIII**, 164 (February 1941), footnote on p. 170. The translation given is rather free, by M. Dresden.
[82] N. van Kampen, thesis, "Contributions to the quantum theory of light scattering," Leiden, January 16, 1952.
[83] W. Pauli and M. Fierz, *Nuovo Cimento* **3**, 167 (1938).
[84] H. A. Kramers, *Ned. Tijdschr. Natuurkd.* **11**, 134–140 (July–August 1944); translated in the *Collected Works*, p. 838.
[85] (a) Letter from Kramers to Heisenberg, dated April 12, 1944, in the possession of Ms. Maartien Kramers. A copy should be in the Kramers file at the Bohr Institute, but it is not there. (b) Translation of this letter by M. Dresden.
[86] H. A. Kramers, "Quantum theory of molecular structures," *Ned. Tijdschr. Natuurkd.* **11**, 42, 57 (1944) (in Dutch).
[87] Private discussion, M. Dresden with Belinfante, March 25, 1976.
[88] *H. A. Kramers Collected Works*, p. 838. What is quoted here is a translation by M. Dresden from the Dutch original, *Ned. Tijdschr. Natuurkd.* **11**, 134–140 (1944), which seems more accurate than that given in the *Collected Works*.
[89] H. A. Kramers, *Ned. Tijdschr. Natuurkd.* **11**, 134–140 (1944). As quoted, this is

again a translation from the Dutch version. This translation is freer than that in the *Collected Works* (p. 838) but makes the point more forcefully. P. A. M. Dirac, *Proc. S. Soc. London Ser. A* **167**, 148 (1938).

[90] *H. A. Kramers Collected Works*, p. 841; *Ned. Tijdschr. Natuurkd.* **11**, 134–140 (1944).

[91] *H. A. Kramers Collected Works*, p. 842; *Ned. Tijdschr. Natuurkd.* **11**, 134–140 (1944).

[92] F. Bopp, *Ann. Phys.* **38**, 345 (1940); **42**, 573 (1942).

[93] E. G. Stueckelberg, *Helv. Phys. Acta* **14**, 51 (1941).

[94] M. Born and L. Infeld, *Proc. R. Soc. London Ser. A* **143**, 210 (1934); **147**, 522 (1934).

[95] P. A. M. Dirac, *Proc. R. Soc. London Ser. A* **167**, 148 (1938).

[96] *H. A. Kramers Collected Works*, p. 842; see *Ned. Tijdschr. Natuurkd.* **11**, 134–140 (1944).

[97] W. Heisenberg, *Z. Phys.* **120**, 513, 673 (1942).

[98] J. A. Wheeler, *Phys. Rev.* **52**, 1107 (1937).

[99] Interview with Suus Kramers, Agnete Kramers, and Maartien Kramers by M. Dresden.

[100] Letter from the Minister of Education to H. A. Kramers, dated July 9, 1946 (Kramers file, Bohr Institute).

[101] Interview with ter Haar by M. Dresden.

[102] Letter from Francis Low to M. Dresden, dated November 3, 1981.

[103] Letter in the Kramers file, Bohr Institute, dated January 14, 1947.

[104] Letter to Kramers, dated June 30, 1947 (Kramers file, Bohr Institute).

[105] S. Schweber, in the Les Houches lectures, 1982.

[106] Letter from Weisskopf to Darrow, dated February 18, 1947. See reference 109.

[107] Letter from Weisskopf to Bethe, dated May 1, 1947.

[108] Letter from Weisskopf to Bethe, dated May 19, 1947, Cornell University Library, Bethe Archives; also quoted by S. Schweber in Les Houches lectures, 1982.

[109] Weisskopf's abstract is contained in the Rockefeller University Archives, 1947, 1948; also quoted by S. Schweber in Les Houches lectures, 1982.

[110] Kramers' outline is contained in the Rockefeller University Archives, 1947. It is also quoted by S. Schweber in Les Houches lectures, 1982. Another copy is in the Kramers file in the Bohr Institute in Copenhagen.

[111] Willis Lamb, in a discussion remark at the Second Shelter Island Conference, June 1983.

[112] *H. A. Kramers Collected Works*, p. 706. Actually, Kramers' reasons for making this claim in 1937 were quite different from the one's mentioned here. It does show that the idea of deviations from the Dirac theory was an old friend of Kramers'.

[113] Private communication from Lloyd Motz. Kramers gave a colloquium on June 10, 1947 at Columbia University, where he presented a derivation of the Lamb shift.

[114] G. E. Uhlenbeck, in a glowing review of Kramers' book, uses this phrase: "Kramers' treatment of Radiation phenomena was 'too original.'" *Ned. Tijdschr. Natuurkd.* 7-1120 (Sept. 1940).

[115] Interview with J. Korringa, N. van Kampen, and F. J. Belinfante by M. Dresden.

[116] H. A. Bethe, *Phys. Rev.* **72**, 339 (1947).

[117] Julian Schwinger, *Phys. Rev.* **73**, 416 (1948).

[118] Richard P. Feynman, *Phys. Rev.* **76**, 749 (1949).

[119] H. A. Kramers, "Non-relativistic quantum electrodynamics and the correspondence principle," *Collected Works*, p. 845.

[120] *H. A. Kramers Collected Works*, p. 831.

[121] H. A. Kramers, "Non-relativistic quantum electrodynamics and the correspondence principle," *Collected Works*, p. 851.

[122] Comment by Dirac, Solvay Conference Reports. Les particules elementaires, *Rapports et discussions du huitieme Conseil de Physique Solvay*, Brussels, September 28–October 2, 1948, published by R. Stoops, Brussels, 1950.

[123] Interviews with Paul Meyer, F. J. Belinfante, and J. Korringa by M. Dresden.
[124] *H. A. Kramers Collected Works*, p. 706.
[125] Interview with G. E. Uhlenbeck by M. Dresden; G. E. Uhlenbeck in "Oude en Nieuwe Vragen der Natuurkunde," inaugural address, Lorentz chair, Leiden, April 1, 1955.
[126] F. Bloch and A. Nordsieck, *Phys. Rev.* **52**, 54 (1937).
[127] W. Pauli and M. Fierz, *Nuovo Cimento* **15**, 167 (1938).
[128] J. Schwinger, *Phys. Rev.* **75**, 651 (1949).
[129] J. Schwinger, *J. Phys.* (*Paris*) *Colloq.* **C8**, Suppl. No. 12, 43 (Dec. 1983).
[130] B. French and V. Weisskopf, *Phys. Rev.* **75**, 1240 (1949).
[131] J. R. Oppenheimer, "Report to the Solvay Conference, 1948."
[132] Interview with A. Pais, April 6, 1981 by M. Dresden.
[133] T. Welton, *Phys. Rev.* **74**, 1157 (1948).
[134] H. A. Bethe, *Phys. Rev.* **72**, 339 (1947); formula 7.
[135] *H. A. Kramers Collected Works*, p. 867.
[136] *H. A. Kramers Collected Works*, p. 868.
[137] N. M. Kroll and W. E. Lamb, *Phys. Rev.* **75**, 388 (1949).
[138] *H. A. Kramers Collected Works*, p. 869.
[139] Interview with Gerda Kramers by M. Dresden.
[140] Written comments on the *Bulletin of the APS 1948*, by M. Dresden.
[141] J. R. Oppenheimer, "Report to the Solvay Conference, Brussels, Belgium, September 27–October 2, 1948," Electron Theory.
[142] Interview with N. van Kampen by M. Dresden.
[143] Interview with J. Korringa by M. Dresden.
[144] Interview with N. van Kampen by M. Dresden.
[145] Interview with S. Wouthuysen by M. Dresden. The "bleak" intellectual desert was the United States.
[146] Weisskopf to Oppenheimer, August 19.
[147] Letter from Bethe to Oppenheimer, dated June 9, 1947, Library of Congress, National Archives, Oppenheimer papers; also quoted by Schweber in the Les Houches lectures, 1982.
[148] V. Weisskopf and B. French, *Phys. Rev.* **75**, 1240 (1949).
[149] A. Pais, *Developments in the Theory of the Electron*, Princeton University Press, Princeton, New Jersey,1948.
[150] Bethe, Schweber, and de Hoffman, *Mesons and Fields*, Vol. 1, Row-Peterson and Co., Evanston, Ill., 1955.
[151] H. A. Bethe, in *From a Life of Physics*, evening lectures at the International Center for Theoretical Physics, Trieste, Italy. IAEA Bulletin, 1968.
[152] Personal communications from Alfred Goldhaber to M. Dresden.
[153] J. Schwinger, *J. Phys.* (*Paris*) **12**, 43 (1982).
[154] S. Schweber, *Some Chapters for a History of Quantum Field Theory 1938–1952*, Les Houches session XL, North-Holland, Amsterdam. See also Les Houches lectures, Summer School 1983.
[155] S. Schweber, *Some Chapters for a History of Quantum Field Theory 1938–1952*, p. 152 in the xeroxed version.
[156] Ibid., p. 166.
[157] See H. A. Kramers, *Theorien des Aufbaues der Materie II*, Akademische Verlagsgesellschaft, Leipzig, 1938, p. 400.
[158] *H. A. Kramers Collected Works*, p. 869.
[159] H. A. Kramers, in the discussion following the Solvay Conference, Rapports du huitieme Conseil de Physique, Solvay, 1950, p. 266.
[160] F. J. Dyson, *Phys. Rev.* **75**, 1736 (1949).
[161] Letter from Kramers to Jost, dated July 13, 1947 (Kramers file, Copenhagen).
[162] In the lecture notes of H. A. Kramers, "Theory of Electrons," 1949–1950, Leiden, notes by C. W. Bentham: (a) p. 102; (b) p. 86.

[163] F. Rohrlich, *Classical Charged Particles*, Addison-Wesley, Reading, MA, 1965, p. 16.
[164] W. Pauli and V. Weisskopf, *Helv. Phys. Acta* **7**, 709 (1934).
[165] Rapports du huitieme Conseil de Physique, Brussels, Belgium, September 27–October 2, 1948: (a) p. 265; (b) pp. 266–267.

PART 4

KRAMERS AS A PERSON AND A SCIENTIST: CONFLICT OR HARMONY?

CHAPTER 17

Personality and Style

I. The Assessments: Public and Personal

There are only a few times when the overall contributions of a scientist are assessed and recorded. Such evaluations are inevitably made whenever a scientist receives an award, an honorary degree, or is otherwise publicly recognized for his or her achievements. Another more common, but unfortunately more melancholy occasion arises on the termination of a career and a life. Indeed, obituaries are often a better source of scientific accomplishments than encyclopedias. For Kramers, the major awards, the Lorentz medal in 1948, the Hughes medal in 1951, came very near the end of his life in April 1952. It is therefore not surprising that the congratulating lecture by van der Waals on the occasion of the award of the Lorentz medal[1] contains the same enumeration of Kramers' scientific achievements as the later eulogies by Casimir,[2] Bohr,[3] Belinfante and ter Haar,[4,5] and John Wheeler.[6] Kramers' scientific works (not counting popular lectures or summarizing talks) contain some 16 papers on the Bohr theory (all written before quantum mechanics) and 13 papers on the mathematical elaborations and formal consequences of quantum mechanics. He published some 21 papers in a different area: magnetic and low-temperature phenomena. In addition, Kramers became very interested in the formal property of the spin, resulting in some 11 papers. Kramers' versatility is further evident in his 15 papers in statistical mechanics, the Ising model, and Brownian motion. Perhaps Kramers made his most profound contribution in his 13 papers on the interaction of matter and the electromagnetic field. To this must be added the two books that Kramers wrote: the book with Holst, a popular exposition of the Bohr theory (see Chapter 12, Section IV), and his monumental Volumes I and II entitled *Quantum Mechanics*. This is a pretty complete compilation of Kramers' contributions to physics.

It is inevitable that different physicists would assess the different parts of Kramers' research in rather different ways. Thus, Kramers' students, Belinfante and ter Haar,[4,5] rate the symbolic method as a significant contribution, while Casimir[2] is more doubtful and Pauli and Wheeler[6] do not mention it at all. Similarly, Belinfante and ter Haar emphasize Kramers' role in renormalization theory, while Casimir is unsure about the eventual significance of Kramers' approach (described extensively in Chapter 16). But there is universal agreement that all Kramers' papers showed a total mastery of the subjects he chose to discuss. He never chose an easy topic to get a quick paper or a cheap success. Quite the opposite: Kramers inevitably selected problems of either great profundity or great complexity or both. "He found no satisfaction in applying his outstanding skill and virtuosity to simple problems [that is simple by his standards] and was glad to leave such problems to others."[2] There are no trivial Kramers papers. It is not surprising that with this large, diverse, and high-caliber scientific body of work Kramers was recognized as one of the major physicists of this time. Kramers' significance was not just recognized by those who eloquently eulogized his passing; it was publicly recognized by a series of awards, some in his native Holland (the most prestigious being the Lorentz medal), others abroad, such as his election as an honorary foreign member of the American Physical Society—a singular and rare honor indeed. Planck and Dirac were examples of the very few outstanding individuals who were so honored. Kramers himself was quite aware that he had been the leading theoretical physicist in The Netherlands for a number of years.[7] He was admired by as formidable a critic as Oppenheimer and highly respected by the uncompromising conscience of physics, Pauli. And yet ....

It is noteworthy that when discussing Kramers' formidable scientific achievements, many of his best friends, close associates, and most devoted pupils nevertheless expressed a certain hesitancy, some slight reservations about the fundamental physical importance of Kramers' work. Such comments were always imbedded in expressions of praise and admiration for his achievements, but even so many voiced some nagging doubts about the ultimate significance of his achievements. Casimir[2] points out that Kramers' work is enormously skillful, and his treatment of the most difficult questions powerful. Still, his work did not contain any major new physical insights, no new physical effects (no prediction of new phenomena, except one). Only rarely did his studies stimulate extensive experimentation or initiate new lines of related investigations.

Typical examples of the type of reservations mentioned are comments by Casimir: "A first group of papers is concerned with the development of the mathematical technique of the new quantum mechanics .... One might be slightly surprised that the number of papers in this field from the years immediately after 1926 is rather small .... Yet to all insiders Kramers was an acknowledged master and his papers never lacked finesse."[2] Later, he made a similar remark: "He [Kramers] is not satisfied until the mathematical problem to which the physical question gave rise is solved in its generality. If this trait has perhaps reduced the total volume of Kramers' work, it gives to his publications a characteristic quality of unselfishness."[2] Wheeler, in a most guarded manner, expresses this same concern, but like Casimir it is phrased as a compliment. Commenting on Kramers' work after his departure from Copenhagen he writes: "His [Kramers'] field of work became wider—whether at any cost to his leadership in theoretical physics is hard to say, his position in any case being outstanding—but certainly showing in ever growing measure his originality and talent."[6] It is clear that both Casimir's and Wheeler's laudatory comments contain noticeable elements of restraint and perhaps even some slight surprise. Perhaps this is because both Casimir and Wheeler shared a rather general sense that in spite of Kramers' acknowledged brilliance, profundity, and major accomplishments, his overall impact on physics as such was rather minor. Weisskopf[8] expressed the view that although the detailed influence of Kramers' work was not that tremendous, he had a great humanizing influence on physics; without him many questions would remain unasked, many ideas would remain undeveloped. That Kramers in a rather unobtrusive way was the source of innumerable studies and investigations of his colleagues, students, and associates was also emphasized by Hugenholz.[9] Discussions, suggestions, questions, inquiries—all are essential for the continued advance of physics, and Kramers excelled in this form of stimulation. The importance of this type of activity for the development of science has not always been sufficiently recognized and Kramers surely suffered from this neglect. This is certainly true, but it does not contradict the basic perception that the overall impact of Kramers' investigations was incommensurate with his talents and abilities. One of Kramers' great admirers, D. ter Haar,[10] indicated that Kramers' major influence on physics was as an exponent of the Copenhagen spirit of quantum theory. Kramers' book, especially Volume I, would certainly bear this out. Of course, the Copenhagen spirit is more identified with Bohr and Heisenberg than with Kramers. So although Kramers' exposition undoubtedly had an important influence, the basic new insights were due primarily to Heisenberg and Bohr, not to Kramers. Interestingly and perhaps surprisingly, it was Bohr himself who commented on the curious discord between Kramers' extraordinary gifts in physics and his ultimate limited influence. At Kramers' funeral, Bohr, by way of explanation to Ubbink, a friend, colleague, and former student of Kramers, exclaimed: "It is a great pity, Kramers was extremely talented, but utterly without ambition."[11] It is pretty sure that Bohr had identified one of the elements, perhaps even an important element, which contributed to Kramers' circumscribed role in physics, but other elements also played an important role. Bohr's explanation, necessarily based on Bohr's perception of Kramers, was just too pat. He did not realize—or in any case did not say—that his own personal relation to Kramers, with its strong elements of dominance, might well have been stifling to Kramers' scientific independence and originality.

Probably more important than these vague allusions—that Kramers with his extraordinary talent might have done more, should have done more, or in any case should be appreciated more—was Kramers' personal dissatisfaction with his own achievements. For Kramers, such feelings were hardly anything very new. From the time he entered the University of Leiden until the end of his career, Kramers had doubts and concerns about his accomplishments. Even his signal success in Copenhagen did not totally eradicate his feelings of fear and uncertainty. The traumatic events, leading from the debacle of the Bohr–Kramers–Slater theory to the formulation of matrix mechanics by Heisenberg, did nothing to bolster his scientific confidence. The professional position in Utrecht conferred academic obligations on him. These as such did not help him to achieve his scientific goals; if anything the opposite is true. Once in Utrecht as a professor of theoretical physics in a prestigious university, any expression of scientific doubt was of course out of the question. A Dutch professor was, by generally accepted perception, an accomplished scholar and a brilliant lecturer, living in a world of pure reason, a world where there is no fear, anxiety, inadequacy, anger, or passion. And although almost everyone and certainly Kramers knew that this popular image was more a caricature than reality, the image nevertheless imposed certain standards of behavior to which Kramers scrupulously adhered. But privately he frequently expressed his personal frustration with the slow progress he was making in his own scientific endeavors. As early as 1927, Kramers writes to his friend Walter Colby at the University of Michigan regarding his projected teaching duties in the Summer School. Kramers explains to Colby why he considers the amount (he believed) he had to teach as extravagant: "You would understand my point of view if you saw the difficulties I have had these first $1\frac{1}{2}$ years in Utrecht, to undertake scientific investigations for myself, because of the great amount of time taken by my lecturing and hundreds of minor duties. My wife always advised me to think, for my own sake of only one thing: scientific research and I suppose she is right in that. You see, Colby, this letter is written by a man who is "weeping" at seeing time pass by without bringing the fulfillment of his scientific ideals."[12] It was impossible to tell whether Kramers was serious or facetious. Still, a concern about his many nonproductive obligations, the passing of time, and his lack of substantive accomplishments is clearly noticeable. These same themes recur, in fact, with increasing frequency in subsequent communications.

Once in a while, Kramers even referred to his own limitations and inadequacies at public occasions. But he did this in his typical half joking, half serious manner so that it is impossible to be totally sure just what he had in mind. Kramers accepted the position of "extraordinary" professor at the Technical University in Delft on October 30, 1931. This meant that he had to give an inaugural address, usually an official, ceremonial occasion. At the end of this thoughtful lecture, which in 1931 must have seemed far from trivial, Kramers, when addressing the students, suddenly exclaimed: "Ladies and gentlemen, students, theoretical physics is a difficult subject, if in my mind's eye, I survey what I know and understand about it [theoretical physics], I am finished in about five minutes."[13] It is impossible to know whether Kramers was serious or facetious when he said this. But it is totally certain that no one in the audience believed him. Kramers' reputation was too great; the respect for the one person in Holland who knew quantum theory was too well established that anyone could seriously entertain the possibility that Kramers' knowledge was not vast or his understanding was not profound. And this Kramers also knew, so whether this remark was merely intended to startle the audience or indeed expressed a deep felt concern is impossible to say. But it is certainly true that in late 1930 and 1940 these self-deprecating remarks increased in frequency. He told S. Wouthuysen in 1939: "I am a burned out flame."[14] There is an illuminating passage in a letter which Kramers writes to Heisenberg.[15] Heisenberg in 1942–1943 had written his early articles on the S matrix, then called the η matrix. He had given some talks about his ideas in Leiden in 1943. Kramers became very excited about Heisenberg's approach, and he made a number of especially important suggestions about the further elaboration of these ideas. Heisenberg in turn was enthralled by the orientation that Kramers had given his η matrix approach: He suggested several times that he and Kramers should prepare a joint paper. But Kramers procrastinated and did not answer. (For some further discussion of this incident see Section II.A.) When he finally did answer, he declined to join Heisenberg in his η matrix venture, although he was interested and convinced that it was important. He tries to give an explanation of his quixotic behavior: "That I did not write you comes from a type of hesitation from which I suffered all my life, when I had the chance to consciously become involved with a really important matter. It is something like deciding to write a beloved—but one still doesn't do it, because one only wishes to do it well, that is with complete commitment. All the worthwhile contributions I have made came about only because I did not make any such conscious decision at all, but rather because I was by chance forced to do these things, through a coalescence of external and internal forces, independent of what I wished to do."[15] This is a remarkable passage; it is reminiscent of Kramers' long ago conflict with Ehrenfest. It will be recalled that Ehrenfest was always doubtful about Kramers' complete commitment to physics. Kramers appears to say here that he would like to be so committed, but that throughout his life he had difficulties in accepting such completely demanding obligations—hence the hesitation. Still it is remarkable that Kramers deliberately withdraws from a cooperation in a topic that fascinated him and that he recognized as particularly significant. In the same letter that Kramers declines Heisenberg's offer to cooperate he writes: "Last fall with your seminar you have given my scientific interest a new lease on life."[15] Still by his own action Kramers retreats from the active pursuit of a promising new research area.

There is more than a little irony in the circumstances that Kramers used the very ceremony when he received his highest honor, the Lorentz medal, to announce his disenchantment and disappointment with his own scientific achievements. There was a selection committee for the selection of the Lorentz medal of which Kramers was a member; however, during the selection meetings in 1947 Kramers was in the United States so he did not participate in those meetings. After Kramers had received the medal, he responded in this fashion: "I want to express my thanks to the Royal Academy of Sciences for the honor of receiving the Lorentz medal. Actually, I should be angry; for this award has become possible through a 'pocket revolution,' I myself am a member of the selection committee, which suggests the choice of the medalists, they have misused my absence last year, for I would surely have voted against the present choice. But my anger dissipates rapidly, for the award gives me a sweet compensation for the sad decline in knowledge and ability of which I have become more and more conscious in these last few years."[1]

This shows, in a most touching manner, Kramers' mood and assessment of his own work at that time. It is also interesting that Kramers, in his acceptance speech, does not talk about his past research or his future plans; he does not talk about himself at all. Instead, he talks about Lorentz's greatness; he discusses physics. He gave an interesting, informative, incisive discussion, but it is evident that he wishes to divert attention from himself and redirect it to Lorentz and to physics in general. Kramers realized at that time that his hopes and dreams for major contributions to physics were and would remain largely unfulfilled. Even though this must have been a deep disappointment to him, he did not blame anyone, he did not complain, and he did not struggle; he accepted the situation with style and gentle resignation. He most certainly did not give up physics or slight his responsibilities as the leading Dutch physicist and a scientist of world standing. But the "holy fire"—the divine inspiration—was gone.[16] His personal judgment of his own scientific accomplishments,[9] even if he did achieve only a small fraction of what he had hoped, still seems cruelly harsh and in stark contrast to the glowing assessment of his colleagues and the scientific community.

II. The Conundrum of the Near Misses

A. The Role of Near Misses

Any physicist actively engaged in research will select problems and develop methods of treatment that are especially suited to his or her style, taste, philosophy, and ability. In this process, value judgments are made all the time; the scientist will concentrate on a certain problem to the exclusion of others.

A scientist generally emphasizes the more fundamental aspects first and leaves the technical details for later discussions. In the typical progression of research, it is commonplace to shift from one problem to another related problem, which seems more treatable, more relevant, or more interesting. Often, the emphasis of the study is changed during the course of an investigation as demanded by the manner in which the material evolves. Thus, active research involves a large number of choices at all stages of the research process.

Kramers made very specific choices; he devoted a great deal of time and effort to questions that he considered as either unanswered or answered too superficially by the then current quantum theory: the relativistic theory of the electrons and radiation theory. He stayed away from all problems in nuclear physics. When in the 1930s the interest generally began to shift toward nuclear problems, the behavior of neutrons, and β-decay theory, Kramers chose not to follow that trend. His fundamental considerations remained focused on electrons and electrodynamics. Even in his more applied studies, Kramers made very particular choices; in the solid-state area, he primarily dealt with low-temperature magnetic phenomena. He never wrote a single paper dealing with conductivity questions.

What Kramers did is totally common. Any productive scientist has to make choices about what to investigate, how to do it, when to pursue and push a line of attack, and when to give it up. If this were not done, research would degenerate into a vague, general set of pronouncements without definite conclusions and without much overall structure. Nevertheless, this mode of operation with its continual choices inevitably means that a number of perfectly plausible research directions are either not pursued at all or not followed for very long. It is not unusual that even in a highly successful scientific career, a number of opportunities are missed, promising avenues remain unexplored, and readily available, important results are either not stressed or overlooked altogether.

What makes Kramers especially unusual, in fact unique, is that the number of his personal "near misses" is so unusually large. Not only is their number especially large, but the further exploration of Kramers' suggestions frequently led to results of quite extraordinary importance. In this sense—which is of course hard to formalize—the importance of Kramers' work is as much determined by what he came *near doing* as by what he *actually* accomplished. In a surprising number of instances, his papers were literally the very last papers written before a major breakthrough occurred. A breakthrough here is defined, not without some arbitrariness, as coinciding with the publication of a seminal paper, which clearly signals a major reorientation in physics. Time and again Kramers published papers full of remarkable insights, technical tricks, and original ideas, often without stressing the novelty of his approach, without emphasizing the decisive changes that his discussion suggested. Often his papers appear as "considerations," reflections, almost ruminations, always interesting, always imaginative, but often lacking a clear-cut physical point. His considerations are only rarely pursued far enough to exhibit the full physical implications of his ideas. As a consequence, many studies of Kramers appear to have an incomplete or inconclusive character; the ideas are there and the formalism is well constructed, but the results are not always stated with either crisp clarity or great confidence. This tendency to deprecate his own results (or at least not to appear boastful about his accomplishments) is also noticeable in Kramers' more formal papers. Often he suggests a highly ingenious original technique in such an off-hand manner that a typical reading would not indicate just how general and how important the procedure actually is. There are instances where the most impressive uses of Kramers' methods were made by others.

It is impossible to escape the feeling that Kramers at times procrastinated on purpose. There are other times where he appears to be deliberately equivocal. He was perfectly capable of making precise, unambiguous statements, but he often did not do it. Many of his conclusions have a somewhat hesitant, tentative character. At times it seems as if Kramers was reluctant to accept the full consequences of his own reasoning. Whether this is related to the difficulty, or the impossibility, of convincing Bohr about his discussion of the Compton effect early in his career (see Chapter 14) is unclear, but it is certain that on that occasion (and there were others) he did not follow his own scientific instincts. Nor did he pursue the Compton effect intensively enough or long enough to enable a definite disposition of its merits to be made. He just gave in and dropped the photon approach altogether. In his later career (after the Bohr–Kramers–Slater theory), Kramers was generally very circumspect and guarded about making radically new physical proposals. Throughout his career, Kramers appeared haunted by the "near miss" syndrome. On many occasions, Kramers came tantalizingly close to significant new insights, to major innovations, but for a variety of different reasons a final, definitive, unequivocal resolution of the attendant problems eluded him. On many occasions, he failed to make the decisive break or take the final crucial step. There is probably no other physicist whose career shows as many important "near misses" as Kramers'.

It is very likely that Kramers' perennial harsh judgment of his own scientific accomplishments is related directly to his realization of the prevalence of the near misses. He must have known, if not at the time certainly in retrospect, that he had come very near the definitive solution of many major problems. In his own terms, "he had been at the threshold of success on many occasions, but he had not crossed that threshold."[17] For a person who valued truly original significant research in theoretical physics above everything else, this must have been especially painful and frustrating. The accompanying disappointment and anger might easily account for the severity of Kramers' judgment about his own activities.

It is even possible that the faint surprise expressed by so many of Kramers' admirers that he did not achieve more in physics might have the same origin. Many of his close associates knew, or sensed, how often Kramers had come exasperatingly close to a major advance only to see his efforts preempted by other physicists. They perceived correctly that Kramers had initiated a promising fruitful direction for research—which is a major contribution in its own right. That he often did not carry out his investigation to a brilliant completion must have caused them much consternation and grief. This might easily be the reason for their expressions of surprise and concern.

These observations raise one obvious and important question. Why of all physicists was Kramers so susceptible to the "missed discovery" syndrome? As will be enumerated in the next section, there are about seven or eight (maybe as many as ten) instances that can legitimately be interpreted as examples of this syndrome. That a major, well-informed, technically superior physicist would miss several major opportunities is easy to believe. It is certainly conceivable that there might be a few additional misses, but once the number gets near seven, eight, or ten (depending a bit on what is counted) it becomes increasingly difficult to attribute this to persistent bad luck or mere coincidence. It is at least reasonable to look for other explanations. This issue of the near misses was central in Kramers' scientific and personal life. It influenced his relation to his fellow scientists and it was a major element in his self-image. So important and so basic is the syndrome of the near misses that it is pertinent and indeed imperative to enumerate the individual occurrences, so that the general pattern can be discerned. Even though this may entail a repetition of some items discussed in earlier chapters, a comprehensive summary of these instances should exhibit the general pattern with convincing clarity.

B. The Litany of Near Misses

The character of the individual occurrences, collectively called "near misses" varies considerably: Sometimes these are specific results that Kramers did not pursue far enough, or on other occasions he had important results but did not publish them, and still other times he had interesting suggestions but was talked out of their publication by others. Whatever the precise details might be in all these occurrences, the scientific work did not come to fruition, the ideas were not exploited to completion, and the initial scientific promise remained unfulfilled.

(1) The first event to be included in this listing has been extensively described in Chapter 14. It may be called the "curious Compton interlude." As explained in Chapter 14, Kramers effectively derived the conservation laws from the Compton effect using the photon notion explicitly. He, in particular, appeared to have the relation between the scattering angles. He certainly did not have (and presumably did not try) to obtain the intensity distribution. In any case, in the subsequent arguments with Bohr, Kramers became concerned that the photon notion was untenable and he dropped his ideas altogether. Even when the Compton effect was experimentally discovered, he did not return to his original formulas, but instead embarked on the Bohr–Kramers–Slater venture (see Chapter 13), which was hardly a success. It is perhaps stretching the point to call this a near miss. Kramers' work was correct, but the scientific atmosphere in which he operated did not accept it as correct and, more important than that, Kramers came to see his work as misdirected. That, if not a near miss, is certainly not a positive outcome of a scientific investigation.

(2) There can be little doubt that Kramers' work in dispersion theory constituted a near miss. As explained in perhaps overly minute detail in Chapter 13, Kramers' contribution to matrix mechanics was major: His investigation led directly to Heisenberg's seminal contributions. It is true in particular that the Kramers–Heisenberg paper (which was primarily Kramers' work) was indeed the last paper written before Heisenberg's breakthrough. (In this instance the word is indeed justified!) This is a typical example where Kramers came to the threshold but did not cross it. He did recognize and state that the beauty of the dispersion relations was that they contained just observables, that all references to orbits had disappeared. But he did not make this observation into a principle the way Heisenberg did. Heisenberg recognized, as Kramers did not, that the laws of physics and especially the equation of motion should be formulated in terms of the matrix elements and not in terms of classical orbital parameters. Still, the line of development which culminated in Heisenberg's matrix elements was unquestionably started by Kramers. As knowledgeable and as critical an expert as Pauli acknowledges that Kramers, through his brilliant replacement of derivatives by differences, prepared the path for quantum mechanics.[18] Pauli further traces the origin of quantum mechanics back to Kramers' initial papers on dispersion theory. He describes, in rather flowery language, the two short notes of Kramers in 1924 as the "acorn" from which the grand structure of quantum theory and wave mechanics grew.[19] The orientation that Kramers gave atomic physics through his study of dispersion phenomena certainly led to matrix mechanics. The pattern of his contributions in this area, with their abrupt cessation after the appearance of Heisenberg's paper, certainly qualify those as a most typical example of a near miss.

(3) A very interesting example of the syndrome which characterized so much of Kramers' research is afforded by the paper mentioned in Chapter 15, Section II, in which Kramers stresses the similarity between commutators and Poisson brackets. Almost all textbooks[20] refer to this most basic connection between classical and quantum mechanics. It is generally and correctly attributed to Dirac. However, as early as November 1925, about the same time as and independent of Dirac, Kramers had obtained the same results. But where Dirac saw that the noncommutativity of the Poisson brackets (and hence the commutators) was the key to the formal structure of quantum mechanics, Kramers used the close connection that he had uncovered only to show the validity of the Bohr correspondence principle. In this case, Kramers really possessed all the technical tools to give a general algebraic formulation of quantum mechanics based on commutators or Poisson brackets (he had been an expert in Hamiltonian mechanics since his early days in Copenhagen) but he certainly did not do it. He clearly had the opportunity to share with Dirac the honor of constructing a general formulation of quantum theory, but he did not seize it. It was clearly an opportunity lost, another miss.

(4) As was mentioned in Chapter 15, Kramers undertook the study of a relativistic spin theory as a response to a challenge by Pauli that such a theory

was impossible. Kramers' extensive investigations and, of course, Dirac's inspired studies showed that Pauli was wrong. By the time Dirac's papers appeared in 1928, Kramers were well on his way toward the formulation of his personal version. Although he never particularly liked the Dirac equation (see Chapter 16) with its austere erudition, he recognized its importance and power and he did not publish any of his own results on the relativistic spin theory until much later. In this subsequent discussion, Kramers could show the equivalence of his procedure with that of Dirac, but his studies did not lead to any new results and no particular recognition. Thus, where Kramers' studies on the Dirac theory perhaps do not exactly belong in the category of near misses, they are again studies leading to correct results, using very interesting techniques, but no new physics emerged and at the time they were published the world of physics showed little interest in or sympathy for his efforts. From a more personal vantage point, Kramers within a span of 2–3 years had come very close to matrix mechanics—he had achieved a formalism equivalent to the Dirac equation—but in terms of published materials he had not obtained either in a definitive, concrete version. As a consequence, he got exactly zero credit for these significant contributions. It is remarkable that in the face of such enormous disappointments, Kramers continued to do physics at all. This surely demonstrates an inordinate intellectual resilience.

(5) Chapter 16 is devoted in its entirety to what is perhaps the most conspicuous of Kramers' near misses. As this long chapter shows, Kramers for many years was literally "hovering" near the renormalization idea. Not only that, he developed the necessary formalism to implement his physical ideas. What he missed was the relativistic version of his procedure, or more significantly, he failed to recognize the new features and the simplifications that the relativistic treatment would bring. But all this was because of a self-imposed limitation. He stated himself: "In the second place it [his treatment] also throws some light on the problem of the divergencies in quantum electrodynamics and on the question connected with the Lamb shift—although, of course, these problems for their exact treatment require an analysis, which goes beyond the approximations which we have imposed upon ourselves."[21] It is this self-imposed limitation to a nonrelativistic description that was his undoing. It is an interesting example of how seemingly innocent choices made in setting up a research program can come back to haunt the investigator. Kramers decided to restrict his renormalization studies to the nonrelativistic domain for what seemed to him good and persuasive reasons. He constructed an effective formalism but the self-imposed restriction was too severe; in demanding a nonrelativistic description, he had eliminated an essential element of the physics; he had by his own design missed the opportunity to construct a relativistic theory.

(6) Kramers' was ambivalent toward symmetries, especially discrete symmetries. Although the results he obtained in this area are generally too fragmentary and sketchy to be strictly counted as near misses, many of the

same features are noticeable. His discussion of the charge conjugation almost stops in mid-air. He had introduced the notion, used it, and studied some of its formal properties—and then left it at that. He certainly did not exploit its formal or physical properties, nor did he give a hint of the possible fundamental significance of this new symmetry operation. Thus, this study again exhibits the incomplete, unfinished character of the near misses. Kramers' comments about the parity inversion are so sketchy and scattered that it is impossible to draw any definite conclusions. It is clear that later in his life he took the discrete symmetries more seriously. He was quite intrigued, for example, with his speculations on neutrinos,[22] but those investigations were never completed. Kramers also changed his thinking about the role of the time reversal transformation. He was at first a bit disenchanted by Wigner's derivation of his result on the Kramers' degeneracy, but as time went by he came to terms with the importance of the time-inversion operation. He did hint—or speculate—on a number of occasions that it was quite possible that the laws of nature would *not* be time-reversal invariant.[23] Whether this was just intended as a rhetorical comment or an announcement of a serious research program is impossible to say. Kramers' studies on the discrete symmetries can probably not be counted as legitimate near misses, but they do again possess the common feature of interesting and important suggestions, which were only partially carried out and at best hinted at future developments.

(7) Quite a different example, if not of a near miss at any rate of missed recognition, is the joint work of Kramers and Klein on the theory of strong electrolytes. This work was started by Klein and Kramers in Copenhagen. The calculation of reaction rates and many other properties of solutions are in principle applications of the Boltzmann kinetic methods to systems of charged particles. Kramers and Klein finished a preliminary study about 1919. In any case, enough had been accomplished so that Klein could obtain his Ph.D. in Stockholm.[24] They more or less surmised the important and new result that the log of the activity coefficient was proportional to the square root of the ionic strength. But they were unhappy with their derivation of that result, and Kramers intended to try to obtain their result using Gibbs' version of statistical mechanics. However, both Klein and Kramers became preoccupied with other problems (primarily in atomic physics), so nothing was done. In 1923, in a justly famous paper, Debye and Hückel obtained exactly the same result earlier surmised by Kramers and Klein. Their method was not systematic but was particularly ingenious. There was nothing left to do for Kramers and Klein. Many years later, Kramers did publish a paper on the statistical mechanics of ionic solutions, using Gibbs' method of the grand ensemble. His treatment was more systematic than the Debye–Hückel treatment; it led to corrections of their formulas. The impact and importance of that result was a great deal less than that of the original Debye–Hückel paper. Kramers' efforts were not exactly a near miss, but surely a missed opportunity.

(8) The paper by Kramers and Wannier on the two-dimensional ferro-

magnet was one of the fundamental contributions to statistical mechanics. It was widely read and recognized as a most significant paper. How could it possibly be considered as a near miss? Kramers, in the discussion of the two-dimensional Ising model, introduced the idea of the transfer matrix. This was of extraordinary importance, but Kramers did not exploit the general formal properties to the utmost. That was done later by Onsager and in a next generation by Baxter. In any case, Kramers used the transfer matrix almost numerically without demonstrating its broad formal implications. In the paper, Kramers *assumed* that there was a phase transition. He had hoped that he could derive the existence of a phase transition but he did not. Finally, Kramers was most anxious to obtain and analyze the *exact* solution of the two-dimensional Ising model, but he was not successful in achieving that either. It is true that the Kramers–Wannier paper made all these significant achievements possible, but, sadly, Kramers did not achieve the goals he had set for himself. He had initiated the approach and indicated the direction; he was *on* the threshold but did not cross it. If not a near miss, it was in Kramers' own terms "only a partial, incomplete success."

C. Interlude: Near Miss Number 9—The *S* Matrix Incident

It is perhaps wrong to call the next incident a near miss; it is perhaps better characterized as a "rejected chance." Kramers had the possibility to participate in a very new, very exciting development and he, after considerable hesitation, chose not to do that. Heisenberg suggested that he and Kramers cooperate in an interesting and very promising project, but Kramers decided against it. This incident is so interesting and unusual that a somewhat more detailed description is very worthwhile—and the slight repetition that this entails is perhaps not too serious.

Heisenberg thought that the various divergencies cropping up in quantum electrodynamics, the Fermi theory of β decay, and the meson theory of nuclear forces, were all manifestations of the existence of a fundamental length l_0. This lengh l_0 was to be as fundamental as $\hbar$ and c; it should limit the applicability of the intuitive notion of distance. Heisenberg then inquired, in the style of Bohr, what features of the current quantum theory of wave fields would survive in this new context. He then—this time repeating himself—asked for the observable quantities in this new theory and concluded that they were the stationary states and the scattering data (probabilities for collision, emission, and absorption). These scattering data can be arranged in a matrix—the scattering matrix. This matrix was to replace the Hamiltonian; Heisenberg's main goal was to find means to calculate or determine the S matrix. To do this, Heisenberg applied general principles, such as Poincaré invariance and conservation of probability, but these principles by themselves were insufficient to determine the S matrix fully. This was roughly the status of the theory in the fall of 1943. The material was later published by Heisenberg in two papers.[25,26] In October of 1943, Heisenberg came to Leiden to give a colloquium on his new theory. This is less simple than it might appear, for Holland at that time was occupied by the Germans and the Germans had closed the University of Leiden. A colloquium visit by a German scientist even as famous as Heisenberg was not appreciated, let alone welcomed by everyone. In fact, Kramers had Heisenberg stay at his house. His talk was arranged as an informal discussion between friends, not an official university function. At the colloquium, Kramers made the penetrating remark that to have any hope that the *S* matrix would be determined by general physical principles (and not by particular forces or potentials), it was essential to consider the *S* matrix as an analytic function of the complex momentum variables. The physical amplitudes would be given by "appropriate" boundary conditions for the real momenta. Heisenberg was particularly impressed by this comment and Kramers was equally impressed by Heisenberg's talk. In fact, after the colloquium, the dinner, and the inevitable music (they played Bach), Kramers stayed up and worked most of the night on Heisenberg's ideas.[27] Kramers realized, as Heisenberg apparently did not, that there was a direct relation between bound-particle states and the occurrence of simple poles in the appropriate matrix elements of the *S* matrix. It is one of the more bizarre aspects of this whole incident that Kramers' student, S. Wouthuysen, who was in hiding from the Germans at that time and clearly could not attend Heisenberg's colloquium, was studying problems leading to such poles in a few, simple nonrelativistic situations. Since it was not without danger for Wouthuysen and Kramers to meet too openly, Kramers' oldest daughter, Suus, was the messenger who usually transported documents, in this case calculations about poles in scattering amplitudes, from Wouthuysen to Kramers.[28] It is evident that Kramers was inordinately interested in the *S* matrix; it is equally clear that his ideas and suggestions were of major significance in the further development of the subject. Heisenberg was totally aware of that. Immediately upon his return to Germany, Heisenberg wrote a very warm letter to Kramers, thanking him and his wife profusely for their hospitality:[29] "It was a consolation to me that human relations can be maintained even in these difficult times." This was the only time Heisenberg ever referred to the war, to the German occupation, to the generally strained relations between the Dutch population and the German authorities. Of course, Heisenberg referred just to the personal, human relations. But in this same letter, Heisenberg reiterates his enthusiasm for Kramers' idea: "Since my return I have thought extensively about your idea to consider the matrix as an analytic function. I am more and more thrilled by your proposal, because I believe that in this way one can really obtain a complete theory of elementary particles (i.e., a theory which contains all the actually existing particles and their interactions, and which perhaps can answer all physically sensible questions). I want to continue to work on this, but I already want to ask you now, are you at all interested in a joint venture. (It is of course also alright with me if you want to publish your idea by yourself.)"[29]

This is a remarkable letter; it shows that Heisenberg was very excited about the prospects of the S matrix theory. Heisenberg uses the German word *begeistert*, translated as "thrilled," to describe his response to Kramers' suggestion. This denotes a very strong, enthusiastic reaction. The letter in the last part also makes unequivocally clear that the idea to consider the S matrix as an analytic function was exclusively Kramers', even to the extent that Heisenberg would understand if Kramers were to publish this idea by himself. Heisenberg's suggestion for further joint work is also explicit and clear; his enthusiasm is evident. Kramers was also interested and excited. In a letter to Clay two weeks after Heisenberg's visit he writes: "I personally got a great deal out of Heisenberg's visit, for me it was a meeting from person to person."[30] It is clear that Kramers was enthusiastic about the scientific results of Heisenberg's visit. He evidently felt that some additional explanation of the visit is called for—probably because of the tense political situation. There is no doubt that Heisenberg's visit was extremely stimulating to Kramers: It lifted him out of a state of depression and general discouragement to return to serious physics.

Even so, Kramers never answered Heisenberg's letter of October 31. Some 5 months later, on March 16, 1944, Heisenberg sent a second letter to Kramers.[31] This letter is more businesslike than the first letter. He mentions that he has continued to work on the S matrix and Kramers' idea. There is a sense of urgency in this letter; Heisenberg has made progress and he feels that something should be published soon. But he repeats his invitation to Kramers to join him. Interestingly enough, he comments in quite an off-hand manner that he visited Copenhagen and that in the Bohr Institute everything runs in the normal fashion. This visit later became somewhat of a cause celebre. It was viewed very differently by the different participants.[32,33] Kramers finally answers the two Heisenberg letters on April 12, 1944. He had not only received Heisenberg's letter but also a preprint of one of Heisenberg's recent papers. Kramers' reply is unusually candid; it reveals a surprising amount about his personal views of his scientific aspirations and approach. The letter,[34] fragments of which were mentioned earlier, starts like a confession: He (Kramers) always had extreme difficulty in becoming totally involved in matters of extreme importance. The difficulty was the greater the more complete a personal surrender was required by this involvement. Kramers compares this with the unwillingness to write a letter to a lover. One wants to do it, but does not, because doing so would demand a depth of involvement that is frightening or that really is not there. After this emotional beginning, Kramers turns abruptly to physics and technical questions. Still, the initial ambivalence remains. He is clearly most interested in—and intrigued by—Heisenberg's results. His comments about Heisenberg's results are enthusiastic; he writes with obvious relish: "I really enjoyed your detailed treatment of a simple two particle model in the first part of your paper. Here one has for the first time a totally relativistically invariant stationary state of a system of interacting particles. It is devilishly unintuitive but the model has a lovely consistency."[34]

Coming from Kramers, this is unusually high praise; he is actually showing more enthusiasm for Heisenberg's paper on the S matrix than he did 18 years earlier for Heisenberg's paper on matrix mechanics, which ushered in the new era. It is also clear that Kramers read Heisenberg's paper in complete detail. He comments on calculations, suggests that it will be necessary to assume something about the time-invariance proportion of the S matrix, and gives additional examples of the use of the S matrix. He obviously spent a great deal of time studying the paper; he understood it completely, controlled the subject, and liked it a great deal.

And yet, in spite of all this interest and enthusiasm, Kramers does not choose to join Heisenberg in pursuing what would surely be an exciting, and likely a successful, venture. In a new section of his letter, which has a totally different tone and mood, Kramers spells out the reasons, or better, the rationale for his refusal. His first reason is that Heisenberg knows the topic so much better than he does, so that he probably would not contribute very much: "You manipulate the examples with such virtuosity that I, for the time being, do not belong in the same ballpark."[34] It is hard to believe that this was written by the same person who suggested new applications and a new invariance property, which incidentally showed considerable virtuosity of his own, just two pages earlier in this same letter. Kramers' arguments do not sound convincing. It appears that he is trying to find excuses for his refusal to participate. Of course, he had already made a major contribution, perhaps in an involuntary fashion, but his suggestion that the analytic character of the S matrix was essential was most important and could not be hidden or ignored. He mentions to Heisenberg that he is pleased that his was a fruitful suggestion, but then he writes: "If you mention in a publication that my suggestion has been useful, then my personal role in the development of the S matrix theory is accurately reported." This may or may not be true, but the issue was not what Kramers had done in the past, but what he was going to do in the future. Kramers' attitude is one of disengagement. He seems to look for reasons why he cannot or should not cooperate with Heisenberg. In this sense, this incident appears as part of the general pattern of near misses or missed opportunities, although in this instance there are factors that make this case more complex.

Kramers had a second reason for deciding not to work with Heisenberg, which sounds more convincing. He feels that the practical difficulties—this was in the middle of World War II—were just too great. It was not easy (although not impossible) to communicate and Kramers evidently felt that he needed a more intensive collaboration. He exclaims in his letter, indicating both interest and frustration: "If only we could talk together more often."[34] How interested he is and how ambivalent he feels can be inferred from a typical Kramers "poly interpretable" remark. "This is not the time for a joint publication—but it is the time for joint work."[34] This seems clear enough, except Kramers for yet another reason claims that he cannot engage in joint work. Kramers was really inspired by Heisenberg's visit to return to active physics and he did a considerable amount of physics from October 1943 on. However, he did *not* do the kind of fundamental physics that he really liked and considered important. Instead, he did quantum chemistry, chemical physics, and a number of minor other problems. Why did Kramers shy away from fundamental questions—of the type that Heisenberg had raised? Here Kramers returns to his original confession—because dealing with these fundamental topics would require a total immersion, a total commitment which he found almost impossible to make. Perhaps, he writes, Heisenberg's actual physical presence would enable him to maintain such a level of total involvement, but just his letters would not be enough.

Kramers invokes another, socially admirable reason why he had to turn down Heisenberg's offer for collaboration. It would be possible for Kramers to be a coequal partner to Heisenberg, only if he would give up all other scientific work. He anticipated, probably correctly, that a relentless pursuit of the *S* matrix ideas would require all his intellectual and emotional energy and leave no time for anything else. Thus, Kramers would of necessity be unable to meet his scientific or university obligations. This was very difficult for Kramers to contemplate. In fact, his sense of duty and obligation was very strong; it was impossible for him to consider a course of action that would cause him to neglect or ignore his responsibilities. He could not even conceive of a mode of action that would give the *appearance* of slighting his obligations.

Thus, there was no way in which he could in good conscience accept Heisenberg's offer of a close serious collaboration. With these arguments and very nearly with these words, Kramers withdraws from active participation in the collaboration of the *S* matrix theory. Heisenberg did continue his investigation and did give Kramers credit in his third paper[35] for pointing out the importance of the analytic properties.

This is an instance where Kramers deliberately chose not to enter a new field, although he had every opportunity to do so. The stated reasons for his refusal are not so impressive. The practical difficulties were surely there, but it was still possible to have contacts. Kramers' devotion to his other duties and obligations are genuine; still, there were no university obligations (the university was closed) and the other scientific responsibilities were of his own choice. He could have concentrated on the *S* matrix theory without seriously handicapping his other work or other people, but he chose not to do it. Perhaps he was somewhat apprehensive about entering so different and unconventional a field (is that why he needed Heisenberg's presence?); perhaps the psychological construction he mentions in his letter is the key: He was constitutionally unable to make a choice that would focus all his intellectual power on a single topic, especially if he felt that this topic was of fundamental significance. But this combination of features is precisely what gives rise to the *near miss syndrome*, so that this *S* matrix incident appears like another legitimate example.

All this seems reasonable and is probably true, but there is one possible factor that might have influenced Kramers' decision in this important matter, which he mentioned and which had nothing to do with near misses. These events all took place in the middle of the war. Kramers had very strong feelings about the war, the German occupation, and the atrocious mistreatment of Jews and political liberals by the Germans. On several occasions, Kramers even asked Heisenberg whether he could intercede with the German authorities on behalf of friends in concentration camps.[34] There is no indication that this had any effect. During his visit to Holland, Heisenberg had several political discussions with Casimir.[36] Political discussions are notoriously vague and it is hard to remember details. But there seems little doubt that Heisenberg considered the Russians as the main enemy which had to be stopped, at all cost, even if this caused unpleasantness and serious problems elsewhere. It is most unlikely that Kramers and Heisenberg talked a great deal about these matters. Kramers would not have argued: Present suffering was real and important to him, and he probably knew Heisenberg's views. It was not in Kramers' character to provoke confrontations or arguments, so he probably ignored the whole topic and concentrated on physics. Still, the political conflicts might well have made Kramers reluctant or even unwilling to cooperate with Heisenberg in a formal manner. Perhaps this is what Kramers had in mind when he said, "this is not the time for joint publication." Complicating this picture still further was Kramers' genuine fondness of Heisenberg. He would not wish to hurt him personally or impute the crimes of the regime to him. So he would tend to stay away from these topics in all his dealings with Heisenberg. Still, his strong ties to Holland might well have been a contributing factor to his ultimate refusal to join Heisenberg in an exciting adventure. So it is not clear that this incident is a pure example of the near miss syndrome; it is impossible to estimate the influence of straight political factors. But it is certainly clear from the material presented here that Kramers had a most unusual opportunity; he evidently was tempted to pursue this chance, but for a variety of reasons he withdrew from active participation and let the chance go by.

D. Near Misses Concluded: Omissions and Style

The last example of a near miss—again interpreted somewhat loosely—is provided by the Kramers–Kronig relations. Both authors had obtained these relations independent of each other, in characteristically different ways. Neither paid much attention to their results for some length of time. It was not until much later that the relation between the principle of causality and the Kramers–Kronig dispersion relations was clearly understood. It is not clear who first realized the important connection between causality and the analytic properties. But it is clear that neither Kramers nor Kronig exploited the full generality of the relations that they had discovered. As such, a number of interesting applications of the dispersion relations were not made until much later, and not by either Kramers or Kronig. In this case, both authors missed opportunities. It is particularly interesting and really surprising that Kramers, when stressing the importance of the analytic continuation of the S matrix elements to Heisenberg, did not immediately relate this to his earlier work on dispersion relations, where the analytic properties had been especially important. It is probably futile to speculate what might have happened if Kramers had made that connection. Neither Kramers nor Heisenberg made the connection, so the basic importance of causality in the S matrix framework was not effectively used until much later. It is clear that the dispersion relations, if not an actual example of a near miss, are yet another instance of results incompletely explored and possibilities not fully recognized.

These ten examples are all different; they occurred at different stages under quite different circumstances in Kramers' career, so any conclusions are bound to be tentative. Nevertheless, all these events show such striking similarities—if not in actual facts then certainly in the attitude and approach they depict—that an understanding of Kramers surely requires a further interpretation of these incidents. A complete clarification of this pattern is hardly to be expected, but it would be intellectually irresponsible not to attempt an explanation of this remarkable prevalence of near misses. Such an explanation—an attempt at gaining some insight—would involve many factors—the delicate interpersonal relations between scientists, purely scientific ability, scientific aspirations, and personal hopes, ambitions, and fears—all counterpoised in an unstable configuration. Perhaps the most accessible element in this myriad of factors is the scientific style. An analysis of Kramers' style is attempted in Section III. Here, as a preliminary to such an analysis, a compilation and discussion of contributions that Kramers did not make, but might well have made, are given. This might well appear as a particularly foolish thing to do. A near miss is already a rather vague, poorly defined, slippery notion. But at least there is some tangible documentary evidence that can serve as a starting point for a further sensible if somewhat speculative discussion. However, there is clearly not much that can be said about research not done, problems not attacked. There are, after all, many reasons, all basically irrelevant, why a scientist might not undertake a particular investigation. He might be tired, busy with other things, have committee meetings, have marital troubles, or be going on a trip around the world. All these are plausible reasons and give little or no information about the person and his science. All this is unquestionably true.

But it is still interesting and illuminating to scrutinize a few of the contributions that Kramers might easily have made and could have been expected to make, given his interests, knowledge, and technical skill, but which for one reason or another he did not make. There are some surprises; in fact, most of them are surprises because many of the investigations would appear to have been natural for Kramers to undertake. That he did not do so clearly shows that he made choices; how deliberate those choices were is hard to know and does not really matter. But selecting certain subjects for intensive study and totally ignoring others shows Kramers' taste and style in actual operation.

This is why a brief examination of these possible contributions is not without its value. It gives a hint of the types of problem that Kramers liked to work at and those he tended to ignore. This information is clearly pertinent for an understanding of Kramers' scientific style.

(1) Perhaps the single greatest surprise is that Kramers never attempted to apply either matrix mechanics or wave mechanics to helium. After all, Kramers had spent close to 8 years in Copenhagen trying to apply the Bohr theory to the helium problem without success. He certainly knew how to treat wave mechanical problems; it would appear quite natural for him to apply the new quantum theory immediately to one of the outstanding failures of the Bohr theory. But there is no indication that he ever tried it. Heisenberg did.

(2) It would also appear reasonable to expect that Kramers might have initiated the electron theory of metals. He certainly was thoroughly conversant with the Lorentz electron theory of metals (as he was familiar in detail with all of Lorentz's work). He had written an article on the kinetic derivation of the Fermi–Dirac distribution. All he would have had to do is put these things together. The technical details leading to the Sommerfeld integrals would have been child's play for Kramers—but he never touched the subject.

(3) Kramers never showed any particular interest in the electromagnetic properties of superconductors. Kramers was totally conversant with all aspects of electrodynamics, so it might have seemed reasonable that the proximity of the experimental results on superconductivity would have stimulated him to investigate the electrodynamics of superconductors. This evidently did not happen; he never wrote a single article on superconductivity. That is remarkable for someone from Leiden.

(4) Kramers throughout his career had been intrigued by the unusual properties of liquid helium. He stayed informed about the strange and new phenomena observed at the Leiden Laboratory. But he never took part in the construction of phenomenological theories or special models. Later in life he became interested in the Landau description via rotons and phonons. He preferred the Landau theory to the two-fluid model; he "did not believe a single two fluid theory."[37] He had thought about, but not published ideas on, second sound. The last paper he wrote in November 1951 was on rotons and phonons.[38] But before that he had never written anything specific on liquid helium. In particular, he never looked into a possible connection between Einstein–Bose condensation and the liquid helium phenomenology. He was intrigued quite early by the Einstein–Bose distribution. In his kinetic derivation of the Fermi distribution,[39] he comments on the "unintuitive" character of the assumptions which would have to be made to obtain the Bose distribution. But he leaves it at that; he does not relate the unintuitive assumptions to a possible condensation phenomenon, let alone to the curious behavior of liquid helium.

(5) It is certainly not inconceivable that Kramers might have originated the Heitler–Landau theory of the chemical bond. He was, of course, totally familiar with the necessary formalism; he was a past master in perturbation

theory and the calculations would not have been an obstacle to him. Especially interesting is that soon after Kramers' arrival in Holland (fall of 1926), van Arkel, a chemist, gave a talk on chemical bonding based on the old Kossel theory (totally non-quantum-mechanical). Kramers was very critical and interrupted many times. Finally, van Arkel getting a little peeved said: "Maybe the day will come when basic physical principles will allow the calculation of chemical properties." To which Kramers answered: "That day is already here." This talk was given some 8 months before the Heitler–Landau paper appeared. Presumably, Kramers was not aware of their work, but he clearly realized the vast possibilities that quantum mechanics opened up for chemistry. Kramers always remained interested in quantum chemistry but he never did any research in this area.

It is also quite surprising that Kramers never invented what became known as the Foldy–Wouthuysen transformation.[40] It was almost second nature for Kramers to look for a suitable, semiclassical description of any quantum system, seek classical analogies, or investigate the classical limit. The Dirac equation is the only instance in which he did not do that. It must have given Kramers a great deal of satisfaction that even though he did not establish the classical limit of the Dirac equation or a classical interpretation of the operators, one of his students (and a collaborator) did. Likely this gave Kramers the feeling that his own understanding and insight would not be lost but would through his students be transferred to the next generation and become a part of physics.

There are, of course, innumerable studies that Kramers could have made but did not. He had an especially profound understanding of the nonequilibrium theory of gases. Several of his papers touch on, or come near, this difficult and unfashionable topic. But although he made some incisive studies of a few special effects and was thinking along these lines in his last paper,[39] he again shied away from a systematic frontal attack on the fundamental questions. His investigations were for all their virtuosity peripheral to the main issues. In this case, the "near misses" and the "might have been" categories almost seem to merge. Together they comprise a characteristic and essential feature of Kramers' scientific (and personal) life, which requires understanding and explanation.

III. Elements of Kramers' Scientific Style

A. The Role of Fear and Courage

It may appear strange to start a section on Kramers' style with a discussion of the role of fear and courage. Actually, scientific achievement is inextricably intertwined with personality traits, so that a serious analysis of personal and social features is indispensable in a discussion of scientific success. Because of this close intermingling of so many disjoint influences, it does not make too much difference where one starts, as long as all factors and their delicate interplay are eventually considered.

Dirac is one of the few truly great geniuses who called attention to the important role of fear in the actual process of scientific research. In a most interesting lecture, Dirac states: "I will try to impart some idea of the feelings of a research worker when he is hot on the trail and has hopes of attaining some important result which will have a profound influence on the development of physics. One might think that a good research worker in this situation would review the situation quite calmly and un-emotionally and with a completely logical mind, and proceed to develop whatever ideas he has in an entirely rational way. But this is far from being the case. The research worker is only human and if he has great hopes he also has great fears. I don't suppose one can ever have great hopes without their being combined with great fears. And as a result, his course of action is very much disturbed. He is not able to fix his attention on the correct logical line of development."[41]

This quotation from Dirac makes two extremely important points. One is the coupling of hopes and fears during the actual research process; the other is that the presence of the fear is very likely to distort the natural, logical development of the investigation. Dirac claims that, although most of the experiences he describes are necessarily his own, he nevertheless believes that the coupling of fear and hope is a general feature affecting all research workers concerned with the foundations of a physical theory. He actually makes a stronger statement: "They are influenced by their fears to quite a dominating extent."[41] Dirac then proceeds to give a number of examples of very great physicists who, he believes, were held back by their fears from the decisive completion of their investigations. Of special importance here is what he had to say about Lorentz. "You must surely have wondered why it was that Lorentz succeeded in getting correctly all the basic equations needed to establish the relativity of space and time but just wasn't able to make the final step establishing relativity. He did all the hard work—all the really necessary mathematics—but he wasn't able to go beyond that and you will ask yourself, why? I think he must have been held back by fears. Some kind of inhibition. He was really afraid to venture into entirely new ground, to question ideas which had been accepted from time immemorial. He preferred to stay on the solid ground of mathematics. So long as he stayed there his position was unassailable. If he had gone further he wouldn't have known what criticism he might have run into. It was the desire to stay on perfectly safe ground which I presume was dominating him."[41] Dirac is well aware that most of this is just conjecture; he gives a number of other examples, which he feels show that the mechanism sketched corresponds very closely to the facts. Even though the examples he picks—Lorentz, Schrödinger, Heisenberg, and Dirac—are hardly a typical foursome of randomly selected physicists, it is a reasonable guess that for almost all active physicists fear and anxiety play an important if not dominant role in the research process. It is probably also true that the more fundamental and far reaching the contemplated changes are, the greater the fear. Correspondingly, the more formal and mathematical, the greater the feeling of confidence and security.

Even though with these simple and straightforward remarks of Dirac the role of fear in scientific investigations is hardly exhausted, it is striking how many of Dirac's observations concerning Lorentz's fears in developing relativity seem to have a direct parallel in Kramers' career. The fear to venture into entirely new ground and to question accepted ideas was very much a part of Kramers' style. Kramers, by his actions, really preferred to stay on the solid ground of mathematics. That his position was unassailable, impervious to criticism if he stayed within a confined formal scheme, was a common motivation in many of Kramers' investigations. There is little doubt that the desire to stay on perfectly safe ground was a stimulus for many of Kramers' specific studies.

Thus, the manifestation of fear that Dirac mentions in his explanation of why Lorentz stopped short of developing relativity is present, noticeable, and even dominant in many of Kramers' studies. It is important to stress that throughout a scientific career the importance of various factors changes continuously. Lorentz might have been kept by fear from developing the radical break leading to relativity. But he had no fear—at least it did not dominate his physics—in the dramatic insight that in electromagnetism, instead of two fundamental, independent electromagnetic fields (E, H) and (D, B) as Maxwell had it, there was only one basic electromagnetic field. Furthermore, where previous to Lorentz electric and magnetic fields were necessarily tied to matter, Lorentz recognized that the seat of the electromagnetic field is empty space. These were most significant and liberating ideas, and here the elements of doubt and fear were overcome.

Thus, although fear itself unquestionably plays a role, its role in turn depends on the scientific surroundings, the personality of the investigator, his courage and independence, his age, and the stage of his scientific development. For Kramers, as for all reflective scientists interested in truly fundamental questions, concern and fear were constant companions. But the role and the influence of fear did vary as his career progressed. In his pre-Bohr stages, Kramers was often moody and concerned; he had severe doubts about his ability and commitment. The Copenhagen years from about 1916 to 1925 witnessed his meteoric rise from an apprentice in atomic physics to heir apparent to Bohr. He was the dominant figure next to Bohr in Copenhagen; Pauli and certainly Heisenberg deferred to Kramers' prestige and status. During that period, he had many successes, few disappointments as with helium, but no failures.

Perhaps the only frustrating experience had been the curious Compton interlude (Chapter 14), but Kramers appeared to have made a satisfactory adjustment. In all these successes, Kramers stayed within the context and confines of either classical physics or quantum theory as laid out by Bohr. These successes must have increased his confidence and bolstered his courage—but his own scientific personality had yet to assert itself. This started with the Bohr–Kramers–Slater theory, which was very close to Kramers' heart. Then there followed in short order three dramatic events, which came close to devastating Kramers' scientific confidence. The failure of the Bohr–Kramers–Slater theory was the failure of a theory to which Kramers felt a personal and philosophical affinity. Heisenberg's success in developing matrix mechanics via the dispersion relations route must have created doubts in Kramers' mind about his ability to recognize and seize scientific opportunities. Dirac's unusual originality and daring in deriving the spin and Einstein's *A* and *B* coefficients would naturally make Kramers worry about the efficacy of his own methods and his own place in physics. None of these events did anything to reinforce Kramers' confidence in his own approach; they almost inevitably must have made him more circumspect, more desirous to stay on perfectly safe ground, as Dirac had said of Lorentz. It is against the background of these events and the ever-present, ever-changing role of doubt, hope, and fear that Kramers' style must be viewed.

B. Characteristics of Kramers' Scientific Style

In addition to the characteristics of Kramers' scientific style (his delight in technical tricks, his great formal power, his penchant for concrete problems, and his almost religious faith in the correspondence principle), Kramers had yet another rather unusual feature. Kramers was one of the few major physicists who did not have a strong a priori belief in the intrinsic simplicity of nature. This belief operates in different ways for different physicists: Einstein certainly was convinced of the existence of an order in the universe which one could aspire to grasp; for Dirac these were mathematical structures capable of expressing fundamental laws in simple ways; for Fermi there were descriptions in terms of concepts and phenomena, allowing a transparent unified understanding of many areas of physics. For Kramers there was none of that. He certainly did not believe that nature *had* to be simple; he had a great *fear* of simplicity and especially of oversimplification. He told his favorite student van Kampen on a number of occasions: "There is no guarantee that nature indeed is simple."[42] His suspicion and mistrust of simplicity ran very deep.

This might well be related to his contempt for cheap success. He had an intense dislike for physics that consisted merely of the straightforward application of known methods and well-understood physical situations, even if in that process new results emerged. To Kramers, the struggle with the unknown, a struggle with an uncertain outcome, was essential. It is *probably* true that Kramers would have preferred to make a contribution which demanded technical virtuosity to settle a point of minor physical significance over a trivial application leading to a novel physical effort. It is *certainly* true, as a survey of his papers shows, that this is exactly what he did most of the time. Kramers did not believe in simplicity on *any* level of description. Such a strong conviction, even if never phrased as a principle, is bound to have a strong influence on the manner in which research is actually carried out. There was another principle, largely unspoken, which had a similar effect on Kramers' scientific style. This principle, which clearly shows Bohr's persuasive influence, was the motto chosen for Kramers' collected works. Since the editors of the *Collected Works* were the persons presumably closest to Kramers and most knowledgeable about his research, this motto represents their perception of Kramers' most outstanding characteristic: "My own pet notion is that in the world of human thought generally and in physical science in particular, the most important and most fruitful concepts are those to which it is impossible to attach a well defined meaning."[43] It is hard to know exactly what to make of this *principle of universal complementarity*; it is not even clear whether Kramers ever used or applied it as such. It certainly sounds like one of those whimsical, provocative statements that Kramers delighted in making. But it does indicate that Kramers was quite willing to tolerate ambiguities; he did not shun contradictions and paradoxes. Whatever the precise role of this pet notion might have been, it is evident that it would *not* propel Kramers to precise, rigorous, sharp formulations of general problems. Since his pet notion comes very near asserting that there is and must be an intrinsic vagueness, no such precision can be expected. If this attitude is combined with a belief in the lack of fundamental simplicity, it becomes understandable why Kramers concentrated so much of his efforts on highly specific questions and why he treated them with such care. This mixture of elements gives Kramers' work a very special and unusual flavor. Casimir in his remarkable autobiography makes a number of incisive comments about Kramers: "He [Kramers] tackled problems because he found them challenging, not primarily because they offered chances of easy success. As a consequence, his work is somewhat lacking in spectacular results that can easily be explained to a layman; but among fellow theoreticians he was universally recognized as one of the great masters."[44] Everything Casimir writes is correct and to the point; it characterizes Kramers' work with great precision. He might still have added that as long as a problem presented a challenge, the actual physical significance of the problem was of minor interest to Kramers. Casimir's reference to Kramers' work clearly shows the respect and admiration he had for Kramers. It is therefore interesting and noteworthy that Kramers and Casimir never did a single piece of research together, in spite of the fact that Casimir of all theorists in The Netherlands at that time was nearest to Kramers in knowledge, sophistication, and ability. Casimir himself explains the reason: "Somehow his approach to problems was different from mine, to me it seemed often somewhat indirect and complicated."[44a] The scientific styles of Casimir and Kramers were altogether different. Casimir's approach was very physical and simple—with a minimum of calculations. The emphasis was definitely on clear and sharp physical formulations. Kramers' approach was definitely not simple:

It had more of a computational character; he stressed the mathematical formalism and used the calculational results as a basis for the physical interpretations. Furthermore, Casimir had more interest in phenomena and experiments than Kramers, who cared more for deduction and logic. It was Casimir's observations that as more experiments became available in low-temperature paramagnetism, Kramers became more interested in ferromagnetism.[45] He shifted to the fundamental problems of phase transitions as modeled by the Ising model. In fact, Kramers asked Casimir to join him in what was to become the Kramers–Wanier paper, but the differences in their approaches and styles were so great that cooperation was never attempted.

This particular character of Kramers' scientific style begins to throw some light on the near miss syndrome. Since Kramers did not believe that there ever would be a basic simplicity, there was really no reason for him to push his ideas to a level where one might expect that the simplicity of the mathematical structure or the transparent character of the physical idea would become apparent. In many of his papers, he was quite willing to stop the discussion at a place where the physics was not completely sharp and the mathematics still mired in approximations. This accounts for the sometimes incomplete character of the investigations; it also explains why it is sometimes difficult to identify the specific contribution of a given paper. Kramers' ideas were wonderful and his instincts excellent, but he lacked the compulsion, perhaps the interest, to push his ideas to total clarity, total precision, or ultimate simplicity. This becomes understandable, at least in part, in terms of his driving principles. He did not believe that a truly important principle could be articulated precisely, nor in any ultimate underlying simplicity. It therefore made little sense to try to extract or abstract general insights from specific examples. To Kramers, such insights would have no other content than the specific features already revealed by his detailed analysis.

C. The Role of Mathematics, Structures, Abstractions

Kramers' attitude toward mathematics, structures, abstractions, and axiomatization in general was of great importance in his research. There was universal agreement that Kramers was a brilliant, powerful mathematician. Kramers not infrequently became as involved in the mathematical intricacies of a problem as in the physics which suggested that mathematical problem. Kramers was a master of analysis; his manipulating power was extraordinary. This was recognized by his students, his colleagues, Pauli, and professional mathematicians.[46] He was especially well versed in complex variable theory, the classical theory of differential equations, and all aspects of the theory of special functions. In a paper with Ittman on the asymmetrical top, he gave a very substantial extension of the theory of Lamé functions.[47] The paper on the dispersion relations (Kramers–Kronig relations) shows an uncommon facility with complex variable theory. According to Kac (who himself was a formidable analyst),[46] Kramers was a great expert on elliptic functions. He had taken a course in elliptic functions as a student, and by reading books and papers in a rather haphazard manner, he had acquired in his own effortless way a powerful technique in that area. Kramers was well aware of his great manipulatory power in analysis; his ability to use this arsenal of powerful methods to obtain specific results was a source of great satisfaction to him. In mathematics as in physics, he enjoyed settling specific questions, summing a series, solving an equation, finding the normalization of a wavefunction in the continuous spectrum. It was there that his manipulatory magic was at its most powerful. He did not have much interest in the investigation of mathematical structures or in abstractions, generalizations, or axiomatizations, all of which are such integral parts of modern mathematics. He was quite convinced (and he said so many times) that he had no need of these abstract general methods; any mathematical problem confronting a physicist (especially him) should be solved using manipulatory methods. The symbolic method that Kramers developed was a case in point. Its specific purpose was to avoid the use of formal group theory, and in this his method was surely successful. The motivation was to replace an abstract, autonomous mathematical discipline with its own concepts, methods, and structures by a well-defined set of computational rules very much in line with Kramers' overall philosophy. He did not like to use formal group theory in physics and he did not really believe that abstract formal mathematical considerations had any significant role in physics.

Nevertheless, Kramers gave considerable thought to the proper relation between mathematics and physics. He considered it an important part of the education of his students that they should appreciate the proper level of rigor for a physicist. Kramers often stated to his students that "mathematics should be used in an artistic manner."[48] It is evident that Kramers, when writing his book, encountered the problem of precisely how much and what type of mathematics was suitable in a physics textbook. He struggled with this question but never arrived at a satisfactory, general answer. His struggles are reflected in his own comments in the preface to his book: "The apparent lack of mathematical sophistication, of which the author is somewhat ashamed, and which is frequently mentioned in the text, is not exclusively due to the incompetence of the author. Sophistication in physics even (or rather especially) in its purest form, i.e., unencumbered by pedagogical considerations, does not live happily with its mathematical counterpart—neither in the restricted mansion of the human mind nor in the equally restricted domain of a monograph. We have not used the theory of groups explicitly. All the same, I am well aware of the fact that the mathematical arguments, especially in the second part, may sometimes put a strain on the reader."[49] This is a somewhat tortuous passage: Kramers is really saying that for a typical theoretical physicist it is difficult and probably impossible to do justice to both the mathematical and physical demands. But there is no doubt that Kramers in all his teaching and public lectures forcefully expressed the view that the physical aspects are fundamental. In his inaugural lecture in Delft, Kramers emphasizes that for a physicist the most important element in the construction of a theory is the introduction and definition of the necessary concepts. It is only in terms of these concepts that it becomes possible to formulate the laws of physics, which then can often be expressed in mathematical term.[50] But the concepts and laws are the true essence of physics. It is curious that even though Kramers' intellectual priorities were clear—the search for physical concepts and physical principles—his own work concentrated very heavily on the mathematical implementation and applications of the laws of physics. But in his assessment of the importance of mathematical ideas in theoretical physics he invariably ascribed a subsidiary role to mathematical considerations. Mathematics might be an enormously useful tool of obvious significance in calculations, but the idea had to come from physics, and there mathematics was no help. In a lecture honoring Debye in 1946 (the lecture was given in 1943 but not published until 1946) Kramers makes the following remarkable comments: "To be a good mathematician it is not sufficient to solve a [physics] problem. It could even be a handicap."[51] It is quite possible that Kramers in his attitude toward mathematics was strongly influenced by his two alter egos—Bohr and Lorentz. Bohr used mathematics very sparingly; Lorentz made many calculations, but he had such an incredible mastery of the problems that the physical results always emerged from fairly elementary calculations. He never used analytic function theory or fancy theorems in differential equations. Kramers admired this trait of Lorentz and surely wished to emulate it. His own approach was mathematically a great deal more sophisticated; even though he might not have desired it, his style possessed extraordinary erudition, much more than Lorentz. It is interesting that Casimir, who was more physical and intuitive than Kramers, also exhibited a remarkable mathematical sophistication. Kramers in a revealing passage told Casimir: "You and I have a certain mathematical vein. I don't want to say it's running very rich but it is there. I do not find it in Lorentz."[44b] It is not exactly clear here whether Kramers is expressing surprise, disappointment, or concern; but it is clear that the proper place of mathematical ideas in a physical theory remained a matter of continual concern.

Kramers' knowledge of mathematics was extremely broad. He was an analyst of world class; he either knew or rapidly mastered matrix techniques. He read Hermann Weyl's book, *Group Theory and Quantum Mechanics*, when he was sick in bed with typhoid fever![52] But he had no taste for and was repelled by abstract mathematical structures. In his book, he deliberately and forcefully avoided group theory. But that was only one manifestation of a very general attitude. He never believed that abstractions could yield anything but trivial recastings of results known through prior computational methods. Kramers was generally antagonistic and slightly contemptuous of abstractions, but he had even less use for axiomatizations. In a letter to Clay he writes: "I do not like the axiomatic method in thermodynamics .... The only conclusion I can draw is that in physics one should avoid axiomatization at all cost."[53] Perhaps Kramers was rather more than usually testy in expressing his dislike of the axiomatic approach to thermodynamics, because he had argued this point ad nauseum with the widow of Paul Ehrenfest—and the mere mention of the topic irritated him.[54] Even in the best of moods, Kramers still had no interest or use for abstractions or axiomatizations. He most likely believed that a mere reordering of the logical sequence of an argument could not produce anything new; it would be like getting something for nothing and his Calvinist soul would rebel against that. What is remarkable is that Casimir, who was a generally much less erudite mathematician than Kramers, was willing to study abstract group theory and he used it with great brilliance and quite extraordinary effectiveness. In fact, to a whole generation of physicists, Casimir the man is probably named after a most important class of operators in group theory—the Casimir operators. Casimir pointed out that a good deal of the involved analysis in Kramers' treatment of the asymmetric top could be simplified enormously by using the properties of the rotation group. But Kramers would have none of it;[55] he insisted on his horrendously complex analytical treatment. In a similar vein, in his treatment of quantum electrodynamics, Kramers did not introduce the longitudinal part of the vector potential because there are no longitudinal photons. That is of course true, but formally the introduction of such a longitudinal field is much more elegant and for the transition to relativity it is essential. Kramers, by insisting on a classical physical interpretation of all his mathematical entities, so obscured the formal structure of his theory that he had made it impossible for himself to meet the equally physical demand of relativistic invariance.

It is impossible not to think back on Dirac's comments about fear in connection with Kramers' insistence on treating specific detailed questions in great mathematical detail. Dirac conjectured that Lorentz never made the decisive break to space–time relativity because he was afraid of the conceptual revolution this would entail. But he felt comfortable and secure with the mathematical treatment which he himself had originated. In that realm he felt immune from criticism. It would appear that this identical conjecture should be applied to Kramers. With formalism, technical treatments, and approximations, he felt totally sure of himself (sometimes even more than just sure). But in dealing with new physical ideas, he was much more tentative and susceptible to criticism. Kramers' mode of operation was, if not determined, at least strongly influenced by his hesitation and unwillingness to contemplate radical changes or major innovations in physics. He was afraid of this jump into the unknown, of facing the problems and paradoxes in an unexplored area. This likely also explains his intense need to seek classical analogies and classical pictures. He was acutely uncomfortable in venturing too far from a classical basis. By leaving a classical basis, he would be alone, intellectually isolated from his secure basis. This was very frightening to Kramers (as it is to most anyone). It is ironic that Kramers himself excluded the one mechanism that would have enabled him to go beyond these classical confines. That would be the road Dirac had traveled by mathematical abstraction. Dirac, however, generally studied structures, only rarely specific problems. Kramers usually studied specific questions where the utility of abstracting the problem is not so apparent. Dirac also believed in an underlying simple, knowable mathematical structure, which would motivate him to investigate the general abstract character of a theory. Kramers believed no such thing. To him it was neither obvious nor necessary that there would be a discernible simplicity in either mathematics or physics or anywhere else. With Kramers' views on mathematics as a purely technical tool not a possible heuristic guide to physical ideas, with his emphasis on problems rather than on structures, with his reluctance to consider abstractions, and with his pessimistic assessment that nature is intrinsically complex, the scope and limitations of his work become generally understandable.

IV. Personal Matters

A. Some Personal Traits

It might seem surprising that personal traits have anything to do with scientific activities. Are not scientific problems objective? Is the treatment not either successful or unsuccessful, right or wrong? So where is the taste, the style, the personality? It must be remembered that in many scientific studies one tries to uncover regularities which are to begin with unknown. If the study is truly fundamental, one always looks for some pattern, some structure. But the perception of this unknown structure varies from one person to another. Even what elements should characterize this unknown order is a personal matter, influenced by private hopes, expectations, and fears. Thus, the choice of the problem and the type of framework envisaged carry an unmistakingly personal stamp. As was said so beautifully by Yang: "The physical universe has a structure and one's perception of this structure, one's partiality to some of its characteristics and aversion to others are precisely the elements that make up one's taste. Thus it is not surprising that taste and style are as important in scientific research as they are in literature, art and music."[56] These remarks by Yang explain the importance of the personality in extremely ambitious and highly sophisticated investigations. But it is important to realize that even on a much more pedestrian level, the personality traits, the choice of scientific problems, the manner of treatment, and the definition of success are entangled in a most complicated skein.

Kramers possessed a variety of personal qualities which, although not directly related to scientific talent, nevertheless influenced his scientific mode of operation and his eventual scientific achievements to a considerable extent. An understanding and appreciation of these traits will go a long way toward an explanation of the otherwise so enigmatic "near miss" syndrome. It must be remembered that Kramers was an extremely controlled, careful, guarded person. He was a most private person, sharing his personal feelings and emotions with only very few—family, old friends like Jan Romeyn, and an occasional intimate friend, such as Toon Kelder (a well-known Dutch painter). Even with them he controlled the level of confidentiality and intimacy. He revealed different aspects of his personality to different people. But *he* defined and set the limits of the relation. This had as a consequence that Kramers was perceived very differently by his students, colleagues, scientific associates, close family, and intimate friends. Of course, this is a rather general situation and not restricted to Kramers. However, because of Kramers' scientific reputation and his unquestioned dominance in Dutch physics (after 1927), this difference in perception became especially pronounced. Persons knowing Kramers primarily or exclusively through scientific contact might well be somewhat surprised even incredulous when learning about some of Kramers' personal characteristics.

Kramers was generally uneasy about making decisions; he often appeared to be tentative in situations calling for specific choices. However, he did reluctantly make decisions. But he was totally unwilling to face conflicts. He would ignore the conflict or try his utmost to explain that the conflict was not really there. He would do almost anything to avoid confrontations. A public confrontation, be it social, personal, scientific, or political, was something to be avoided at all costs. His strategy of life—that of course includes science—was designed to bypass situations where conflicts or confrontations might appear. He would go out of his way, employing all the resources of his formidable intellect, to look for alternatives and compromises, however contrived and artificial, just to avoid an open conflict. Another aspect of this general unwillingness to face conflicts was that Kramers in both his scientific and personal life was loathe to make sharp breaks, sudden radical changes. He never would want to precipitate a conflict or create a situation where irreconcilable choices became mandatory. His attitude can best be summarized by stating that he resisted and cringed at the necessity of having to make choices which would limit his options in the future in any way whatsoever. He fervently wished to maintain all possible options at all stages of his personal and scientific life. It is now clear why Kramers had such a very difficult time making intense commitments; this would impose a long-lasting, binding obligation on him, automatically restricting his options—which is just what he did not wish to do. By the same token, his unwillingness to make choices or face conflicts—or phrased positively, his willingness to entertain compromises and bypass conflicts—severely restricted the range of actions that could be taken. It is overwhelmingly likely that with such constraints one becomes an observer, an interpreter, a critic, a commentator, perhaps a developer, but certainly not an originator or revolutionary. The desire to evade conflicts and to avoid confrontations at all costs, the hesitancy to make option-reducing choices, the extreme reluctance to take decisive irreversible actions, all these were life-long characteristics of Kramers. They were not just congenital properties of Kramers himself, but many members of the Kramers' extended family had strikingly similar characteristics. Kramers' mother, Mrs. Breukelman, was a gentle, kind-hearted lady who was known for her remarkable talent to make feuding members of the family get along at last in minimal harmony.[57] There was a cousin who showed these traits in a very pronounced manner: He was extremely intelligent, very knowledgeable, but he never did anything. He just let the world go by, without actively participating, "he would be happy to sit, watch and just stir the porridge."[58]

Another family member, H. C. Kramers, a nephew of H. A. Kramers, again showed many of these same traits. H. C. Kramers came to Leiden in 1939 to study physics. Kramers was most helpful to H. C. all during the war time. But after the war, H. C. was very unsure about what he wanted to do; he could not really make up his mind. He vacillated for a long time between theory, experiments, or no physics at all. Kramers had many talks with him and he told him about the congenital Kramers' trouble: They did not know how to be forceful, they did not know how to insist on their rights, and they had a hard time making decisions. He ended his characterization of the Kramers family by what almost amounts to a slogan: "A Kramers must fight to fight."[59] It is quite understandable that these character traits, with their ensuing gentleness, would give the appearance of lack of ambition. It was mentioned earlier that Bohr was convinced that Kramers did not have a grain of ambition. That is not exactly true; Kramers was interested in understanding, anxious to make significant contributions, and avid to be recognized for these efforts. But he was not personally vain; if personal recognition would entail a priority fight or a public confrontation, he would withdraw altogether. If the possibility of an open conflict arose, Kramers would say: "What does it really matter, as long as the work gets done."[58] (*Wat doet het er eigenlyk toe.*) With that he would gracefully withdraw from the competition and possible conflict. No wonder he was considered such a kind person. He always gave in!

Kramers' life is full of specific instances, all illustrating the personality traits mentioned here. He had a terrible time, as was discussed so extensively in Chapter 10, deciding to concentrate on physics—thereby giving up his many other options such as music and literature. He had an even more difficult time deciding to marry Storm, even though he knew she was pregnant (Chapter 11). After all, marriage is the extreme option-eliminating action. Kramers' talent in avoiding confrontations was legendary. After Kramers' promotion, there was a small party at Romeyn's room which—although large for a student room—had only two chairs. Two brash young students of Kramers' circle came in early and sat down in the chairs. After the promotion, Kramers, Bohr, and Lorentz came in, but the students, in what seems a surprisingly contemporary attitude, just remained seated. Kramers approached them with a big smile accepting their congratulations. He pulled the two students up by their ears, smiling all the time, and offered the chairs to Bohr and Lorentz. There was no fight, no incident, no confrontation.[58] On another occasion, there was a somewhat official dinner party at the home of Kramers' father and his new wife (Kramers' stepmother). Kramers at that time was unofficially engaged to Storm. She was invited—not very graciously; Kramers' stepmother did not have much use for Storm. Soon after the dinner had started, "Ma Kramers" suggested that they speak French, supposedly to give the party class, but more likely to embarrass or exclude Storm. Kramers did not object, but he spoke such silly French and feigned such a terrible pronunciation with such exaggerated Dutch colloquialisms that the whole thing was quickly dropped. Again, Kramers had very smoothly averted a possibly unpleasant scene; as it was, nobody was very embarrassed and there was no conflict. This was very much Kramers' style, to avoid confrontations by humor, slyness, or if necessary, by obfuscation. And it did not matter much whether the conflicts were minor domestic quarrels, scientific issues of great profundity, or political controversies.

Kramers' unwillingness to participate in definite actions was especially noticeable in the political sphere. Although he was abstractly interested in politics, he never took political action: He never signed a petition in his life and he never took a strong political position.[58] (He made an exception during the German occupation.) Very often he saw or claimed he saw some merit in the opposing position. This must have been hard on his close friends, the Romeyns, who were political activists. Annie Romeyn in an interview said almost in exasperation: "He [Hans Kramers] would never fight for anything."[58]

The fact that Kramers, because of his extraordinary broad range of abilities and his unusual intellect, had so many options and saw so many alternatives may ultimately have been a serious handicap. It kept him from evolving strong attitudes or dominating convictions in his scientific, personal, and political lives. He did not like to make decisions anyhow, but to have to contend with a mind capable of making compelling arguments for all sides of an issue must have made the decision-making process that much harder. Because he had and saw so many options, none of which he wanted to pursue or ignore, he had frequent periods of restlessness and indecision.[60] During such periods, his behavior vacillated between extremes; he could be taciturn, withdrawn, morose and depressed, or he could be loquacious, holding forth at great length about almost any topic from religion to rowing. Because of Kramers' vast knowledge, he would even in ordinary conversations allude to persons or events in literature and mythology. Perhaps because he had thought so much more and so much deeper than most, and because he saw ambiguity in so many places, many of Kramers' statements have an oraclelike quality. He seemed to wish to project a sphinxlike image; he liked statements that had multiple meaning, whose converse (à la Bohr) might also be true.[61] It is not clear whether these statements were to be taken seriously, merely intended as thought provoking, or should just be ignored. Perhaps the intent was not to mystify or obfuscate, but to express Kramers' own doubts and vacillations.

It is important to realize that the personality traits and characteristics described here pervaded Kramers' entire life. The illustrations presented may appear as only occasional incidents which must not be given an exaggerated significance. However, this would be quite wrong. The crucial characteristics—the unwillingness to provoke or face confrontations and the doubts and hesitancy about making decisions and radical breaks—all were an integral part of his personality. When one of Kramers' daughters was asked what she considered her father's most typical, most characteristic feature, she said without hesitation: "He always had great trouble in crossing a threshold."[62] She knew little about physics; she had never heard of "near misses" (and she was not told).

There is one other factor which also had a considerable influence on Kramers—as indeed it would on anyone—and that is the status of his physical health. For a person generally in good health, he was sick a lot. Ever since his sojourns in the hospital in Rotterdam in 1919 and later in Copenhagen in 1920 he worried a lot about his health. Even though Bohr was some 10 years older than Kramers, Bohr could "outwalk" Kramers anytime. In a number of letters, Kramers complained about one physical ailment or another. He suffered from a bad back, gout, and a number of other annoying debilitating minor illnesses. He was very frequently tired; he complained more about lack of energy, exhaustion, and the inevitable depression than anything else. He actually managed to work even when his health was not the best—when he was tired and worn out. Kramers' insistence on working when he was not in the best physical shape, no doubt took its toll. But his complex personality would not let him do otherwise; he had to continue physics. To make radical, daring, new contributions takes courage, brashness, and physical strength, and Kramers had problems with all of these. That he achieved so much, operating under the constraints of his own personality, is a miracle.

B. The Surroundings

As was described in detail in Chapter 10, Kramers grew up and was educated in The Netherlands. It is important to repeat that he was very much a part of the Dutch culture at that time. He liked the country, knew its literature and history, and had deep and strong roots in Holland. Even though he spent many years in Denmark and traveled a great deal, he always retained this special fondness for The Netherlands. Kramers was a member of the well-educated upper middle class; he had strong ties to the professional, intellectual, and artistic Dutch community. Even though early in his career there might have been some confusion about his personal role in that community, there was never any doubt that he would be an academic, a scholar, or an artist of some kind and not a plumber, athlete, or politician. After he had definitely decided to pursue a career in physics, his peer group was circumscribed even further: It consisted primarily of young scientists aspiring to become productive full-fledged, respected members of the Dutch physics establishment. The physics establishment in The Netherlands at that time was not a particularly well-defined entity; operationally, it consisted of the full professors in physics at the major universities and technical high schools. Since at that time there were four universities in Holland and two technical high schools, the total number of scientists actively participating in scientific investigations was quite small—probably somewhere between 50 and 100. Each one of these professors had a very considerable amount of freedom in deciding what to teach and how to teach it. If there was any overall direction or general trend, it was due to the scientific dominance of a few outstanding individuals, such as Lorentz in Leiden and van der Waals in Amsterdam. Obviously, both the teaching and research were carried out in the midst of the Dutch culture. It is therefore not surprising that certain characteristic elements of that culture would seep into physics. It is understandable that these cultural influences would be more conspicuous in education than in research. Care, precision, soundness, steadiness, and accuracy were some of the central, most valued elements of the Dutch science. For a student to be called careless or sloppy was considerably worse than to be called inept or of limited intelligence. The educational system, curriculum, teaching methods, and examination procedures in the high schools all mirrored these same values. In the universities the structure was a great deal less formal and rigid, but the values were very much the same. To give a correct, well-grounded, thorough discussion; to carry out a detailed, exhaustive, painstaking experiment; this was recognized as the essence of good and sound science.

Kamerlingh Onnes, the famous low-temperature physicist who discovered superconductivity, expressed the prevailing view of the experimental physicists in his inaugural address in Leiden, November 11, 1882, entitled "The Significance of Quantitative Research in Physics." In this lecture, Onnes suggested that every physics laboratory should have a motto affixed over the entrance: "From measurement to knowledge."[63] This is quite an unequivocal pronouncement; to Onnes the only road to knowledge was via careful, persistent, systematic measurements. It might well be more extreme than most experimentalists would believe; but it did indicate a very common and pervasive trend. [It is not without interest, and certainly not without relevance, that this same Kamerlingh Onnes, while talking to G. E. Uhlenbeck, a favorite student of Ehrenfest, asked him (in 1924): "These Maxwell equations, what good are they?!"[64]] There were no comparable pronouncements emanating from the theoretical physicists in Holland; Lorentz was too wise a man to make such categorical pronouncements. Ehrenfest (see Chapter 10, Section II) was of course Austrian, not Dutch, but his style was not at variance with the Dutch tradition. He liked specific problems and he was especially intrigued by models and examples, which exhibited specific physical points. In many of his studies, he constructed systems and equations, which expressed the physical ideas and no more. The treatment given was invariably precise and unusually lucid. Ehrenfest would not tolerate carelessness, ambiguity, and vagueness; in this way he just became a chain in the long Dutch tradition of precision, clarity, and exactitude. This had as a byproduct that Ehrenfest and his students, especially Uhlenbeck and Kramers, became extremely critical of much that passed for theoretical physics. They both became past masters at asking embarrassing, but pertinent, questions at colloquia.

This was the scientific atmosphere during Kramers' formative years in Leiden. He made this aspect of Ehrenfest's attitude very much his own. He never put up with slipshod thinking; he hated careless, superficial, amateurish physics. He believed in a thorough technical grounding before going on to more fashionable topics. He was firmly convinced that a proper, systematic ordering of the topics was necessary for a true appreciation of physics. When his nephew, H. C. Kramers, decided to embark on the study of physics, Kramers advised him how he might best go about that: He should start with fundamentals, then progress to harder topics. He gave him several books to read, such as Maxwell's book on thermodynamics. However, H.C. wanted to read books by Jeans on the expanding universe and by Eddington on the new pathways in science. These books, although written by experts, were effectively popular expositions of current topics. Kramers objected strenuously. He did not approve of those books; they talked too much "around the subject." He then gave a typical Dutch characterization: "These books are not sufficiently sound and solid."[65] So H.C. read these books without telling his uncle. It is most interesting that when Nico van Kampen (who became a major scientist in his own right) became interested in physics as a young man, he asked his uncle, Fritz Zernike, also an eminent physicist, for advice about what to read. Zernike gave van Kampen a traditional list. When van Kampen wanted to read Jeans and Eddington, Zernike also discouraged his nephew for exactly the same reasons as Kramers had given his nephew, using exactly the same words and even the identical grammar. Rather than just considering this as an amusing coincidence, it is more reasonable to suppose that to a Dutch professor of physics, the Dutch virtues of solidity, soundness, and meticulous care, were an integral and indispensable part of physics. There is of course no doubt that they were right. Yet it is certainly possible that an undue emphasis on these very qualities would make it impossible to pursue speculative ventures or explore alternate ideas. Whether this was the case with Kramers is hard to know. He was certainly very scholarly and erudite, but his work was often lacking in daring and originality. Whether his affinity for the Dutch virtues made him less adventurous, less speculative than he might have been, given his insight and sophistication, is of course a matter of conjecture. But it is certain that the perennial insistence on clarity, precision, soundness, and completeness, can have a severely inhibiting effect on a person and on the progress of an investigation, especially in the early stages of a research venture. Throughout Kramers' work there are occasional speculative glimpses, but these are usually rapidly suppressed or dismissed as mere logical possibilities. At times it seems as if Kramers quite suddenly encounters an obstacle which keeps him from pursuing his radical ideas further, as if he were getting too bold. There is only one notable and obvious counterexample—the Bohr–Kramers–Slater theory. Unfortunately, that ended in a dismal failure. But in all his other investigations, Kramers' critical faculties far outstripped his creative abilities. Whether this is at all related to his early exposure to the Dutch surroundings is impossible to say, but it is certain that these experiences could not have stimulated him to daring new exploits. These circumstances may not explain Kramers' timidity in striking out on his own, but it is hard to believe that they were utterly without effect.

C. The Interpersonal Relations

Kramers' personal relations with his scientific peers played an unusually important role in his life. Because of the importance of physics in Kramers' life, almost all his social contacts involved physicists. With the possible exceptions of Jan Romeyn, the historian and friend of his youth, and Toon Kelder, the painter and a very close friend in later life, Kramers had only a few friends not associated with physics. The Kramers home in Leiden was a social center for Dutch physicists: It was there that Kronig met his wife, and the physicist Fokker would come and sing at the many musical evenings. Kramers was generally well liked; his relation to most of the physicists was one of great congeniality and cordiality. Of course, most of those relations were not of great depth or great intimacy. Kramers' controlled, guarded, very private personality monitored the intensity and the range of each and every one of these relations. Naturally, they were all different. The physicists of greatest importance to Kramers, both personally and scientifically, came in two groups of three: the older group, Lorentz, Ehrenfest, and Bohr; and the Copenhagen group, Heisenberg, Pauli, and Klein. Although Kramers knew many other physicists personally and studied their papers, the two triumvirates were closest to him in approach, scientific style, and scientific goals. As time went on, Kramers paid less and less attention to the investigations of other physicists; by 1948, he no longer read books and he only glanced at an occasional paper.[66] He maintained whatever scientific contacts he desired via letter or personal conversations with a very small number of physicists. Among these were Heisenberg, Pauli, Klein, and Bohr, the four surviving members of the original triumvirates. It appeared that Kramers was firmly convinced that even at that time these persons were at the frontier of physics. He clearly did not feel that in maintaining these contacts to the exclusion of most others he would miss any significant ongoing developments. So he effectively stuck to the personal and scientific associations of his youth, ignoring new people and new developments.

Of course, the nature of these personal involvements varied enormously. Kramers had an extraordinary admiration for Lorentz. He had immense respect for Lorentz's style, for the scope of his activities, and for his supreme mastery of physics, but he did not know him very well and their direct personal contact was probably negligible. Their total correspondence consisted of a few letters, one written by Kramers to Lorentz[67] in 1918 in connection with a paper he had written on general relativity (Kramers called it a "little paper"), and a few others dealing with arrangements for the Solvay Conference.[68] Still, Lorentz had an enormous impact on Kramers; there can be little doubt that Kramers desired to emulate Lorentz to the extent that he could. He succeeded at least in part; he was for a considerable period of time the leading physicist in Holland. Perhaps the most interesting and indirect way of Lorentz's influence is that Kramers often found it necessary to explain or justify his interests in matters outside physics by the contrived claim that there was a hidden, somewhat mysterious, deep connection between these matters and theoretical physics. Lorentz, a man of universal talent, deliberately restricted his own activities to physics. In one of his popular lectures he stated: "The physicist, and this holds for all of us, must restrict himself to reading in his way in the book of the world."[69] This was precisely the kind of restriction that was most difficult for Kramers to accept. In fact he did not accept it; he read widely and voraciously. However, he justified his reading of Baudelaire's *Les fleurs du mal* to his friend Casimir by insisting that its spirit is somehow akin to that of theoretical physics. If Kramers was serious—and that is impossible to know—this comes near the silliest argument that Kramers ever gave, but it does show that facetiously or not he felt, even at this stage of his career, the need to justify his literary interest in terms acceptable to Lorentz.

The relation between Ehrenfest and Kramers was analyzed in great detail in Chapter 10. That relation went through several stages; it was personal and intense. But in this relation, intense as it was, Kramers really reveals very little of himself. Especially in the later correspondence (1928), there is a feeling of frenzy and desperation on Ehrenfest's part, which Kramers tries to assuage by stressing Ehrenfest's great contributions and important role. But Kramers' own emotions, fears, and concerns are nowhere noticeable. In two letters, Ehrenfest writes: "Practically all new theoretical papers stand before me as a completely ununderstandable wall and I am at a total loss. I no longer know the symbols or the language and I no longer know what the problems are."[70] "Please help me to understand things the way a school teacher should and not a wisdom producer or a copublisher. This I must postpone for the foreseeable future, maybe for ever."[71] In answering these letters Kramers thanks Ehrenfest for making him study the theory of Lamé functions in 1915 so that he was properly oriented for his papers with Ittman on the asymmetrical top. "I find it ironic that you can give others so much and yourself so little."[72] Kramers continues for six pages; he is understanding and compassionate, and one learns about his trip to the United States, about Ornstein's laboratory, but nothing at all about Kramers. The net, direct influence of Ehrenfest on Kramers' scientific style is very minor, this in spite of the very strong personal involvement. The influence of Lorentz was much more pronounced even though there was almost no personal involvement.

Unquestionably, the strongest personal influence on Kramers' life was Bohr. Bohr's influence covers the complete spectrum, scientific, intellectual, emotional, and personal. Kac, in his remarkable autobiography, comments on the difference between learning physics and mathematics in isolation.

"Learning physics by oneself is essentially impossible, or to quote Uhlenbeck, One must follow a matter."[73] If anyone ever took that admonition to heart it was Kramers; he was a true disciple of Bohr. The scientific aspects of the relation between Bohr and Kramers were discussed in Chapter 11. It is important to realize that the personal and indeed emotional aspects of that relation are of equal importance. It was in his relation to Bohr that Kramers' personality traits—his unwillingness to fight and his abhorrence of conflicts—became especially important. This cannot be separated from Kramers' totally genuine and effusive admiration of Bohr's physics. In almost all popular and general lectures, Kramers emphasizes the dominant role of Bohr and Bohr's style. "In Bohr's work nothing is clear and unambiguous—there are no conclusions—no definitions—it is more a continuous struggle against the unknown, which paradoxically has led to great innovations which have left Bohr's imprint on this period."[74] Kramers' complete acceptance of the Bohr ideology and his reluctance and inability to confront Bohr on any level whatsoever clearly made it difficult for Kramers to develop his own independent scientific identity in a forceful manner. The fluidity of the Bohr philosophy, combined with Kramers' scientific dependence, gives a major clue to the *near miss syndrome*. The Bohr–Kramers relation was continually changing without materially altering their mutual admiration and devotion. Kramers on occasion complained that he was exploited by Bohr[75]—something his wife had told him for many years, but when she said it, Kramers always denied it. Nevertheless, when Bohr asked Kramers to do something, he told Korringa, "This really goes too far."[75] During the war time, Kramers started to have regular and frightening dreams about Bohr. This worried him to such an extent that very soon after the war was over, he took a tiny plane, a two-seater, to travel to Copenhagen to see that everything was alright. Bohr on at least two occasions tried seriously to lure Kramers back to Copenhagen. Of course, Kramers had visited Copenhagen regularly, but it appears that Bohr had more permanent arrangements in mind: "I am writing on behalf of our institute, where we never forget what we owe to your collaboration over so many years to ask if it would be possible for you in the course of the coming autumn to visit us and to give a series of lectures on your recent work .... This would indeed be a most valuable inspiration to our whole group and an extreme pleasure to all your friends and old pupils in Copenhagen."[76] This is certainly a warm letter; it appears that Kramers' answer is lost, likely because of the turbulent circumstances just before World War II. In the middle of the war, to be precise, July 1, 1942, Bohr repeats his invitation, this time there is a direct suggestion that Kramers would permanently be associated with the Bohr Institute.[77] Bohr refers explicitly to Kramers' paper on Brownian motion and to his work on the classical theory of the electron (Chapter 16). There is something puzzling about this invitation. July 1, 1942 was right in the middle of World War II and it is hard to see how Kramers could have moved on a permanent basis from one German-occupied country to another. Possibly, Kramers might have obtained permission for a short visit, but a major move seems impossible. It is also unlikely that Kramers as a major figure in Dutch physics would just pack up and leave in the middle of a disruptive war. Perhaps Bohr was just worried about Kramers in Holland and this was his way to offer him asylum. Nothing came of it, but it is certainly another indication of the closeness between Bohr and Kramers. In his relation to Bohr, Kramers revealed more of himself than he did to anyone else with the possible exception of Toon Kelder and Jan Romeyn.

The degree to which Kramers confided in Bohr can be appreciated from two letters Kramers sent to Bohr. The first letter was written January 8, 1948, about half a year after he had returned from the United States, just after he had recovered from his stroke.

> Thank you for your Christmas letter, which in its mood of warm friendship had an uplifting effect on me at this time, when the worries of the world and one's soul weigh heavily on me—probably more than under normal conditions of my health .... Regarding the situation of the world at the moment I am doomed to be satisfied to develop dialectical points of view, both with respect to world politics as it is (and used to be) with whatever feeling of hope and despair of a spectator. I am pleased to know that you will do what you can in America: you are one of the few persons who do not bend himself to the different realities, both in physics and in politics—also one of the few who know which values play a role in the "culture." Salvation of the world lies in a combination of such an attitude with strength of character. I have fantasized many thoughts about strength of character these last months and the interplay between this on the one side and intelligence and ability on the other side. My difficulty is that I do not have any talents to be a participant in the necessary drama which this combination requires; unfortunately, I must sit back as a spectator, like Shakespeare, but he could write it down, and I am only his "gentleman." [78]

This is a most remarkable letter: It shows Kramers' undiminished adulation of Bohr and also expresses his deep concern for the world and for humanity. But perhaps most surprising is Kramers' explicit and candid assessment of his own strength of character. Kramers never doubted his intelligence and ability, but he often expressed doubts about his combativeness. In this letter, he states unequivocally that he can only be a spectator because he lacks sufficient strength to do otherwise. Kramers wrote another letter to Bohr, perhaps not quite as revealing as the earlier one, but still divulging an interesting aspect of his concerns. "As far as my soul is concerned (both with regard to mankind and science and society), I am experiencing feelings that I need to talk to you soon about, even if it is only to be momentarily free from speculations about happiness and misfortune and fate." [79] This passage conveys a melancholic mood. Clearly, Kramers would only write this to a very close friend. Later in the letter, Kramers recounts an interesting dream: "In Switzerland I recently dreamt that Margrete [Bohr's wife] whom I met sitting in a park—offered me support and help to escape from the claws of Nazism, which in a large meeting just before (when there were only Russians present) had *sworn* that I was guilty and had added—when I wanted to defend myself—that I was not to have permission to reply, but only express a hope. Anyhow, I got away; first from the female Nazi guard who was assigned to me, later from the dream itself."[79] This is how the letter ends, without further comment. It would be presumptuous to try to interpret this dream, but it does show that 4 years after the German occupation of Holland and 3 years after Kramers dealt with the Russians in the U.N. (see Chapter 18, Section IV.B), the impressions had not faded and his fears had not disappeared. This, too, he confided just to Bohr. These letters, better than any analysis, give a complete and perceptive picture of the Kramers–Bohr relation.

Klein was a particularly good friend of Kramers. Their relation was comfortable and relaxed—not beset by conflicts, tensions, or antagonisms (scientific or otherwise). This no doubt was because Klein like Kramers was a kind, generous, and compassionate person, who lacked the vicious critical faculties of Pauli and the intense competitiveness of Heisenberg. Just as Kramers had come very near matrix mechanics and did not discover it, so Klein had come very near wave mechanics and did not quite discover it either. (The only remnant of his work is the K in the Klein–Gordon–Schrödinger equation.) Klein and Kramers tried to console each other but they mainly commiserated. Klein like Kramers was neither a fighter nor a pugnacious advocate of his position. When Klein tried to help in the deep rift between Heisenberg and Bohr concerning the uncertainty relation in 1926–1927, he only got into trouble. Kramers, who was aware of the conflict, wanted no part of it (he was in Utrecht at the time) and gave Klein advice, which characterizes both Klein and Kramers: "Do no enter this conflict, we are both much too kind and gentle to participate in that kind of struggle. Both Bohr and Heisenberg are tough, hard nosed, uncompromising and indefatigable. We [Kramers and Klein] would just be crushed in that juggernaut."[80] It is interesting to record that both Klein and his wife insisted that Klein did his most original and daring work when Bohr was away from Copenhagen.[80] The friendship between Klein and Kramers was so deep and satisfying that Kramers could be relaxed about his limitations and weaknesses, because he knew that these would be unquestioned, understood, and completely accepted.

The curious Pauli–Kramers–Heisenberg triangle was analyzed in Chapter 13, Section V. That relation did not change a great deal after Kramers had left Copenhagen. It is interesting that later in his career Kramers tended to defer to Pauli and Heisenberg. He seemed inclined to believe that they continued to dominate physics to the same extent they had from 1924 to 1932. For example, Kramers always refers to the Dirac–Pauli matrices and to the Dirac–Heisenberg hole theory. He clearly had a much greater affinity for Pauli and Heisenberg than for Dirac. The personal relation between Pauli and Kramers was warm and pleasant and centered on physics. There was an easy congeniality, typical of persons who have known and liked each other for a long time. Pauli accepted Kramers' home life—he liked the children—but in true Pauli fashion on his own terms. He often brought the children gifts, but since he felt two children were enough he would bring two presents and no more.[81] Kramers and Pauli could have public disagreements, even caustic arguments, without any effect on their relation. In fact, they rather enjoyed these performances. In the summer school in 1931, in Ann Arbor, Michigan, Oppenheimer was giving a seminar. Both Pauli and Kramers were in the audience. Pauli kept objecting and interrupting, so that it was hard for Oppenheimer to say anything. Kramers got up and said: "Shut up, Pauli, and let us hear what Oppenheimer has to say. You can explain how wrong it is afterwards."[82] (This is, by the way, exactly what happened.) This type of half-joking, half-serious exchange was typical for Kramers and Pauli. There was great mutual respect, considerable warmth, and deep affection.

The relation between Kramers and Heisenberg was in many ways more complicated. There is no doubt that Kramers had an enormous and long-lasting admiration for Heisenberg. They maintained close contact for many years. Yet there was something guarded about their relation. It was polite, civil, even warm, but it lacked intensity and depth. Most noticable is the absence of humor or lightheartedness. Everything is very serious and business-like, with polite and proper references to the family, health, and other "impersonal" personal matters. Kramers wrote a very pleasant and generous letter congratulating Heisenberg on getting the Nobel prize. In this letter there is no trace of jealousy or disappointment; quite the contrary, Kramers writes that "this time the Nobel Committee had an easy choice to make," a most complementary thing to say.[83] After Kramers had decided to leave Utrecht and go to Leiden, he was instrumental in trying to get Heisenberg to come to Utrecht. After a long series of cordial exchanges, Heisenberg finally decided that he did not wish to leave Germany. But all this was done in a most pleasant atmosphere. The cordial relation between Heisenberg and Kramers was unaffected by the war. Kramers went out of his way to arrange a visit for Heisenberg in the middle of the war. Kramers was enthused and stimulated by Heisenberg's visit. In turn, Heisenberg tried to be helpful in keeping the low-temperature instruments in the Leiden university instead of having them sent to Germany, and Kramers was very grateful for his efforts. This is recorded in a letter dated July 5, 1944.[84] From then on there is no further communication between Kramers and Heisenberg for some time. The next contact was a card sent by Heisenberg to Kramers on March 23, 1947, from Göttingen.[85] The occasion was rather remarkable; Kramers evidently had sent the Heisenberg family a "care" package containing lots of food, and Heisenberg thanks him profusely. Heisenberg writes that he was worried about Kramers during the "hunger winter" in 1945; he had not known that Kramers had survived the war until he was in England. Heisenberg had also heard that Kramers had accepted an important and responsible position in the boundary are between politics and physics. Then he adds: "I hope that in this difficult field you will be as successful as you were earlier in physics." Actually, Kramers' gift of food in early 1947 to someone in Germany was quite impressive. The Netherlands and the Kramers family were literally ravaged by hunger: Things were better in 1947 but far from perfect. To send food was an act of real compassion. Kramers always considered Heisenberg an extraordinary genius, almost a young "Wagnerian hero," and a good if not intimate friend, even though there was a certain stiffness in the relation. Part of this had to do with the Nobel prize. Heisenberg at a dinner in Leiden sat next to a world famous Latin scholar, Professor Wassink.[86] Heisenberg told Wassink that Kramers was deeply disappointed, while Mrs. Kramers had suffered for many years because Kramers never got the Nobel prize. Evidently, the Nobel prize was mentioned often at the Kramers home; Kramers said that if he got the Nobel prize he would buy many paintings, but Mrs. Kramers was not so sure she wanted to spend the money that way.[87] Thus, although Kramers never said, wrote, or hinted anything about the Heisenberg Nobel prize, it is evident that Heisenberg himself was not so sure that the topic was totally absent from Kramers' daydreams. Complicating the Heisenberg–Kramers relation even further was the fact that Mrs. Kramers did not like Heisenberg at all. She felt that Heisenberg used and exploited Kramers' talents and gave credit reluctantly if at all. Storm was very ambitious for her husband; she believed that he had been exploited by many physicists, starting with Bohr. On the other hand, she knew how fond Kramers was of Heisenberg and how special Bohr was. So she never was too explicit in her criticism, but she did all she could to encourage Kramers to stand up for his rights and see that he got the recognition and rewards that were rightfully his. After Kramers died, when Heisenberg told Mrs. Kramers that her husband's role in the creation of quantum mechanics was vastly underrated, she was especially unhappy. She felt that Heisenberg should tell the world of physics. She already knew but she could not tell that world. Storm throughout her life struggled to get Kramers to fight for himself, but it never had any effect. More precisely, the effect was negative: Kramers' attachment to his colleagues was great and any hint that they were using him or his work would make him particularly angry. The suggestions or the demands by Storm that he aggressively pursue a policy which would guarantee that he receive all the credit due, clashed head on with the most basic canon of his character—his unwillingness to engage in violent confrontations. Thus, ironically enough, Storm's ambition for her husband's reputation became part of a gradual estrangement. Storm's heroic efforts at an accommodation with those she believed exploited her husband netted her nothing, except possibly his displeasure. Kramers carried his unwillingness to engage in arguments so far that even when he was subject to rather unpleasant and abusive treatment he still withdrew rather than fight.

W. J. de Haas, a professor of experimental physics in Leiden, an imaginative and whimsical person, and Kramers had rather strained relations. The precise reason is not obvious; de Haas was very anxious to get the Nobel prize, which he felt he deserved. (Perhaps he felt he had to get it; his father-in-law was the immortal Lorentz.) In innumerable letters he asks for Kramers' help to get the coveted award and when Kramers could not or would not help him, he became bitterly resentful. This anger spilled over into many areas. On July 20, 1945, he writes an angry letter to Kramers; he is extremely critical of Casimir. The situation did not improve and 3 years later de Haas writes: "You

[Kramers] do not know what experiments are or what experimental physics is. You got bad advice from Casimir."[88] There are literally dozens of these letters, complaining, cajoling, accusing. Still, Kramers remained courteous. The continual badgering of de Haas did irritate Kramers, but he never took any action to end this eternal petulant whining. Storm told Kramers a hundred times to stand up to de Haas and get rid of this nuisance, but so attached was Kramers to his personal nonbelligerency that even in this most deserving case he suffered instead of provoking confrontation.

D. Fear and Decency

It would be quite wrong to say that Kramers' unwillingness to face conflicts was an exclusively negative quality. His reluctance to engage in controversy was rooted in profoundly human compassion. He perhaps more than anything was scared and unwilling to hurt or insult other people. He may have carried this to an extreme, by not contradicting, not disagreeing with potentially objectionable and controversial opinions, but his underlying motivation was surely a desire to be kind, understanding, and compassionate. Kramers was unusually intelligent; he knew perfectly well that not every person is sensible, not all arguments are correct, not all actions are admirable. This combination of compassion and insight often caused him to appear whimsical and capricious, so that it was hard to know when he was serious, teasing, or indeed profound. But underlying it all was a profound decency, a great gentleness, and a respect for others. One of Kramers' admirers, a student of John Wheeler, John S. Toll, whose thesis was an extensive elaboration of Kramers' work on dispersion relations, claimed that he could describe Kramers' attitude in one phrase: "He did not have the killer instinct."[89] And Toll was quite right: Gentleness, kindness, respect for others are not the qualities that make for a relentless, single-minded, ruthless pursuit of *one* idea of *one* person to the explicit exclusion of other ideas and other persons. The attitudes of kindness, decency, and understanding pervaded his life (as Kramers himself would say, "his soul"). This, coupled with his recognition of the existence of real suffering, actual misery, and rampant stupidity, left him often in a state of inactive melancholy. The "human condition" was a tangible reality to Kramers, not a philosophical or sociological abstraction. The day that Paris fell to the Germans in the beginning of World War II, Pais and Kramers were together.[90] They sat in silence, mourning the irretrievable loss of a civilization. Neither the world nor Kramers quite recovered from that loss. Kramers was sensitive, perhaps even overly sensitive to all human problems, personal and social. But what made him such a difficult person to understand and appreciate was his own compulsion to deny, hide, or even ridicule his own sensitivity. Kramers was especially close to his brother Jan, a most renowned Arab scholar, a colleague at the University of Leiden. They often played bridge together and shared an ironical sense of humor.[91] An elaborate party had been planned for Kramers' birthday, December 17, 1951, with the usual games and festivities. Jan Kramers died during that day. When Hans Kramers heard of the death of his brother, who, of course, was supposed to be at the party, the first and only thing he said was: "Let us get somebody else for the party."[91,92] It is hard to believe that this casual, callous comment expressed his true feelings, but he of course was totally incapable of showing his true emotions. Sadness, grief, weakness, and despair were not for the public view, not even for his family.

In spite of Kramers' attempts at concealing his sensitivity and compassion, these were crucial elements in his makeup. This became very evident in the education of his children. They were not allowed to think or care about money. Any kind of competition was strongly discouraged; in fact, competitive sports were practically forbidden. To appreciate the complexity of Kramers' personality, it must be realized that these genuine feelings coexisted with a number of quite different and barely compatible attitudes. Everyday matters were of a supreme unconcern to Kramers. He was not interested in them and did not wish to talk about them. His thoughts were intellectual, cultural, and analytical. Trivialities irritated him; he possessed a certain degree of intellectual arrogance. If people had nothing interesting to discuss, he would say: "Let us talk about the immaculate conception."[93] He could be a little sharp and mildly unpleasant; he had little tolerance for stupid people. Yet all this was imbedded in a person of great kindness, understanding, and unusual sensitivity. He was at all times aware of the human condition and the limitations of his own mind. He was a curious mixture of self-confidence, conceit, arrogance, abject humility, and paralyzing self-doubt. This is not such an usual combination, but in Kramers' case they seemed mixed in a very destructive way. In really crucial matters, his apprehensions and doubt would always win out, and he would gracefully withdraw from the fray.

Kramers had definite and even strong opinions on many issues; these had to be expressed without provoking an open conflict. He became remarkably adroit at managing this balancing act. While in Utrecht, Kramers had to contend with the director of the physical laboratory, L. S. Ornstein. Ornstein was a strong, opinionated person who had utterly no use for quantum mechanics. Kramers had just one room, and he never managed to get a place for his assistant. Ornstein ran the laboratory as a true director. He decided everything without consultation, often without communication. Altogether, he was a difficult person with whom to deal. Kramers during Ornstein's life never tangled with him, never insisted on having sufficient space to work, and generally avoided any public clash. But when he delivered Ornstein's obituary he pretty much told everything "the way it really was." He was candid and did not omit Ornstein's less admirable qualities. Yet the obituary as a whole made a thoughtful, compassionate impression—a typical Kramers performance.

It is interesting, although slightly beside the point, that Lorentz also had trouble getting the proper space from his director of the physical laboratory in Leiden, Kammerlingh Onnes. In a new addition to the Leiden Laboratory in 1906, Lorentz had hoped to get a room for himself, one for his assistant, and two small laboratories for his personal use. (By that time Lorentz was probably the world's leading physicist!) But through a mistake, an omission, or perhaps intentionally, these rooms were added to the main laboratories and Lorentz never got to use them. Lorentz's daughter in her biography of Lorentz writes: "I remember well how disappointed my father was as a result of this administrative measure. However, the matter was not discussed. My father preferred rightly or wrongly to keep his peace of mind rather than create a disturbance unless it were strictly necessary."[94] These very words could be applied to Kramers. Very few things in life were worth making a public fuss about for either Kramers or Lorentz. Having a suitable place to work was not one of them!

Kramers once told his nephew H. C. that to do something genuinely new makes one feel as "if one is kissed by an angel." He told him in the same conversation that to achieve something so original "one needs holy courage."[95] It is clear from Kramers' work and style, but above all from his diaries, that he only rarely managed to muster the necessary courage. He was more often torn by doubts and beset by fears, which often guided him in paths which led nowhere. Fear and anxiety about his role in physics were his constant companions. When he left Bohr in 1926 to go to Utrecht, he wrote a letter to Kronig in which he expressed great misgivings about his ability to continue physics all by himself. "How can I do anything without Bohr?"[96] But in the same letter he wrote that he wishes to be independent. This pattern of conflicting desires continued in one form or another throughout his life. He literally ached to do something important and novel. When the opportunity presented itself in the form of collaboration with Heisenberg on the S matrix theory, Kramers was first enthusiastic, then willing, next reluctant, and eventually refused. His excuse was: "You manipulate these concepts so brilliantly that I do not belong in the same ballpark." Perhaps Kramers was afraid to be scientifically isolated and adrift. Without Bohr—or someone like Bohr—there was no one to guide him in the complete unknown and this frightened him. But he also did not wish to be guided by Bohr or anyone else; he insisted that he should be independent—penetratingly original. This whole series of contradictory wishes and his consequent inability to live up to his own standards must have caused Kramers great frustration and anguish. It is not surprising that he often would be exhausted and melancholy. Even his choice of scientific problems becomes understandable: They had to be difficult and involved to be worthy of his ability, but could not be too far a departure from the existing framework, for then the intellectual loneliness would set in. Only in his popular lectures, where he felt less scientific responsibility, does Kramers speculate—but even there in a guarded and controlled manner. Kramers was totally conscious of his inability to withstand the fulminations of Pauli, the denunciations of Bohr, or the objections of Heisenberg. He was unable in the face of such criticism to go his own way, although he desperately wished to do so and he knew he should. Kramers told his friend and colleague Ubbink: "If you do something that looks promising and you are convinced that there is something to it, pursue it, push it, develop it, do not be dissuaded. Be stubborn, pugnacious, belligerent, combative. I never was."[97] This was unusually emphatic advice for Kramers to give; usually he just hinted or suggested. This strong language shows just how intensely Kramers felt about his lack of persistence in pushing and completing his own promising research ventures.

An enumeration of a number of personality traits does not explain a person or his behavior. But it is nevertheless striking how many parallels there are between Kramers' scientific career and his personal life. His unwillingness to make sharp, radical, or irreversible breaks and his intense desire to avoid conflicts operated in his emotional life in exactly the same way it did in science. His aversion to simplicity and directness colored his social and political philosophy as it did his approach to physics. His tendency to withdraw from a dispute is manifest throughout his personal life and academic life. It is interesting and perhaps surprising that Kramers as a reflective person knew all this very well. This knowledge did not make him very happy, nor did it do him much good. He was aware of his own brilliance but it was of no help in overcoming his perennial self-doubt. His fervent wish for penetrating and profound originality clashed with his equally deep need for continuity, certainty, and security. The complex interplay of all these conflicting personality characteristics certainly circumscribed his scientific objectives and thereby limited his scientific accomplishments.

CHAPTER 18

Obligation and Duty

I. The Equivocal Role of Religion

The Kramers family was deeply Calvinistic. As a young boy, Hans Kramers went to church regularly, received additional religious training, and was confirmed when he was 14 years old. His mother had arranged a party (a supper) for this special occasion. But young Hans was quite unhappy. He felt that a religious ceremony did not call for parties or other frivolous entertainment, but should be the occasion for serious, thoughtful contemplation of human life.[98] (His mother did manage to persuade him to come to the supper!) Kramers' later attitude toward religion became ambiguous. He liked to discuss problems of Christian theology. Ever since his youth, he had read extensively in the religious literature, and was especially well versed in the letters of St. Paul. The Swiss theologist Karl Barth was one of his favorite authors. Kramers' religious perorations had the same mixture of knowledge, brilliance, and irony as his discussions in other areas. After his adolescence, he never went to church, and he ignored any kind of organized dogmatic religious system. Actually, he did go to church once in a while, to escape the nagging of his sister-in-law, Mrs. Jan Kramers. But he then took his revenge by sleeping soundly through the whole service. He was certainly not religious in any traditional sense; but he was not a dogmatic atheist like Dirac either. Pauli beautifully summarized Dirac's religious views: "Our friend Dirac has a religion and its main tenet is 'There is no God and Dirac is his prophet.'"[99] Kramers' original religious views evolved in two quite distinct directions, neither having any relation to traditional religious practices. Kramers developed a very strong sense of duty, obligation, and responsibility, directly from his earlier formal religious beliefs. It is perhaps in this domain that the Dutch culture and the Calvinist religion are connected most directly. Religion imposes obligations, duties, and responsibilities—and these remain, even if the religious basis itself disappears. As another consequence of his earlier religious training, he became deeply concerned with questions of ethics, good and evil, and redemption, but especially suffering and compassion. Perhaps he became so interested in these questions because he sought a deeper understanding of the intrinsically vulnerable and tragic human condition. None of this had anything to do with a formalized religion, and even less with rituals and church attendance. The typical organized religious activities became a matter of supreme indifference to Kramers. His finance, Storm, educated by nuns, wished to get married in the Catholic Church. That was fine by Kramers. The nuns wanted to pray for him; that was also fine. However, Kramers did not tell his father and stepmother that he had gotten married in the Catholic Church. He also told Storm not to tell people in Holland that she was Catholic, for this would bring the priest all over the house and this Kramers did not want. These were not religious objections on Kramer's part, but means to avoid awkward social situations. Storm complied and gradually her Catholic faith was eroded by Kramers' active lack of interest—not active opposition! Their daughters, Suus and Agnete, were baptized as Catholics; this again was a matter of indifference to Kramers.

It is therefore somewhat surprising, and again shows the complex character of Kramers, that in spite of his almost hostile lethargy to traditional religious practices, he was deeply concerned with abstract theological problems and also with personal religiosity. In the perceptive and touching eulogy for Kramers, delivered in Oegstgeest on April 29, the minister states: "From the outside it might appear that he [Kramers] looked down on theology. A church funeral for Professor Kramers who rarely if ever came to church is however not unreasonable for those who knew Kramers more intimately. Kramers has lived and struggled more with God than most people knew."[100] Kramers' religious ruminations are unexpected and profound. "You theologists speak about the miracle when you no longer understand it, but I consider it a miracle when I sometimes do understand a little of it."[100] He expressed his own feelings and his deep compassion shortly before he died: "I believe in a God who made the flowers, but where is the God who dries the tears?"[100] He asked this question perhaps for the first time after the funeral service for his brother Jan. Later, he tried to comfort a friend who was frightened in the face of imminent death even though Kramers knew that he himself would not have long to live. The question he raised then, and time and time again, is the anguished cry of a truly compassionate person: "Where is the God who dries the tears?" It is remarkable that the minister, in his eulogy, hints that it is through people like Kramers that God dries the tears. This is strikingly similar to Arnold Siegert who believed that in Kramers he saw one of the 32 genuinely decent persons that God had sent down to earth to make the miserable fate of the Jews more bearable.[101] Kramers himself was not afraid to die. When he heard that he had lung cancer he told his wife: "I will suffer my fate with dignity and resignation."[102] He did not struggle against life; he would not struggle against the prospect of death. But he did struggle, to understand and to comprehend; in this respect he was the same in religion, science, and life. He asked unexpected questions and detested facile trivializations of deep problems. He struggled valiantly to come to terms with injustice, suffering, stupidity, his own demons, and his own Gods. That was his own personal religion.

II. Obligation and Duty

Kramers' highly developed sense of duty and responsibility, which can be traced back to his early Calvinist training, was of unusual significance in his later life. According to his oldest daughter (Suus), he maintained a serious, thoughtful attitude toward religious matters all during his life.[103] One of the manifestations of this attitude was that he perceived his role as a physicist and an educator as a mission with a religious tinge. When Kramers gave lectures in the early quantum theory, he often described his task as "spreading the gospel according to Copenhagen."[104] But this was really more than a clever turn of phrase; Kramers felt very strongly that he was duty-bound to convey his understanding and insight to the physics community. In the same vein, it was a sense of duty that kept him going in physics in spite of many disappointments. He knew he possessed knowledge and technical skill; he was obliged to continue and not give up. He even postponed his operation for lung cancer several times because he wanted to finish some papers.[104]

Not only did Kramers take his responsibility to physics extremely seriously, he was equally conscientious about all other obligations he undertook. In the early days in Copenhagen, he wrote the letters, arranged for visitors, taught the classes, and even carried books from one building to another. Later, as a professor in Utrecht and Leiden, he accepted his many administrative responsibilities—which he really did not like—with extraordinary good grace. (But he never asked for and indeed never got a personal secretary.) In this devotion to duty, the Dutch spirit and the Calvinist doctrine have totally merged; it was one of Kramers' dominant traits. The feelings of duty and obligation were not restricted to the academic life or a particular university, although Kramers was quite attached to Holland. Right after World War II, Kramers was offered the Chair of Theoretical Physics at Cambridge, which had been vacant since the death of the famous physicist Fowler. At that time, Kramers was not all that happy at Leiden: The perennial bickering with de Haas left him exhausted and irritated and his marital situation was far from perfect. A new environment might be just the stimulation he needed. Furthermore, it was a considerable honor to be offered this prestigious chair. But taking the Cambridge position would entail a change; he would have to leave Holland and make a decision. Even though Kramers was now older, he was no better at making decisions. He changed his mind innumerable times; each change was accompanied by a new compelling reason. He talked to his wife and children; they obviously argued that it was his decision. He finally decided to take the position; he told this in great confidence to Casimir.[105] He visited Cambridge to make all the final arrangements for his move. But after his wonderful visit to Cambridge, he changed his mind again. He had one of his daughters send a telegram to Professor Bragg declining the offer. Kramers' uncertainty and vacillation are still apparent in the memo he wrote for his daughter. The complete text including changes is as follows:[106]

> Professor Bragg
> Cavendish Laboratory
> Cambridge
> Back Leiden Sunday. After extensive discussion authorities and others decided to ~~accept~~ stay in Leiden, although sad memory of beautiful Cambridge dreams.
> Kramers

It is evident that Kramers made his decision with the utmost difficulty. Even when he had made it, it was hard to give up "the beautiful Cambridge dream." Probably his sense of duty toward The Netherlands kept him from accepting this challenging and stimulating new beginning. Casimir lost a great opportunity because of Kramers' procrastinations. He would have been offered the Cambridge chair had Kramers refused it. At pretty much the same time, he was offered a major position at the Philips Company.[107] If Kramers had made a quick decision, Casimir would have had the choice of two major positions.

After Kramers declined, the position was offered to Casimir, but by that time Casimir had accepted the Philips position and could not go back on that decision, even if he had wished to do so. Kramers had a number of other possibilities for new positions. On March 3, 1947, he was approached by the California Institute of Technology;[108] only a few days later, on March 7, 1947, he received a serious inquiry from Brown University.[109] On November 3, 1949, Slater (from the Bohr–Kramers–Slater theory), not known as a close friend or scientific admirer of Kramers (see Chapter 13, Section III.B), offered Kramers a position at M.I.T. as a replacement of Weisskopf who was going to CERN.[110] Kramers also received an offer to join the Institute for Mathematical Sciences at New York University, now known as the Courant Institute.[111] They were anxious to have Kramers to start a collaboration with a mathematical group, directed by the brilliant young mathematician K. Friedrichs. None of these offers caused Kramers quite the trouble that the Cambridge offer had. But in the end he turned them all down. His attachment to the culture which had nurtured him was too great; his sense of duty would not allow him to abandon a country ravaged by war, no matter how alluring new opportunities might be. "One does one's duty till the end."

III. The Special Role of Teaching

Nowhere is Kramers devotion to duty more conspicuous than in his total dedication to teaching. He considered the transmission of knowledge—the conscientious instruction of students—as one of his primary responsibilities.

He took it extremely seriously, no matter where he was. He prepared his lectures and classes with equal care whether he was at his home university or was a short-time visitor at another place. So seriously did he take his teaching obligation that he reworked every lecture, for every class he taught—even if he had done so many times before.[112] Kramers was quite explicit about the importance of teaching; in his inaugural address at Leiden he said: "I have not accepted the position of an incarnated hand and yearbook, or of an abstracting journal of theoretical physics, but rather the position of a physicist who is going to instruct the youth .... It will be my duty to help develop in each student as close a relationship between him and physics as possible."[113] And Kramers did exactly that. He did not just help students by his inspiring lectures and brilliant and careful explanations; he prepared research topics, gave innumerable suggestions, and functioned as an all around father confessor. The teaching profession was actually a tradition in the Kramers family. Hans Kramers' grandfather was unhappy with the pre-university teaching in The Netherlands. He organized a school near the Hague where subjects were taught very carefully in a pleasant, somewhat relaxed manner. Academically, the school was a great success; financially it was a disaster. Kramers' grandfather refused to compromise the quality of the school's teaching for a more disciplined approach, so eventually the school was disbanded.[114] Kramers not only felt his teaching as an obligation, but he really loved explaining ideas and information to others, and it did not really matter whether this was done in a formal classroom, a lecture, a small seminar, or a private discussion. It was a great pleasure to him to communicate new information to any audience. He had indeed the rare gift to captivate any audience with talks about physics or science, irrespective of the audience's knowledge of the topic.

In Copenhagen, at the Bohr Institute, one of Kramers important functions was to organize and arrange the teaching program. Although Bohr was of course a professor and had the primary responsibility for the lectures, Kramers frequently substituted for Bohr in his absence. At the occasion of the opening of the Institute for Theoretical Physics on March 3, 1921, Bohr specifically mentions the great help that he had received from Kramers, not only in scientific work but also in the teaching and guidance of students: "This leads me to emphasize once more that this Institute is not intended solely for scientific research but is also to be a homestead for the teaching of physicists and others with special interest in physics ... the combination of this task is of the greatest importance."[115] A few years later in January 1923, Bohr asked the ministry of education for two positions so that he could devote his time to research, the other position would have primary responsibility for the teaching. This was turned down by the Minister of Education on July 17, 1923, as it happened exactly the same date that Jeans in England offered Bohr a position in Cambridge. By a mechanism not unknown even today, the Cambridge offer managed to produce the two positions that Bohr had requested. Kramers was appointed to the second position and from that time on (in 1923) he carried the main teaching responsibility in the Bohr Institute.

Kramers' pedagogical capacities were actually recognized at an earlier occasion. A movie house in Copenhagen, the Palace Cinema, decided to show a popular show on the Einstein theory of relativity. But it was necessary to have some explanation and commentary in the film. Kramers, who was described as "Bohr's principal assistant" was selected to do this. He was also charged with choosing a proper musical introduction to the evening. He picked Beethoven's Egmont Overture—why, no one knows. Evidently, Kramers did well. According to the newspaper reports, he knew all the requisite science and his presentation "was not a burden on anyone's intellect." He explained everything clearly and cleverly, with a sense of humor. So Kramers' debut as a commentator was a great success; he played for several weeks to packed houses. After all, the theory of relativity, according to the advertisements, was the most spectacular product of this century. With the money that Kramers made, he bought a sewing machine for his wife![116] It was after this popular success that Kramers decided to write a popular book on the Bohr theory (see Chapter 12, Section IV).

Kramers prepared his lectures with meticulous care and he delivered them with verve and wit. It was clear to the students that they were in the presence of a real active physicist, who had direct contact with many of the living masters of the field and who participated in the ongoing developments. It was not unheard of that students in engineering at the technical high school in Delft switched to physics, actually to hear Kramers in Leiden.

So conscientious was Kramers about teaching that as he became more and more experienced, he lectured slower and slower and consequently covered less and less material. When some of his more advanced and ambitious students complained about this, he defended himself: "Everytime I look at Miss L. [a student taking some of Kramers' courses in physics; she majored in chemistry and minored in physics] sitting in the front row and staring at me with her puzzled eyes, I feel I am going too fast and I pity the poor girl, so I add another explanation. Later when you'll be teaching yourself, you'll have the same experience. One just can't teach slowly enough. You'll also find out how important it is to add sufficient historical details."[117]

In colloquia and seminars he showed his concern for the students in another way; he often asked simple and even naive-appearing questions to slow down the speaker and give the students a chance to catch up. Quite often, his naive questions were not so naive either! He did everything he could to see to it that students would profit from the seminars, sometimes by giving a brilliant summary of what the speaker had said (or should have said), sometimes by relating the topic to other ongoing investigations.

Kramers loved to teach. He did this on all occasions, when going for a walk, at a picnic, even during an exam. In giving the typical oral examinations, Kramers would spend a whole morning (10–12:30) with every student. He wanted to probe the boundaries of a student's knowledge. He very quietly asked harder and harder questions, until the student either did not know the answer or did not know what to do. Then he would say: "Let me help you."

In this way the examination actually became an instructional tool. It generally worked, but not always. Sometimes the students, fearful of the examinations, asked Kramers' assistant Belinfante what the professor was likely to ask. Belinfante then coached the students and helped them with the likely topics; they managed to take and pass the examinations without learning much at all.[118] It is unlikely that Kramers knew about this; it is certain that he would not have cared one way or another. Although he was kind to all students, even to the extent of accepting them for a Ph.D. thesis, he only spent time with those he considered worthwhile. He left the others to shift for themselves or just flounder.

But even his method of dealing with his promising students was unorthodox. When his assistant Belinfante planned to take his oral doctoral exam, Kramers told him exactly what to study, that is, what would be on the exam. He specifically told him not to study quantum theory since that would be his thesis area. However, at the examination the only questions that Kramers asked were precisely on quantum mechanics. When Belinfante later asked (presumably complained to) Kramers about this, he said: "I just wanted to see how much quantum mechanics you knew, without studying it."[118] This somewhat whimsical behavior of Kramers, in relation to his students, was not all that uncommon. One of Kramers' last students was N. Hugenholz (he actually obtained his Ph.D. degree with L. van Hove), who became a major scientist in his own right. Kramers evidently had high expectations of him and treated him with great kindness. He encouraged Hugenholz to visit him often and talk about physics: "As long as I live you must make use of it."[119] After a while, Kramers told Hugenholz: "You should take your doctoral examination pretty soon." However, Hugenholz had not taken a single one of the required oral examinations. When he mentioned this, Kramers said: "We talked a lot, I know, what you know."[119] This totally personal approach to education and training was very typical for Kramers.

Apart from teaching, Kramers was also helpful in guiding students through the bureaucratic maze of a university. Another one of Kramers' favorite students, S. Wouthuysen, came to Leiden on the recommendation of Rutgers. Wouthuysen was studying chemical engineering, in Belgium, but hoped and wished to study physics. He had solved some problems for Rutgers, who was impressed and took him to Leiden and introduced him to Kramers. At the first meeting, Kramers looked at Wouthuysen after he finished playing his cello and said "Is this now the miracle of the Century?"[120] But he liked Wouthuysen and certainly thought he was an excellent prospect to study physics. However, Wouthuysen had taken few courses and had no diplomas; it was even hard for him to be formally allowed to the first, the "candidates exam." However, Kramers exploited a loophole in the rigid rules. If a person by experience or background could be allowed to omit part of the candidates exam (he would get advanced credit for it), he automatically qualified to take the rest. So Wouthuysen gave a colloqium on thermodynamics; Mrs. de Haas, who taught that subject (she was Lorentz's daughter and de Haas' wife), gave him the credit and Wouthuysen was free to take the rest. He advanced rapidly from that point on. When he had to take his doctoral examination, Kramers once again came to his rescue. Wouthuysen was a little worried about his mathematics. So in the exam, which was run by Kramers, he let Wouthugsen talk for such a long time about physics (which he did brilliantly) that there was little or no time for mathematics—yet another example Kramers' consideration for all aspects of his students' welfare.

Kramers had officially 27 thesis students. They were of widely varying quality, and Kramers' personal relations to his students varied enormously. He was very fond of some and helped them a lot; others he accepted, while still others he tolerated. But in general Kramers was not too demanding of his students; he would occasionally suggest an idea or method of approach, but generally there was not the day-in, day-out, frantic activity so typical of his own work with Bohr. Usually, he left his students pretty much alone for long stretches of time. For example, Kramers suggested in 1938 that Belinfante investigate possible Diraclike equations for meson theory. At that time, Kramers had a great deal of interest in the formal properties of such equations, but Belinfante worked out all the details. Kramers suggested only general directions; the research itself was carried out by Belinfante.[118] This was pretty much the typical pattern. With a few notable exceptions, Kramers was not all that happy with his Ph.D. students. Probably his most famous student was Koopmans, who received the Nobel prize in economics. Koopmans wrote exactly one paper in physics. So although he was unquestionably successful, it is hard to attribute this success totally to Kramers' teaching. Only a few of Kramers' students made a real impact on physics. That was in part because Kramers was not particularly selective in the students he accepted. Ehrenfest only accepted outstanding students, with obvious talents, whereas Kramers would reject only students with obvious deficiencies. This meant that the continual instruction of these weaker students was yet another drain on Kramers' intellectual energy. Only on rare occasions was the scientific exchange with his students of direct use to Kramers. His family thought that Kramers spent too much time with his students anyhow. They, like everyone else, exploited Kramers' good nature and gentleness. At one time van Kampen was working with Kramers at his home. This was sometime after Kramers had suffered his stroke, and the family was still worried that Kramers worked too hard. First, the youngest daughter, Maartien, would come into his study and ask her father to stop working, but he would ignore her or tell her to go away. A little later, Mrs. Kramers would tell him that the time had come to stop; he would tell her he would in a little while, but he did not. Eventually, his oldest daughter, Suus, would tell him to stop, and then he would.[121] Even though Kramers enjoyed a number of his students enormously, looked at as a whole, it was not an extraordinary group. But with almost no exceptions, all students from the best to the weakest recognized that working with Kramers, was an incredible stroke of good fortune and they deeply appreciated this privilege.

Kramers was well aware that the existing University curriculum for physics students was far from ideal. Freshman students had to take thermodynamics and kinetic theory well before they had taken, let alone understood, analysis or differential equations. The whole program, consisting officially of 2-year lecture courses, was disorganized and unstructured. Kramers suggested that some of the graduate assistants get together to set up a tighter, more sensible, more appropriate curriculum. This committee came up with a proposal, which would make the study before the candidate exam take about 3 years, with a particular ordering of the courses. In that way, the necessary mathematical tools would have been presented and digested before they had to be used. Although Kramers had suggested something like this, he never said anything about it publicly in either faculty or senate meetings. Changes in curriculum of this type had to be submitted to a sequence of faculty committees. To get such a proposal to the policy-making body, it had to be approved by several such committees. This proposal was turned down by the very first committee. Only two faculty members voted in favor. The conservative trend in the Leiden faculty was too strong. *Any* change of any kind was greeted with skepticism; a change that appeared to lower the requirements or that would make it easier for the students would have no chance at all.[118] Although in sympathy with the proposed change, Kramers was unwilling to disagree publicly with his peers. Privately, he disagreed strongly with their decision, but he did not use his considerable influence to try to change their minds. That was not his style.

In a rather different vein, Kramers, in 1947, pushed very hard for the establishment of a chair at the University of Leiden in "The Methodology and Conceptual Analysis of Exact Sciences." Such a chair in the methodology of science was eventually established. The first person to hold this chair was Dr. J. B. Ubbink. He had obtained his Ph.D. degree in physics with Kramers, on a topic in low-temperature physics. Ubbink had always been interested in the more general, almost philosophical questions, and after his degree, strongly encouraged by Kramers, he returned to these questions. His scholarly work deals almost exclusively with the methodology and philosophical analysis of science.

Not only did Kramers urge the establishment of such a chair, he also was an active participant in setting up a new program, presumably for all students, to provide a unified view of all science. This program was called "colloquium generalum" (general college). Kramers was one of the main lecturers in that program. Kramers' letters urging the establishment of that program give an interesting insight into his views on science and education at that time (1947). He especially stresses the need for reintegration and reunification which, he claims, experience has shown is necessary for many students: "One can almost say the students are 'homesick' for such integration."[122] Kramers indeed started this colloquium generalum with great enthusiasm and expectations. The audiences were large and very interested. Kramers began with an apology: "It is impossible to understand physics from the outside, what we really should do is study some real physics. But this is outside of the scope of this course.

So we have to use words. But words are so limited. With words alone, we are almost helpless."[123] Kramers' own notes contain several versions of his opening lecture. They are interesting because they almost read like his personal philosophy of the nature of physics. "Physics is in many ways the easiest but also the most advanced branch of the natural sciences. Today I want to speak to you about the nature of these advances. On the basis of my considerations, it should become clear to you how incredibly difficult it has been to achieve these advances, how powerless the human mind is to create something that is genuinely new, that is, something that is really and truly new, something that can be characterized as a step forward on the path toward the development of the human mind and spirit. I would so very much have you appreciate, how much greater a miracle it is, that a human being understands something, than that he does not understand it."[124] It is likely but not certain that Kramers mentioned this intensely personal approach to science in his opening lecture. From then on, his lectures have a slightly whimsical character. Each lecture has a slogan. The initial slogan is "No understanding without action." He tries to explain that interpretations and understanding must be based on experimentation and observation. He spends a lot of time on the next slogan: "What is separated is unified, what is together is separated." In the subsequent discussion, which is rather cursory, he explains that apparently unrelated phenomena such as electricity and magnetism, or terrestrial and celestial phenomena, nevertheless are closely related—in fact identical. Evidently, these lectures did not go over that well, for he comments in his notes: "I know how I should continue but I don't have the strength to write it all down." It is not surprising, but still very much in character that Kramers stresses that although the second slogan is similar to dialectical materialism (he refers to the "damned reconciliation of differences"), Bohr's notion of complementarity is much more significant and much deeper. The last slogan that Kramers enunciates is again a very personal view: "Simplicity is not the hallmark of the essential." From then on, Kramers' notes become more and more fragmentary; instead of sentences as quoted before he just jots down a reminder, a brief comment, or a name. In many ways, Kramers' notes appear as random notes taken from a stream of consciousness. It is clear that one can learn a great deal about Kramers from these remarks. "We feel it as a blessing if we succeed in introducing a useful new concept. And how often does it, even then, not happen, that it is not given to the first discoverer of this new concept to really understand the essence of his discovery, to really plumb the depth of his result."[125] This sounds very much like Kramers; how much this meant to the students is not clear, but Kramers probably did not intend his notes for the students. Kramers then states that the greatest and deepest change in human thinking is the quantum theory. He characterized Planck as a man of compromise and Einstein as a person who would pursue an idea consistently, stubbornly without the slightest hesitation. Bohr is a person with a mysterious intuition; Heisenberg is a young hero. Kramers finally adds another slogan and some advice. The slogan is: "Technique is to be able to do what you want and to want what you can do."[123] His advice is: "If you want to steal nature's secret by experiment—then you must cajole her like an excellent but devious husband." But it is clear that Kramers' interest was waning. He writes in his notes: "This was done poorly in the lecture." The later lectures are not as carefully prepared as the earlier ones. In fact, apart from a few sudden insights, they are quite ordinary and even a little belabored. His initial enthusiasm had all but disappeared and the class was becoming an onerous duty. The attendance at his lectures started to decline and after a few months it decreased precipitously. Whether the students after all were not homesick for a unification of all sciences or whether Kramers' highly personal and sophisticated philosophy was too erudite for an average class is not a bit clear. But it is certain that the experiment was unsuccessful, and Kramers never again participated in this program. Even so, it was at his instigation that the program started and that the chair was established. It shows that Kramers' interest in science, education, and students was intense, broad, and undiminished by the suffering of the war.

IV. Duty: The War and Its Aftermath

A. The Wartime

It is a little silly to rate individuals in terms of their degree of suffering during a time of war and occupation by a foreign power. But it is certain that Kramers as a decent and compassionate person must have recoiled in horror at the new wave of barbarism which swept over Europe during World War II. Kramers, infused as he was in the Western culture, was deeply affected by the disintegration of European civilization. There is always the obvious suffering due to hunger and lack of heat. Kramers was a large man; the lack of food was very hard on him and it affected him more than many others. In addition, Kramers felt a very special responsibility toward his students. As the leading physicist in The Netherlands, Kramers felt an obligation to continue the educational and scientific program as long as possible. The cares for his wife and children added to his enormous burden during the war time. And that was not all: The gradually deteriorating condition of Jewish colleagues and students caused Kramers time and again to intercede for them or provide discrete help. Soon after The Netherlands were occupied (May 1940), the Germans started their efforts to make the most renowned Dutch university into a Nazi-dominated institution. On November 26, 1940, the Jewish professors were dismissed. Professor R. P. Cleveringa, the dean of the faculty of law, took over the lecture of one of his dismissed Jewish colleagues. Cleveringa was promptly jailed; he had taken the precaution to pack a suitcase. A student strike followed (the next day) and the university was closed on November 27, 1940. On April 30, 1941, the university was reopened by the Germans but only for examinations and thesis defenses. The Germans hoped eventually to reopen the university; they expected that they could gradually dismiss undesirable faculty members and appoint new professors sympathetic to the new regime, so that the university would become a Nazi showplace. However, the faculty, resolutely and vigorously opposed all these moves. When it became known that on March 16 and March 20, 1942 two distinguished faculty members were fired, the bulk of the remaining faculty (56 in all) handed in their resignations. Kramers was one of the 53 professors who resigned. That was a highly principled and courageous action to take. Resignation clearly implies complete loss of salary; even though the university fund of Leiden was especially generous in coming to the aid of the professors, it could not for long supply the needed funds. Kramers literally eked out a livelihood by lecturing on theoretical chemistry to Royal Dutch Shell officials. The university was closed in April 1942 and remained closed. No further attempts were made to reopen the university or to "nazify" it.

Kramers was a genuinely loyal Dutch citizen and he violently disliked the German oppression and the cruel German antisemitism. That must have caused him great difficulties at the time of Heisenberg's visit in the fall of 1943. Kramers was happy to accept Heisenberg as a friend—he stayed overnight at his home—but not as a representative of the German government. He certainly did not want to have this visit interpreted as tacit approval of German policies. All this is perhaps a little touchy but not yet a real conflict. However, in a long walk with Casimir, Heisenberg and Casimir discussed politics and history.[126] Heisenberg claimed that it always had been Germany's historic mission to defend the Western culture against the Eastern barbarians and the present war was just another instance of Germany's historical destiny. He concluded: "A Europe under German leadership might be the lesser evil."[107,126] It is doubtful that Kramers would have that type of discussion. Like Casimir, Kramers would be upset, most uncomfortable, and extremely uneasy about the trend of the discussion. It was precisely the kind of controversy that he always tried to avoid. Even without talking with Heisenberg about these matters, Kramers might have had an inkling of Hesienberg's views. Kramers admired Heisenberg tremendously; they understood each other and shared a deep interest in music. They were both admirable examples of their national cultures. It must have shocked and hurt Kramers, who was an extremely perceptive and sensitive person, to realize that Heisenberg, in spite of all this background, did not appreciate the feelings of an oppressed and tortured people. Heisenberg was certainly not a Nazi. He did not condone the excesses of the regime and he even tried in his own way to be of help to some of the victims, but a chasm existed between Kramers and Heisenberg. This may well be the real reason why Kramers did not choose to work with Heisenberg on the *S* matrix, although other factors surely played a role. This "Heisenberg incident" must have created profound conflicts for Kramers; it could only have strengthened his expectation of an imminent disintegration of the culture.

Kramers did a great deal to try to maintain some semblance of a scientific life. The university of course was closed, but Kramers gave a number of clandestine classes. van Kampen[128] remembers a number of such classes in rather unlikely places, and ter Haar mentions in the preface of his well-known book *Statistical Mechanics*[127] that parts of it are based on lectures given by Kramers in 1944–1945. Since the university was officially closed, these were "underground classes." He brought his nephew H. C. Kramers, who had gone into hiding, the well-known textbook on theoretical physics by Joos. He kept in close contact with A. Pais. At one time when Kramers was visiting Pais, the place was raided by the Gestapo. Pais went to his cramped quarters (a closet) upstairs. Kramers also went up, sat down in the adjoining room and read aloud from a commentary on Shakespeare, all the time the Gestapo were looking around. Eventually, the Gestapo left; Kramers left shortly thereafter and Pais changed his address. Pais thought the whole incident took some 15 minutes; it actually took $2\frac{1}{2}$ hours. It was a most remarkable, kind, and courageous thing for Kramers to do.[90]

The German authorities generally left Kramers more or less alone. Whether this was because of his international eminence is not clear; the Germans generally did not let their actions be influenced by international condemnation. Even so, Kramers did not escape his share of irritating and frightening experiences. He once was told by the German authorities that he had to leave his house which would be occupied by German officers. It took three students a full day to move his books to the astronomical library. After the Kramers' family left their home, but before the Germans moved in, the angry neighbors had cut down the fruit trees in the garden and cut up the carpet in the house, rather than have the Germans have them. Soon afterward the Kramers family was allowed to return to their home; evidently the German officers feared that Kramers' large house, surrounded by meadows, presumably with a conspicuous Nazi flag, would make a tempting target for allied bombers. Whatever the reason, the Kramers family was given permission to reoccupy their house and they were left in relative peace.[118,129] Even though this incident caused no lasting harm, it must have been an unnerving experience. As the war ground on, the suffering increased. Many students, sought by the Germans and facing deportation, went underground in various places in the physics laboratory (the basements). Such close confinement, the fear, and the unrelieved tension had its inevitable effect and tempers flared. Kramers changed his place of work to the laboratory; his presence had a calming effect on the hard-pressed students. He worked and the students tried to work; even if no great physics was done, they retained their sanity. One of the graduate students managed to smuggle a load of potatoes from his father's farm from the "Zuiderzee polder" into Leiden. Kramers got a cart and a horse, but he did not know how to manage a horse. Kramers and his colleague Taconis,[130] who did know how to manage a cart and horse, dressed up like vegetable peddlers and distributed the potatoes to students and laboratory personnel most in need of food. Kramers, all during the war, literally was confronted by one personal tragedy after another, by one crisis after another. He handled all problems with sympathy and understanding. But this, together with the physical hardships and the unrelenting pressure of ever present responsibilities, took its toll. Kramers never really was the same after the war.[131] His health, never all that robust, deteriorated sharply. The scope of his responsibilities increased enormously. But even in the middle of these new demands on his time and energy, he did not forget the personal suffering and private tragedies that the war had brought to so many individuals. One of Kramers' students, Wouthuysen, who later became a well-known professor at the University of Amsterdam, was forced to go underground in Belgium. He of course could do very little physics during his underground existence. After the war, Kramers went out of his way to see to it that Wouthuysen would get back into physics. A very select and important conference was planned by Bohr and Kramers in 1951 to discuss and argue about the future of European physics. The organizer of the conference was Rosenthal, a member of the Bohr Institute in Copenhagen. To have serious, concentrated, and intelligent discussions, attendance was to be by invitation only. Kramers implored Rosenthal to invite Wouthuysen. By that time Wouthuysen was not so well known, so Rosenthal demurred and asked Kramers why he insisted that Wouthuysen be invited. Kramers answered "Because I love him."[132] Rosenthal was speechless, but Wouthuysen was invited. Right after the war Kramers organized the Foundation for Fundamental Research of Matter (F.O.M.). This was part of his concerted effort to revive and strengthen Dutch physics after the war. Kramers was the first chairman of that organization; he made Wouthuysen secretary. This was yet another example of Kramers' efforts to reintegrate Wouthuysen into physics. He introduced Wouthuysen to Oppenheimer and started him on his way toward a distinguished career in physics.

Thus, Kramers not only was instrumental in saving Dutch physics during World War II, but in his profound humanity he saved and rebuilt many individual lives. He accepted all his responsibilities as a physicist, a human being, and a Dutch citizen during and after the war, often at the expense of his own peace of mind, his health, and his cherished dreams.

B. Kramers' Work in the United Nations

The Atomic Energy Commission (AEC) of the United Nations was created by a resolution of the General Assembly of the United Nations, during the 17th plenary session of the United Nations on January 24, 1946. Its mandate was to investigate all problems associated with nuclear energy. The specific charge to the AEC included:[133]

> The commission shall provide with the utmost dispatch and inquire into all phases of the problem and make such recommendations from time to time with respect to them as it finds possible. In particular the Commission shall make specific proposals

(a) For extending between all nations the exchange of basic scientific information for peaceful ends
(b) For control of atomic energy to the extent necessary to ensure its use only for peaceful ends
(c) For the elimination from national armaments of atomic weapons and of all other major weapons adaptable to mass destruction
(d) For effective safeguards by way of inspection and other means to protect complying States against the hazards of violations and evasions.

Clearly, the AEC would deal with some of the most fundamental, most important, and most difficult questions in the postwar world. The resolution stipulated further that members of the Security Council, in 1940, would have representatives in the AEC. Since Holland that year was a member of the Security Council, there would be Dutch representatives in the AEC. The Minister of Foreign affairs, Mr. E. N. van Kleffens, was appointed as the representative, while Kramers was selected as scientific adviser and alternate representative. It was of course a great, if not altogether unexpected, honor for Kramers to be chosen for a position of such world responsibility. There is no doubt that Kramers accepted this appointment only because of a strong sense of obligation and duty. He did not enjoy political infighting and he really hated to go to meetings. The only positive aspect of the innumerable and interminable meetings was that Kramers developed a renewed respect for Oppenheimer, who could tolerate and run meetings much better than Kramers could. Oppenheimer could summarize and synthesize what had been accomplished, what were the areas of conflict, and what remained to be done. But to Kramers his acceptance of the position with all its concomitant obligations was pure duty, with no redeeming features. The AEC met for the first time on June 14, 1946; the last meeting was on December 30, 1946. The AEC and its subcommittees met 82 times during that period, and Kramers was faithful in his attendance of the meetings.

So important did Kramers consider this task that, in spite of the serious financial sacrifice this would entail, he left for New York soon after he had received this generous letter from the Minister of Education.[134] The discussion in the AEC followed the usual contentious pattern of the East–West arguments. The American plan tied the prohibition of nuclear energy for military purposes directly to a system of detailed international inspection, to be controlled by an international agency. This was not too popular with the Eastern bloc. They were most sensitive to on-sight inspection; but much more serious was the stipulation that if a country had violated the international agreements regarding nuclear weapons, the country in question would lose its veto power in the Security Council. To this the Russians objected vehemently. The Russian plan would call for an international convention which would outlaw the use of nuclear energy for military purposes. This would be the first step; controls and sanctions would be decided later by committees which were yet to be established. The majority favored the American plan, but the Russians remained adamant in their objections and after 2 months of meetings, no agreement of any kind was reached. In this impasse it was decided that the commission should make a study of the technical and scientific basis of nuclear energy. Once this was accomplished, one could more rationally decide just what the control of nuclear energy actually would entail. It might even be possible to develop a system of controls based on a rational assessment of the relevant factors. A subcommittee of the AEC was established to study these questions. Its official title was Scientific and Technological Subcommittee of the Atomic Energy Commission of the United Nations. This 12-nation United Nation agency had representatives of Australia, Brazil, Canada, China, Egypt, France, Mexico, The Netherlands, Great Britain, the United States, Poland, and the Soviet Union. The representative of the Soviet Union was Andrei A. Gromyko. He started his discussion by insisting that "nobody is asking for secrets." All the Russians were interested in, he claimed, was a method of exchange of scientific data for the peaceful use of nuclear energy. But Gromyko immediately provoked a fight about the chairmanship of this subcommittee. He suggested Dr. Stefan Pienkowski, Vice Chancellor of the University of Warsaw, but in the voting he lost to Kramers. Then Gromyko argued that there should be two alternating chairmen; he specifically proposed Kramers (because of his eminence and moderation) and the Polish representative. But this proposal was voted down by a vote of 10 to 2. There was to be just one chairman, and Kramers was it.

One could argue—although Kramers would disagree—that this was the international high point of his career. Because of this selection as chairman, he became more visible on the international scene. He was even interviewed by the *New York Herald Tribune*. Kramers was described as "a quiet spectacled physicist who worked with Niels Bohr." It mentioned that Kramers had achieved world fame through his work on atomic structure and low-temperature physics. Especially interesting is the editorial comment that atomic physics is of practical significance as in nuclear fission, but low-temperature physics is pure science that has not figured in practical applications. Some of the quotes attributed to Kramers are interesting. "After five years under Nazi occupation, we are no longer idealists; we are realists." In connection with his move to Copenhagen in 1916, he said: "My blood urged me to go abroad." But most important are his observations regarding his role as chairman of the AEC's subcommission. "I have never seen a big scientific discovery which did not lead to some discovery for humanity.... I will undertake my duties with no preconceived opinions—I will try to understand what the others say."[135] This last quote is of course vintage Kramers; it explains why he was trusted and why he was a perfect chairman. He would listen and try to understand; he would moderate and not insist on imposing his private opinions. The charge to the scientific and technological subcommittee was very much in Kramers' line. The sharp separation of the technological and the political aspects of the control of nuclear energy was a first essential step for the eventual development of a rational policy. One should first know what it is precisely that has to be controlled and monitored, what technical means are available for control and detection, before any system of control measures can be suggested. The political implementations of such controls is of course of later (but dominant) concern. Kramers' committee worked extremely hard to identify the technical aspects necessary for the effective control of nuclear energy. Included are such items as the production, manufacture, and storage of nuclear fuel. The committee had two obvious difficulties: (1) A good deal of the relevant information was classified and unavailable. (2) Although in principle the problems were scientific and technical, and not political, they had such obvious political relevance that it was often difficult to separate the political from the technological features. If there is not a great deal of confidence and trust to begin with, there may be deep suspicions that a priori political attitudes are hidden beneath allegedly objective scientific claims. It is to Kramers' immortal credit that, in spite of these almost insuperable difficulties, he was able to hammer out a report that was approved unanimously. This took prolonged and arduous negotiations by the subcommittee; on Kramers' part it took infinite patience, a complete understanding of all details, and a sympathetic appreciation of all viewpoints. This was one of the few occasions that the East and West managed to come to a unanimous agreement. Kramers was overjoyed. This report was written during September 1946 and officially approved by October 1, 1946. Its fundamental conclusion is simple: "Effective control of atomic energy is technologically feasible."[136] At no time was Kramers' sense of decency, moderation, and fairness more effective and more important. Unfortunately, this was not the end of the story. The report of the subcommittee was presented to the full AEC. There, of course, the political considerations, which were deliberately omitted from the discussions of the subcommittee, became of paramount importance. An innumerable number of meetings were held, during October, November, and December 1946. The problems considered were always the same; how could one on the basis of the report of the subcommittee develop suitable and effective control mechanisms which would prevent the misuse of nuclear energy for military purposes. In November, the AEC decided that in spite of the still unsettled status of their deliberations it would still be useful to make a report to the Security Council, especially because of the unanimous acceptance of Kramers' subcommittee's report. The subsequent political discussions in the Security Council offered no hope for agreement. Kramers' valiant effort had come to naught. He had desperately tried to create a rational basis for the effective international control of nuclear energy, he succeeded in that difficult and thankless task and remarkably enough he obtained the unanimous agreement of both East and West. It was a devastating blow for him—and for the whole world—that the politicians of the East and West were unable to reach an agreement on that same basis. In the increasingly acrimonious and self-serving debates in the United Nations, the unanimous acceptance of Kramers' report, with its optimistic message that nuclear energy after all could be controlled, was all but forgotten. It must have been a shattering disappointment for Kramers. He never again met with the subcommittee or the AEC. On January

31, 1947, Kramers received a terse note from The Netherlands government, thanking him in the most perfunctory manner for his "meritorious service."[137] It seems unbelievable that a country as civilized as The Netherlands, would acknowledge the outstanding service of one of its most illustrious scientists, in the pursuit of one of the most important questions facing the human race, in so shabby a manner. When Kramers returned home in August 1947, there were no bands welcoming him, neither the minister nor a representative came, no one in fact paid any attention to him on his return. One can only conclude that the government did not know how important were the issues with which Kramers dealt nor how important his contribution was.

His home university did not show a great deal more understanding. On January 14, 1947, Kramers wrote to the authorities in Leiden to ask for an extension of his leave.[138] This was eventually approved on the same financial terms as his earlier leave. That is why, although Kramers lived in Princeton, he taught at Columbia University in New York to supplement his income. In Princeton, after his many months with the AEC, Kramers appeared totally exhausted.[90] While in New York, Kramers stayed for some time with an uncle of his student Wouthuysen. Wouthuysen was also in New York and was looking forward to long relaxed, informal talks with Kramers. However, Kramers was so exhausted from his daily talks with Gromyko that after dinner he would fall asleep immediately. Wouthuysen did not have any chance to have a serious talk with Kramers.

It is hard to know whether Kramers had any great expectations or high hopes when he accepted the appointment to the AEC. It is certain that he agreed to undertake this task out of a sense of duty to The Netherlands, obligation to the world, and devotion to science. Since he was an experienced, brilliant, and sophisticated person, he must have known that the chances of success were slight at best. But to a duty-bound person that does not matter; the injurious effects on his health were a minor matter. The obligation was there and it had to be fulfilled.

CHAPTER 19

Kramers' Self-Image

I. Kramers and the Psychology of Physicists

On March 23, 1943, Kramers gave a lecture for which he had selected the title "Psychological Aspects of Physics." The occasion was a symposium, organized by the Dutch Physical Society, on the general topic "Aspects and Insights of Physics." The symposium was held in Utrecht. It was not unusual that Kramers as the leading Dutch physicist would be asked to give a major address at that symposium, nor that the choice of topic would be left largely up to him. It was suggested that he might possibly discuss some psychological and personal aspects of research in physics.[139] What was surprising was that Kramers agreed to give the lecture in the first place. Kramers' knowledge in philosophy, religion, and music and his erudition in literature were widely known, but it was not generally known whether he knew or cared a great deal about psychology. Kramers was literally bombarded with requests to give general talks, which he usually declined; but he accepted the invitation to talk about some psychological aspects of physics at this symposium with alacrity. Evidently, he felt that he had something pertinent to say about the topic. The talk showed that Kramers had a considerable familiarity with the current psychological literature and the general mode of psychological thinking, but surprisingly he did not talk about that at all. He mentions in passing some of the methods and ideas of psychology and he hints that these might be of interest to physicists, but he does not relate these aspects of psychology to physics in any way. Instead, he deliberately restricts himself to the psychology of *physicists* as persons, not of physics as a field. The discussion of the psychology of physicists is presented in a highly personal manner; Kramers does not present the results of tests or scientific psychological analysis; he merely records and organizes personal experiences. He states this explicitly:

"Today we will reflect on psychological facts and phenomena well known to all of us because of our activities in physics, perhaps without being able to suggest experiments which would deepen our understanding [of these phenomena]."[140] He ends this introductory statement with a rather typical sardonic remark: "If one were to be unpleasant, one could say that I will just engage in idle chatter."[140]

Kramers starts out his discussion by enumerating what he feels are some general characteristics of investigations in physics. But the characteristics that he singles out are hardly well-defined (or even poorly defined) accepted psychological notions; instead, they are properties which, in Kramers' view, are significant elements in every individual's scientific activities. He presents these properties in dialectical pairs. Thus, he mentions that *reasoning* and *intuition* both play essential complementary roles in any scientific investigation. He even gives a definition of reasoning as a process that purposely seeks out interrelations, based on experience, knowledge, and memory. Intuition by contrast, is the process whereby, without recognizable conscious effort, a result or insight suddenly appears. The language that Kramers uses to describe this process is strikingly similar to that used so often by Bohr to describe quantum processes. The intuitive recognition "happens uncontrolled by and even uncontrollable by the mind."[140] Much more typical for Kramers own feelings is his statement that an intuitive understanding "is not something discovered by the person, but that instead it was whispered to him by an angel."[140] But the main point of this discussion of the dichotomy of reason and intuition is Kramers' claim that the difference between these two indispensable ingredients is not all that great. Kramers views the distinction between the two as quantitative rather than qualitative. Both the reasoning and intuitive understanding result from very similar—maybe even identical—types of mental activities, some of which are conscious and controllable, while others are not. These others might even appear irrational or random, but all are intertwined is such a complex pattern that a strict separation is not possible.

Another dialectical pair of properties, which Kramers argues is of special importance in the psychology of scientific invention, consists of *logic* and *representations.* Physics only becomes possible because of vast simplifications, idealizations, auxiliary representations, and images which all physicists, consciously or unconsciously, inevitably make. Physics selects from all the amorphous materials—sense data, experiences, feelings, and images in the psyche—a few elements that are presumably those relevant for the physical world. It is, according to Kramers, a common element in the psychological makeup of all physicists that each one attributes a picture or visualization to the elements so selected. That is the representation aspect. The interrelation between these elements so selected has to do with the logic. This psychological reduction and representation process, which Kramers claims makes it possible to do physics in the first place, varies from person to person, but the process itself is universal. Even so, the process is extremely complex and intensely personal. "What strange and foolish associations, what curious mnemonics characterize our work and make it possible?"[140] Kramers stresses that to him a cosine is black and a sine is white; this helps him to analyze and remember trigonometric identities. An electron was to Kramers a bluish, watery, tapioca sphere. He asserts unequivocally that the processes of verbalization and visualization are universal for all analytical and creative activities in physics: "Those who claim that their scientific thinking is *not* imbedded in a complex of largely irrelevant pictures and contrived relations, can be correct only to the extent that to them the emphasis is not on spatial–visual representations."[140] And with this flourish Kramers ends the discussion and never returns to it. It appears pretty evident that whatever universal validity Kramers description of the psychology of doing physics might have, it certainly represented his private understanding of his own activities. It is interesting that Feynman, when developing his brilliant description of phenomena in space–time, also associated a visual picture with these processes: A positron is an electron going backward in time and "getting blacker all the time."[141] Perhaps the kind of visual representation that Kramers suggested is indeed common, although a sample consisting of Feynman and Kramers is hardly compelling statistical evidence.

The final elements, which Kramers asserts play a dominant role in the intellectual activities of physicists, are the contrast or even the tension between *belief* and *unbelief* or between faith and skepticism. This tension that Kramers feels is of purely psychological origin and it determines the nature and type of scientific activity. The belief and unbelief refer to the credence or doubt attributed to the currently accepted maxims and methods of physics, the confidence or skepticism toward the traditional concepts and goals of physics. Many years later, Pauli made a related distinction between what he called *maximum* and *minimum* physicists. He called a *maximum physicist* one who only believes those results that are inevitable consequences of a theoretical structure, while a *minimum physicist* believes everything that is not explicitly prohibited by that structure.[142] To Kramers, these maximum and minimum attitudes are simultaneously present in the psyche of every scientist. The continual shifting between these attitudes causes great tension; this Kramers feels is a creative tension, responsible for most original work. Without such tension, Kramers claims that physics rapidly would become sterile. One of the conclusions that Kramers draws from this conflict between belief and unbelief is that a really "fruitful, living theory cannot and should not be perfect; it should never be finished or complete it may never exclude the emergence of new horizons." This may well explain Kramers reluctance, if not his opposition to axiomatizations of any kind in theoretical physics. It may explain his hesitation to make an unqualified commitment to any physical principle even one as universal as relativistic invariance or charge conservation. In any case, to Kramers, the dichotomy between belief and unbelief was a basic, almost determining factor in research in physics. Kramers does not exactly come out and say that, but he comes very close: "Belief and Unbelief provide a psychological definition of our passive and active attitude toward our methods and working hypothesis."[140] He even refers to the almost religious power such beliefs might have, but he warns, no doubt alluding to his own scientific career, how potent, influential, and ultimately destructive a lack of faith or an overly skeptical view can be. The three dialectical pairs—*reasoning and intuition, logic and representations*, and *belief and unbelief*—are, according to Kramers, the mental elements that play a basic role in creative physics. It is evident that these are hardly well-established psychological characteristics. They are instead Kramers' private attempt to formalize and organize his personal reactions to his own experiences in the world of physics.

In discussing physicists rather then physics, Kramers introduces a number of distinct categories. Of course the classification of scientists into various types is a rather old, well-studied subject. It is an active field of research up to this day. However, Kramers' categories do not correspond in any direct way to the standard psychological or sociological classification schemes. In his lecture, Kramers first reminds the audience of some of the standard, well-known types, such as the scholar and encyclopedist—scientists who are extremely well informed but often without a great deal of originality. Their main role is to summarize the status of the field in textbooks or handbooks. This category is distinct from that of the critical thinkers who tend to give a logical, critical analysis of materials believed to be well known and well established. Like the encyclopedists, the critics are important in the general development of science; they often point out insufficiencies and contradictions in existing schemes and in so doing they pave the way for new approaches. As a final known category, Kramers mentions the contrast between the *romantic* and *classical* types of scientist, first introduced by Ostwald.[143] But most interesting are the categories again dialectically presented, which appear to express the subdivision of physicists as Kramers personally sensed them. A classification scheme obviously depends on the criteria selected for such a classification, and Kramers evidently felt that his criteria were more suitable for an understanding of the psychological processes of physicists than the more traditional psychological yardsticks.

Kramers first introduces the dual categories of realists and formalists. The former are more attuned to phenomena and experiments, while the latter lean toward formalism and mathematical structures. This among theoretical physicists is a well known and conspicuous distinction. Kramers states emphatically that these physicists, who are (or wish to be) exclusively formalists or realists, are usually not particularly good physicists. He asserts that Einstein is a first-rate formalist precisely because he is also an excellent realist; of "Bohr, one could say just the reverse."[140] No one would argue with the statement about Einstein, but the formal power of Bohr implied by this quotation is less obvious and clearly represents Kramers' personal views. However, the distinction between formalists and realists is certainly a useful means to categorize theoretical physicists.

Kramers' next category, although not totally standard, is still not all that unusual. In the dialectical mode which he adopted throughout this lecture, he distinguishes between *absolute thinkers* and *linguistic thinkers*. For the absolute thinkers, the creative process—the development of their ideas and the execution of their calculations—appears to be largely independent of any kind of verbalization. After intense thought or extensive computations, their papers are rapidly written and require little or no revision. Dirac, Fermi, and Yang are obvious examples. The linguistic thinkers, a category that was undoubtedly introduced to accommodate Bohr, are those who use the verbal articulation as a means of defining and refining their ideas and thoughts. Their papers are written, rewritten, revised, corrected linguistically, and scrutinized almost without limit. The remnants of these many struggles and changes are quite often still noticeable in the final version. Kramers sees a parallel in the way composers operate: "Mozart's music is absolute, it enters from heaven without any appearance of conscious struggles, while in Beethoven's music the signs of a conscious struggle are often quite noticeable."[140]

However, by far the most unusual category introduced is that of *independents* and *dependents*. Kramers takes great pains to explain that to him a *dependent* physicist is not just a person who dutifully carries out tasks assigned by someone else, or who elaborates or embroiders suggestions of teachers or colleagues. Quite the contrary, to Kramers a dependent physicist can be creative, original, and innovative, often more so than an independent physicist. What makes him dependent is his inability to be completely convinced of the correctness and importance of what he actually does. The dependents in Kramers classification lack the absolute certainty that their work is of supreme significance. They continually worry about the validity, relevance, and above all significance of their work. To become successful, they may very easily need the help and support of a less talented but more forceful person. The encouragement that these dependent physicists continuously need is clearly of a personal and emotional, as well as a scientific, character. It is obvious that in such personal encounters, factors other than purely scientific will become particularly relevant. All aspects of the personalities will become involved. That clearly creates very complex and tangled situations involving both psychological and scientific factors. These complicated interrelations have a great bearing on the eventual scientific productivity of the dependent physicists, but Kramers does not comment any further on these matters. He evidently found the contrast between dependent and independent physicists so striking that he felt justified in devoting a complete category to it in his classification.

The last element in Kramers' classification also carries a personal imprint; he distinguishes between *serious* and *playful* physicists. The playful physicists consider physics as an amusing game or a fascinating sport, while the serious physicists tend to be hyperserious and ponderous. It is hard to make these delineations precise. One can be deadly serious about a game and facetious about weighty matters, still Kramers feels that this subdivison corresponds to a real distinction among practicing physicists. Gamow, Feynman, Glashow, and possibly Schrödinger belong to the playful types, while Mach, Born, Bohr, and Planck can be classified as serious thinkers. Kramers' classifications end here; the complete scheme includes the formalists and the realists, the dependents and the independents, the absolute thinkers and the linguistic thinkers, the serious and the playful thinkers. This is surely a highly personal and very unorthodox classification scheme. It is pretty easy to guess where Kramers would place himself in his own classification: he would be a playful, dependent, formalist and realist (he did think of himself as a major physicist, so he had to be both a formalist and realist). He was probably more nearly an absolute thinker than a linguistic thinker, although Bohr's influence cannot be underestimated.

In the final section, Kramers raises the question to what extent the other everyday activities of a physicist carry the imprint of his professional preoccupation. In attempting to analyze this question, Kramers argues that the answer will be very different depending on whether one concentrates on internal personal questions or on external public behavior.

As far as the external behavior is concerned, a physicist will probably behave in much the same way as any other person of the same social and economic class. However, the inner life can be influenced profoundly by physics. Kramers writes: "It [physics] can give a color—an ominous color at that—to the background of his humanity."[140] The last section of this lecture —where Kramers discusses the impact of physics on life—is written in a somewhat turgid, almost oraclelike fashion. The last quotation is a good example of that style. It is therefore not always easy to fathom just what Kramers had in mind. Perhaps he was just enumerating the personal reactions of physicists he had known to their physics; perhaps he was just expressing a number of personal concerns that had bothered him over the years. For example, he mentions that some physicists wished that they were not so involved with physics: "it is too difficult for me" or "it leaves me totally cold" were some common complaints. Both Einstein and Pauli made occasional remarks like that. But on other occasions, these same people often exclaim: "The understanding of physics is the only thing which makes a miserable life worthwhile." Still others seek to incorporate their physics within a larger philosophical or religious framework. "Physical truth to them becomes a reflection of a higher, deeper—the deepest truth."[140]

The last passage of Kramers' lecture deals with the relation between truth and value. Evidently, Kramers feels that this is a central problem that all physicists face continuously. It is unfortunately not very easy to guess what kind of value he is referring to. The examples suggest that he uses *value* in a very narrow sense, as value for physics. For example, he writes: "of what value is this experiment, or this consideration." That clearly indicates its utility for physics, not its social or human value. But it may very well be that he later used *value* in a more general sense. The last sentence in his lecture is particularly enigmatic: "The struggle for truth and value and the tension between them characterize the inner life of many physicists. How it [the struggle] is fought is largely a matter of character. Often he dies before a victory of any significance has been achieved. But if he has achieved such a victory, thanks to God and his friends, then the results of the victory often remains a public secret."[144]

In spite of a great deal of thought, it has been impossible to come up with a rational, let alone compelling, interpretation of this sentence.

What can be learned from this remarkable lecture? There can be very little doubt that in spite of its well-organized structure, Kramers delivered a lecture about himself. His way of describing the personal elements in physics (belief and unbelief) and his classification scheme have an unmistakable personal stamp. His introduction of the category of dependent physicists seems tailor-made for himself. The description of an original, inventive, and intellectually powerful physicist, who lacks the assertiveness and conviction of his own importance to achieve ultimate success, is after all a portrait of Kramers. But here this self-portrait is hidden in a rather arcane classification scheme. It was also in connection with this category of the dependents that Kramers admitted, for the first time in print, how great the influence of personal interactions actually was, even if these were of a nonscientific nature. Other intensely personal elements are scattered throughout this lecture, often as side remarks or offhand comments. In his description of encyclopedists and critical thinkers, he comments that it must be remembered that a great deal of novelty comes to physics because of a pronounced lack of knowledge and a marked lack of critical understanding. The idea that physics progresses because of naive misdirection was a favorite theme of Kramers. In reviewing a book on Einstein's life he wrote: "To think of something new and creative is so difficult that new discoveries have often been reached by the queerest of detours and by the guidance of definitely misleading signposts."[145] One gets the impression from this lecture—and also from Kramers' diaries[146] that he almost feels that there is an element of unfairness that this haphazard procedure would lead to such successes. On the other hand, one cannot escape the feeling that Kramers would have liked to have the courage to follow definitely misleading signposts. This type of dichotomy occurs over and over again in Kramers' work and life. It is striking how Kramers' lecture in psychology emphasizes conflicts, contrasts, and struggles. His whole classification scheme is organized in terms of contrasts.

The conclusion that in this lecture he primarily talked about his own attitudes and reactions seems inescapable. The very last part of the lecture where he talks about the influence of physics on the inner person and the conflict between value and truth appears quite contorted. In spite of the somewhat convoluted arguments, this lecture gives as complete a picture of Kramers' personal views on life and his role in physics as his nature would allow.

II Kramers' Lecture on Debye: What Is an Ideal Scientific Biography?

In 1946, a book was published in The Netherlands which carried the somewhat pretentious title *Dutch Heroes in the Natural Sciences.*"[147] This book contained a collection of sketches of famous Dutch scientists. One of these essays was about the physical chemist Peter Debye, born in 1884 in the southern province of Limburg, who received the *Nobel* prize in chemistry in 1936. The essay was written in 1941 by Kramers, but because of the wartime stringencies the publication of the book was delayed until 1946. Kramers' essay is particularly interesting not only because of the biographical material it contains about Debye, but because it reveals a great deal about Kramers' attitude toward science and scientists. This is one of the very few documents where for some unknown reason Kramers "lets his hair down." He is surprisingly candid about his feelings about science; however, he is considerably more guarded about his feelings about Debye.

The biographical part of the essay is rather standard; it contains the customary laudatory enumeration of Debye's accomplishments. Kramers recalls that Debye, in a lecture given in 1912 in Leiden, primarily if not exclusively addressed Lorentz. Intelligent and bright as Debye was, he recognized "a higher judge" and the success of his lecture would be measured by Lorentz's approval. That is how Kramers in 1941 remembered the events of 1912. Lorentz's superiority was and remained unquestioned and unchallenged.

But more interesting and more illuminating are Kramers' general comments about science. These comments are intensely personal. He starts out by talking about science and scientists: "Science is for those who study it, a source of exaltation." [148] The Dutch word that Kramers uses here is *verrukking*, a strong word denoting a very powerful emotion. It was a rather quaint, stilted word; its emotional content is best approximated by a linear combination of elation, exultation, and euphoria. Kramers uses it a great deal, and throughout this discussion it is translated as "exultation." For example, in his lecture on psychology (Section I), Kramers uses the same word to describe the emotional reaction to understanding and appreciating something genuinely new. He interprets it as being "pulled away" from "everyday" understanding with an accompanying euphoria. Kramers continues his explanation of science:

> Science is for those who study it a source of exultations. That is old news—but often it is not understood what this word implies. Sometimes one just gets pleasure out of of the activity. But besides that, the interaction between nature and the human spirit, which is called science, give rise to the phenomenon that, while involved in hard and difficult work, one becomes suddenly conscious of the fact that one is torn away from the ordinary, from the well known, out of the daily character of consciousness. Something "new" has happened to the investigator or the learner—and this lies on a higher plane of our human understanding—and this "something new" has assumed a concrete form in the mind of the investigator. I point here to the creation and recreation which takes place in science, which if it happens to some one, places as it were, this person on the next rung of the ladder we all try to climb. Yes and it is then indeed so that it does not appear that we climbed at all, but as if a force from outside ourselves, say an angel, pulled us away from our previous level and brought us in his mercy, to that higher level in an indescribable incomprehensible manner." [149]

This is obviously an intensely emotional description of a process of scientific creation as Kramers must have experienced it. This passage has really nothing to do with Debye (and Kramers actually forgets about Debye for several pages); it has everything to do with Kramers. The remarkable intervention of "the angel" as the motive force behind the emergence of scientific insight is vintage Kramers. This essay is about the only place where his feelings are expressed so completely and explicitly.

Kramers continues his discussion of the exultation associated with true, unselfish, and profound research for some time, but he recognizes that this essay is written for a rather general audience, so sooner or later he has to explain the emotions that accompany scientific creation. He makes this connection in an unusual manner. Throughout this essay, Kramers stresses that the true significance of Debye's work, especially its original, assured, and decisive character, can only be appreciated by professional physicists and chemists. But this is not the audience for whom this essay is intended. To gear this essay to the proper audience, Kramers argues that the exultation and elation experienced by the scientific researcher are in fact not restricted to the participating researcher, but are indeed available to all human beings and can be sensed and appreciated by scientists or lay people. "I want to see this exaltation as something all humans strive toward—you, the reader, wish to experience this exaltation yourself by participating in the ongoing process of the expansion and development of the human spirit, to which science bears witness."[150] It is the emotional impact of scientific research and scientific understanding which, according to Kramers, is of overriding significance. Kramers considers this essay on Debye as a miniature biography. In this connection, he lists what he considers the essential elements in a scientific biography. It is somewhat surprising that Kramers considers what the scientist discovered, what he or she saw that others did not, and just how he or she discovered it as of minor importance. Nor does he feel that it is of special interest to know what kind of a person the scientist was, or just what special events and encounters shaped his or her life and career. All this, according to Kramers, is of interest only to professional scientists or psychology students and of little relevance to the general audience for which the biography is intended. Instead, Kramers suggests that a scientific biography should emphasize the emotional aspects of the intellectual adventure that is science. He asserts that even if a person knows little about a science and even less than that about the scientist, there still will be emough shared human experiences so that an appreciation of the emotional impact, generated by creative scientific activities, can be enriching and even fulfilling to a general audience. Kramers addresses the audience in an intense and flowery manner:

> When reading a scientific biography, which this essay is intended to be in miniature form, you do not especially want to learn what Debye found out, what nobody knew before, you are not specially curious what kind of a person Debye is and what sensational things happened to him. Oh yes you might be interested in these things but not for the information (unless you are a physicist or a chemist or a student of human behavior)—for this information is only of little use in your own life. No you want to know about the scientist because your human

> and patriotic intuition tells you that there is an exaltation to be experienced which in turn can produce an exaltation into yourself, because even with all the uncertainty about the man and his field—there is so much in both of you—that is near to you, that it reflects something of your being and your aspirations."[150]

In spite of Kramers' convoluted language, the idea is clear; it is the emotional involvement of science, with its associated moments of ecstasy, which constitutes the essence of a scientist's life. This is what Kramers claims a good biographer must convey. Although there can be no doubt that Kramers was totally honest and genuine when he described the feelings and emotions accompanying research, there is still something idealized and single-minded about Kramers' description. He knew very well that these moments of pure elation and perfect joy are rare and can certainly not be produced by conscious effort. Most of the time a researcher is confused, frustrated, and beleagued by conflicting ideas, limited technical abilities, and competing colleagues. Perhaps Kramers meant to say that the *possibility* of such moments is what keeps researchers going through a muddle of grubby details and thorough confusion. But the feelings expressed by Kramers in this essay, come as close to explaining Kramers' perennial search for the "scientific grail" as anything he ever wrote.

The part of the essay that deals directly with Debye reveals quite another aspect of Kramers' personal attitude. In this essay, Kramers is quite complimentary about Debye—yet his enthusiasm is definitely controlled. It is clear that Kramers sees Debye as an important and productive first-rate scientist, but he was definitely not in the class of Lorentz and Bohr or Heisenberg and Pauli. (Remember that Kramers aspired to be in the class of Heisenberg and Pauli!) "Debye did not resolve fundamental riddles—he did not speak the liberating and initially poorly understood words, which alter the basic conceptions of the problems and which mark a milestone on ourliberation from bondage."[150]

This assessment of Debye's science is not particularly critical yet it is still not an enthusiastic endorsement of Debye as a scientist. Certainly Kramers appreciated and admired Debye, but he insisted that he was not of world class. The problems he tackled were significant and important, but not fundamental. His methods were clever, singularly suited to the problem at hand, incisive, and original, but not profound, universal, or deep. Debye generally did the type of physics and chemistry that Kramers thought was on the trivial side, but he did it so well that he achieved impressive results. Kramers acknowledged this without rancor, but he reserved his unqualified scientific admiration and his scientific loyalty for those who dared to tackle more basic questions.

So behind all the customary accolades, there was a distinct coolness in Kramers' essay. It is not at all impossible that part of the coolness had a political basis. Debye studied in Germany and he stayed there for the greater part of his scientific career. He stayed even after the Hitler takeover in 1933 and did not leave until January 1940, after World War II had exploded on the world. Debye, both scientifically and politically, was an opportunist; he did not concern himself about things that could not be changed—and that obviously included the Nazi regime. When that regime would not give him what he needed in 1940, he managed to emigrate to the United States where he accepted a distinguished professorship at Cornell University. But it appears that to Kramers, Debye stayed in Germany way too long. In any case, when Kramers visited Cornell University, he very pointedly ignored Debye.[151] Holland had suffered grievously under the German occupation and Kramers made no secret of his personal dislike of those who sympathized or tolerated the German excesses in the 1930s. Apart from some personal recollections, there is no documentary evidence for the political basis of Kramers' attitude, although it remains a distinct possibility.

The main interest of this essay is not what it says about Debye or about the relation between Kramers and Debye, but the surprising candor which Kramers displays in describing the intense emotions experienced in the process of doing scientific research. Kramers, in this essay, gave a general description of these emotions, which are in fact common to all committed researchers. In this connection, it is appropriate to quote another mathematical physicist, Mark Kac, who eloquently and movingly described his reactions to the discovery of a result in statistical mechanics: "I would find it difficult to convey to the reader the thrill which I experienced, when together with Uhlenbeck and Hemmer we discovered that the coexistence of the gaseous and liquid phases in a simple model was reflected in the double degeneracy of an eigenvalue of an integral operator. That nature should know so much mathematics bounded on the miraculous."[152] Kramers never did record his own reactions to a specific piece of research, but it is overwhelmingly likely that he would express the same excitement, the same thrill, the same exaltation, and the identical feeling of wonder and euphoria in having contributed to the understanding of the mysteries of nature.

III. Fragments from Notebooks and Diaries

A. From the Notebooks: Science and Scientists

The Dutch high schools used to have a quaint custom of the "school agenda"; this was an official school datebook or appointment book. Each student in a given school was required to buy one of those appointment books which were really no more than elaborate calendars, with every day divided into hours to correspond to different classes. The students were obliged to note down the assigned homework for the particular day and hour; they also had to record the dates and hours of major and minor exams in each class. One person was selected to maintain a second such notebook, called a "class agenda," which always had to remain in the school building so that the complete record of the homework for an entire year was always available—presumably for examination and scrutiny by the principal of the high school.

In the agenda for the gymnasium, the most prestigious school, each day had a motto, a famous saying or maxim by a Greek or Latin author; the other schools had to be satisfied with English, French, German, or Dutch quotations. Eventually, these agendas contained mathematical formulas and conversion tables. Many persons became quite attached to the type of agenda for their particular school. Kramers used this kind of notebook for many years to keep track of his lectures, seminars, classes, appointments, and other obligations. But instead of noting down the homework assignments as students did in their agendas, he got in the habit of writing down his reactions to meetings attended or to students he had met. He used the ample space provided in the agenda to make or indicate calculations or comment on matters of physics. Often he made sarcastic remarks about the ineffectiveness of the meetings or the pomposity of the participants. In spite of the brevity of the comments, these notebooks are extremely valuable as an unedited record of Kramers' spontaneous reaction to persons, issues, and events.[153] From time to time, Kramers started more formal diaries, usually written in Danish, in which he recorded candid observations about his students, colleagues, life, hopes, and expectations. Some of these diaries are lost; apparently Kramers at times disposed of all or part of these diaries, so that only scattered fragments remain.[154] But there is one diary which is completely preserved.[155] It was started on May 5, 1945, the day of liberation of Holland from the German occupation; he recorded his activities and thoughts from May 5 until August, in great detail. After August there are fewer entries; its last entry is December 15, 1945. Kramers left for a trip to Stockholm the next day and it appears that he never reverted to a systematic accounting of his thoughts and activities. The period that this diary covers is a most important one. With the war over, everyone was eagerly looking forward to a new beginning, with grand expectations of what the future could and should bring. Kramers' hopes, future plans, anticipations, fears, and misgivings are all recorded in this diary. The diary is brutally frank and at times intensely personal. This diary and the remaining notebooks provide an unusual and illuminating insight into Kramers' state of mind at that period of his life. There is something rather upsetting and even embarrassing about reading, analyzing, and reporting on materials that were so personal and so private. Nevertheless, it would appear unconscionable to ignore the unconstrained utterances of the protagonist of this biography even when in so doing one has to deal with profoundly personal feelings, guarded jealously and always hidden from unwanted intruders.

Even a rather superficial examination of the comments that Kramers made in his notebooks about other physicists would show that the opinions were frequently critical and quite harsh. About de Broglie he writes numerous times that "he is really a stupid character," he was especially critical of Urey: "He is a nice person but I doubt that I can ever get any physics through his head, but I will keep on trying."[153] Actually, Kramers had expressed similar feelings in letters to Clay and Bohr (to Clay about de Broglie, to Bohr about Urey). When referring to the activities of a rather well-known physicist in Princeton, Kramers always used the expression (in his notebook[153] and also in private discussions) *Doch Wieder Dumm* ("dumb again"). In Kramers' notebooks, there are also frequent comments about the results of his meetings with students. A typical entry would read: "Met for 2 hours with X, nothing again—can he do anything?"[153] It was generally believed that on those occasions when Kramers was critical in public, he did not really mean it; it was only semi-serious. Because of Kramers' jocular but also ambiguous manner of phrasing his objections, it was easy to believe this. However, when these public critical utterances are combined with his private notes, it becomes quite clear that he was often uncompromisingly serious in his criticism. He just used this ambiguous mode of expression to avoid public embarrassments or public conflict. An interesting example of this phenomenon occurred in connection with some remarks made by Goudsmit. At one time the great Arthur Eddington gave a lecture about his alleged derivation of the fine structure constant from fundamental theory. Goudsmit and Kramers were both in the audience. Goudsmit understood little but recognized it as farfetched nonsense. Kramers understood a great deal and recognized it as total nonsense. After the discussion, Goudsmit went to his friend and mentor Kramers and asked him: "Do all physicists go off on crazy tangents when they grow older. I am afraid." Kramers answered "No Sam, you don't have to be scared. A genius like Eddington may perhaps go nuts but a fellow like you just gets dumber and dumber."[156] Goudsmit thought this was a great joke, and he quoted it on many occasions.[156] Actually, the situation is a bit more involved. When Kramers quipped his answer to Goudsmit, he surely considered it as a good joke; yet, underlying this facetious comment about Goudsmit was something quite serious. In his letters of recommendations and in his diaries, Kramers is very doubtful about Goudsmit's role in physics. He states explicitly: "Goudsmit is not of the broad caliber to be demanded of a Dutch University Professor."[157] Thus, when Kramers jokes that a person like Goudsmit was getting dumber and dumber and did not have to worry about his future deterioration, it was not just a joke to Kramers; it did contain a good share of his actual feelings, although to the world at large, it could always be interpreted as a clever wisecrack. In this way, the diaries provide a setting, a basis for the deeper understanding of the public statements; of course the diaries contain a great deal of material never made public at all.

It is perhaps not surprising that while writing to himself in the privacy of his study Kramers exhibits the same fascination and passion for physics that he did in public. Clearly, his interest in science was not a passing fancy, a means to become famous, or a public relations gimmick; it was the central part of his being. The first thing he wrote in his diary right after the war was over was a list of physics ideas "to be done immediately." He referred to the prospect of implementing these ideas as "infinitely fascinating." All the "purple prose" that Kramers used in his lecture on Debye, describing the near ecstasy of scientific creation, recurs in his personal diaries. That theoretical physics was Kramers' life was not a slogan or a fancy phrase to impress others; his personal diaries show once again that physics was his overriding concern. But his diary shows equally clearly that matters of obligation and duty—even when they conflicted with his scientific goals—were a perennial worry. This conflict is stated in one of the very first sentences in this diary: "What a great deal of work awaits me now. For me science? or the organization of science? The latter means intense preoccupation with the University and the Kamerlingh Onnes Laboratory, the other is my calling."[155] This clash between what Kramers wished to do and what he felt he was obliged to do pervades the diary, and the conflict remained unresolved. Kramers was asked to become the acting director of the Kamerlingh Onnes Laboratory, to replace the director W. J. de Haas, whose behavior during the war had been sufficiently questionable to warrant such a move. But Kramers did not really want to be a director—or acting director—and his diary entries record his daily struggles to decide just what to do. Even his list of "things to do in physics" is in part determined by his responsibility toward his students.

This list is actually quite a curious list: He mentions three problems related to liquid helium, two quite offbeat classical problems, three popular, semiphilosophical papers which he had promised to write, and some as yet unspecified problems in astrophysics. He mentions that he "dreams of writing a short book on thermodynamics;" he hopes to continue work on statistical mechanics, order–disorder phenomena, and the stopping power of α particles. Almost all these topics were associated with the scientific activities of his students at that time. The only surprising subject on Kramers' list is symbolic logic. He evidently read the basic books—Hilbert and Ackerman, a Dutch book by Beth, a treatise by Tarski—with great interest. His diary records the daily progress he made, almost in the manner of the original school agenda. Why Kramers had become so interested in this most formal of mathematical disciplines is not at all clear. But he certainly studied it most seriously, and after digesting these basic books, it appears that he dropped the subject. He did not mention it in any of his popular talks; he never even so much as hinted that it had anything to do with physics. Apparently, he was just curious enough to study the basic books with care, so he knew in some detail what it was all about—and that was enough!

The conflict between Kramers' obligation toward physics in Leiden and in Holland and his own intense interest in and love for physics was never resolved; it pervades the diary. Kramers can hardly contain his excitement when, on June 2, 1945, Sieg Wouthuysen brought him the first two issues of *The Physical Review.* He writes that there was so much thrilling material in the journal that he stayed up all night reading it. At the same time he was so perturbed by the teaching of subjects like thermodynamics and statistical mechanics, "that something should be done about it." He observes numerous times that the research hegemony had shifted from Europe to the United States and he felt that Holland's research talent had been drained by the war.

In response to these conflicting demands, Kramers effectively decided to do everything he could to restore the research vitality of The Netherlands. In his diaries he complains that he cannot do it all perfectly—but given his inability or unwillingness to limit his options or restrict his activities—he did a most remarkable job of everything.

B. From the Diaries: The Person and His Demons

Kramers kept his diary with great regularity from May 5 to August 10 and sporadically until December 25, 1945. It is understandable that, when writing practically every day, not all the entries deal with profound, scientific, or soul-searching questions but include a great deal about the daily life of the writer and of his reactions to quite ordinary events. But of course that information is important to round out the image of the person described. A few minor things stand out. Evidently, Kramers was very concerned about his health. He describes his aches and pains in great detail, and he dwells at length on the progress of his illnesses. He often had trouble with his stomach; he frequently complained (in his diaries) of weakness, tiredness, and a general lack of energy. More often than not, he would lie down in the middle of the afternoon and he almost always would go to bed early. He had a pronounced weakness for cigars; he smoked between 10 and 15 cigars a day. When they were not available during the war time, he suffered withdrawal symptoms. He was almost as grateful to Sieg Wouthuysen for bringing him cigars as for bringing him the copies of *The Physical Review*.

Evidently, the Kramers' social life was very busy. Almost everyday, someone would come for dinner or "to drink coffee" (a venerable Dutch institution, where the drinking of the coffee is merely the excuse for lengthy visits and extended conversations). In addition, there were the traditional birthday celebrations of the extended family (father, mother, aunts, uncles, cousins), the wedding anniversaries, and the commemoration of special events (such as the anniversaries of degrees obtained) so that their social life was full. It appears that Kramers generally enjoyed these activities. In his diaries he often comments on the pleasant, agreeable evenings he spent as, for example, with his brother Jan and his wife Gerda on the occasion of their 30th wedding anniversary.[158]

The diary also shows unmistakably how central music was in Kramers' and indeed the Kramers family's life. Many of the social events centered around music. Only rarely did they go to formal concerts (although that did happen); most of the time there were small private gatherings, where "people made music." They played chamber music, sonatas. Storm very frequently sang, often Bach, but other composers were well represented. For special concerts, Storm would often take voice lessons, with usually spectacular results; her voice was large and beautiful, and with training and coaching the results were impressive, as when she sang a Bach cantata on July 3, 1945. Kramers records this with obvious pleasure in his diary. Many entries in the diary refer to his need for more or more sustained cello practice. It is quite evident that he practiced nearly every day for at least an hour. He clearly was an accomplished musician, but he suffered the frustrations and irritations of all extremely talented amateurs. After not practicing for a few days he exclaims that "he can no longer trill, the vibrato is also off." He was capable of playing truly great music—and he came very near playing it extremely well. But to maintain this level required constant practice; to improve, it was nearly impossible, given all the other demands on his time and energy. That was the source of his constant frustration with music. Kramers was especially pleased when playing the Brahms piano quartet (July 12, 1945); the other three musicians all complimented him on his performance. He notes this in his diary, with the comment that he really should not be so vain, that public praise should not be so important to him. These scattered comments about his social and musical life give a vivid picture of Kramers' activities. What is perhaps most interesting is that frequently, right in the middle of describing a concert or a conversation, Kramers returns to physics. Sometimes he jots down an idea to think about or a reference to look up; other times his references to physics are quite detailed. Thus, experimental curves, giving the heat of evaporation of helium with a question mark somewhere, may appear out of nowhere. Still other times his diary contains "miniderivations" of new results. All this shows again that whatever Kramers was thinking, dreaming, or fantasizing, physics was always there and refused to stay in the background.

Kramers was an introspective person, and it is therefore to be expected that there would be a great deal of self-analysis and self-contemplation in his private writings. He often wrote that he was preoccupied with "searching his soul." These discussions are invariably brief and disorganized; they appear more like meanderings through his private world. Since these were his own private reflections, there was clearly no need for any coherence. Thus, Kramers wonders whether his friends or acquaintances were of "noble character." Kramers' close friend, the painter Tom Kelder, was considered as especially noble; his colleagues in the Kamerlingh Onnes Laboratory, such as de Haas and Keesom, were most definitely "not noble." He also wonders about the "nondramatic nature of his soul." It is true that Kramers in his daily life often made a detached, even cold, impersonal impression. Kramers was evidently aware of this presumed lack of warmth and involvement in his everyday association, and it seemed to have bothered him somewhat. He tries to give some explanation; "All my conflicts and tensions die out, without any struggle with other humans."[155] He adds almost in the nature of a confession: "The smallest dramatic effect—whether in life or in the theatre—that I really comprehend emotionally, makes a tremendous impression on me." Perhaps Kramers is saying that he is so sensitive that if he did not put up this barrier of detachment, life would become impossible for him, since he would be tossed around from one powerful emotion to another.

Most—perhaps even all—of the diary was written at night. During the summer of 1945 there were a number of beautiful June nights (this is not all that often the case in The Netherlands) and Kramers stayed up the better part of these nights to enjoy the unusually mild, clear weather. He wrote in his diaries in the very early morning when the birds just started to sing. In a somewhat nostalgic mood he wrote (this was a month after the war ended): "I am rather happy; but it is the happiness of the former springs of my dreams."[155] In such moods Kramers was especially likely to write and quote all kinds of poetry. He quoted some French poetry but said of his French literature that "he no longer knew a thing about it," and that he made himself a little ridiculous using it. Since there was no audience, it did not make too much difference what he said. With the multiple roles so often assumed by the authors of a diary, he wrote in defense of himself and also laughed at himself.

Kramers discusses his vanity, pride, and arrogance in many diary entries. That may seem strange, for Kramers surely never made the impression of a proud, boastful man. Yet, as the diary entries make clear, he chided himself frequently and felt guilty about what he perceived as his personal vanity. After Kramers had been praised by his colleagues for his performance in the Brahms piano quartet, he felt particularly happy and pleased. But he could not leave it at that. In his diary entry that night he commented at length that it was wrong to be so pleased. It seemed only proper to him to struggle against or deny such feelings of pleasure. Kramers then proceeds to analyze just why he feels so strongly that one should fight against such feelings, which encourage self-importance. He attributes this in part to the effect of his Christian education. He was taught since time immemorial that vanity is a sin to be atoned for or eliminated. The true Christian is humble and self-effacing. Praise and compliments are true instruments of the devil. Even though Kramers as a mature person no longer believed the Christian doctrine in a literal sense, he certainly accepted the philosophical principles embodied in the Christian theology. He states explicitly: "On the road to self purification one should not strive toward fame and honor."[155] This might have been Kramers' goal; he strongly feels that this indeed is the proper interpretation of Christ's injunction: "Not my will but thine be done." But he had real trouble with the interpretation and consistent implementation of these ideas. Quite the contrary, in his diary he struggles time and again against the overpowering tendency to enjoy his status and to relish his fame. It was easy for Kramers not to strive toward authority; he never was an "empire builder", and the acquisition of power was of no interest to him. But he did have real trouble, in his own sense, being properly humble. Kramers was delighted when a passer by or student pointed him out as the "Dean of Dutch Scientists,"[155] but he felt deeply guilty about the pleasure this recognition gave him. Kramers was evidently very bothered by the conflict between his self-imposed austere humility and the intense pleasure he derived from public appreciation.

Not only did Kramers enjoy the adulation of his scientific public, he actually needed public appreciation to function properly—this in spite of his protestations that one should not pursue fame, glory, and renown. His need for public approval became quite apparent in a minor incident described in his diary. Evidently, Kramers was criticized in a faculty meeting for certain suggestions and proposed actions. The actual issue is of no particular significance. A typical faculty controversy in the senate of the University of Leiden in 1945 is unlikely to be of lasting significance. As always, it was an issue of prerogatives, consultations, and egos. But Kramers was very upset by these public rejections, and he wrote extensively about it in his diary. He feels that this public action by the senate was very nearly an insult and certainly a blow to his pride and vanity. He then writes what is probably the most remarkable phrase in his whole diary: "I so much want to be liked and admired; I want to be seen as brilliant, intelligent, kind, and understanding, whether I am or not."[155] No doubt many people would find it difficult to recognize Kramers in that exclamation. There is no doubt, however, that when Kramers wrote this, this is exactly the way he felt, because he expands and elaborates on this same theme. He struggles with his inability to accept the senate's critical judgment with the proper humility. He argues that "true humility" would be for him to admit his errors to the full faculty. But this he cannot and will not do: "I will not submit to the members of the faculty, only to God."[155] And he continues this emotional outburst by writing: "This means for me, I will only apologize to humans who are truly God's witnesses—I will bow down only for those who truly love me, such as a father or mother loves an errant child."[155] After this paroxysm Kramers clearly quieted down; he writes that there is no point in his continuing anger, but with his usual perceptive insight, he recognizes that he will bow down—which really means "show his sensitivity and great personal vulnerability"—only to those who not only love him but understand him as well. And this he felt was almost never true. Kramers, like everyone else, had his hopes, fears, anxieties, and anger. It is perhaps not surprising that he had to struggle with his own personal demons. He was desperately afraid that he was not liked, loved, understood, or admired; but unlike others, he kept these feelings and struggles to himself. Unearthing these feelings after so many years should make Kramers more understandable and more human.

C. From the Diaries: Personal Entanglements

Scattered throughout the diary are comments and complaints about Professor W. J. de Haas. Right after World War II was over, Professor de Haas was removed as the director of the Kamerlingh Onnes Laboratory, pending an investigation of the Reparations and Cleansing Committee,[159] of his behavior during the war. Kramers was effectively in charge of the laboratory. Kramers and de Haas never got along particularly well before these unpleasant events; de Haas' firing did not make it much better, although de Haas in a most transparent manner tried to "makeup" to Kramers. Kramers was well aware of this, but in his customary way he avoided public confrontations. In his diary he is quite frank: "When will I have to tell him what I think of him."[160] Although Kramers clearly saw de Haas for what he was, he never did tell him what he thought of him, in spite of the insistent urging of his wife. But many of the diary entries testify to his annoyance and irritation with de Haas and with the whole situation. Many times when de Haas wanted to come over in the evening and talk—always because he was misunderstood and under-appreciated—Kramers called or wrote him that he was tired or busy, so the visit was postponed or canceled. This happened many times. In his diary, he recorded these same evenings that he really could not stomach another "complaining session" with de Haas. The very last entry, written December 15, 1945, reads: "Further again these thoroughly unappetizing cockfights with W. J."[155]

The discussion of more intense personal matters, such as private hopes and fears—what were called the personal demons in the previous section—is sensitive. But there is still another even more intimate level of experiences which is yet more difficult to analyze and assess. In most scientific biographies there is almost a complete taboo against the frank discussion of the private lives of scientists. It very much appears like a "conspiracy of silence." The taboo against discussions of the sex life of a scientist is effectively complete. Such discussions are considered to be in very poor taste and surely completely irrelevant. A rather well-known Dutch scientist stated categorically: "It is impossible to imagine that the great Dutch academic scholars before the 1950's would have any interest in sex whatsoever."[161] Whether intended as facetious or not, as a complaint or compliment for the younger generation, this quote surely shows the great reluctance and severe misgivings of many scientists to acknowledge romantic love, let alone sexual interest as significant elements in the life of a scientist. But in Kramers' case the diaries and notebooks, supplemented by stories, letters, and poems (see Section III), provide remarkable first-hand sources revealing a complete human being—one who touched greatness, who struggled with anger and compassion, who felt misunderstood—with unmet emotional and sexual needs.

The relation between Kramers and his wife, Storm, was one of the major components of his emotional life. Superficially, their life and marriage was comfortable and pleasant. There were no major conflicts and no conspicuous problems. Kramers' character would of course avoid even the slightest indication of a public confrontation anyhow. The children were well cared for and had a happy, pleasant, and warm childhood. Even so, the marriage was far from perfect. It was suspected by many and known to a few close friends that the relation between Storm and Kramers was not particularly fulfilling.[162] There was little closeness, no deep understanding, no tenderness in the relationship. Kramers writes precisely this in his diary: "It is not only that I should be loved—but also that I should be understood that I impose as a condition (for my love) and this I feel is never satisfied by Storm."[155] Kramers remarks numerous times that he misses genuine and deep contacts with Storm. But

Kramers was a very smart and perceptive person, and he realized full well that where he did miss an intense intimate association with Storm, he was at the same time quite fearful of such a deep and demanding contact. Thus, even if Storm had been willing and able to initiate and sustain a relation of this intensity, it is likely that Kramers would have passively resisted and might have sabotaged such a relation. Kramers, in all phases of his life, had great trouble making unconditional, open-ended commitments. This affected all aspects of his life, but it influenced his relations with women in a particularly direct way. Kramers was always a little diffident with women. He grew up in a family of all boys; his first association with women did not come until he was a student at Leiden. As all adolescents, he developed a healthy interest in sex, but as was typical for that epoch in Holland, the interest was restricted to fantasies and verbal activities. As a student, Kramers' most important relation with a young woman, Waldi van Eck, again had this very indecisive and unsettled character. It was on again, off again; no commitments were made, no decisions were taken, the relationship was never clearly defined, it was certainly not consummated, nor ever terminated. After Kramers returned from Copenhagen, some 10 years later, with a wife and children, Waldi van Eck became again, or remained, a friend of the family for many years. The desk in Kramers' study belonged to her. Kramers' youngest daughter was well aware of the past relation between her father and Waldi van Eck and referred to her as "my prenatal friend."[163] The actual relationship between Kramers and Waldi van Eck remained static; it did not evolve, it did nor reach a crisis, and it did not mature—but it did not stop either.

Although the relationship between Storm and Kramers was infinitely more intense and obviously much more important, there were nevertheless striking similarities. As described in Chapter 11, Section V, the love affair between Kramers and Storm was turbulent, intense, and full of caprice. After Kramers had left Holland for Copenhagen, his interest in sex matured and deepened. In his diary in 1945,[155] he recalls that in the early days in Copenhagen (1916–1919) he was preoccupied with sex. From his notebooks and especially his letters to Romeyn at that time,[164] it can be inferred that this interest had shifted from the purely theoretical to a more operational level. This in particular meant that in the relationship with Storm, sex played a very important role. Although hardly startling, this still seems to have confused Kramers to some extent, because he seemed unable to distinguish clearly between purely physical attraction and a deep, companionable love. Thus, it surprised and bothered him that when he was with Storm he still missed the emotional fulfillment that came from his intellectual contacts in Holland. This was no doubt one of the reasons for his unwillingness to make definite long-lasting commitments as explained in Chapter 11. The decision to get married was forced on Kramers by Storm's pregnancy. Given Kramers' background, his deep sense of responsibility, his profound decency, and his Dutch Calvinist upbringing, he could do nothing else. He did struggle with that inevitable decision for the summer of 1920, but he knew all along that he did not have a choice. But he did have a choice in following Storm to Paris—when it appeared that their relation was just finished (see Chapter 11). He could have let the relation end there. But instead, he joined Storm for an unplanned week in Paris, which led directly to the birth of their oldest daughter.[165] But there should be no misunderstanding, the decision to get married was Kramers' alone: Storm was surely willing—but definitely not insisting. This was the basis of the marriage; Kramers was forced to do what he disliked and feared most in the world, to make an irrevocable decision.

It is interesting that the relation between Storm and Kramers did not change a great deal after this turbulent start. Kramers' feelings of an incompleteness in the relationship intensified as time developed. This is often the case in an unsuccessful love relationship; he felt frequently lonely and misunderstood. He began to withdraw as early as 1924, and he refers to how "parsimonious" his emotions had become. Consequently, the relationship, incomplete from the start, become more inpersonal and detached as time progressed. Kramers confided less and less in his family; he concentrated his time and energy, including emotional energy, on physics. Storm, no doubt aware of the situation, tried her best to ameliorate the situation. But she was severely limited; she felt intellectually inferior to Kramers. She often said: "It is too bad that you won't have children with a truly intelligent woman, just imagine how brilliant they would be."[166] It is of course true that Storm intellectually was no match for Kramers, but she was very bright and had a great deal of common sense. In fact, her judgment of students and colleagues of Kramers was usually a good deal better than his own. She was extremely protective of Kramers' rights and prerogatives and had an excellent sense when Kramers did not get, or did not insist on getting, his "due." She often tried to coerce Kramers into taking assertive or aggressive actions. These actions, intended obviously to help and support Kramers, had actually the opposite effect. Kramers would become impatient, defensive, and sometimes downright angry. He would inevitably withdraw further, making communication effectively impossible. Storm, continuously frustrated in her efforts, became at times a bit too insistent and a trifle shrill—exacerbating the situation even further. Thus, the relationship, whose initial basis was somewhat suspect, deteriorated steadily in the course of time. It was a quiet estrangement very much in Kramers' style, not a sharp conflict or a public break. Still by the late 1930s, it was known among the close friends that the marriage was no longer very satisfying to Kramers. He told his close and intimate friend Tom Kelder that he felt that "Storm and I no longer communicate."[167] Although not quite as explicit, many other people were also aware that the relationship had deteriorated sharply (e.g., Weisskopf and Wouthuysen). Kramers himself writes in his diary on May 8, 1945, only 3 days after the liberation from the German occupation, that he is increasingly irritated with the kind of things Storm talks about. It is of course not surprising that after a tense and frightening period of enforced togetherness, small annoyances would become major irritations. But Kramers' unhappiness is more than that. "If I were not strongly opposed, it would soon be the end of this relationship—but for that

I am too easy going—or perhaps too respectful of what others would not understand."[155] It is a little sad to contrast Kramers' own words with those spoken by Bohr on the occasion of Kramers' funeral. "It was in Danish artistic circles that Kramers soon after his arrival met his wife, who was to bring him so much happiness and beome so great a support to him."[168] These are the traditional things to say in a public eulogy. It is a melancholy reflection on all relationships that a candid public assessment is impossible. The relationship between Kramers and Storm was anchored on music. That started it and it was the ultimate cause of its persistence. Surely, Kramers must have gotten other pleasures from the relationship, but for many years, it functioned mainly as an effective inhibition against deep emotional contacts. Deprived from those contacts, Kramers felt emotionally isolated, misunderstood, melancholy, and unfulfilled.

It is hardly surprising that, sooner or later, Kramers would meet a woman, who he believed would meet the emotional needs left unfulfilled by his marriage. At the summer school held at Ann Arbor, Michigan in 1938, Kramers met a young woman physicist, Rose M., and an unusual but quite strong relationship developed between them. It came very near being a love affair—it may even have been a full-fledged love affair—but it was strangely muted. It certainly was not a casual flirtation; it had moments of great intimacy, recorded in their letters.[169] It is obvious that they talked a great deal to each other about intensely personal matters and also about physics, science, and scientists. Kramers found in that relationship the understanding, tenderness, and responsiveness he had craved for these many years. But it was a strange relationship nevertheless. Kramers brought Rose to his home in Leiden for lunch: Storm and the children were well aware of what was going on. The two younger girls (Maartien and Agnete) were puzzled how their father could be "in love" with this woman who had red nails and chewed with her mouth open.[170] As young girls, they were surprised, a little intrigued, and giggled a lot. Storm handled Rose as just another guest, with charm, grace, and hospitality. Sometime after the visit to Leiden, Rose and Kramers went for a weekend (or longer) to Domburg in the province of Zeeland—Kramers' favorite place for solitude and reflection. It was to this sojourn that Kramers referred in his diary. One of the entries (July 13, 1945) was written in English. It starts out: "I rose at 5 A.M." Then Kramers writes immediately "I and Rose at 5 A.M." Such comments are scattered throughout the diary, sufficiently often to show that the relationship with Rose was important, deep, and long lasting. The diary entry was written in 1945; the initial meeting in Ann Arbor was in the summer of 1938. During Kramers' visit to the United States in connection with his work for the AEC., he visited Rose several times. At the same time they maintained an extensive correspondence. It is very hard to know how satisfying that relationship ultimately was. It clearly met some deep need in Kramers; but there is in his letters, in his approach to the affair, a certain reluctance and hesitation. As always in Kramers' relationships, he held back and did not surrender; neither in his marriage nor in his love affairs could he bring himself to the unqualified acceptance of another person with the corresponding loss of personal identity. However, the relationship with Rose was not merely an innocent flirtation or diversion. Both took their relationship quite seriously; both were heavily involved. After a visit from Kramers, Rose writes him: "I miss you in a new way."[171] The affair ended in a rather bizarre way. Rose had not been married during this period, but with a coincidence worthy of a soap opera in contemporary television, in 1948 she married John C. Slater, Kramers long-time adversary from the Bohr–Kramers–Slater theory. At that time, the letters stop abruptly. Nevertheless, traces of the relationship between Rose and Kramers remained even then. It will be recalled that Slater had little use for Bohr and if possible even less for Kramers (see Chapter 13 Section III.B). Yet Slater writes in his autobiography "My old friend Kramers was at that time president of IUPAP."[172a] This was written in 1949—after he had married Rose. He comments later that "his mistrust of Bohr was well satisfied" (July 1951)[172b] and he expresses his sadness at not seeing Kramers: "I was sorry not to have a final chance to see my friend Kramers" (July 1952).[172c] It is of course impossible to know in how far Slater's radical change toward Kramers and his explicit lack of change toward Bohr was influenced by his marriage to Rose. But it is hard to believe that this complete reversal in attitude could be unrelated to Rose's relationship to Kramers. It is interesting and remarkable that on November 3, 1949, Slater made an offer to Kramers, asking him to spend some time as a visiting professor at M.I.T.[173] Whether Rose had anything at all to do with this invitation is of course a totally unanswerable question, but it is pushing innocence a bit too far to believe that there is no connection at all. This incident was the last recorded evidence of any possible traces of the relationship which had preoccupied Kramers for such a long time. It was a relationship in which once again personal and scientific interests were intertwined in an extricable mesh. Later—very near the end of his life—Kramers had yet another involvement, a Mrs. J., the wife of a visiting professor at Princeton. This was a much more traditional love affair (if such a thing exists). In this, Kramers' commitment to science played no role whatsoever. The many letters accompanying this affair are quite ordinary letters, generally light, at times facetious, but never terribly serious.[174] It appears that Kramers had a congenial, pleasant time in this relationship—but it had neither the depth nor the depression nor the elation that characterized his relationship with Rose.

All his involvements were guarded and controlled; and undoubtedly met some deep-felt needs. But it is unlikely that they made him deliriously happy, or indeed resolved some of his deep conflicts. Most likely, his new personal entanglements created more problems than they resolved.

D. The Story of Camilla

Kramers was an editor and frequent contributer to a Dutch journal *Het Kouter*, according to the motto on its frontispiece "An Independent Journal for Religion and Culture." The articles in that journal were typically short essays—generally very erudite and usually written by university professors or those who aspired to that kind of position. In the July 1940 issue, an article appeared—ostensibly written by a woman referred to as Camilla—under the title "Science, Friendship, Comfort." [175] The story is little more than a description of the conversation of this woman Camilla and her intimate friend, the scientist Antonius. In the beginning of the story, Antonious expresses extreme unhappiness with the role of science in human life. In the ensuing discussion it becomes transparently clear that Antonius is no one else but Kramers. The arguments made, the ideas expressed, the slightly convoluted language—all point unmistakingly to Kramers. This was first noted by Opechowski—a close collaborator and friend of Kramers.[176] He mentioned this to Casimir who, after reading the article, wholeheartedly agreed that Antonius was certainly Kramers. He guessed (as it turned out correctly) that the article was in fact written by Kramers.[177] But both Casimir and Opechowski were puzzled by the woman, Camilla, in the story. Actually, the woman was Rose, Kramers' long-time paramour. This could be inferred from the nature of the story told, which followed some of their correspondence to the letter, from the nature of the relationship it depicts, and from a number of tiny telltale details. But none of this inferential material is really necessary. Kramers had confided in his friend, the painter Tom Kelder, about his relationship with Rose. In the frequent correspondence on this topic Tom Kelder always referred to Kramers' relationship with Rose M. as "You and your Camilla." [178] These references occur sufficiently often that no doubt can remain that Camilla is Rose.

Kramers wrote this article, from what he believed to be the vantage of Rose M. He describes that relationship and the role physics played in it in considerable detail. Even a casual reading of the story would show that Camilla herself was a scientist—which is of course exactly correct; Rose and Kramers met at a summer school and their personal intimacy was inextricably intertwined with their attitudes toward physics.

In the story, Antonius (Kramers) drops by to see Camilla (Rose) and even before he has sat down he blurts out: "I am sick of this striving toward harmony and unity in my considerations about science, the world and humanity." [175] Antonius (Kramers) continues: "I notice this striving in myself and others.... I would really like to characterize it as a sick excess of the human spirit." [175] These are obviously harsh words. To Kramers the search for unity, understanding, and harmony had always been a dominant, life-giving goal. To judge it so harshly, even with words given to Antonius, means that he frequently had severe doubts about the major purpose of his life. To deal with this he needed understanding, comfort, and encouragement. In the story, Camilla recognizes that right away; she contradicts Antonius: "It is no sick excess at all, it is the essence of the method whereby you think and whereby I think—and using it you have been able to create new results in your scientific investigations." [175] It is to be noted that in this phrase Camilla reveals her scientific background: the words, "the method whereby you think and whereby *I* think," show that explicitly.

There is a certain contrapuntal quality about this story. Kramers wrote it, but from the orientation of Rose. Thus, Antonius indeed speaks directly for Kramers, but Camilla speaks for what Kramers hoped, sensed, or expected that Rose would express. Thus, Camilla, recognizing the depressed mood of Antonius because of his opening exclamation, sets out to console, help, and comfort Antonius. (It is precisely the absence of this kind of comfort or help with intellectual conflicts that Kramers felt so strongly in his relationship to Storm.) Thus, Camilla tries to convince Antonius that his striving for unity was not "a sick excess," but instead an important goal. She said this because "she was proud of him," but he "was not the kind of person who finds consolation in self satisfaction."[175] In this passage the contrapuntal character becomes very pronounced. By this construction, Camilla becomes an almost perfect Greek chorus, understanding all shades, in Kramers' mood. When Kramers has Camilla say "he was not the kind of person who finds consolation in self satisfaction," she expresses one of Kramers' strongly held convictions. This complete and perfect understanding which Kramers attributed to Camilla was one of the features he missed so desperately in his marriage.

The story itself revolves around Camilla's efforts to comfort Antonius, to help him through his mood of disenchantment and depression when all the goals and purposes to which he had devoted his life appeared as pointless, pathetic efforts toward self-justification or efforts to justify his own self-importance. While making some miserable coffee (in The Netherlands it is utterly unthinkable to have any confidential discussions without coffee), Camilla wonders just how she can go about consoling Antonius. It is an interesting sidelight that while thinking about this, she realizes that once she starts to explain something she becomes very animated and enthusiastic—which "her friends attribute to her southern background."[175] This comment is completely incomprehensible unless it is realized that Rose came from South Carolina. She taught in an exclusive, Southern girls school "Sophie Newcomb" in New Orleans.[179] In the "deep south" such lengthy verbal outbursts by women are very much part of the culture and it shows Kramers' sensitive understanding of Rose; he recognized this feature in her and gave it to Camilla to verbalize.

The actual discussion of Antonius' depression is not all that unusual. Camilla explains that after concentrated work, she too becomes frequently unhappy, disappointed, and depressed. However, she does not blame science per sé as the cause for unhappiness, and sooner or later she forgets her unhappiness. At this point, Antonius interrupts with a typical Kramers' commet: "I don't want to forget my unhappiness, I wish to overcome it, by conscious effort."[175] It was very much a part of Kramers world view that weaknesses and fear should not be condoned, that hard work and unflinching devotion to duty were the only way to conquer depression. In principle, that might be an admirable attitude which should be adopted. But in the middle of a debilitating depression such admonitions hinder more than they help. Camilla tried to overcome Antonius' despondency by very ordinary means, by talking, by waiting, by understanding, by getting Antonius to talk. And true to character, Antonius did start to talk, with an unmistakably evangelistic fervor. He appeared unable to stop his stream of consciousness and he really delivers a lengthy peroration on friendship and consolation as divine mercy. His phraseology has a theological strain: "Let a friend comfort you he then becomes automatically your incarnate heavenly friend. In the relation between friend and friend, your soul naturally adopts the true, humble attitude towards the merciful character of consolation."[175] This is a rather characteristic Kramers' sermon. It shows Kramers' subtlety and finesse that while he (Antonius) is delivering this oration, he recognizes that Camilla does not really listen (she lets her mind wander) and worry about the continual difficulties that come up in any relationship and the struggle that must be fought to maintain the relationship honestly and faithfully. Antonius continues to talk, but he shifts from friendship and consolation to science—its claims, purposes, and pretenses: "Let me stick with science—that is the only thing I know something about."[175] Many of Antonius' discontents are strangely reminiscent of Kramers' doubts about universal laws in physics, when Kramers was in the throes of the Bohr–Kramers–Slater theory. At that time, he was not a bit sad that the conservation laws would have a limited validity and that no strictly causal description was possible. In this confession to Camilla, he has his alter ego say: "I will only note that we do not understand the essential basis of our scientific work at all. The work however is so fascinating that we can not help but look for a slogan, which we imagine gives a reasonable explanation of creation. We say there is unity, harmony. But what we picture, or think, is either trivial or patently false. But it would be equally possible to say that there is no unity, but multiplicity."[175] It is interesting to recall that in connection with the Bohr–Kramers–Slater theory, Kramers insisted that God could have created the world in an altogether different mode. And so Antonius continues to talk about science, its inadequacies, pretenses, and irresistible fascination. But in this process his anger is gradually spent; he admits that in spite of everything he too searches for a unifying slogan. This search gives great beauty and coherence to science, which he loves to explain to students and colleagues. His anger has abated and his depression has lifted; he turns to Camilla to thank her for her sensible "down to earth" approach to science and life.

In the story, Kramers has Camilla deny that she was the agent helping him through this depression. Instead she says: "As answer I will only quote something, a poem, you wrote yourself—but perhaps you have forgotten it—I knew very well that he had not forgotten."[175]

Of human things our intellect
can only ten percent detect
The rest our stupid heart may know
When its mysterious sources glow

So ends the story of Camilla on "Science, Friendship, Comfort." This is a remarkable and illuminating story. It does not reveal any new and different aspects of Kramers' personality. It does show the emotional intensity with which he was caught up in science; it demonstrates once again the inextricable mixing of his scientific and personal life. It also reemphasizes his pervasive scepticism of the universal principles of science. Perhaps most of all it shows his extraordinary sensitivity to the moods and perception of Rose—even in relation to himself. The story provides yet another unexpected insight into the personal world of an intensely private person.

It is not so clear why Kramers wrote this story. He must have realized that his own identity was not well protected. The phraseology and philosophy of Antonius would immediately identify him as Kramers to his intimates as Opechovski and Casimir did right away. It is unlikely that Kramers gave much thought to the recognition of Camilla. Although his family (wife and children) knew about Rose, he no doubt guessed (correctly) that they would never read an article in *Het Kouter*. Thus, Camilla remained an unidentified, mysterious woman. It is reasonable to guess that writing the Camilla story was to Kramers yet another mode of self-expression. The despair and anger expressed by Antonius' about the role of science were recurring themes in his life. The hope that through the intervention of a "perfect woman" all these conflicts could be resolved was a rather common fantasy. (Kramers was fond of the Wagnerian opera "The Flying Dutchman," where this is the central theme.)

E. The Poems

Literature played a most important role throughout Kramers' life. This started as early as high school and continued, and even intensified, during his college career, when he became editor-in-chief of the student literary magazine *Minerva*. Even after he had made the reluctant decision that physics, not literature or music, would be his life's work, he still maintained his passionate interest in literature. He read voraciously in French, German, Dutch, Danish, English, Swedish, Latin, and Greek. Not only did he read the literature, it became part of his being. It was not uncommon that in his diaries he quoted a Swedish poet, only to have his train of thought change to French poetry, which in turn was followed by a commentary on Horace (this exact sequence occurred in his diaries). Thus, literature, or perhaps better the cultural vista's opened up by the world's literature, became an integral part of Kramers' personality. Kramers had a particular interest in poetry. He read it regularly and quoted it frequently, often to the embarrassment of his students and colleagues, who rarely knew to what he was referring. But he was not just passively interested in poetry; he frequently used poems as a medium of communication. He wrote poems for all kinds of occasions, some rather trivial and lighthearted, as for the Santa Claus festival. Other times, he found that writing a poem was exactly the appropriate way to express feelings of friendship, admiration, or encouragement and hope. When Pais was in Holland, during World War II, in obviously difficult and harrowing circumstances (underground), Kramers wrote a poem for him in which the phrase "The best is still to come" was prominent.[181] It almost seems that in the poems Kramers felt that he could allow himself an explicit degree of emotion and involvement that his reticent and controlled nature would keep him from exhibiting in normal "nonpoetic" circumstances. There are a number of other poems which are totally and deeply personal. In many of those he really "bared his soul." They can occasionally be found in the diaries or notebooks; frequently, they are just handwritten on a loose piece of paper. They are surely not collected together. They have no central theme and they are just the deeply felt poetic expressions of the emotions of the moment. Sometimes, he used the poems of others to express his own feelings. Thus, there is one diary entry, which reads in its entirety:

> My life closed twice before its close
> it yet remains to see
> if immortality unveil
> a third event to me
> so huge, so hopeless to conceive
> as these that twice befell
> Parting is all we know of heaven
> and all we need of hell

It is likely that Kramers quoted this poem by Emily Dickinson because it, better than anything else, captured his mood. It is also likely that he quoted this from memory; it was most unlikely that, in the middle of the night when this entry was written, he would look through the collected works of a poet or consult an anthology in the hope that he would find a poem suitable to express his exact mood. It is interesting that Kramers used poetry to express his deep emotions and not music or composition. Kramers was a fine and perceptive musician, who strongly responded to music. It is a little puzzling that, with all his musicianship, he never used original music as a means of emotional release. There is no evidence that Kramers with all his technical mastery of the cello and the piano ever composed anything—even half-way serious—for either instrument. He did not even like to improvise very much. But poetry was different: He wrote poetry at any time, under any pretext. It is particularly remarkable that Kramers spent so much of his time writing and refining, beautiful, sensitive translations of poems in foreign languages. Thus, in 1941, he wrote a very perceptive translation of "Les fenêtres," a poem by Mallarmé and one of Kramers' favorites. This Dutch translation is reproduced in the Appendix. There hardy seemed any point in making an English translation of the Dutch translation. Actually, Kramers engaged in some correspondence with Dr. de Boer, a famous French scholar, in Leiden, regarding his translation. This correspondence is largely lost, but even the few remaining fragments indicate how seriously Kramers took these translations. He wanted to capture the poetic essence of the poem and recast that in poetic Dutch.[182] Dr. de Boer was somewhat critical of the liberties that Kramers took—but nonetheless he was quite impressed that a mere physicist would take the time and have the ability to engage in such serious literary ventures. In a very similar vein, Kramers in 1942 started a beautiful translation of Shelley's "To a sky-lark." The translation remained imcomplete, but what little is available of the Dutch version is also reproduced in the Appendix. Again it made no sense to retranslate Kramers' Dutch translation of an English poem back into English. Consequently, just the Dutch stage in the translation process is presented.

These translations show Kramers' deep abiding interest in literature and his sensitivity to poetry. But they do not reveal a great deal about Kramers the person. There are many, many poems, which are direct images of Kramers' moods. Such poems do not define or characterize a person, yet they, like diaries, give an untrammeled distinct view of otherwise inaccessible aspects of a personality. Without pretending to enter on a literary analysis of Kramers' poems, it is still pertinent to quote some examples of Kramers' efforts. These more "private" poems were generally (almost without exception) written in Dutch but a translation—with all the known limitations thereof—still gives an uncommon insight into Kramers' state of mind, his moods, and brooding anguish. English versions of four of these poems, translated by the American–Dutch poet Claire Nicolas White,[183] are presented here. The Dutch originals are quoted for comparison in the Appendix.

Rotterdam

For seven centuries all told
this town that worshipped God and gold
grew, then one day destroyed it
and as the blazing flames devoured it
the fire lashed at the greedy dreams
and vomited flame spitting streams.
Even the storm fed flame and fire,
sparing the merchant nor the poor
man's tulips. House on house did burn.
St. Lawrence church then had its turn.
The organ died, so silencing
The praise that once to God did ring.
The stock market now gray with ash
the splendor of the ancient port,
lively and fair, all came to nought.
The storm blew fierce, the fire raged on.
This was the end of Rotterdam.

Marriage has not always brought
All that we at first had thought
Often sun but sometimes rain
And for that I'll take the blame
Faults I have like any sport
Though I'm a devoted sort
And I still can't understand
How flesh and soul go hand in hand

And how our spirits celebrate
What our senses consummate
For happiness we have been born
Yet at times we felt forlorn
And difference of temperament
Often caused you discontent.

November 20, 1942

On this bitterest of bitter days
he came to me as reticent as always,
asking the timeless question, why, what for,
well-known to me but never voiced before.

What is eternity? How to behave?
I begged him to desist, too tired by half
and all that I with desperate nerve could go
was ask him back, did anyone tell you?

Once you were, master, my faithful companion,
taught me to see while I followed you blind.
Now the creator's dream is over, I'm awake.
Though you returned, I'm no longer your child.

(A farewell offering; come, let me show you
how woman's revelation can console you.)

The Enemy

With anxious hope the runaway reaches for the brambles.
His foot gropes for a solid rock. Painfully he hoists himself up.
It is night and the wind raw. Anyone watching
would know that he'll be safe reaching the summit.
But does he trust his rescue? Ach!
A firm hand reaches for him out of the darkness
and plunges him back into the depth where he is wanted.

So also does the Enemy pull me down,
this Enemy whom I've known and never denied.
At times his hand is just a word, a sign.
One word, no more. How can such a sign harm me?

But though one word, one signal thus can wound me,
where is the word whose strength can lift me up,
the sign that throughout the heavy journey
calls me on high and gives support?
Oh would I could believe such word or sign.

CHAPTER 20

Epilogue: Does One Know Better—Understand More?

I. The Objective Recognition: Prizes and Awards

The most obvious and direct way to assess the significance of a scientist is by the kind of public and professional honors he receives. Even if this does not tell the whole story, it has the advantage that there is visible documentation of the esteem in which the scientist is held by his peers. It has the disadvantage that what is public and even conspicuous is not necessarily felt as most important either by the profession or the individual. Thus, the zenith of Kramers' career was undoubtedly his election in 1946 as Chairman of the Scientific and Technological Subcommittee of the United Nations Atomic Energy Commission. Still, this election came at a time when Kramers had doubts and worries about his scientific status and future. He himself attributed this election to his linguistic skill rather than his scientific importance. Important as this election was, it was to him most assuredly not the highlight of his scientific career. Still, something can be learned about a person and his life by examining the recognition and rewards he received for his professional activities. In chess terms, Kramers would undoubtedly be classified as a grandmaster. His many awards and the important functions that the profession entrusted to him all testify to that classification.

He received honorary degrees from a number of eminent universities, such as Oslo, The Sorbonne, Lund, and Stockholm. He was elected to membership in a number of academies. The first academy to bestow this honor was the Royal Danish Academy. He was elected in 1925, at the uncommonly young age of 31, a year before he returned to The Netherlands. He was also a member of the academies in The Netherlands, Belgium, France, and Norway, as well as the Royal Society of Edinburgh. He received an unusual honor on May 21, 1948, when he was elected an honorary foreign member of the American

Physical Society. That year, the Society elected an unusually large number of foreign members (to make up for the interruption caused by World War II), but even so only five additional members were elected: Chadwick, de Broglie, von Laue, Sommerfeld, and Dirac. Kramers was obviously in excellent company: All except for Sommerfeld (who probably should be) were Nobel prize winners.

Kramers' eminence as a major physicist was also evident from the important functions which the profession literally thrust upon him. It is noteworthy that Kramers accepted these responsibilities only after the end of World War II. Evidently, the dramatic events of these terrible years made Kramers more willing to accept social, organizational obligations which were only indirectly related to research in physics. In The Netherlands, Kramers was instrumental in establishing a new research organization: The Investigation of the Fundamental Properties of Matter (in Dutch, *Fundamenteel onderzoek der materie*). This organization played a most important role in the remarkable revival of Dutch science after World War II, and Kramers encouraged and supported the organization for many years. He often suggested young and developing physicists as members of the various boards and committees, so the young scientists could be reintegrated in the ongoing research in physics. Kramers also played an active role in the revitalization of the Institute for Nuclear Research in Amsterdam. This is remarkable because Kramers never wrote a single paper in nuclear physics; he never taught a course in that field. He nevertheless felt that Holland needed a nuclear research institute and he considered it his duty to help and encourage the establishment of such a facility—even though he himself would not have a great deal of use for such an institute. Kramers also played a major role in a Dutch–Norwegian cooperative venture, which resulted in The Joint Establishment for Nuclear Energy Research. As a consequence of this cooperation, a nuclear pile was built in Kjellar, Norway. In the fall of 1951, Kramers visited Princeton, but he cut this visit short so that he could be present at the inauguration of this atomic pile. Had he lived (this was only a few months before his death), he would surely have played a major role in planning the future activities of this facility.

Perhaps the most prestigious and conspicuous position that Kramers held in the international organization of physics was the presidency of IUPAP (International Union of Pure and Applied Physics). He officially accepted this position on February 11, 1947, and he remained president until 1950. In this capacity, he organized (and attended) 26 international conferences in 11 countries. In 1950, when Kramers was still president and Amaldi was secretary general, a number of meetings were held in Geneva in The European Cultural Center concerning the establishment of a major nuclear high-energy facility. This was the beginning of CERN, the fabulous European high-energy laboratory.[184] These activities clearly demonstrate Kramers' deep involvement with all aspects of the world of international physics.

Quite distinct from his recognition as a physicist were the honors that

Kramers received as a highly erudite and cultured member of the Dutch intellectual society. He was one of the editors of the great *Dutch Encyclopedia*. In his diaries, he often chides himself that he does not pay as much attention to this function as he should. He also was an editor and contributor to *Het Kouter*, the Dutch journal for religion and culture mentioned in Chapter 19 in connection with the story of Camilla. Not only was he a member of the Dutch Society for Literature, but he was especially honored by having his obituary included in the annual review of deceased members—a recognition usually reserved for well-established literary figures.

Where the scientific and intellectual world recognized and rewarded Kramers' greatness, the university bureaucracy and the government on the other hand were much more muted in their praise. Kramers had to argue and fight to obtain permission to leave for a semester or to accept a visiting position. On November 27, 1947, the government publicly rewarded Kramers by making him a Knight in the Order of the Dutch Lion. In the Dutch society, there are different categories of rewards for faithful and distinguished service. The system is very structured. Almost all university professors will receive such a distinction upon the completion of their teaching obligations. Kramers no doubt received his honor because of his activities in connection with his work for the AEC. Although an honor, it is not an especially outstanding honor. The government behaved correctly; it recognized his services in a suitable public manner, but there was no realization and certainly no acknowledgment that Kramers was a most unusual scientist—a brilliant and unselfish representative of the Dutch scientific culture. At no time did the Dutch government or the Dutch University confer on Kramers the special treatment he deserved.

The scientific world did better, awarding him a number of coveted and prestigious medals. In 1948, he received the Lorentz medal, a singular honor indeed. In 1951, he was awarded the Hughes medal of the Royal Society. Very shortly before he died, on March 5, 1952, he received the James Scott prize, another coveted award. But the "honor of honors"—the Nobel prize—he never received. It was something that he did think about; as time went on he gradually became resigned to the fact that he would never receive this honor. It did bother him, however; in his diary there are frequent entries that "he talked a little Nobel" last night.[185] There are persistent, widely believed rumors that Kramers would get the Nobel prize in 1952. He was certainly one of the candidates, but he had been that for some time. All rumors can be traced back to information supplied by Rosenthal, a friend of Kramers in Copenhagen.[186] But unfortunately, the situation is not that clear-cut. Kramers had been considered as a possible recipient of the Nobel prize for some time but no definite decision appeared to have been taken at the time of his death. Kramers himself had expected or certainly hoped that some time he would receive this "honor of honors." But it appears that after his work on renormalization theory was superseded so rapidly, he had all but given up hope that he would be so honored. As with so many promising expectations and high hopes, Kramers had to make the melancholy adjustment to yet another disappointment.

II. Finale

Writing the last section of a biography is very similar to a melancholy farewell from a close personal friend one is never to see again. There will be no further communications possible. On such an occasion there is an understandable desire to make sure that everything of importance has been said. There will be no further chance to rectify, explain, or reinterpret, nor will it be possible to remedy inadvertent omissions or unintended misunderstandings. Because of the finality of such leave-taking, there is a tendency to try to be particularly deep, serious, and profound; no trivialities should mar such a solemn occasion. And yet, in actual life, the protagonists during such a final encounter usually discuss very ordinary matters in a most offhand and pedestrain way. It is more the other way around: It is the occasion that confers a special poignancy on what is discussed; the subject matter is often incidental.

And so on completing this biography of Kramers' life and work it is pertinent to ask whether, after all these lengthy discussions, there is more appreciation of Kramers the scientist, more understanding of Kramers the man. Kramers was surely a man of contrasts and contradictions. He had a definite style in science, but he was not a maker of taste and fashion. He was extremely articulate but he doubted the effectiveness of language. "My own pet notion is that in the world of human thought generally and in physical science in particular, the most important and most fruitful concepts are those to which it is impossible to attach a well-defined meaning."[187] Recognizing Bohr's influence on Kramers, this is still a remarkable motto for a person whose expository skill was legendary. He was a brilliant, clear lecturer; yet at times there was an oraclelike quality about his pronouncements. He was intrinsically, genuinely Dutch; yet he was very much a citizen of the world. There was a mocking lightness, a whimsical strain about him; yet he admired the typically Dutch virtues of soundness and solidity. He always appeared as composed, confident, poised, and in complete control; yet he often felt severe doubts and on occasion woefully inadequate. None of the listeners to his masterful lecture, "Fundamental Difficulties of a Theory of Particles," delivered on April 14, 1944, would have guessed that only a day before he wrote to Heisenberg that he had no idea just what he was going to say. In spite of the confidence he exuded on all scientific matters, to himself he often seemed ill-prepared and deficient. It is remarkable that his lectures and seminars were so polished and so well-delivered; his scientific control was so manifest that no trace of these worries or insecurities remained in his public presentations. But privately, as his diaries and letters make abundantly clear, these concerns were his perennial companions. This much has been learned.

But these apparently contradictory and conflicting features were all essential parts of a single human being: a person who had not only enormous intelligence but interests in almost all facets of human existence. He had a deep and abiding interest in the classical literature of many countries, but he was also fascinated by detective stories and he read them regularly. This was no contradiction or conflict, but merely another example of his unusual breadth. It shows once again that Kramers was reluctant to restrict himself in any way. The conflict with Ehrenfest as a student, which caused him so much pain, originated from this very inability to confine himself rigidly to just physics. This restriction made him uneasy and restless. And even though the romantic ideals of his youth (to be a scientist of Lorentz's stature, a literary critic, a poet, a sensitive interpreter of music, an isolated, lonely genius, who feels, understands, and ameliorates the suffering of the world) had necessarily faded, substantial traces of these fantasies remained throughout his life. He emotionally was never quite reconciled to the obvious impossibility of being all things to all people. He was even reluctant to give up his efforts to understand all of science and mathematics, although he knew that to be a fantasy as well. A tiny but telling example is his serious intensive study of symbolic logic right after the end of World War II. In terms of his scientific career, this was a silly thing to do. He never used it, it was of no relevance to his teaching or research, and he was overburdened with scientific obligations; yet he took the time out of his demanding schedule to study this scientifically remote and esoteric topic, just because he enjoyed it. It almost seemed that with the physical liberation of The Netherlands and the end of the war, Kramers experienced a momentary personal liberation, which would allow him to follow his natural, if secret, inclination to learn, study, and investigate whatever he felt like, for the sheer pleasure of doing it, without having to account to his colleagues, his students, or indeed himself.

So many of the quixotic features in Kramers' behavior were really consequences of the contrasting demands which he continually made on himself. His youthful image as a lonely genius, who senses the suffering of the world (perhaps even religiously inspired), was of course rarely verbalized, but it remained a very important part of Kramers' personality. The human condition was to him not an empty phrase, but a deeply felt emotion. He was acutely sensitive to individual suffering. More than most people, he was aware of the deep-seated conflicts and horrible unfairness inherent in the human condition. The inevitable suffering this caused was one of the main sources of his persistent melancholy. In a lecture honoring the friend of his youth, Dirk Coster, on the occasion of the 25th anniversary of his professorship, Kramers exclaimed: "Life is beautiful but not for everybody."[188] This is hardly a very original observation, although it is perhaps somewhat unusual to stress it on a festive occasion. But to Kramers the many problems of the world, from minor irritations to major calamities, were totally real and he never lost sight of individual tragedies. On numerous occasions, he did what he could to alleviate these personal problems, whether they were articulated or not. However, in what became a pattern, he did this in the most unobtrusive way imaginable. It seemed at times as if he was almost embarrassed by his personal sensitivity to the problems of others. He did a great deal for many people: He encouraged students, found jobs for colleagues, and helped refugee scientists, but he always tended to minimize his own contribution. He was most anxious to help, but he did not want to make a public spectacle of his assistance.

Kramers was deeply convinced that the problems and the ugliness of human existence could be redeemed by the beauty of science. To him, physics was not an activity where a few problems are solved by clever tricks. Through physics, it became possible to transcend the boundaries of the obvious, the commonplace, the limitations of a humdrum world. Through physics, it became possible to appreciate the unanticipated unity and the extraordinary structure of the world around us. The processes leading to this understanding are so subtle and refined that they belong to the greatest creative achievements of the human species. The appreciation of the beauty of science can lift the human spirit far above the suffering and agonies of everyday existence. Science was to Kramers an integral and indispensable part of human culture, which made it possible to survive in a harsh and unfriendly world. It was not a personal sacrifice to devote a life to science—it was a privilege. Of course, a life devoted to science does not just consist of exciting, exhilarating experiences. Much of the time there is confusion, frustration, and fear. In the same lecture honoring Dirk Coster, Kramers remarks: "Only through our own struggles do we make our life."[188] Perhaps this is Kramers' most profound legacy. Our own personal efforts and deep involvement are essential to give us peace and personal fulfillment. And this will inevitably require unrelenting efforts and continual struggles. But this is yet another legacy of Kramers. His life was an unending struggle for beauty, compassion, decency, and understanding. Although thwarted and disillusioned, suffering the cruel disappointment of discovering and seeing, but not entering, the promised land, he never gave up; he never wavered in his intellectual integrity or his deep humanity. Perhaps Kramers' life was the answer to the question he so often asked: "Where is the God who dries the tears?"

APPENDIX

Poems

Dutch translation of "Les fenetres" by Mallarmé

Beu van het gasthuis en zyn lucht die waart
Omhoog langs vlakke vaalte van gordynen
Tot waar aan leege muur het groote kruisbeeld staart
Richt dien de dood reeds merkte, zich op uit zyn pynen

En schuifelt aan, niet te warmen 't oud gewricht
Maar om, stil aan, wat zon te zien daarbuiten
En drukt de witte haren, het maagre gezicht
Aan't raam, waar eenh heldere straal komt kleuren op de ruiten.

Al't lauw kon hy wel slokken, die mond door koorts verzengd
Zoo ging hy eens, nog jong, zich aan zyn lief verdwazen
En laven zich aan teere huid_hy plengt
Een lange bit're kus op lauwe gulden glazen
Hy leeft, in roes, vergeet wat schrik hem 't sterven doet
Vergeet 't gedwongenbed, de klok, de medicynen
De hoest en als de avond over de daken bloedt
Ontwaardt zyn oog, daarginds inlichtvervuld deinen

Gulden galeien, die zwanen gelyk
Dom'len op purperen, geurige stroomen
Wieg'lende vormen, warm glanzend en ryk
Die achteloos grootsch in herinnering droomen.

Zoo greep ook my de walg van wie, de ziel als steen
zich wentelt in geluk, zoals zyn lusten't vragen
En ander, niet en die styfhoofdig gaart by een
Dat slyk dat hy zyn vrouw met boorling aan komt dragen.

Ik ben gevlucht em klampm'aan iedere vensterschouw
waarlik, denrug gekeerd naar't leven, in genade
turen mag in haar glazen bespreukeld door eeuwigen dauw
Die in den kuischen morgen van het eindelooze baden/

Spiegelend, zie ik my engel; ik sterf, maar hunker in schroom
wenkt hier de kunst, of een diepste, geheimste waarde
dat ik her boren, mag dragen in luister myn droom
Op naar dien hemel, een bloeiende schoonheidsgaarde
Maar helaas!. 't Hier beneden is Meester, en vaak
Dringt het walgelyk spook tot myn plaats by de ruiten.
En Domheid en Laagheid, met onrein gebraak
My dwingen den neus, onder "t staren, te sluiten

Is er geen middel, myn ziel, al dat bit're gewaar
Om het glas, door het monster bezwadderd, te kloven
En te ontvluchten_ach met veerenloos wiekenpaar
Al dreig ik te tuimelen, op weg naar de eeuwige hoven/

Translated by H. A. Kramers, late 1941

The Marriage†

De echt heeft vaak niet dat gebracht
wat je'r oorspronkelyk van dacht
Vaak was erzon en ook soms regen
Dat heeft wel aan myzelf gelegen
Voor fouten bleef ik niet gespaard
Al ben ik toegewyd van aard
En 'k heb nog altyd niet verstaan
Hoe geest en vleesch te zamen gaan
En hoe vervoering in den geest
gevierd wordt als een zinnenfeest
'k wil in verbeelding alles geven
Maar schiet te kort in't daagelyksch leven
Wel zyn wy voor geluk geboren
Maar't ging een beetje soms verloren
En onzer beider constitutie
Bracht je wel eens tot revolutie

November 20, 1942

Op deze droefste van veel droeve dagen
kwam ingehuld, als steeds hy op my toe
De eeuwen oude vraag van wat en hoe
kwam, dien ik weet maar nimmer konde vragen

Wat is het al, hoe moet gy U gedragen
Ik Weer hem smekend af, ik ben zoo moe
Wat ik alleen, in moed uit angst, nog doe
is Wedervraag: Wie zyn het die U zagen

Eens waart gy, meester, my een trouwe makker
Leerend my zien, wyl ik U volgde blind
Thans is die scheppings droom voorby, 'k ben wakker
wel kwaamt gy weer, maar 'kben niet meer Uw kind

Een bee ten afscheid nog, dat ik U schouwe
toen openbaarde zich de liefste vrouwe

† Untitled in original.

Rotterdam 1940

In zeven eeuwen wies de stad
De stad, die geld en God aanbad
In eenen dag is zy vergaan
Het vuur heeft onze stad verdaan
't Veer streek de grauwe gierendroom
en braakte vlammenspuwende bron
Wee 'tstorm schiet de vlam te hulp
spaart koopmansslot noch armentulp
En huis op huis de vonken ving
Sint Laurens kerkeburcht verging
En 'torgel stierf, zoo als weleer
zyn zang verstierf die zong God's eer
De gryze beurs 't werd alles asch
Waar eens het geroes van" koopvolk was
Der oude havens toevalspracht
Zoo schoon zoo levend, zyn ontkracht
Neer smakte 't vuur, de storm blies aan
Zoo is ons Rotterdam vergaan

De Leeuwerik

Heil, genie van blijheid! Vogel zijt gij niet
Uit hemelsche nabijheid, g'uw vol hart vergiet
In zangvloed van Uw hoog, onvoorberaden lied.

Hooger, hooger steeds wilt van d'aard' gij springen.
Als een vuurwolk. Reeds gaat gij 't blauw doordringen
En zingend stijgt g' aldoor en stijgend blijft gij zingen.

In de gulden glans van de zon, aan 't dalen
Onder wolkentrans blijft gij gieren, dwalen,
Een blijdschap hulselloos, weldra niet in te halen.

Purpurbleeke schemer wischt reeds uit Uw baan.
Als ster aan den hemel in daglicht schuilgegaan
Verdweent g' en nog hoor ik Uw schrille juichkreet aan.

Pijleflits, als 't waar' van zilversikkels schijnen
Blanke uchtend klaar moog zijn glans doen kwijnen
Tot nauwelijks wij meer zien, wij voelen dat zeer zijn.
(niet voelen meer zijn deinen)

Aard' en lucht bevolkt alom Uw geluid;
Zoo vanuit een wolk, elders ongestuit,
Giet in den ijlen nacht de maan zijn lichtfloers uit.

Wat gij zijt; wien thans 'k U gelijken moge,
Weet ik niet. Geen glans drupt van regenbogen
Zoo heerlijk als hetlied, dat gij stort uit den hoogen.

Zoo zingt, zelf, vervaagd in 't lichtrijk der gedachten,
'n Dichter, ongevraagd, hymnen die ons brachten,
Roers'len van hope en angst, die wij voordien niet achtten.

(Zoo, een dichter zingt, vervaagd in 't lichtrijk der gedacht
Hymnen, ongevraagd, die der wereld brachten
Roers'len van hope en angst, die wij voordien niet achtten.)

Zoo een jonkvrouw rein in een burchtetoren
Met zoetste melodij, in eenzaam uur geboren,
Haar ziel in minne troost, en 't welt naar aller ooren.

Zoo ook, niet te schouwen, gouden glimworm gloeit,
Die, in del van dauw, zijn ijle lichtwaas sproeit,
Tussen gras en bloemen door geen blik gemoeid.

Zoo een open roos, schuil in 't eigen loover:
't Lauwe windje loos komt haar blaadjes rooven
Tot zoetste geur'n dien loomgewiekten dief verdooven.

De Vyand

In angstig hopen grypt de vluchteling naar de struiken
zyn voet tast naar een vasten steen, hy haalt zich moeizaam op
't is nacht en ruwe wind wie zou hem zien
weet zich gered als hy den top bereikt
Maar gelooft hy aan zyn redding ach'
een vaste hand grypt uit de duisternis hem aan
en stort hem naar omlaag waar men hem zoekt
Zoo stoot ook my de vyand naar omlaag
de Vyand dien ik kende en niet ontzag
een woord een beeld soms is zyn hand
een woord niets meer; hoe kan een beeld wel schaden
Maar zoo een woord een beeld my schaden kon
waar is het woord, welks kracht myheft
het beeld, dat in den zwaren tocht
omhoog my wenkt, en steunt
O' geloove ik zulk een woord, zulk beeld

Notes for Part 4

[1] van der Waals, Jr., in "Koninklyke Nederlandsche Akademie van Wetenschappen Meeting," October 30, 1948.
[2] H. B. G. Casimir, *Ned. Tijdschr. Natuurkd.* **18**, 167 (1952).
[3] N. Bohr, *Ned. Tijdschr. Natuurkd.* **18**, 161 (1952).
[4] F. J. Belinfante and D. ter Haar, *Science* **116**, 3021 (Nov. 21, 1952).
[5] F. J. Belinfante and D. ter Haar, early version of his obituary. I am grateful to Dr. Belinfante for providing me with an early draft of this obituary.
[6] John A. Wheeler, *The Year Book of the American Philosophical Society*, 1953, p. 355.
[7] H. A. Kramers, personal diary, in the possession of Ms. Maartien Kramers.
[8] Interview with V. Weisskopf by M. Dresden.
[9] Interview with N. Hugenholz by M. Dresden.
[10] Interview with D. ter Haar by M. Dresden.
[11] Interview with Ubbink by M. Dresden.
[12] Letter from Kramers to Colby, dated February 17, 1927, Bohr file, Copenhagen.
[13] Kramers' inaugural address, Delft, October 30, 1931; translation by M. Dresden.
[14] Interview with S. Wouthuysen by M. Dresden.
[15] Letter from Kramers to Heisenberg, dated April 12, 1944; translation by M. Dresden.
[16] Interview with G. E. Uhlenbeck by M. Dresden.
[17] Interview with Ubbink (paraphrased) by M. Dresden.
[18] Letter from Pauli to Fokker, dated February 11, 1947.
[19] Letter from Pauli to Fokker, dated February 11, 1947. This is nowhere near the language Pauli uses. It is believed that the paraphrase given in the text is about what Pauli wished to convey.
[20] L. Schiff, *Quantum Mechanics*, McGraw-Hill, New York, 1949; A. Messiah, *Quantum Mechanics*, Wiley, New York, 1958.
[21] *H. A. Kramers Collected Works*, p. 847.
[22] Interview with Jan Korringa by M. Dresden.
[23] H. A. Kramers lecture on April 20, 1933, "Past and Future," not included in the *Collected Works*.

24 Interview with O. Klein by M. Dresden.
25 W. Heisenberg, *Z. Phys.* **120**, 513 (1943).
26 W. Heisenberg, *Z. Phys.* **120**, 673 (1943).
27 Interview with Agnete Kramers by M. Dresden.
28 Interview with S. Wouthuysen by M. Dresden.
29 Letter from Heisenberg to Kramers, dated October 31, 1943, Kramers file; translation by M. Dresden.
30 Letter from Kramers to Clay, dated November 18, 1943, Boerhave Institute, Leiden, The Netherlands.
31 Letter from Heisenberg to Kramers, dated March 16, 1944, Kramers family file.
32 W. Heisenberg, in *Physics and Beyond*: *World Perspectives*, Vol. 42, Harper & Row, New York, 1971.
33 Interview with S. Hellmann by M. Dresden.
34 Letter from Kramers to Heisenberg, dated April 12, 1944, in the Kramers family file; translation by M. Dresden.
35 W. Heisenberg, *Z. Phys.* **123**, 93 (1944).
36 Interview with H. B. G. Casimir by M. Dresden.
37 Interview with H. C. Kramers by M. Dresden.
38 H. A. Kramers, "Reflections on Rotons and Phonons," *Physica* **18**, 653–654 (1950); *Collected Works*, p. 953.
39 *H. A. Kramers Collected Works*, p. 376.
40 L. Foldy and S. A. Wouthuysen, *Phys. Rev.* **78**, 29 (1950).
41 P. A. M. Dirac, "Some of the early developments of quantum theory," Coral Gables Center for Theoretical Studies, CTS-HS-69-1, March 1969.
42 Interview with van Kampen by M. Dresden.
43 H. A. Kramers, in *Physical Science and Human Values*, Princeton University Press, Princeton, N.J., 1947; also the *motto* in the *Collected Works.*
44 H. B. G. Casimir, *Haphazard Reality*, Harper & Row, New York, 1983: (a) p. 155; (b) p. 153.
45 H. B. G. Casimir, *Ned. Tijdschr. Natuurkd.* **18** (7), 167 (July 1952).
46 Conversation between Mark Kac and M. Dresden.
47 H. A. Kramers and G. P. Ittmann, Z. Phys. **53**, 553–565 (1929).
48 Interview with S. Wouthuysen by M. Dresden.
49 H. A. Kramers, *Quantumtheorie des electrons und der Strahlung*, Preface V, Akademische Verlagsgesellschaft, Leipzig, 1938. Translations from the German edition by M. Dresden. The translation of this passage in the English edition is contrived and does not quite catch the mood.
50 Inaugural address, Delft, October 30, 1931, "Reality and the Formation of Concepts."
51 H. A. Kramers, lecture on Peter Debye in Nederlandsche Helden der Wetenschap, 1943 (1946).
52 Discussion remark of H. A. Kramers at a Wednesday night colloquium, fall 1938.
53 Letter from Kramers to Clay, dated November 18, 1943.
54 Interview with Opechovsky by M. Dresden.
55 Interview with Casimir by M. Dresden.
56 C. N. Yang, "*Autobiography 1982*," in Selected Papers 1945–1980, Freeman, San Francisco, 1983.
57 Interview with Gerda Kramers by M. Dresden.
58 Interview with Annie Romeyn-Verschoor by M. Dresden; also in *Omzien in Verwondering*, Uitgevery de Arbeiderspers, Amsterdam, 1970.
59 Interview with H. C. Kramers by M. Dresden.
60 Interview with Maartien Kramers by M. Dresden.
61 Interview with Paul and Marianne Meyer by M. Dresden.

[62] Interview with Agnete Kramers-Kuiper by M. Dresden.
[63] H. B. G. Casimir, *Haphazard Reality*, Harper & Row, New York, 1983, p. 101. Dutch translation: *Door meten tot weten.*
[64] G. E. Uhlenbeck, private conversation. Dutch translation: *Die Maxwell vergelykingen, wat heb je daar nou aan?*
[65] Interview with H. C. Kramers by M. Dresden. Dutch translation: *Die bocken zyn niet degelyk genoeg.*
[66] H. A. Kramers, personal communication to S. Dresden, spring of 1948.
[67] Letter from H. A. Kramers to Lorentz, dated January 1, 1919; Kramers calls his paper a *Verhandelingetje*, Lorentz Collection, The Hague.
[68] Letter from H. A. Kramers to Lorentz, dated February 14, 1927, Lorentz Collection, The Hague.
[69] *Lorentz Collected Papers IX*, p. 181.
[70] Letter from Ehrenfest to Kramers, dated August 24, 1928, Boerhave Institute, Leiden, The Netherlands.
[71] Letter from Ehrenfest to Kramers, dated November 4, 1928, Boerhave Institute, Leiden, The Netherlands.
[72] Letter from Kramers to Ehrenfest, dated January 18, 1929, Boerhave Institute, Leiden, The Netherlands.
[73] Mark Kac, *Probability Number Theory and Statistical Physics*, K. Baclawski and M. D. Donskerd (eds.), M.I.T. Press, Cambridge, MA, 1979.
[74] H. A. Kramers, Leiden inaugural address, September 28, 1934; "Physics and Physicists."
[75] Interview with Korringa by M. Dresden. Dutch translation: *Dit is toch wel te bar.*
[76] Letter from Bohr to Kramers, dated July 4, 1939, Bohr Institute, Kramers file.
[77] Letter from Bohr to Kramers, dated July 1, 1942, Bohr Institute, Kramers file.
[78] Letter from Kramers to Bohr, dated January 8, 1948, Bohr Institute. I am grateful to Dr. Rüdinger for bringing this letter to my attention.
[79] Letter from Kramers to Bohr, dated May 11, 1949, Bohr Institute. I am grateful to Dr. Rüdinger for providing me with a copy of this letter.
[80] Interview with Klein by M. Dresden; memo undated from Kramers to Klein.
[81] Interview with Agnete Kramers-Kuiper and Maartien Kramers by M. Dresden.
[82] Recollection of S. A. Goudsmit on H. A. Kramers.
[83] Letter from Kramers to Heisenberg, dated December 15, 1933, at the Heisenberg Document Library, Munich.
[84] Letter from Kramers to Heisenberg, dated July 5, 1944, Munich.
[85] Postcard from Heisenberg to Kramers, dated March 23, 1947, Kramers file, Bohr Institute.
[86] Communication from Wassink to S. Dresden.
[87] Interview with Agnete Kramers-Kuiper by M. Dresden.
[88] Letter from de Haas to Kramers, dated October 11, 1948, Kramers file, Copenhagen.
[89] John S. Toll, private communication.
[90] Interview with A. Pais by M. Dresden.
[91] Interview with Mrs. Gerda Kramers by M. Dresden.
[92] Interview with N. Hugenholz by M. Dresden.
[93] Interview with Korringa by M. Dresden.
[94] *H. A. Lorentz—Impressions of His Life and Work*, North-Holland, Amsterdam, 1957.
[95] Interview with H. C. Kramers by M. Dresden.
[96] Interview with R. de L. Kronig by M. Dresden.
[97] Interview with Ubbink by M. Dresden.
[98] Interview with Suus Kramers-Perk and G. C. Kramers by M. Dresden.
[99] Compare Heisenberg, *Physics and Beyond: World Perspectives*, Vol. 42, Harper & Row, New York, 1971.

[100] This is a translation and rearrangement of the sermon held in the Green Church in Oegstgeest on April 29, 1952. All the ideas expressed in this quote are contained explicitly in that sermon.
[101] Arnold Siegert, private communication.
[102] Interview with Annie Romeyn by M. Dresden.
[103] Interview with Suus Kramers-Perk by M. Dresden.
[104] Interview with Korringa by M. Dresden.
[105] Interview with Casimir by M. Dresden.
[106] Original text in the Kramers family file.
[107] H. B. G. Casimir, *Haphazard Reality*, Harper & Row, New York, 1983.
[108] March 3, 1947, letter from CalTech, Kramers file.
[109] Letter from Lindsay to Kramers, dated March 7, 1947, Kramers file.
[110] Letter from Slater to Kramers, dated November 3, 1949, Kramers file, Bohr Institute.
[111] Letter from Courant to Kramers, dated February 10, 1949 Kramers file, Bohr Institute.
[112] Interview with Suus Kramers-Perk by M. Dresden.
[113] Kramers inaugural address, Leiden, September 28, 1934, "Physics and Physicists," translation by M. Dresden.
[114] Interview with H. C. Kramers and Mrs. Jan Kramers by M. Dresden.
[115] *Bohr Collected Works*, Vol. III, p. 25.
[116] Interview with Annie Romeyn by M. Dresden.
[117] Interview with ter Haar by M. Dresden; ter Haar letter to Belinfante; private conversation between Kramers and M. Dresden.
[118] Interview with Belinfante by M. Dresden.
[119] Interview with Hugenholz by M. Dresden.
[120] Interview with S. Wouthuysen by M. Dresden.
[121] Interview with van Kampen by M. Dresden.
[122] Letters, Kramers correspondence, in possession of Agnete Kramers-Kuiper.
[123] Kramers' notes on "Colloquium Generalum." These are very sketchy, incomplete lecture notes.
[124] This is a translation by M. Dresden from Kramers' notes on "Colloquium Generalum." The translation is quite free. Kramers' language is somewhat contorted.
[125] Kramers' notes on "Colloquium Generalum," by M. Dresden.
[126] Interview with Casimir by M. Dresden.
[127] ter Haar, *Statistical Mechanics*, Rhinehart & Co., New York, 1954.
[128] Interview with van Kampen by M. Dresden.
[129] Interview with ter Haar by M. Dresden.
[130] Interview with Taconis by M. Dresden.
[131] Interview with G. E. Uhlenbeck by M. Dresden.
[132] Interview with Rosenthal by M. Dresden.
[133] First report of the AEC to the Security Council, January 3, 1947.
[134] Letter from the Minister of Education to Kramers, dated July 9, 1946, Kramers file, Copenhagen.
[135] Almost all the information in this section comes from the private papers of Kramers in the Kramers file of Mrs. Agnete Kramers-Kuiper. It contains many other things, for example, Kramers' outline of the AEC's activities, written February 8, 1947, presumably submitted to the Dutch Foreign Minister, since the text is in Dutch. The quotes given here are from an interview in the *New York Herald Tribune*, September 23, 1946.
[136] ter Haar and Belinfante, *Science* **116** (3021), 555 (Nov. 21, 1952).
[137] Letter from the ministry to H. A. Kramers, dated January 31, 1947, Kramers file.
[138] Letter from H. A. Kramers to the Minister of Education, dated January 14, 1947, Kramers file.

[139] H. A. Kramers, "aanzicht der Natuurkunde," lecture, March 13, 1943, in *Ned. Tijdschr. Natuurkd.* **10**, 228–235 (1943).
[140] H. A. Kramers, "aanzicht der Natuurkunde," lecture, March 13, 1943; translation by M. Dresden.
[141] Feynman, lecture, University of Iowa, 1959.
[142] Pauli, lecture, Institute for Advanced Study, 1957; lecture, American Physical Society, New York, 1957.
[143] W. Ostwald, "Grosse Manner," *Akad. Verlagsgesellschaft*, Leipzig.
[144] H. A. Kramers, "aanzicht der Natuurkunde," lecture, March 13, 1943. Since this passage is hard to understand or interpret, the translation provided is rather precise.
[145] H. A. Kramers, book review of Frank's book on Einstein, *Saturday Review*, March 8, 1947.
[146] H. A. Kramers diaries, in the possession of Maartien Kramers.
[147] T. P. Sevensma, *Nederlands Helden in the Natuurwetenschappen*, Kosser (ed.), Elsevier, Dordrecht, 1946.
[148] Reference 147, literal translation by M. Dresden.
[149] Ibid.
[150] Reference 147, free translation by M. Dresden.
[151] Interview with E. Montroll by M. Dresden.
[152] Mark Kac, *Probability, Number Theory and Statistical Physics*, Selected Papers, K. Badawski and M. D. Donsker (eds.), MIT Press, Cambridge, MA, 1979.
[153] H. A. Kramers "note books" or "agendas" are in the possession of Ms. Maartien Kramers.
[154] Interview with Klein by M. Dresden.
[155] A complete copy of this diary is in the possession of Ms. Maartien Kramers. Translations of this diary quoted in text are by M. Dresden.
[156] S. A. Goudsmit, "It might as well be spin," *Physics Today* (June 1976), p. 40.
[157] Letter from Kramers to Uhlenbeck, dated October 5, 1934.
[158] Kramers often uses the Dutch word *gezellig*, which is really quite untranslatable. It contains elements of pleasant, cozy, warm, homey; it denotes a feeling of gentle euphoria.
[159] This awkward title is an exact translation.
[160] See reference 155, no translation. Occasionally, Kramers wrote part of his diary entries in English (or French or Latin).
[161] Good taste forbids me to publish the name of the person who made this statement.
[162] Interview with Annie Romeyn by M. Dresden.
[163] Interview with Maartien Kramers by M. Dresden.
[164] Romeyn correspondence.
[165] Interview with Suus Kramers-Perk by M. Dresden.
[166] Interview with Maartien Kramers and Agnete Kramers-Kuiper by M. Dresden.
[167] Interview with Casimir by M. Dresden.
[168] Niels Bohr, *Ned. Tijdschr. Natuurkd.* **18**, 166 (July 1952).
[169] These letters are in the possession of Mrs. Agnete Kramers-Kuiper.
[170] Conversations with Maartien Kramers and Agnete Kramers-Kuiper.
[171] Letter from Rose M. to Kramers, dated June 20, 1947, Bohr file, Copenhagen.
[172] J. C. Slater, *Solid State and Molecular Theory*, Wiley, New York, 1975: (a) p. 233; (b) p. 240; (c) p. 242.
[173] Letter from Slater to Kramers, dated November 3, 1949, Kramers file, Bohr Institute.
[174] These letters are in the possession of Mrs. Agnete Kramers-Kuiper.
[175] *Het Kouter*, 5de Jaargang No. 7, July 1940. Translations appearing in the text are by M. Dresden.
[176] Interview with Opechovski by M. Dresden. I am especially grateful to Professor Opechovski for calling this article to my attention.

[177] Letter from Casimir to Opechovski, dated October 8, 1977.
[178] Letter from Tom Kelder to Kramers, dated June 1944, Bohr file, Copenhagen.
[179] Interview Clemson University, South Carolina.
[180] Reference 175; this is written in English in the original.
[181] Interview with A. Pais by M. Dresden.
[182] Letter from Kramers to de Boer, dated July 14, 1942, written on the stationery of the Kamerlingh Onnes Laboratory. Probably the only time that Mallarmé's name was ever written on paper with the letterhead of a physics laboratory. This is the only letter that appears to have survived; even it is damaged. Original is in the possession of Ms. Martien Kramers.
[183] Usually, the translations from the Dutch are carried out by M. Dresden. My control and understanding of both languages (English and Dutch) is sufficiently good that not too much is lost or added in the translation. But the understanding and translation of poetry is an altogether different matter. For this reason, these poems were translated by a well-known American author and poet, Claire Nicolas White. She was born in The Netherlands and lived there for about 20 years. Her control of the language is excellent. She has written plays and poetry in both English and Dutch. She was clearly an ideal person to translate these poems. The translations presented here are due to her.
[184] Discussion with and information from Amaldi (May 1985, Fermi Laboratory).
[185] Diaries in the possession of Maartien Kramers; entries 1948, 1950, 1951.
[186] Interview with Rosenthal by M. Dresden.
[187] H. A. Kramers in a discussion at Princeton in 1946; *Physical Science and Human Values*, Princeton University Press, Princeton, NJ, 1947.
[188] Lecture by H. A. Kramers, in the Kramers file, Copenhagen.

Index

Numbers refer to pages, numbers in parentheses refer to the references of that part of the book: Part 1, pp. 3–84; Part 2, pp. 85–308; Part 3, pp. 311–438; Part 4, pp. 441–551.

A coefficients
 used by Kramers and Bohr, 105, 106, 107 (39)
Amplitudes
 phases of, 229, 230
 relation to Fourier coefficients, 226, 227
Andrade, reference to BKS theory, 190 (173)
Atomic Energy Commission—scientific and technological subcommittee
 agreement reached, 504
 charge to the AFC, 501 (133)
 establishment of a subcommittee, 503 (135)
 Kramers, delegate and chairman, 380
Atomic model—Bohr's
 book about, 133 (89)
 early reception, 110 (50) (51)

Balmer series, 17, 18
Belinfante, F. J., 326 (36)
 eulogy, 441 (4), 151, 442
 exam, 494 (118)
Bethe, H. A.
 calculation of the Lamb shift, 390 (116)
 recollections of Kramers' influence, 424 (151)
 at Shelter Island, 344
Bohr, Niels, 20–27
 answer to Pauli's criticism of the correspondence principle, 270 (31)
 different approach to atomic physics, 21
 difficulty of expressions/letter to Sommerfeld, 43 (52), 29 (24)
 doubts about Dirac's hole theory, 340 (64)
 eulogy by, 441 (3)
 influence on Kramers, 33
 invitations to Kramers during World War II, 479 (76)
 Kramers' ambition, 443 (11)
 Kramers as first collaborator, 27
 Kramers' help with the revision of the 1916 paper, 102, 133 (34)
 letters of recommendation for Kramers, 312 (8) (9)
 mathematical elegance, 30 (28)

Bohr, Niels (*cont.*)
the 1916 paper withdrawn, expanded, 101, 102
and the photon, 31, 32 (32) (33) (34) (36)
postulates, 21, 22, 23
prediction: the Pickering–Fowler series is due to helium, 23
provisional character of, 26, 27 (23)
reactions to Bohr's theory, 23, 24, 25
reactions to the failure of BKS, 211, 210 (213) (214) 215)
reaction to outcome of Geiger–Bothe experiment, 47 (67) 48 (68) (69)
style of physics, 28, 29, 30 (24) (25) (26) (27)
urges joint publication of Kramers–Heisenberg paper, 225 (246)
Bohr hydrogen model
comment in *Nature*, 24 (15)
reaction of Sommerfeld, 23 (13)
reception by Jeans, 24 (14)
reception—criticism
von Laue, 24 (16)
Einstein, 24 (17) (18) (19)
Bohr Institute, The
organization and early members, 34–40, 80 (37)
style of physics, 36
Bohr–Kramers–Slater Theory
abandons conservation laws, 175 (157)
background: Slater's ideas, 159–163
basic physical ideas, 171, 172, 173, 174
Born's reaction to, 46 (64), 201, 211 (188) (212)
Colby's query, 183
the Compton effect, 185–190
cost of the photon elimination, 178
distant and near collisions, 184
Einstein's reaction, 47, (65), 204, 205 (193) (194)
estrangement of authors, 166–171 (133) (134)
first mention of the BKS theory, 31, 32 (33) (36) (35)
Klein's reaction, 183 (165)
Kramers' exposition to Sommerfeld, 203 (192)
Kramers' involvement, 46 (62) (63)
Ladenburg's reaction to Einstein's reaction, 204 (196)
method of construction, 164, 165 (146) (147)
Pauli's reaction, 205, 206 (198)
in retrospect, 165 (149) (150)
Schrödinger's reaction, 182 (161)
second thoughts, 178
Bohr–Sommerfeld Theory
conflicts, 40, 41, 42 (48) (49), 103 (37)
Boltzmann, Ludwig
molecular theory, 10
probability, 11
Born, Max
letter to Einstein about quanta, 42 (47)
reactions to the BKS theory, 46 (64)
troubles with helium, 122 (75)
where does the Bohr theory fail?, 42, 43
Bothe, Walther-Wilhelm
experiments on BKS theory, 47, 208 (207) (206)
Broglie, Louis de
"wave" idea, 65 (109), 66
Brownian motion
Kramers' "escape" probability, 323 (34)

Canonical transformation
in structure elimination, 341, 342, 366 (75) (79) (80)
use of variation principle, 397, 398
van Kampen's derivation, 406
Canonical variables of radiation theory, 346, 348
impossibility, 349 (69), 353
proper set, 369, 370 (83) (84) (82)
Casimir, H. B. G.
eulogy by, 441 (2), 442 (2)
and Heisenberg's visit, 499 (107) (126) (127)
story of Camilla, 529 (177)
Chemical bonds
Kramers' interest in, 461
Classical physics
conflicts in, 18, 19
scope—confidence in, 9

Colloquium Generalum, 496 (122) (123)
  decline, 497, 498
Commutation relation
  Born–Jordan obtained by Kramers, 315 (20)
Compton, Arthur Holly
  experiment on the BKS theory, 47, 209 (211)
Compton Effect
  Kramers' semidiscovery, 290–298
  Sommerfeld's assessment, 160
Correspondence principle
  description, 106 (32)
  insistence on, 392
  use by Kramers, 129, 375 (90)

Darrow, Karl, 380
  role in organizing Shelter Island conference
Darwin, Charles Galton
  on the conflict between electromagnetism and quantum theory, 42 (51)
Debye, Pieter
  Kramers and Debye Hückel theory, 452 (24)
  Kramers' assessment, 515 (150)
Decomposition of fields, 351
Dirac equation
  modification due to self-energies, 388, 389 (112)
Dirac, P. A. M.
  on Bohr's style, 29 (24)
  contributions to radiation theory, 332 (47) (48)
  doubts about the Dirac equation, 334 (50) (51)
  effect on Kramers, 333, 334
  on the role of fear in research, 462 (41)
  role in quantum mechanics, 58, 59, 60
  talk on Kramers' quantum conditions, 388, 389 (112)
Dispersion relations
  as basis for a fundamental theory, 245 (256)
  chronology of publication, 144
  Heisenberg's attempt to use, 46 (60), 245 (255)
  Kramers' assessment, 46 (61)
  Kramers' greatest achievement, 45 (58)
  Lorentz view of, 147 (117)
  no reference to orbits, 44, 45
  Pauli's assessment, 45 (59)
  replacement of derivatives by differences, 44 (57)
  role in the Bohr theory, 44, 45
Doublet splitting, 61, 62
Duty, 490, 492
Dyson, F. J., 337
  Kramers' choice for expository lecture, 420 (144)

Eddington, A. S.
  congratulations to Kramers, on the X-ray papers, 137 (95)
Ehrenfest
  on the Bohr theory, 25 (21) (22)
  Burgers' relations with, 93 (13), 111 (54) (53)
  Chairman search committee; Julius' successor, 311
  dependence on Kramers, 313 (11) (12), 478 (70) (71)
  Kramers' appreciations, 92 (11)
  Kramers' difficulties with, 92, 93 94 (21)
  Kramers' letters of recommendation, 312 (7)
  letter to Bohr, suggesting Kramers for the Utrecht position, 311 (3)
  relations to Bohr, 112 (56)
  students' appreciation of Ehrenfest and Kramers, 6
  style of teaching, 92
Einstein
  Einstein's photons—disliked, 15, 16 (90)
  fluctuations and the material quantum gas, 66 (112)
  ghost field, 66 (110)
  incompatibility of quantum and classical physics, 15
  influence on Schrödinger, 66 (113)
  letter of recommendation for Kramers, 312 (4) (15)

Einstein (*cont.*)
on light quanta (photons), 31 (29) (30)
reaction to BKS theory, 47 (65)
reaction to the Bohr theory, 24 (16) (17) (18) (19)
Electromagnetic mass, 328, 329 (42), 330, 364, 388
Electron
unanticipated discovery, 11
Electron radius
only structure dependent remnant, 412
Electrons
as mathematical points, 377 (94) (95)
Energy conservation laws
abandoned in BKS, 175 (157)
Energy fluctuations
in BKS theory, 197 (184)
Experimental consequence of the BKS theory, 190, 191
Experimental mass, 330, 331, 343, 360
Extended electron
Lorentz picture, 328 (42)

Fear
in research, 462 (41)
Fermi, Enrico
style, 60
Feynman, Richard
on the emotional attachment to a theory, 191 (174), 337
quantum electrodynamic formalism, 390 (118)
Field, external and self, 359
divergencies at particle locations (74) (75) (76)
Fine structure
constant, 335
Formal theory, 171
formal character of the BKS theory, 171, 226
"Future More Complete Theory", 26, 42
and BKS theory, 46

Geiger, Hans Wilhelm
experiments on the BKS theory, 47
letter to Bohr on BKS theory, 47 (66)
Ghost field
Born's recollection, (111) 143
reference to Einstein's, 143 (110) (111) (112)
Wigner's recollection, 143 (112)
Göttingen University
Göttinger gelehrsamkeit, 37
"School of Physics", 37
Goudsmit, S. A.
Kramers' evaluation, 518 (156) (157)
spin idea, 60 (100)
Gravitation
comment in Kramers' dissertation, 113

ter Haar, R.
interview, recollection of Compton incident, 293 (356)
obituary, 441 (4) (5)
de Haas, W. J., 483, 484 (88)
conflicts with, 519, 523, 524 (160) (155)
Half-integer quantum numbers, 42 (49)
Hamiltonian adopted by Kramers, 399, 400
comparison with Dirac Hamiltonian, 400
Dirac's comment, 403 (122)
final version, 406
Hamiltonian mechanics, Kramers' mastery, 110 (49)
Heisenberg, Werner
"all models are wrong", 42 (50)
ambition, 263 (293)
attitude towards physics, 39
"Bohr is the pope", 268 (300)
on Bohr's style, 262 (290) (291)
competition with Kramers, 264 (292), 265
on criteria for publication, 261 (283)
differences with Bohr, 262 (285), 295 (362)
different attitude toward Bohr, 130, 131
disagreement with Schrödinger, 69, 70 (123) (125)
η matrix theory and Kramers, 453–458

first discussion with Bohr, 125, 126, 128 (86) (87)
first meeting with Bohr, 39 (45)
on half-integer quantization, 42 (49)
the Heisenberg–Dirac bet about the spin, 64
influence of Bohr on, 253 (267) (268)
lack of communication, matrix mechanics with Kramers, 278
later appreciation of Kramers, 283 (343)
Pauli and Heisenberg's contrasting styles, 261
reaction to Pauli's reaction to matrix mechanics, 58 (94) (95)
use of dispersion ideas to construct matrix mechanics, 50, 51 (76)
visit to Leiden, 1943, 454
Heitler, Walter
incorrect definition of fields, 331
Helium,
Kramers' efforts, 120, 121 (71) (72) (73)
Holst, Helge
joint author with Kramers, 132–134 (89), 441
Honors and awards
academies, Royal Danish, The Netherlands, Belgium, France, Norway, Edinburgh, 536
honorary degrees, Oslo, Sorbonne, Lund, Stockholm, 536
medals, Hughes medal, Lorentz medal, James Scott prize, 538
Hugenholz, N., 443 (9), 494 (119)
Hughes medal, received, 1951, 441
Human condition, 484, 540

Induced dipole moment
calculation, 220
classical result, 217
coherent part, 219
the false resonances, 242
form suggested by BKS, 226
incoherent part, 231, 232, 236
Intensity calculations, 107, 108 (41), 109 (43)
International union of pure and applied physics
Kramers president of, 3
Invariance principles
relativistic, importance, 377
Ising Model
Kramers' contributions to, 323 (33)
as a "near miss", 453
work at, 379

Jost, Res
lecture on Kramers and the Compton effect, 295 (361)
on Planck and contradictions, 14 (7)

Kac, Marc
on "following a Master", 479 (73)
on Kramers and elliptic functions, 467 (46)
on "Mathematics", 516 (452)
Kamerlingh, Onnes H.
in Leiden, 475 (63)
on Maxwell, 475 (64)
Klein, Oskar, Support of Kramers, 1925, 48 (71) (72)
attempt to console Kramers, 277 (324)
discussion with Bohr, on Kramers–Heisenberg relation, 273 (321)
discussions on Compton effect, 294 (359)
friendship with Kramers, 481
second Institute visitor, 35
Korringa, Jan
on Bohr's demands, 479 (75)
interview, recollection of Compton effect, 293 (357)
interview, "Research is as if one is kissed by an angel", 4, 79 (1)
Kouter, The story of Camilla, 528–532 (176) (177) (175)
Kramers, Gerda, 484 (91)
Kramers, H. A.
"a Kramers must fight to fight", 274 (324), 473 (58)
"appendage to Bohr", 44 (55)
appreciation of Ehrenfest, 92 (11)
appreciation of Heisenberg's explanations, 252 (262)

Kramers, H. A. (*cont.*)
arguments with Pais; are all divergencies classical?, 427 (149)
"be pugnacious", 487 (97)
on Bohr's style, 479 (74)
chairman of subcommittee, AEC, 503 (135)
characterization of, 7, 79 (5)
classical approach in 1950 lecture course, 428 (162a)
consolidation of relations with Pauli, 272
criticism of the Dirac theory, 336 (53) (54) (55) (56), 338 (58), 385, 386, 387
departure for Utrecht, 281
derivation of the dispersion relations, 150–154
detailed objections to the Dirac theory, 339, 340
difficulties of the Bohr theory, 43, 44 (54) (56)
disappointment, lack of recognition for renormalization, 419, 422 (142) (143)
distinction between his and Dirac's Hamiltonian, 358 (72)
distribution of potatoes, 500 (130)
on doing research, 4, 79 (1)
elaborations of BKS theory, 196, 197 (185), 198
the exultation of science, 514 (150)
feelings about German occupation, 458
first encounter with Lorentz, 91 (9)
first postcard to Bohr, 35 (38)
first visit to Copenhagen, 27
and the German authorities, 500 (118) (129)
guiding ideas, 194 (178), 195 (179)
Heisenberg as a "young hero", 252 (263)
Heisenberg's 1943 visit, 455 (30) (34)
influence of Bohr on, 28, 29
interest in the Utrecht position, 311 (1) (2)
interlude in Arnhem, 94, 95, 96
Kramers and the photon, 32 (36)
the Kramers–Pauli bet about the spin, 63 (107) (108)
lack of personal ambitions, 255
on the law of causality, 195 (176)
lecture on Debye, 512–516 (147) (148)
lecture on particle physics, April 14, 1944, 372 (84)
lecture on psychology, 506–512 (139) (140)
letters to Ehrenfest, 478 (72)
marital problems, 524, 525 (155) (164) (168)
marriage and courtship, 116, 117
misgivings, leaving Copenhagen, 313 (14)
Pais and the Gestapo, 500 (90)
Pauli's consolation, 257 (272)
personal dissatisfaction, 444, 445 (12) (13) (14) (15)
personal doubts, 95, 96 (20) (21), 98, 99 (24) (25) (26), 318 (30)
on personality of, 6, 7 (79), 4
on photons, card to Fowler, 289 (335)
poems and poetry, 532–535 (181) (182)
postwar letters to Bohr, 480 (78) (79)
preserved diary, 517 (155)
program for the theory of electrons, 341 342
promotion in Leiden, 113, 114
quadratic stark effect, 127 (82)
rapid rise in the Bohr inner circle, 36, 80 (38) (39) (40) (41)
rapporteur at Shelter Island, 383 (107) (108)
reaction to Bethe's Lamb shift calculation, 416, 417 (138)
reaction to brother's death, 485 (91) (92)
reaction to Heisenberg's *S* matrix, 373 (85)
reaction to the Lorentz medal, 446 (1) (16)
reactions to experiments in the BKS theory, 48 (70) (71) (72) (73)
reactions to Heisenberg's paper, 252 (261)
reactions to Kramers' presentation, 390 (114)

reactions to the failure of BKS, 212, 213
relation to Heisenberg, 482 (83) (85) (86)
relation to Pauli, 482 (82)
relation with Rose M., 527 (170) (169) (172)
resignation from Leiden, 499
retrospective assessment of BKS, 198
role of fear, 463, 469, 470
scientific attitude, 193 (177), 194
scientific faith, 192 (175a)
scientific isolation, 486, 487
scientific style, 464–466 (42) (43) (44)
silence about Heisenberg meeting, 214 (225) (226)
state of mind, summer 1925, 53, 54 (90) (91), 277 (323), 279 (332)
style and the nondiscovery of matrix mechanics, 285–288
suggests Dyson for lectures in Eindhoven, 420 (144)
ties to Holland, 279, 280 (333)
tiredness, 44
trip with Bohr to England, 256 (270)
on understanding, 5, 79 (3)
veneration of Lorentz, 327, 374 (39) (40) (41), 477 (67)
visit to Göttingen, 213 (222)
why not work relativistically?, 431 (165)
Kramers–Heisenberg dispersion relations, 215 (228)
chronology, 221
credit for, 273 (319) (322)
methodology, 230
rapid acceptance, 242
support of BKS, 244
Kramers (Mrs. Anna Petersen, Storm)
attitude toward Heisenberg, 282
first meeting, 114 (59), 115 (60) (61)
general character, 526 (166)
on Heisenberg–Kramers collaboration, 282, 483
on Kramers' naivete, 282 (340)
protective of Kramers, 526 (167)
Kramers, H. C.
discussion with H. A. Kramers, 274 (324), 472
"to do something new", 486 (95)
what to study, 476 (65)
Kramers–Kronig relations
as a near miss, 458
Kronig, R. de Laer
recollection of Compton effect, 293 (358)
spin idea, 62

Ladenburg, N., 44, 148, 149, 150 (119) (120)
Lamb shift
calculation by Bethe, 414 (134)
calculation by French and Weisskopf, 411 (130)
calculation by van Kampen, 413
first announcement, 344 (66) (67) 389
needed to stimulate theory, 411 (131)
Leiden University
Kramers' student years, 90–94, 379 (100)
Light: two theories, 199 (187)
Einstein–Bohr conflict, 207 (202)
Sommerfeld, 206 (200)
Lorentz, H. A., 8–12
on black body radiation, 10
characteristics, 9, 10
electromagnetic mass, 330, 331 (42)
letter of recommendation, 312 (6)
picture of an electron, 328, 329, 332
the problems of the mechanics of quanta, 42, 81 (46)
"spirit of the theory", 377
Lorentz medal, 441 (1)
Low, Francis Eugene
letter about Kramers and Lorentz, 327 (40)

Mach, Ernst
attitude towards molecules, 10
Mac Innes, Duncan Arthur
role in organizing Shelter Island, 381
Mass renormalization
classical, 330, 363, 365, 377
Mathematics
role of in Kramers' work, 466–470 (44) (48) (49) (51) (52) (53)

Matrix
from amplitudes to matrices, 228, 229
matrix elements of dipole, 235
Matrix mechanics
Bohr's reaction, 51 (80) (81) (82)
Born's reaction, 51 (78)
Dirac's reaction, 51 (83)
Einstein's reaction, 51 (84) (85)
Heisenberg's construction, 50, 51
Kramers' reaction, 51 (77)
Pauli's reaction, 51 (79)
transition to, 56, 57
who told Bohr?, 51, 52 (26) (87) (88) (89), 276, 277
Wigner's reaction to, 57 (92)
Maxwell, James Clerk
electrodynamic laws, 9
Michelson, A. A.
"on future discoveries", 9, 79 (6)
Millikan, Robert
on photons, 16, 79 (10), 79 (11)
Molecules
conflicts generated by, 10

Near misses, 446–449
enumeration, 449–453 (18) (19) (20)
not crossed a threshold, 448 (17)
Nernst, Herman Walter
assessment of quanta, 16, 79 (12)
Newton, Isaac
confidence in Newtonian concepts, 8–12
Null vector
Kramers' usage, 345

Obituary, 6 (4)
Occupation
wartime, 498–501
Opechovski, W., 325, 326
application of Kramers' formalism, 360 (75), 368
story of Camilla, 529 (176)
Oppenheimer, J. Robert
assessement of particle physics, 384
dominant role at Shelter Island, 385
frustration with Dirac theory, 335 (51)
on the heroic time, 4, 79 (2)
"renormalization is a new synthesis", 418
in Shelter Island, 381, 382, 383
Ornstein, L. S., 485
Oscillator in a radiation field, 360 (51) (77) (78)
Kramers-type treatment, 367

Pais, Abraham
the day Paris fell, 484 (90)
domain of anticorrespondence, 429
principle of "disparity", 429 (149)
references to Kramers, 422 (149)
Pauli, Wolfgang
appreciation of Kramers' X-ray paper, 137 (94)
assessment of Kramers' contributions, 284 (349) (350)
attitude towards physics, 38 (44)
"on belief in authority", 206 (198)
Bohr's invitation to Copenhagen, 38, 80 (44)
continual respect for Bohr, 257, 258 (273) (274)
critical attitude, 39
criticism of BKS, 269
criticism of Kramers and orbits, 258, 259 (276) (278) (303), 271
criticism of the correspondence principle, 270 (311) (312), 271 (314)
criticism of the Dirac theory, 335 (52), 340 (64)
on Dirac's religious views, 488 (99)
doubts about the orbit notion, 49, 81 (74) (75)
first meeting with Bohr, 38, 80 (43), 253 (266)
frustration with the Bohr theory, 43, 81 (53)
on Heisenberg, 260 (279)
on Heisenberg's 1/2-integer quanta, 42, 81 (48)
on Kramers, 36 (40) (41)
Kramers as "his eminence", 268 (301), 269 (303)
on Kramers' spirits in Utrecht, 284 (347)

letter to Kramers on BKS, 197 (184)
on maximum and minimum physicists, 508 (142)
reaction to matrix mechanics, 57, 58
reaction to the spin, 63, 64 (107)
on the reality of photons, 289 (306)
recollections of Compton effect, 294
superficiality of an electron theory not giving α, 335
unhappiness with Kramers–Bohr trip, 256 (270)
Pauli–Weisskopf paper: The anti-Dirac paper, 431 (164)
Ph.D.'s
Kramers, 113
Phase transitions
discontinuities as limit properties, 323
Photon
Bohr's antagonism toward, 139–142 (103) (109)
Einstein's introduction, reception, 15, 16 (90), 28–33
Planck, Max
conflict with Mach, 10
on quanta, 15, 79 (8, 9)
on quantization of energy, 14, 79 (7)
Probability
needed to explain thermodynamics, 11
Probability field, 173
Probability in quantum theory, 74 (131) (132)
and BKS theory, 74, 75 (133) (135) (136)
discussions in Copenhagen, 76, 77

Quantization, half-integer, 122, 123 (79)
Quantum electrodynamics
status of, 382, 383
Quantum jump, 25, 26
Quantum level, 26

Raman spectrum, 224 (243)
Reservations
Casimir, 442 (2), 443
ter Haar, 443 (10)
Weisskopf, 443 (8)
Wheeler, 443 (6)
Rohrlich, Fritz
calculation of electromagnetic mass, 330 (44)
Romeyn, Jan
advice how to combat exhaustion, 319 (31)
early letters, 88, 89 (2) (3) (4)
inquiry about the BKS theory, 207 (203)
letter about the X-ray paper, 137 (94)
letters about Bohr and Copenhagen (54) (90)
preparation for "the exam of state", 90
visit to his home, 6
Rutherford
structure of the atom, 18, 19

S matrix, 378 (97) (98)
Schrödinger, Erwin
description of goals, 66, 69 (114) (121)
on quantum jumps, 25, 80 (20)
Schrödinger wave equation, 67 (114)
Dirac's reaction, 71 (129)
Einstein's reaction, 68 (115)
influence of, 67, 68
Lorentz's comments, 72 (130)
Pauli's reaction, 68 (119)
Planck's reaction, 68 (117)
statistical conservation laws, 196 (182)
Schweber, Sylvan
references to Kramers, 423, 424 (150) (154) (155)
Schwinger, Julian, 337
quantum electrodynamics, 384 (117)
refers to Kramers, 409 (128) (129), 422 (117), 425
at Shelter Island, 344
Self-energy
possible observable effect, 344
Self-field
definition, 350, 351
example, 362 (77)
splitting, 253, 355, 356
splitting adopted, 393, 394, 395
van Kampen splitting, 407, 408

Serpe, J., 325, 326
application of Kramers' ideas, 360 (77)
Shelter Island Conference
Kramers' presentation, 344, 379–391
Siegert, A. J. F.
characterization of Kramers, 7, 79 (5)
Slater, John Clarke
on conservation laws, 32 (35)
ideas on quanta, 161 (142)
Slater's assumptions, 163 (139) (140)
Smekal–Raman effect, 224 (242)
Solvay Conferences
report, 391–402 (19)
Sommerfeld, Arnold
on BKS theory, 203 (191)
letter to Bohr on the Compton effect, 160 (134) (135) (136)
quantum conditions, 103 (37)
reaction to Bohr's theory, 23
students, 37, 38, 39
style of physics, 80 (40) (42), 37, 79 (13)
Spectral analysis
discussion, 17, 18
Spin idea, 60, 61
Kramers' approach to, 317 (54)
Kramers' bet with Pauli, 316 (25) (26)
Spin orbit coupling
Einstein's help, 62 (102) (103)
Heisenberg's calculation, 62
Spinor invariants
Kramers' formalism, 317 (29)
Stark effect intensity calculation, 97, 107 (39), 125, 126 (81)
Statistical mechanics, 321–324
Steller energy
Kramers' idea in BKS, 197
Storm, *see* Kramers, Mrs. Anna Petersen
Structure-independent formulation, 342, 343, 369, 396
Stuewer, Roger, 302 (134)
Symbolic method, 317, 321 (29) (54)
Symmetries
Kramers' ambivalence, 451, 452 (22) (23)

Taconis, K., 500 (130)
Teaching, 491–498
in Copenhagen, 492 (115)
Theology, 5 (3), 488, 489 (100) (101)
Thermal spectrum
calculation, 13
Planck's concerns, 14, 15
Toll, John S.
Kramers and the killer instinct, 484 (89)
Thomas factor, 62, 63
Pauli's reaction and capitulation (106)
Transcriptions
classical-quantum, 247, 248, 249 (259) (229)

Ubbink, J.
Kramers' advice to be stubborn, 297
Uhlenbeck, G. E.
Kramers' letter to, 518 (157)
spin idea, 60 (100), 326 (38)
Ultraviolet, catastrophe, 14
Urey, Harold
Kramers' assessment, 517 (153)
Kramers' letter to, 53 (88), 277

van der Waals, J. D.
congratulatory lecture by, 441
van Kampen, N., 325 326
advice from his uncle, 476
Hamiltonian, 404 (82)
Virtual oscillators, virtual field, 172, 173, 174, 182
introduced in the BKS theory, 45
suggested by dispersion formulae, 155, 156, 157
virtual processes, to replace virtual oscillators, 235, 238
virtual states, sum over, instead of sum over oscillators, 233
von Laue, M.
book review of Kramers and Holst, 134

W. K. B. paper, 16 (316)
Weisskopf, Victor
assessment of particle physics, 384 (109)

interest in basic questions, 383 (106)
at Shelter Island, 344
Wheeler, John A.
on Kramers' style of questioning, 5, 79
obituary, 441 (6), 442
organization of Shelter Island Conference, 381
Wolfskehl lectures
origin and influences, 38, 124
Wouthuysen, S., 454
brings *Phys. Rev.*, 519
and *S* matrix (28)
transformation, 461 (40), 494 (120)

X rays, Kramers' paper, 134–138 (93)

Yang, C. N.
on scientific taste, 470 (56)

Zernike, Frits
advice to van Kampen, 476

MIX
Papier aus verantwortungsvollen Quellen
Paper from responsible sources
FSC® C105338
FSC
www.fsc.org

If you have any concerns about our products,
you can contact us on
ProductSafety@springernature.com

In case Publisher is established outside the EU,
the EU authorized representative is:
Springer Nature Customer Service Center GmbH
Europaplatz 3, 69115 Heidelberg, Germany

Printed by Libri Plureos GmbH
in Hamburg, Germany